Lecture Notes in Artificial Intelligence 10449

Subseries of Lecture Notes in Computer Science

LNAI Series Editors

Randy Goebel
University of Alberta, Edmonton, Canada
Yuzuru Tanaka
Hokkaido University, Sapporo, Japan
Wolfgang Wahlster
DFKI and Saarland University, Saarbrücken, Germany

LNAI Founding Series Editor

Joerg Siekmann
DFKI and Saarland University, Saarbrücken, Germany

More information about this series at http://www.springer.com/series/1244

Ngoc Thanh Nguyen · George A. Papadopoulos
Piotr Jędrzejowicz · Bogdan Trawiński
Gottfried Vossen (Eds.)

Computational
Collective Intelligence

9th International Conference, ICCCI 2017
Nicosia, Cyprus, September 27–29, 2017
Proceedings, Part II

 Springer

Editors
Ngoc Thanh Nguyen
Department of Information Systems,
 Faculty of Computer Science
 and Management
Wrocław University of Science
 and Technology
Wrocław
Poland

George A. Papadopoulos
Department of Computer Science
University of Cyprus
Nicosia
Cyprus

Piotr Jędrzejowicz
Department of Information Systems
Gdynia Maritime University
Gdynia
Poland

Bogdan Trawiński
Department of Information Systems,
 Faculty of Computer Science
 and Management
Wrocław University of Science
 and Technology
Wrocław
Poland

Gottfried Vossen
Department of Information Systems
University of Münster
Münster
Germany

ISSN 0302-9743   ISSN 1611-3349   (electronic)
Lecture Notes in Artificial Intelligence
ISBN 978-3-319-67076-8   ISBN 978-3-319-67077-5   (eBook)
DOI 10.1007/978-3-319-67077-5

Library of Congress Control Number: 2017952854

LNCS Sublibrary: SL7 – Artificial Intelligence

Printed on acid-free paper

This Springer imprint is published by Springer Nature
The registered company is Springer International Publishing AG
The registered company address is: Gewerbestrasse 11, 6330 Cham, Switzerland

Preface

This volume contains the proceedings of the 9th International Conference on Computational Collective Intelligence (ICCCI 2017), held in Nicosia, Cyprus, September 27–29, 2017. The conference was co-organized by the University of Cyprus, Cyprus and the Wrocław University of Science and Technology, Poland. The conference was run under the patronage of the IEEE SMC Technical Committee on Computational Collective Intelligence.

Following the successes of the 1st ICCCI (2009), held in Wrocław, Poland, the 2nd ICCCI (2010), in Kaohsiung, Taiwan, the 3rd ICCCI (2011), in Gdynia, Poland, the 4th ICCCI (2012), in Ho Chi Minh City, Vietnam, the 5th ICCCI (2013), in Craiova, Romania, the 6th ICCCI (2014), in Seoul, South Korea, the 7th ICCCI (2015), in Madrid, Spain, and the 8th ICCCI (2016), in Halkidiki, Greece, this conference continued to provide an internationally respected forum for scientific research in the computer-based methods of collective intelligence and their applications.

Computational Collective Intelligence (CCI) is most often understood as a sub-field of Artificial Intelligence (AI) dealing with soft computing methods that enable making group decisions or processing knowledge among autonomous units acting in distributed environments. Methodological, theoretical, and practical aspects of computational collective intelligence are considered as the form of intelligence that emerges from the collaboration and competition of many individuals (artificial and/or natural). The application of multiple computational intelligence technologies, such as fuzzy systems, evolutionary computation, neural systems, consensus theory, etc., can support human and other collective intelligence, and create new forms of CCI in natural and/or artificial systems. Three subfields of the application of computational intelligence technologies to support various forms of collective intelligence are of special interest but are not exclusive: semantic web (as an advanced tool for increasing collective intelligence), social network analysis (as the field targeted to the emergence of new forms of CCI), and multi-agent systems (as a computational and modeling paradigm especially tailored to capture the nature of CCI emergence in populations of autonomous individuals).

The ICCCI 2017 conference featured a number of keynote talks and oral presentations, closely aligned to the theme of the conference. The conference attracted a substantial number of researchers and practitioners from all over the world, who submitted their papers for the main track and seven special sessions.

The main track, covering the methodology and applications of computational collective intelligence, included: multi-agent systems, knowledge engineering and semantic web, social networks and recommender systems, text processing and information retrieval, data mining methods and applications, sensor networks and internet of things, decision support and control systems, and computer vision techniques. The special sessions, covering some specific topics of particular interest, included cooperative strategies for decision making and optimization, computational swarm intelligence, machine learning in medicine and biometrics, cyber physical systems in automotive area, internet of things - its relations and consequences, low resource language processing, and intelligent processing of multimedia in web systems.

We received in total over 240 submissions from 39 countries all over the world. Each paper was reviewed by 2–4 members of the International Program Committee of either the main track or one of the special sessions. We selected the 114 best papers for oral presentation and publication in two volumes of the Lecture Notes in Artificial Intelligence series.

We would like to express our thanks to the keynote speakers: Yannis Manolopoulos from the Aristotle University of Thessaloniki, Greece; Andreas Nürnberger from the Otto-von-Guericke University Magdeburg, Germany; Constantinos S. Pattichis from the University of Cyprus, Cyprus; and Sławomir Zadrożny from the Systems Research Institute of the Polish Academy of Sciences, Poland, for their world-class plenary speeches.

Many people contributed towards the success of the conference. First, we would like to recognize the work of the Program Committee co-chairs and special sessions organizers for taking good care of the organization of the reviewing process, an essential stage in ensuring the high quality of the accepted papers. The chairs of the workshops and special sessions deserve a special mention for the evaluation of the proposals and the organization and coordination of the work of seven special sessions. In addition, we would like to thank the PC members, of the main track and of the special sessions, for performing their reviewing work with diligence. We thank the Local Organizing Committee chairs, the publicity chair, the web chair, and the technical support chair for their fantastic work before and during the conference. Finally, we cordially thank all the authors, presenters, and delegates for their valuable contribution to this successful event. The conference would not have been possible without their support.

It is our pleasure to announce that the conferences of the ICCCI series continue a close cooperation with the Springer journal Transactions on Computational Collective Intelligence, and the IEEE SMC Technical Committee on Transactions on Computational Collective Intelligence.

Finally, we hope and intend that ICCCI 2017 will significantly contribute to the academic excellence of the field and lead to the even greater success of ICCCI events in the future.

September 2017

Ngoc Thanh Nguyen
George A. Papadopoulos
Piotr Jędrzejowicz
Bogdan Trawiński
Gottfried Vossen

Organization

Honorary Chairs

Pierre Lévy — University of Ottawa, Canada
Cezary Madryas — Wrocław University of Science and Technology, Poland
Costas Christophides — University of Cyprus, Cyprus

General Chairs

Ngoc Thanh Nguyen — Wrocław University of Science and Technology, Poland
George A. Papadopoulos — University of Cyprus, Cyprus

Program Chairs

Costin Badica — University of Craiova, Romania
Kazumi Nakamatsu — University of Hyogo, Japan
Piotr Jędrzejowicz — Gdynia Maritime University, Poland
Gottfried Vossen — University of Münster, Germany

Special Session Chairs

Bogdan Trawiński — Wrocław University of Science and Technology, Poland
Achilleas Achilleos — University of Cyprus, Cyprus

Doctoral Track Chair

George Pallis — University of Cyprus, Cyprus

Organizing Chair

Georgia Kapitsaki — University of Cyprus, Cyprus

Publicity Chair

Christos Mettouris — University of Cyprus, Cyprus

Keynote Speakers

Andreas Nürnberger	Otto von Guericke University Magdeburg, Germany
Yannis Manolopoulos	Aristotle University of Thessaloniki, Greece
Constantinos S. Pattichis	University of Cyprus, Cyprus
Sławomir Zadrożny	Systems Research Institute of the Polish Academy of Sciences, Poland

Special Sessions Organizers

1. *CSI 2017: Special Session on Computational Swarm Intelligence*

Urszula Boryczka	University of Silesia, Poland
Tomasz Gwizdałła	University of Lodz, Poland
Jarosław Wąs	AGH University of Science and Technology, Poland

2. *WebSys 2017: Special Session on Intelligent Processing of Multimedia in Web Systems*

Kazimierz Choroś	Wrocław University of Science and Technology, Poland
Maria Trocan	Institut Supérieur d'Électronique de Paris, France

3. *CPSiA 2017: Special Session on Cyber-Physical Systems in the Automotive Area*

Adam Ziębiński	Silesian University of Technology, Poland
Markus Bregulla	Technische Hochschule Ingolstadt, Germany
Rafal Cupek	Silesian University of Technology, Poland
Hueseyin Erdogan	Continental Ingolstadt, Germany
Daniel Grossman	Technische Hochschule Ingolstadt, Germany

4. *LRLP 2017: Special Session on Low-Resource Languages Processing*

Ualsher Tukeyev	al-Farabi Kazakh National University, Kazakhstan
Zhandos Zhumanov	al-Farabi Kazakh National University, Kazakhstan

5. *CSDMO 2017: Special Session on Cooperative Strategies for Decision-Making and Optimization*

Piotr Jędrzejowicz	Gdynia Maritime University, Poland
Dariusz Barbucha	Gdynia Maritime University, Poland

6. *IoT-RC 2017: Special Session on Internet of Things – Its Relations and Consequences*

Vladimir Sobeslav	University of Hradec Kralove, Czech Republic
Ondrej Krejcar	University of Hradec Kralove, Czech Republic

Peter Brida   University of Žilina, Slovakia
Peter Mikulecky   University of Hradec Kralove, Czech Republic

7. *MLMB 2017: Special Session on Machine Learning in Medicine and Biometrics*

Piotr Porwik   University of Silesia, Poland
Alicja Wakulicz-Deja   University of Silesia, Poland
Agnieszka   University of Silesia, Poland
  Nowak-Brzezińska

International Program Committee

Muhammad Abulaish   Jamia Millia Islamia (A Central University), India
Sharat Akhoury   University of Cape Town, South Africa
Ana Almeida   GECAD-ISEP-IPP, Portugal
Orcan Alpar   University of Hradec Králové, Czech Republic
Bashar Al-Shboul   University of Jordan, Jordan
Thierry Badard   Laval University, Canada
Amelia Badica   University of Craiova, Romania
Costin Badica   University of Craiova, Romania
Hassan Badir   Ecole Nationale des Sciences Appliquées de Tanger,
                          Morocco
Dariusz Barbucha   Gdynia Maritime University, Poland
Nick Bassiliades   Aristotle University of Thessaloniki, Greece
Maria Bielikova   Slovak University of Technology in Bratislava,
                          Slovakia
Leon Bobrowski   Bialystok University of Technology, Poland
Mariusz Boryczka   University of Silesia, Poland
Urszula Boryczka   University of Silesia, Poland
Abdelhamid Bouchachia   Bournemouth University, UK
Peter Brida   University of Žilina, Slovakia
Robert Burduk   Wrocław University of Science and Technology,
                          Poland
Krisztian Buza   Budapest University of Technology and Economics,
                          Hungary
Aleksander Byrski   AGH University of Science and Technology, Poland
Jose Luis Calvo-Rolle   University of A Coruna, Spain
David Camacho   Universidad Autonoma de Madrid, Spain
Alberto Cano   Virginia Commonwealth University, USA
Frantisek Capkovic   Slovak Academy of Sciences, Slovakia
Richard Chbeir   LIUPPA Laboratory, France
Shyi-Ming Chen   National Taiwan University of Science
                          and Technology, Taiwan
Amine Chohra   Paris-East University (UPEC), France

Kazimierz Choroś	Wrocław University of Science and Technology, Poland
Mihaela Colhon	University of Craiova, Romania
Jose Alfredo Ferreira Costa	Universidade Federal do Rio Grande do Norte, Brazil
Boguslaw Cyganek	AGH University of Science and Technology, Poland
Ireneusz Czarnowski	Gdynia Maritime University, Poland
Paul Davidsson	Malmö University, Sweden
Tien V. Do	Budapest University of Technology and Economics, Hungary
Vadim Ermolayev	Zaporozhye National University, Ukraine
Nadia Essoussi	University of Carthage, Tunisia
Rim Faiz	University of Carthage, Tunisia
Faiez Gargouri	University of Sfax, Tunisia
Mauro Gaspari	University of Bologna, Italy
Janusz Getta	University of Wollongong, Australia
Daniela Gifu	University "Alexandru Ioan Cuza" of Iasi, Romania
Daniela Godoy	ISISTAN Research Institute, Argentina
Antonio Gonzalez-Pardo	Universidad Autonoma de Madrid, Spain
Manuel Grana	University of the Basque Country, Spain
Foteini Grivokostopoulou	University of Patras, Greece
Marcin Hernes	Wrocław University of Economics, Poland
Huu Hanh Hoang	Hue University, Vietnam
Tzung-Pei Hong	National University of Kaohsiung, Taiwan
Mong-Fong Horng	National Kaohsiung University of Applied Sciences, Taiwan
Frederic Hubert	Laval University, Canada
Maciej Huk	Wrocław University of Science and Technology, Poland
Dosam Hwang	Yeungnam University, South Korea
Lazaros Iliadis	Democritus University of Thrace, Greece
Agnieszka Indyka-Piasecka	Wrocław University of Science and Technology, Poland
Dan Istrate	Université de Technologie de Compiégne, France
Mirjana Ivanovic	University of Novi Sad, Serbia
Jaroslaw Jankowski	West Pomeranian University of Technology, Poland
Joanna Jędrzejowicz	University of Gdańsk, Poland
Piotr Jędrzejowicz	Gdynia Maritime University, Poland
Gordan Jezic	University of Zagreb, Croatia
Geun Sik Jo	Inha University, South Korea
Kang-Hyun Jo	University of Ulsan, South Korea
Jason Jung	Chung-Ang University, South Korea
Tomasz Kajdanowicz	Wrocław University of Science and Technology, Poland
Petros Kefalas	University of Sheffield International Faculty, CITY College, Greece

Rafał Kern	Wrocław University of Science and Technology, Poland
Marek Kisiel-Dorohinicki	AGH University of Science and Technology, Poland
Attila Kiss	Eotvos Lorand University, Hungary
Marek Kopel	Wrocław University of Science and Technology, Poland
Jerzy Korczak	Wrocław University of Economics, Poland
Jacek Koronacki	Polish Academy of Sciences, Poland
Leszek Kotulski	AGH University of Science and Technology, Poland
Ivan Koychev	University of Sofia "St. Kliment Ohridski", Bulgaria
Jan Kozak	University of Economics in Katowice, Poland
Adrianna Kozierkiewicz-Hetmańska	Wrocław University of Science and Technology, Poland
Bartosz Krawczyk	Virginia Commonwealth University, USA
Ondrej Krejcar	University of Hradec Králové, Czech Republic
Dalia Kriksciuniene	Vilnius University, Lithuania
Dariusz Król	Wrocław University of Science and Technology, Poland
Elzbieta Kukla	Wrocław University of Science and Technology, Poland
Julita Kulbacka	Wrocław Medical University, Poland
Marek Kulbacki	Polish-Japanese Academy of Information Technology, Poland
Piotr Kulczycki	Polish Academy of Sciences, Poland
Kazuhiro Kuwabara	Ritsumeikan University, Japan
Halina Kwaśnicka	Wrocław University of Science and Technology, Poland
Mark Last	Ben-Gurion University of the Negev, Israel
Nguyen Le Minh	Japan Advanced Institute of Science and Technology, Japan
Hoai An Le Thi	Université de Lorraine, France
Florin Leon	"Gheorghe Asachi" Technical University of Iasi, Romania
Edwin Lughofer	Johannes Kepler University Linz, Austria
Juraj Machaj	University of Žilina, Slovakia
Bernadetta Maleszka	Wrocław University of Science and Technology, Poland
Marcin Maleszka	Wrocław University of Science and Technology, Poland
Yannis Manolopoulos	Aristotle University of Thessaloniki, Greece
Urszula Markowska-Kaczmar	Wrocław University of Science and Technology, Poland
Adam Meissner	Poznań University of Technology, Poland
Ernestina Menasalvas	Universidad Politecnica de Madrid, Spain
Hector Menendez	Universidad Autonoma de Madrid, Spain

Jacek Mercik	WSB University in Wrocław, Poland
Radoslaw Michalski	Wrocław University of Science and Technology, Poland
Peter Mikulecky	University of Hradec Králové, Czech Republic
Alin Moldoveanu	University Politehnica of Bucharest, Romania
Javier Montero	Universidad Complutense de Madrid, Spain
Ahmed Moussa	Université Abdelmalek Essaadi, Morocco
Dariusz Mrozek	Silesian University of Technology, Poland
Kazumi Nakamatsu	University of Hyogo, Japan
Grzegorz J. Nalepa	AGH University of Science and Technology, Poland
Fulufhelo Nelwamondo	Council for Scientific and Industrial Research, South Africa
Filippo Neri	University of Napoli Federico II, Italy
Linh Anh Nguyen	University of Warsaw, Poland
Loan T.T. Nguyen	University of Warsaw, Poland
Adam Niewiadomski	Lodz University of Technology, Poland
Agnieszka Nowak-Brzezinska	University of Silesia, Poland
Alberto Nunez	Universidad Complutense de Madrid, Spain
Manuel Nunez	Universidad Complutense de Madrid, Spain
Tarkko Oksala	Aalto University, Finland
Mieczyslaw Owoc	Wrocław University of Economics, Poland
Marcin Paprzycki	Polish Academy of Sciences, Poland
Rafael Parpinelli	Santa Catarina State University, Brazil
Marek Penhaker	VSB -Technical University of Ostrava, Czech Republic
Isidoros Perikos	University of Patras, Greece
Marcin Pietranik	Wrocław University of Science and Technology, Poland
Elias Pimenidis	University of the West of England, UK
Piotr Porwik	University of Silesia, Poland
Radu-Emil Precup	Politehnica University of Timisoara, Romania
Ales Prochazka	University of Chemistry and Technology, Czech Republic
Paulo Quaresma	Universidade de Evora, Portugal
Mohammad Rashedur Rahman	North South University, Bangladesh
Ewa Ratajczak-Ropel	Gdynia Maritime University, Poland
Tomasz Rutkowski	University of Tokyo, Japan
Jose A. Saez	University of Salamanca, Spain
Virgilijus Sakalauskas	Vilnius University, Lithuania
Jose L. Salmeron	University Pablo de Olavide, Spain
Ali Selamat	Universiti Teknologi Malaysia, Malaysia
Natalya Shakhovska	Lviv Polytechnic National University, Ukraine
Andrzej Siemiński	Wrocław University of Science and Technology, Poland
Dragan Simic	University of Novi Sad, Serbia

Vladimir Sobeslav	University of Hradec Králové, Czech Republic
Stanimir Stoyanov	University of Plovdiv "Paisii Hilendarski", Bulgaria
Yasufumi Takama	Tokyo Metropolitan University, Japan
Zbigniew Telec	Wrocław University of Science and Technology, Poland
Diana Trandabat	University "Alexandru Ioan Cuza" of Iasi, Romania
Bogdan Trawinski	Wrocław University of Science and Technology, Poland
Jan Treur	Vrije Universiteit Amsterdam, Netherlands
Maria Trocan	Institut Supérieur d'Électronique de Paris, France
Krzysztof Trojanowski	Cardinal Stefan Wyszyński University in Warsaw, Poland
Ualsher Tukeyev	al-Farabi Kazakh National University, Kazakhstan
Olgierd Unold	Wrocław University of Science and Technology, Poland
Ventzeslav Valev	Bulgarian Academy of Sciences, Bulgaria
Bay Vo	Ho Chi Minh City University of Technology, Vietnam
Gottfried Vossen	ERCIS Muenster, Germany
Lipo Wang	Nanyang Technological University, Singapore
Izabela Wierzbowska	Gdynia Maritime University, Poland
Michal Wozniak	Wrocław University of Science and Technology, Poland
Sławomir Zadrożny	Polish Academy of Sciences, Poland
Drago Zagar	University of Osijek, Croatia
Danuta Zakrzewska	Lodz University of Technology, Poland
Constantin-Bala Zamfirescu	"Lucian Blaga" University of Sibiu, Romania
Katerina Zdravkova	Ss. Cyril and Methodius University in Skopje, Macedonia
Aleksander Zgrzywa	Wrocław University of Science and Technology, Poland
Adam Ziębiński	Silesian University of Technology, Poland

Program Committees of Special Sessions

CSI 2017: Special Session on Computational Swarm Intelligence

Urszula Boryczka	University of Silesia, Poland
Tomasz Gwizdałła	University of Łódź, Poland
Jarosław Wąs	AGH University of Science and Technology, Poland
Ajith Abraham	Scientific Network for Innovation and Research Excellence, USA
Andrew Adamatzky	University of the West of England, UK
Costin Badica	University of Craiova, Romania
Jan Baetens	Ghent University, Belgium
Mariusz Boryczka	University of Silesia, Poland
Wojciech Froelich	University of Silesia, Poland

Rolf Hoffmann   Technische Universität Darmstadt, Germany
Genaro Martínez   Computer Science Laboratory IPN, Mexico
Dariusz Pierzchała   Military University of Technology, Poland
Franciszek Seredyński   Cardinal Stefan Wyszyński University in Warsaw,
                              Poland
Georgios Sirakoulis   Democritus University of Thrace, Greece
Rafał Skinderowicz   University of Silesia, Poland
Mirosław Szaban   Siedlce University of Science and Humanities, Poland
William Spataro   University of Calabria, Italy
Krzysztof Trojanowski   Cardinal Stefan Wyszyński University in Warsaw,
                              Poland
Barbara Wolnik   University of Gdansk, Poland
Wojciech Wieczorek   University of Silesia, Poland

WebSys 2017: Special Session on Intelligent Processing
of Multimedia in Web Systems

Kazimierz Choroś   Wrocław University of Science and Technology,
                              Poland
Jarosław Jankowski   West Pomeranian University of Technology, Poland
Ondřej Krejcar   University of Hradec Kralove, Czech Republic
Tarkko Oksala   Helsinki University of Technology, Finland
Andrzej Siemiński   Wrocław University of Science and Technology,
                              Poland
Jérémie Sublime   Institut Supérieur d'Électronique de Paris, France
Maria Trocan   Institut Supérieur d'Électronique de Paris, France

CPSiA 2017: Special Session on Cyber-Physical Systems in the Automotive Area

Markus Bregulla   Technische Hochschule Ingolstadt, Germany
Daniel Grossman   Technische Hochschule Ingolstadt, Germany
Dariusz Kania   Silesian University of Technology, Poland
Rafał Cupek   Silesian University of Technology Poland
Hueseyin Erdogan   Continental Ingolstadt, Germany
Damian Grzechca   Silesian University of Technology, Poland
Sebastian Budzan   Silesian University of Technology, Poland
Roman Wyżgolik   Silesian University of Technology, Poland
Krzysztof Tokarz   Silesian University of Technology, Poland
Marcin Fojcik   Western Norway University of Applied Sciences,
                              Norway
Mirosław Łazoryszczak   West Pomeranian University of Technology, Poland
Grzegorz Ulacha   West Pomeranian University of Technology, Poland
Krzysztof Małecki   West Pomeranian University of Technology, Poland
Grzegorz Andrzejewski   University of Zielona Góra, Poland
Marek Drewniak   Aiut Sp. z o.o., Poland
Adam Ziębiński   Silesian University of Technology, Poland

LRLP 2017: Special Session on Low-Resource Languages Processing

Ualsher Tukeyev	al-Farabi Kazakh National University, Kazakhstan
Zhandos Zhumanov	al-Farabi Kazakh National University, Kazakhstan
Madina Mansurova	al-Farabi Kazakh National University, Kazakhstan
Altynbek Sharipbay	L.N. Gumilyov Eurasian National University, Kazakhstan
Rustam Musabayev	Institute of Information and Computational Technologies, Kazakhstan
Zhenisbek Assylbekov	Nazarbayev University, Kazakhstan
Jonathan Washington	Swarthmore College, USA
Djavdet Suleimanov	Institute of Applied Semiotics, Russia
Alimzhanov Yermek	al-Farabi Kazakh National University, Kazakhstan

CSDMO 2017: Special Session on Cooperative Strategies
for Decision-Making and Optimization

Dariusz Barbucha	Gdynia Maritime University, Poland
Vincenzo Cutello	University of Catania, Italy
Ireneusz Czarnowski	Gdynia Maritime University, Poland
Joanna Jędrzejowicz	Gdansk University, Poland
Piotr Jędrzejowicz	Gdynia Maritime University, Poland
Edyta Kucharska	AGH University of Science and Technology, Poland
Antonio D. Masegosa	University of Deusto, Spain
Javier Montero	Complutense University, Spain
Ewa Ratajczak-Ropel	Gdynia Maritime University, Poland
Iza Wierzbowska	Gdynia Maritime University, Poland
Mahdi Zargayouna	IFSTTAR, France

IoT-RC 2017: Special Session on Internet of Things – Its Relations and Consequences

Ana Almeida	Porto Superior Institute of Engineering, Portugal
Jorge Bernardino	Polytechnical Institute of Coimbra, Spain
Peter Brida	University of Žilina, Slovakia
Ivan Dolnak	University of Žilina, Slovakia
Josef Horalek	University of Hradec Kralove, Czech Republic
Ondrej Krejcar	University of Hradec Kralove, Czech Republic
Goreti Marreiros	Porto Superior Institute of Engineering, Portugal
Peter Mikulecký	University of Hradec Kralove, Czech Republic
Juraj Machaj	University of Žilina, Slovakia
Marek Penhaker	VSB Technical University of Ostrava, Czech Republic
José Salmeron	Universidad Pablo de Olavide of Seville, Spain
Ali Selamat	Universiti Teknologi Malaysia, Malaysia
Vladimir Sobeslav	University of Hradec Kralove, Czech Republic
Stylianakis Vassilis	University of Patras, Greece
Petr Tucnik	University of Hradec Kralove, Czech Republic

MLMB 2017: Special Session on Machine Learning in Medicine and Biometrics

Nabendu Chaki	University of Calcutta, India
Robert Czabański	University of Silesia, Poland
Rafał Deja	Academy of Business in Dąbrowa Górnicza, Poland
Michał Dramiński	Polish Academy of Sciences, Poland
Adam Gacek	Institute of Medical Technology and Equipment, Poland
Marina Gavrilova	University of Calgary, Canada
Manuel Grana	Computer Intelligence Group, Spain
Michał Kozielski	Silesian University of Technology, Poland
Marek Kurzyński	Wrocław University of Technology, Poland
Dariusz Mrozek	Silesian University of Technology, Poland
Bożena Małysiak-Mrozek	Silesian University of Technology, Poland
Agnieszka Nowak-Brzezińska	University of Silesia, Poland
Nobuyuki Nishiuchi	Tokyo Metropolitan University, Japan
Piotr Porwik	University of Silesia, Poland
Małgorzata Przybyła-Kasperek	University of Silesia, Poland
Roman Simiński	University of Silesia, Poland
Dragan Simic	University of Novi Sad, Serbia
Ewaryst Tkacz	Silesian University of Technology, Poland
Alicja Wakulicz-Deja	University of Silesia, Katowice, Poland

Additional Reviewers

Ben Brahim, Afef	Meditskos, Georgios	Piasny, Lukasz
Filonenko, Alexander	Mihailescu, Radu-Casian	Schomm, Fabian
Holmgren, Johan	Mls, Karel	Thilakarathne, Dilhan
Le, Hoai Minh	Montero, Javier	Vascak, Jan
Liutvinavicius, Marius	Phan, Duy Nhat	Zając, Wojciech

Contents – Part II

Cooperative Strategies for Decision Making and Optimization

Gene Expression Programming Ensemble for Classifying Big Datasets   3
Joanna Jędrzejowicz and Piotr Jędrzejowicz

Shapley Value in a Priori Measuring of Intellectual Capital Flows.   13
Jacek Mercik

MDBR: Mobile Driving Behavior Recognition Using Smartphone Sensors. . .   22
Dang-Nhac Lu, Thi-Thu-Trang Ngo, Hong-Quang Le,
Thi-Thu-Hien Tran, and Manh-Hai Nguyen

Adaptive Motivation System Under Modular Reinforcement Learning
for Agent Decision-Making Modeling of Biological Regulation   32
Amine Chohra and Kurosh Madani

Computational Swarm Intelligence

Simulated Annealing for Finding TSP Lower Bound   45
Łukasz Strąk, Wojciech Wieczorek, and Arkadiusz Nowakowski

A Cellular Automaton Based System for Traffic Analyses on the
Roundabout   56
Krzysztof Małecki, Jarosław Wątróbski, and Waldemar Wolski

The Swarm-Like Update Scheme for Opinion Formation   66
Tomasz M. Gwizdałła

A Comparative Study of Different Variants of a Memetic Algorithm
for ATSP   76
Krzysztof Szwarc and Urszula Boryczka

Improving ACO Convergence with Parallel Tempering   87
Rafał Skinderowicz

Modeling Skiers' Dynamics and Behaviors.   97
Dariusz Pałka and Jarosław Wąs

Genetic Algorithm as Optimization Tool for Differential Cryptanalysis
of DES6   107
Kamil Dworak and Urszula Boryczka

Machine Learning in Medicine and Biometrics

Edge Real-Time Medical Data Segmentation for IoT Devices
with Computational and Memory Constrains 119
  Marcin Bernas, Bartłomiej Płaczek, and Alicja Sapek

A Privacy Preserving and Safety-Aware Semi-supervised Model
for Dissecting Cancer Samples 129
  P.S. Deepthi and Sabu M. Thampi

Decision Fusion Methods in a Dispersed Decision System - A Comparison
on Medical Data.. 139
  *Małgorzata Przybyła-Kasperek, Agnieszka Nowak-Brzezińska,
  and Roman Simiński*

Knowledge Exploration in Medical Rule-Based Knowledge Bases......... 150
  *Agnieszka Nowak-Brzezińska, Tomasz Rybotycki, Roman Simiński,
  and Małgorzata Przybyła-Kasperek*

Computer User Verification Based on Typing Habits
and Finger-Knuckle Analysis 161
  *Hossein Safaverdi, Tomasz Emanuel Wesolowski, Rafal Doroz,
  Krzysztof Wrobel, and Piotr Porwik*

ANN and GMDH Algorithms in QSAR Analyses of Reactivation Potency
for Acetylcholinesterase Inhibited by VX Warfare Agent 171
  *Rafael Dolezal, Jiri Krenek, Veronika Racakova, Natalie Karaskova,
  Nadezhda V. Maltsevskaya, Michaela Melikova, Karel Kolar,
  Jan Trejbal, and Kamil Kuca*

Multiregional Segmentation Modeling in Medical Ultrasonography:
Extraction, Modeling and Quantification of Skin Layers
and Hypertrophic Scars 182
  *Iveta Bryjova, Jan Kubicek, Kristyna Molnarova, Lukas Peter,
  Marek Penhaker, and Kamil Kuca*

Cyber Physical Systems in Automotive Area

Model of a Production Stand Used for Digital Factory Purposes 195
  *Markus Bregulla, Sebastian Schrittenloher, Jakub Piekarz,
  and Marek Drewniak*

Improving the Engineering Process in the Automotive Field Through
AutomationML 205
  Markus Bregulla and Flavian Meltzer

Enhanced Reliability of ADAS Sensors Based on the Observation
of the Power Supply Current and Neural Network Application   215
  Damian Grzechca, Adam Ziębiński, and Paweł Rybka

ADAS Device Operated on CAN Bus Using PiCAN Module
for Raspberry Pi   227
  Marek Drewniak, Krzysztof Tokarz, and Michał Rędziński

Obstacle Avoidance by a Mobile Platform Using an Ultrasound Sensor   238
  Adam Ziebinski, Rafal Cupek, and Marek Nalepa

Monitoring and Controlling Speed for an Autonomous Mobile Platform
Based on the Hall Sensor   249
  Adam Ziebinski, Markus Bregulla, Marcin Fojcik, and Sebastian Kłak

Using MEMS Sensors to Enhance Positioning When
the GPS Signal Disappears   260
  *Damian Grzechca, Krzysztof Tokarz, Krzysztof Paszek,
  and Dawid Poloczek*

Application of OPC UA Protocol for the Internet of Vehicles   272
  Rafał Cupek, Adam Ziębiński, Marek Drewniak, and Marcin Fojcik

Feasibility Study of the Application of OPC UA Protocol
for the Vehicle-to-Vehicle Communication   282
  Rafał Cupek, Adam Ziębiński, Marek Drewniak, and Marcin Fojcik

Data Mining Techniques for Energy Efficiency Analysis of Discrete
Production Lines   292
  *Rafal Cupek, Jakub Duda, Dariusz Zonenberg, Łukasz Chłopaś,
  Grzegorz Dziędziel, and Marek Drewniak*

Internet of Things - Its Relations and Consequences

Different Approaches to Indoor Localization Based on Bluetooth
Low Energy Beacons and Wi-Fi   305
  Radek Bruha and Pavel Kriz

Towards Device Interoperability in an Heterogeneous
Internet of Things Environment   315
  Pavel Pscheidl, Richard Cimler, and Hana Tomášková

Hardware Layer of Ambient Intelligence Environment Implementation   325
  Ales Komarek, Jakub Pavlik, Lubos Mercl, and Vladimir Sobeslav

Lightweight Protocol for M2M Communication   335
  *Jan Stepan, Richard Cimler, Jan Matyska, David Sec,
  and Ondrej Krejcar*

Wildlife Presence Detection Using the Affordable Hardware Solution
and an IR Movement Detector   345
    Jan Stepan, Matej Danicek, Richard Cimler, Jan Matyska,
    and Ondrej Krejcar

Text Processing and Information Retrieval

Word Embeddings Versus LDA for Topic Assignment in Documents   357
    Joanna Jędrzejowicz and Magdalena Zakrzewska

Development of a Sustainable Design Lexicon. Towards Understanding
the Relationship Between Sentiments, Attitudes and Behaviours   367
    Vargas Meza Xanat and Yamanaka Toshimasa

Analysing Cultural Events on Twitter   376
    Brigitte Juanals and Jean-Luc Minel

A Temporal-Causal Model for Spread of Messages in Disasters   386
    Eric Fernandes de Mello Araújo, Annelore Franke,
    and Rukshar Wagid Hosain

One Approach to the Description of Linguistic Uncertainties   398
    Nikita Ogorodnikov

Complex Search Queries in the Corpus Management System   407
    Damir Mukhamedshin, Olga Nevzorova, and Aidar Khusainov

Entropy-Based Model for Estimating Veracity of Topics from Tweets   417
    Jyotsna Paryani, Ashwin Kumar T.K., and K.M. George

On Some Approach to Integrating User Profiles in Document Retrieval
System Using Bayesian Networks   428
    Bernadetta Maleszka

Analysis of Denoising Autoencoder Properties Through Misspelling
Correction Task   438
    Karol Draszawka and Julian Szymański

New Ontological Approach for Opinion Polarity Extraction from Twitter   448
    Ammar Mars, Sihem Hamem, and Mohamed Salah Gouider

Study for Automatic Classification of Arabic Spoken Documents   459
    Mohamed Labidi, Mohsen Maraoui, and Mounir Zrigui

"Come Together!": Interactions of Language Networks and Multilingual
Communities on Twitter.   469
    Nabeel Albishry, Tom Crick, and Theo Tryfonas

Bangla News Summarization   479
  Anirudha Paul, Mir Tahsin Imtiaz, Asiful Haque Latif, Muyeed Ahmed,
  Foysal Amin Adnan, Raiyan Khan, Ivan Kadery,
  and Rashedur M. Rahman

Low Resource Language Processing

Combined Technology of Lexical Selection in Rule-Based
Machine Translation   491
  Ualsher Tukeyev, Dina Amirova, Aidana Karibayeva, Aida Sundetova,
  and Balzhan Abduali

New Kazakh Parallel Text Corpora with On-line Access   501
  Zhandos Zhumanov, Aigerim Madiyeva, and Diana Rakhimova

Design and Development of Media-Corpus of the Kazakh Language   509
  Madina Mansurova, Gulmira Madiyeva, Sanzhar Aubakirov,
  Zhantemir Yermekov, and Yermek Alimzhanov

Morphological Analysis System of the Tatar Language   519
  Rinat Gilmullin and Ramil Gataullin

Context-Based Rules for Grammatical Disambiguation
in the Tatar Language   529
  Ramil Gataullin, Bulat Khakimov, Dzhavdet Suleymanov,
  and Rinat Gilmullin

Computer Vision Techniques

Evaluation of Gama Analysis Results Significance Within Verification
of Radiation IMRT Plans in Radiotherapy   541
  Jan Kubicek, Iveta Bryjova, Kamila Faltynova, Marek Penhaker,
  Martin Augustynek, and Petra Maresova

Shape Classification Using Combined Features   549
  Laksono Kurnianggoro, Wahyono, Alexander Filonenko,
  and Kang-Hyun Jo

Smoke Detection on Video Sequences Using Convolutional and Recurrent
Neural Networks   558
  Alexander Filonenko, Laksono Kurnianggoro, and Kang-Hyun Jo

Intelligent Processing of Multimedia in Web Systems

Improved Partitioned Shadow Volumes Method of Real-Time Rendering
Using Balanced Trees   569
  Kazimierz Choroś and Tomasz Suder

Online Comparison System with Certain and Uncertain Criteria Based
on Multi-criteria Decision Analysis Method   579
    Paweł Ziemba, Jarosław Jankowski, and Jarosław Wątróbski

Assessing and Improving Sensors Data Quality in Streaming Context   590
    *Rayane El Sibai, Yousra Chabchoub, Raja Chiky, Jacques Demerjian,
    and Kablan Barbar*

Neural Network Based Eye Tracking   600
    Pavel Morozkin, Marc Swynghedauw, and Maria Trocan

Author Index   611

Contents – Part I

Knowledge Engineering and Semantic Web

Mapping the Territory for a Knowledge-Based System   3
   Ulrich Schmitt

A Bidirectional-Based Spreading Activation Method for Human
Diseases Relatedness Detection Using Disease Ontology   14
   Said Fathalla and Yaman Kannot

Semantic Networks Modeling with Operand-Operator Structures
in Association-Oriented Metamodel   24
   Marek Krótkiewicz, Marcin Jodłowiec, and Krystian Wojtkiewicz

Knowledge Integration in a Manufacturing Planning Module
of a Cognitive Integrated Management Information System   34
   Marcin Hernes and Andrzej Bytniewski

The Knowledge Increase Estimation Framework for Ontology
Integration on the Relation Level   44
   Adrianna Kozierkiewicz-Hetmańska and Marcin Pietranik

Particle Swarm of Agents for Heterogenous Knowledge Integration.   54
   Marcin Maleszka

Design Proposal of the Corporate Knowledge Management System.   63
   Ivan Soukal and Aneta Bartuskova

Dipolar Data Integration Through Univariate, Binary Classifiers   73
   Leon Bobrowski

Intelligent Collective: The Role of Diversity and Collective Cardinality   83
   Van Du Nguyen, Mercedes G. Merayo, and Ngoc Thanh Nguyen

RuQAR: Querying OWL 2 RL Ontologies with Rule Engines
and Relational Databases   93
   Jarosław Bąk and Michał Blinkiewicz

The Efficiency Analysis of the Multi-level Consensus
Determination Method. 103
   Adrianna Kozierkiewicz-Hetmańska and Mateusz Sitarczyk

Collective Intelligence Supporting Trading Decisions on FOREX Market. . . . 113
   Jerzy Korczak, Marcin Hernes, and Maciej Bac

Social Networks and Recommender Systems

Testing the Acceptability of Social Support Agents in Online Communities . . .   125
    Lenin Medeiros and Tibor Bosse

Enhancing New User Cold-Start Based on Decision Trees Active Learning
by Using Past Warm-Users Predictions   137
    Manuel Pozo, Raja Chiky, Farid Meziane, and Elisabeth Métais

An Efficient Parallel Method for Performing Concurrent Operations
on Social Networks.   148
    Phuong-Hanh Du, Hai-Dang Pham, and Ngoc-Hoa Nguyen

Simulating Collective Evacuations with Social Elements   160
    Daniel Formolo and C. Natalie van der Wal

Social Networks Based Framework for Recommending Touristic Locations . . .   172
    Mehdi Ellouze, Slim Turki, Younes Djaghloul, and Muriel Foulonneau

Social Network-Based Event Recommendation   182
    Dinh Tuyen Hoang, Van Cuong Tran, and Dosam Hwang

Deep Neural Networks for Matching Online Social Networking Profiles   192
    Vicentiu-Marian Ciorbaru and Traian Rebedea

Effect of Network Topology on Neighbourhood-Aided Collective Learning . . .   202
    Lise-Marie Veillon, Gauvain Bourgne, and Henry Soldano

A Generic Approach to Evaluate the Success of Online Communities   212
    Raoudha Chebil, Wided Lejouad Chaari, and Stefano A. Cerri

Considerations in Analyzing Ecological Dependent Populations
in a Changing Environment   223
    Kristiyan Balabanov, Robinson Guerra Fietz, and Doina Logofătu

Automatic Deduction of Learners' Profiling Rules Based on Behavioral
Analysis.   233
    Fedia Hlioui, Nadia Aloui, and Faiez Gargouri

Predicting the Evolution of Scientific Output   244
    Antonia Gogoglou and Yannis Manolopoulos

Data Mining Methods and Applications

Enhanced Hybrid Component-Based Face Recognition   257
    Andile M. Gumede, Serestina Viriri, and Mandlenkosi V. Gwetu

Enhancing Cholera Outbreaks Prediction Performance in Hanoi,
Vietnam Using Solar Terms and Resampling Data   266
  Nguyen Hai Chau

Solving Dynamic Traveling Salesman Problem with Ant
Colony Communities.   277
  Andrzej Siemiński

Improved Stock Price Prediction by Integrating Data Mining Algorithms
and Technical Indicators: A Case Study on Dhaka Stock Exchange.   288
  Syeda Shabnam Hasan, Rashida Rahman, Noel Mannan,
  Haymontee Khan, Jebun Nahar Moni, and Rashedur M. Rahman

A Data Mining Approach to Improve Remittance
by Job Placement in Overseas.   298
  Ahsan Habib Himel, Tonmoy Sikder, Sheikh Faisal Basher,
  Ruhul Mashbu, Nusrat Jahan Tamanna, Mahmudul Abedin,
  and Rashedur M. Rahman

Determining Murder Prone Areas Using Modified Watershed Model   307
  Joytu Khisha, Naushaba Zerin, Deboshree Choudhury,
  and Rashedur M. Rahman

Comparison of Ensemble Learning Models with Expert
Algorithms Designed for a Property Valuation System.   317
  Bogdan Trawiński, Tadeusz Lasota, Olgierd Kempa, Zbigniew Telec,
  and Marcin Kutrzyński

Multi-agent Systems

Multiagent Coalition Structure Optimization by Quantum Annealing   331
  Florin Leon, Andrei-Ştefan Lupu, and Costin Bădică

External Environment Scanning Using Cognitive Agents   342
  Marcin Hernes, Anna Chojnacka-Komorowska, and Kamal Matouk

OpenCL for Large-Scale Agent-Based Simulations   351
  Jan Procházka and Kamila Štekerová

A Novel Space Filling Curves Based Approach to PSO Algorithms
for Autonomous Agents.   361
  Doina Logofătu, Gil Sobol, Daniel Stamate, and Kristiyan Balabanov

Multiplant Production Design in Agent-Based Artificial Economic System. . .   371
  Petr Tucnik, Zuzana Nemcova, and Tomas Nachazel

Role of Non-Axiomatic Logic in a Distributed Reasoning Environment   381
  Mirjana Ivanović, Jovana Ivković, and Costin Bădică

Agent Having Quantum Properties: The Superposition States
and the Entanglement   389
  Alain-Jérôme Fougères

Sensor Networks and Internet of Things

A Profile-Based Fast Port Scan Detection Method.   401
  Katalin Hajdú-Szücs, Sándor Laki, and Attila Kiss

Sensor Network Coverage Problem: A Hypergraph Model Approach.   411
  Krzysztof Trojanowski, Artur Mikitiuk, and Mateusz Kowalczyk

Heuristic Optimization of a Sensor Network Lifetime
Under Coverage Constraint   422
  *Krzysztof Trojanowski, Artur Mikitiuk, Frédéric Guinand,
  and Michał Wypych*

Methods of Training of Neural Networks for Short Term Load
Forecasting in Smart Grids.   433
  Robert Lis, Artem Vanin, and Anastasiia Kotelnikova

Scheduling Sensors Activity in Wireless Sensor Networks   442
  Antonina Tretyakova, Franciszek Seredynski, and Frederic Guinand

Application of Smart Multidimensional Navigation in Web-Based Systems. . . .   452
  Ivan Soukal and Aneta Bartuskova

WINE: Web Integrated Navigation Extension; Conceptual Design,
Model and Interface   462
  Ivan Soukal and Aneta Bartuskova

Real-Life Validation of Methods for Detecting Locations, Transition
Periods and Travel Modes Using Phone-Based GPS and Activity
Tracker Data   473
  *Adnan Manzoor, Julia S. Mollee, Aart T. van Halteren,
  and Michel C.A. Klein*

Adaptive Runtime Middleware: Everything as a Service   484
  *Achilleas P. Achilleos, Kyriaki Georgiou, Christos Markides,
  Andreas Konstantinidis, and George A. Papadopoulos*

Decision Support & Control Systems

Adaptive Neuro Integral Sliding Mode Control on Synchronization
of Two Robot Manipulators   497
  Parvaneh Esmaili and Habibollah Haron

Ant-Inspired, Invisible-Hand-Controlled Robotic System to Support
Rescue Works After Earthquake   507
   Tadeusz Szuba

Estimation of Delays for Individual Trams to Monitor Issues in Public
Transport Infrastructure   518
   Marcin Luckner and Jan Karwowski

Novel Effective Algorithm for Synchronization Problem
in Directed Graph   528
   Richard Cimler, Dalibor Cimr, Jitka Kuhnova, and Hana Tomaskova

Bimodal Biometric Method Fusing Hand Shape and Palmprint
Modalities at Rank Level   538
   Nesrine Charfi, Hanene Trichili, and Basel Solaiman

Adaptation to Market Development Through Price Setting Strategies
in Agent-Based Artificial Economic Model   548
   Petr Tucnik, Petr Blecha, and Jaroslav Kovarnik

Efficacy and Planning in Ophthalmic Surgery – A Vision
of Logical Programming   558
   Nuno Maia, Manuel Mariano, Goreti Marreiros, Henrique Vicente,
   and José Neves

A Methodological Approach Towards Crisis Simulations:
Qualifying CI-Enabled Information Systems   569
   Chrysostomi Maria Diakou, Angelika I. Kokkinaki,
   and Styliani Kleanthous

Multicriteria Transportation Problems with Fuzzy Parameters   579
   Barbara Gładysz

Author Index   589

Cooperative Strategies for Decision Making and Optimization

Gene Expression Programming Ensemble for Classifying Big Datasets

Joanna Jędrzejowicz[1(✉)] and Piotr Jędrzejowicz[2]

[1] Institute of Informatics, Faculty of Mathematics, Physics and Informatics,
University of Gdańsk, 80-308 Gdańsk, Poland
`jj@inf.ug.edu.pl`
[2] Department of Information Systems, Gdynia Maritime University,
Morska 83, 81-225 Gdynia, Poland
`pj@am.gdynia.pl`

Abstract. The paper proposes a new GEP-based batch ensemble classifier constructed using the stacked generalization concept. In our approach combination of base classifiers involves evolving the meta-gene using genes induced by GEP from randomly generated combinations of instances with randomly selected subsets of attributes. The main property of the discussed classifier is its scalability allowing adaptation to the size of the dataset under consideration. To validate the proposed classifier, we have carried-out computational experiment involving a number of publicly available benchmark datasets. Experiment results show that the approach assures good performance, scalability and robustness.

Keywords: Gene expression · Classification · Big data sets

1 Introduction

The goal of classification is to predict unknown objects or concepts. Classification requires inducing the model (or function) that describes and distinguishes data, classes or concepts. Among numerous approaches, models and learners there are classifiers based on Gene Expression Programming (GEP). Gene Expression Programming was introduced by Ferreira [5]. In GEP programs are represented as linear character strings of fixed length called chromosomes which, in the subsequent fitness evaluation, evolve into expression trees without any user intervention. This property makes GEP induced expression trees a useful tool for constructing classifiers [6]. Among several example applications of GEP to solving classification problems one can mention GEPCLASS system allowing for the automatic discovery of flexible rules, better fitted to data [23]. Zeng with co-authors [26] proposed a novel Immune Gene Expression Programming as a tool for rule mining. A different example of GEP application to classification was proposed by Li and co-authors [15]. They proposed a new representation scheme

© Springer International Publishing AG 2017
N.T. Nguyen et al. (Eds.): ICCCI 2017, Part II, LNAI 10449, pp. 3–12, 2017.
DOI: 10.1007/978-3-319-67077-5_1

based on prefix notation. Karakasis and Stafylopatis [13] proposed a hybrid evolutionary technique for data mining tasks, which combines the Clonal Selection Principle with Gene Expression Programming. Avila with co-authors proposed GEP algorithm for multi-label classification [3].

Expression Trees induced by Gene Expression Programming can be used as base classifiers in the ensemble of classifiers. The idea was proposed by Jędrzejowicz and Jędrzejowicz [8] and further developed in [9–11]. In these papers several approaches to combining expression trees including majority-voting, boosting and clustering were suggested.

In this paper our goal is to construct an ensemble of GEP-based base classifiers suitable to classify large datasets. For combining base classifiers, we use the idea of stacked generalization proposed by Wolpert [24]. Stacked generalization is a general method of using a high-level model to achieve greater predictive accuracy [21]. In the vast literature on the stacked generalization there are two basic approaches to the manner in which base classifiers are combined. The first one assumes combining somehow outputs from the base classifiers at a higher level to obtain classification decision. Alternatively, at a higher level, models (here base classifiers) are integrated into a meta-model, subsequently used to predict unknown class labels. In the paper we propose a solution based on the second approach where meta-gene is evolved from genes, that is expression trees, developed at a lower level.

To enable using the proposed scheme for dealing with the big datasets we suggest to take advantage of the inherent information redundancy present in data and often exploited through applying various data reduction techniques. Example of the data reduction applied to stacked generalization can be found in [18]. Another interesting approach to reducing data dimensions is a selection of base classifiers as, for example, proposed in [1]. Instead of using some sophisticated data reduction techniques we decided to enable users to control the number of base classifiers, the number of instances from the training set used to induce base classifiers, and the number of attributes used to construct them. The above settings are expected to remain related to the number of instances in the full training dataset to assure obtaining predictions in a reasonable time. Number of instances set by the user is used to randomly select instances from the training set. Number of attributes set by the user is used to randomly select attributes for each base classifier. Number of base classifiers should be set with a view to assure high probability of having all attributes represented within base classifiers. Subsequently, genes evolved from randomly selected sets of data are used to induce a final meta-gene.

The rest of the paper is organized as follows. Section 2 contains the detailed description of the proposed ensemble classifier in two versions. The first for classifying data with two classes and the second for multi-class classification. Section 3 contains results of the validating computational experiment. Final section contains conclusions and suggestions for future research.

2 GEP Ensemble Based on Stack Generalization Concept

The idea of gene expression programming (GEP) is based on evolving the population of genes which is subject to genetic variation according to fitness calculation. Each gene is divided into two parts as in the original head-tail method, see [5]. As usual in GEP, the tail part of a gene always contains terminals and head can have both, terminals and functions. The size of the head (h) is determined by the user and for classification task the suggested size is not less than the number of attributes in the dataset. The size of the tail (t) is computed as $t = h(m-1)+1$ where m is the largest arity found in the function set. In the computational experiments the functions are: logical AND, OR, XOR, NOR and NOT. Thus $m = 2$ and the size of the gene is $h + t = 2h + 1$. The terminal set contains triples $(op, attrib, const)$ where op is one of relational operators $<, \leq, >, \geq, =,$ $\neq$, $attrib$ is the attribute number, and finally $const$ is a value belonging to the domain of the attribute $attrib$. Thus for a fixed gene g and fixed instance from the data set r the value $g(r)$ is boolean and for binary classification each gene naturally differentiates between two classes. Attaching an expression tree to a gene is done in exactly the same manner as in all GEP systems. Consider the gene g with head $= 6$, defined below.

0	1	2	3	4	5	6	7	8	9
OR	AND	AND	AND	$(=,2,0)$	$(>,150)$	$(>,8,10)$	$(>,1,0)$	$(<,1,10)$	$\cdots$

The start position (position 0) in the chromosome corresponds to the root of the expression tree (OR, in the example). Then, below each function branches are attached and there are as many of them as the arity of the function - 2 in our case. The following symbols in the chromosome are attached to the branches on a given level. The process is complete when each branch is completed with a terminal. The number of symbols from the chromosome to form the expression tree is denoted as the termination point. For the discussed example, the termination point is 8. For the attribute vector $rw = (8.0, 1.0, 0.0, 20.0, 1.0, \cdots)$ the value of the above gene g is *true*.

Metagenes are representing ensemble classifiers. Similarly as above, the set of functions contains logical ones and terminals are identifiers of genes from the trained population. Algorithm 1 is used for binary classification. The input parameters contain those usual for genetic learning (size of population, probability of each of considered genetic operations) as well as parameters for data reduction: numbers of instances chosen for learning, number of attributes and number of base classifiers.

For the multi-class classification two approaches, OVA or OVO, can be applied. In the first case OVA (one-versus-all) the number of training steps is proportional to $|C|$ - the number of classes, as training takes place separately

---

Algorithm 1. Binary classification with meta-genes

Input: data $D = TD \cup TS$, split into training and testing, in - number of
instances, an number of attributes, ng - number of base classifiers
Output: qc - quality of classification
1 **for** $i = 1$ *to* ng **do**
2   filter training data TD to TD_i by choosing randomly in instances from TD
3   randomly choose subset AT of attributes of size an
4   use AT to find best gene g_i for training set TD_i
5   add g_i to the population PG of best genes (base classifiers)
6 filter training data TD to TD_M by choosing randomly in instances
7 generate population PM of meta-genes using genes from PG and training set
TD_M
8 train PM using genetic operations
9 return best fitted meta-gene mg from PM
10 use mg to perform classification of all instances from testing set TS
11 calculate qc - proportion of correctly classified instances
12 **return** qc

Algorithm 2. Value of instance row for gene classifier

Input: vector $\boldsymbol{g} = (g_1, \cdots, g_{|C|})$ of $|C|$ genes, instance r
Output: class $c \in C$
1 initialize integer counters $ct_1, \cdots, ct_{|C|}$ to 0
2 **for** $c = 1$ *to* $|C|$ **do**
3   **if** $g_c(r) = true$ **then**
4         $ct_c + +$
5   **else**
6         $ct_i + +$ for $i \neq c$
7 $c = \arg max_{c \in C} ct_c$
8 **return** c

---

for each class. For OVO (one-versus-one) approach the training is computationally demanding, as for each pair of classes $c1, c2 \in C$ the population of genes
which separates best instances from $c1$ and $c2$ is evolved. In this case, the cost
of training is proportional to $|C|^2$.

In the experiments OVA approach was applied and the gene classifier was a
vector of length $|C|$ containing genes specialized for distinguishing instances for
each separate class. For the testing stage the value of $\boldsymbol{g} = (g_1, \cdots, g_{|C|})$ for a
given instance r is a majority vote of $|C|$ counters as shown in Algorithm 2. The
details of multi-class learning are given in Algorithm 3.

Algorithm 3. Multi-class classification

Input: data $D = TD \cup TS$, split into training and testing, in - number of
instances, an number of attributes, gn - number of base classifiers
Output: qc - quality of classification

1 **for** $i = 1$ to gn **do**
2   filter training data TD to TD_i by choosing randomly in instances from TD
3   randomly choose subset AT of attributes of size an
4   **for** $c \in C$ **do**
5   use AT to find best gene g_c which differentiates best instances from
           class c
6   add $g = (g_1, \cdots, g_{|C|})$ to the population PG of gene classifiers
7 set $qc = 0$
8 **for** $r \in TS$ **do**
9   **for** $g \in PG$ **do**
10   use Algorithm 2 to calculate class for $g(r)$
11   **if** $g(r)$ *is correct* **then**
12              $qc + +$

13 **return** $qc/|TS|$

---

3   Computational Experiment Results

To validate the proposed ensemble classifiers computational experiment has been carried-out. Experiment involved 10 data sets. Luxembourg dataset has been taken from [27]. The remaining datasets have been taken from UCI Repository [16]. Numbers of instances, attributes and classes for each considered datasets are shown in Table 1. The table includes also recent literature reported accuracies obtained using different classifiers.

Accuracy of classification has been calculated from 10 repetitions of the 10-cross-validation scheme for each dataset. Area Under ROC Curve (AUC) has been calculated in a similar manner for each dataset with 2 classes. Percent of instances used to construct base classifiers as well as the number of base classifiers have been set to assure that the computation time needed to run all 10 rounds of the 10-cross-validation scheme does not exceed 1 min on a notebook with Intel i7 processor. Number of the selected base classifiers depended also on both - the number of attributes in each dataset and the number of attributes allocated to each gene. Experiment results including accuracies, standard deviations, and parameter settings for all considered datasets are shown in Table 2.

Example relation between the number of attributes selected randomly to induce each base classifier and the respective classifier performance (all other settings remaining as shown in Table 2 for Banknote Authentication and Wilt dataset) are shown in Figs. 1 and 2, respectively. To better evaluate generalization ability of the approach we propose to use the robustness index calculated as follows:

$$R = \frac{Acc(40\%) - Acc(1\%)}{Acc(40\%)}$$

Table 1. Datasets used in the reported experiment

Dataset	Instances	Attributes	Classes	Literature reported accuracy
Amazon commerce	1500	10000	50	0.805 (Synergetic NN [17])
Bank marketing	45211	17	2	0.844 (ANN [14])
Banknote authentication	1372	5	2	0.959 (Multilayer perceptron [2])
Cover type	581012	54	7	0.784 (Multi-class SVM [4])
Credit card	30000	24	2	0.820 (K-nearest neighbour [25])
Electricity	45312	6	2	0.698 (Non-parametric stream classifier [7])
KDD99	494012	42	2	0.892 (OzaBagADWIN [19])
Luxembourg	1901	32	2	0.856 (Logistic regression [22])
Poker hand	1025010	11	10	0.774 (EDDM [20])
Wilt	4839	6	2	0.900 (Hybrid pansharpening [12])

Table 2. Experiment results and algorithm settings

Dataset	Acc.	+−	AUC	+−	% instances	No. base	noAtt
Amazon commerce	0.8149	0.0043	-	-	50	50	4000
Bank marketing	0.8976	0.0113	0.7617	0.0088	20	30	15
Banknote authentication	0.9682	0.0127	0.9658	0.0103	40	10	3
Cover type	0.7911	0.0097	-	-	0.1	5	50
Credit card	0.8154	0.0175	0.7443	0.0194	20	15	20
Electricity	0.9051	0.0332	0.8869	0.0296	2	15	4
KDD99	0.8083	0.0234	0.8345	0.0137	5	5	38
Luxembourg	1.000	0	1.000	0	10	30	16
Poker hand	0.8026	0.0031	-	-	10	20	9
Wilt	0.9431	0.0186	0.7988	0.0098	20	20	4

where $Acc(x)$ denotes average accuracy calculated over 10-cross-validation scheme with x percent of the training set used to induce base classifiers. Values of the robustness index for the considered datasets, with all other settings as shown in Table 2, are shown in Table 3.

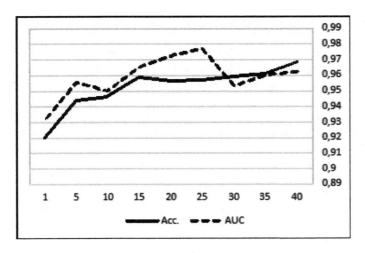

Fig. 1. Accuracy and AUC versus the percentage of attributes used to induce base classifiers for the Banknote Authentication dataset

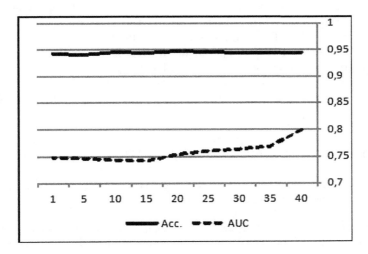

Fig. 2. Accuracy and AUC versus the percentage of attributes used to induce base classifiers for the Wilt dataset

Table 3. Robustness index for the considered datasets

Am.C.	B.Mark.	B.Aut.	Cov.T.	Cr.C.	Elect.	KDD	Lux.	P.H.	Wilt
0.0033	0.0815	0.0502	0.0111	0.011	0.0064	0.035	0.0081	0.0131	0.0008

4   Conlusions

The main contribution of the paper is proposing a new GEP-based batch ensemble classifier. As in all ensemble classifiers, final classification decision comes from combining base classifiers. In the proposed approach such combination involves evolving the meta-gene using genes induced by GEP from randomly generated combinations of instances with randomly selected subsets of attributes. Main property of the discussed classifier is its scalability. The user controls the size and number of genes as well as the number of instances in the subset of the training dataset used to induce base classifiers. The above feature allows to deal with big datasets and obtaining classification decisions in a reasonable time. Performance of the proposed classifier, in terms of classification accuracy and area under the ROC curve, is quite satisfactory as compared with the performance of some leading, literature reported, classifiers. One should however note that such comparisons are only approximate, since some of the datasets used in the reported computational experiment have been so far used only to evaluate adaptive or incremental classifiers. Another important property of the proposed approach is robustness with respect to user decisions as to the above described control parameters. This finding has been supported in our experiment through calculation of the proposed robustness index. In the worst case (bank marketing dataset) using only 1% of the available instances in the training dataset instead of using 40% of them, decreases classification accuracy by only 8%. In all other cases such a decrease was considerably smaller.

Future research will aim at supporting user control decisions on parameter values through analytical tools identifying structural properties of the dataset at hand.

References

1. Álvarez, A., Sierra, B., Arruti, A., Gil, J.M.L., Garay-Vitoria, N.: Classifier subset selection for the stacked generalization method applied to emotion recognition in speech. Sensors **16**(1), 21 (2016)
2. Awwalu, J., Ghazvini, A., Bakar, A.A.: Comparative analysis of algorithms in supervised classification: a case study of bank notes dataset. Int. J. Comput. Trends Technol. **17**(1), 38–43 (2014)
3. Ávila-Jiménez, J.L., Gibaja Galindo, E.L., Zafra, A., Ventura, S.: A gene expression programming algorithm for multi-label classification. Multiple-Valued Logic Soft Comput. **17**(2–3), 183–206 (2011)
4. Crain, K., Davis, G.: Classifying forest cover type using cartographic features. Stanford University (2014)
5. Ferreira, C.: Gene expression programming: a new adaptive algorithm for solving problems. CoRR, cs.AI/0102027 (2001)
6. Ferreira, C.: Gene Expression Programming: Mathematical Modeling by an Artificial Intelligence. Studies in Computational Intelligence, vol. 21. Springer, Heidelberg (2006). doi:10.1007/3-540-32849-1

7. Hosseini, S.A., Rabiee, H.R., Hafez, H., Soltani-Farani, A.: Classifying a stream of infinite concepts: a Bayesian non-parametric approach. In: Calders, T., Esposito, F., Hüllermeier, E., Meo, R. (eds.) ECML PKDD 2014. LNCS, vol. 8724, pp. 1–16. Springer, Heidelberg (2014). doi:10.1007/978-3-662-44848-9_1
8. Jędrzejowicz, J., Jędrzejowicz, P.: GEP-induced expression trees as weak classifiers. In: Perner, P. (ed.) ICDM 2008. LNCS, vol. 5077, pp. 129–141. Springer, Heidelberg (2008). doi:10.1007/978-3-540-70720-2_10
9. Jędrzejowicz, J., Jędrzejowicz, P.: A family of GEP-induced ensemble classifiers. In: Nguyen, N.T., Kowalczyk, R., Chen, S.-M. (eds.) ICCCI 2009. LNCS, vol. 5796, pp. 641–652. Springer, Heidelberg (2009). doi:10.1007/978-3-642-04441-0_56
10. Jędrzejowicz, J., Jędrzejowicz, P.: Experimental evaluation of two new GEP-based ensemble classifiers. Expert Syst. Appl. **38**(9), 10932–10939 (2011)
11. Jędrzejowicz, J., Jędrzejowicz, P.: Combining expression trees. In: 2013 IEEE International Conference on Cybernetics, CYBCONF 2013, Lausanne, Switzerland, 13–15 June 2013, pp. 80–85. IEEE (2013)
12. Johnson, B.A., Tateishi, R., Thanh, H.N.: A hybrid pansharpening approach and multiscale object-based image analysis for mapping diseased pine and oak trees. Int. J. Remote Sens. **34**(20), 6969–6982 (2013)
13. Karakasis, V., Stafylopatis, A.: Data mining based on gene expression programming and Clonal selection. In: IEEE International Conference on Evolutionary Computation, CEC 2006, part of WCCI 2006, Vancouver, BC, Canada, 16–21 July 2006, pp. 514–521. IEEE (2006)
14. Koc, A.A., Yeniay, O.: A comparative study of artificial neural networks and logistic regression for classification of marketing campaign results. Math. Comput. Appl. **18**(3), 392–398 (2013)
15. Li, X., Zhou, C., Xiao, W., Nelson, P.C.: Prefix gene expression programming. In: Rothlauf, F. (ed.) Late Breaking Paper at Genetic and Evolutionary Computation Conference (GECCO 2005), Washington, D.C., USA, pp. 25–29, June 2005
16. Lichman, M.: UCI machine learning repository (2013)
17. Liu, S., Liu, Z., Sun, J., Liu, L.: Application of synergetic neural network in online writeprint identification. Int. J. Digit. Content Technol. Appl. **5**(3), 126–135 (2011)
18. Mertayak, C.: Utilization of dimensionality reduction in stacked generalization architecture. In: The 24th International Symposium on Computer and Information Sciences, ISCIS 2009, 14–16 September 2009, North Cyprus, pp. 88–93. IEEE (2009)
19. Olorunnimbe, M.K., Viktor, H.L., Paquet, E.: Intelligent adaptive ensembles for data stream mining: a high return on investment approach. In: Ceci, M., Loglisci, C., Manco, G., Masciari, E., Ras, Z.W. (eds.) NFMCP 2015. LNCS, vol. 9607, pp. 61–75. Springer, Cham (2016). doi:10.1007/978-3-319-39315-5_5
20. Pesaranghader, A., Viktor, H.L.: Fast hoeffding drift detection method for evolving data streams. In: Frasconi, P., Landwehr, N., Manco, G., Vreeken, J. (eds.) ECML PKDD 2016. LNCS, vol. 9852, pp. 96–111. Springer, Cham (2016). doi:10.1007/978-3-319-46227-1_7
21. Ting, K.M., Witten, I.H.: Issues in stacked generalization. J. Artif. Intell. Res. (JAIR) **10**, 271–289 (1999)
22. Turkov, P., Krasotkina, O., Mottl, V.: Dynamic programming for bayesian logistic regression learning under concept drift. In: Maji, P., Ghosh, A., Murty, M.N., Ghosh, K., Pal, S.K. (eds.) PReMI 2013. LNCS, vol. 8251, pp. 190–195. Springer, Heidelberg (2013). doi:10.1007/978-3-642-45062-4_26

23. Weinert, W.R., Lopes, H.S.: GEPCLASS: a classification rule discovery tool using gene expression programming. In: Li, X., Zaïane, O.R., Li, Z. (eds.) ADMA 2006. LNCS, vol. 4093, pp. 871–880. Springer, Heidelberg (2006). doi:10.1007/11811305_95
24. Wolpert, D.H.: Stacked generalization. Neural Netw. **5**(2), 241–259 (1992)
25. Yeh, I.-C., Lien, C.H.: The comparisons of data mining techniques for the predictive accuracy of probability of default of credit card clients. Expert Syst. Appl. **36**(2, Part 1), 2473–2480 (2009)
26. Zeng, T., Tang, C., Xiang, Y., Chen, P., Liu, Y.: A model of immune gene expression programming for rule mining. J. Univ. Comput. Sci. **13**(10), 1484–1497 (2007). http://www.jucs.org/jucs_13_10/a_model_of_immune
27. Zliobaite, I.: Controlled permutations for testing adaptive classifiers. In: Discovery Science, pp. 365–379 (2011)

Shapley Value in a Priori Measuring of Intellectual Capital Flows

Jacek Mercik$^{(\boxtimes)}$ (iD)

WSB University in Wroclaw, Wrocław, Poland
jacek.mercik@wsb.wroclaw.pl

Abstract. Analysis of transmission of intellectual capital as a specific types of information requires consequently different models. The graph presentation maybe in use but it needs more complicated structure including logical conditions and multi connections between the same nodes. In the process of evaluation of the role of each nodes of such graph the concepts of Shapley value and taxonomy dendrite were used. The obtained results let to evaluate the role of nodes not only as a separate element containing the intellectual capital but also as an element of much bigger structure.

Keywords: Group decisions · Shapley value · Taxonomy dendrite · Intellectual capital

1 Introduction

In this paper we deal with a specific types of information, i.e. intellectual capital by preparing a model with so called non-fast transmission. The model shares the basic assumption that each element of the system of non-fast transmission of knowledge takes part in it to some extent. The article presents a way of measuring this participation based on the Shapley value of cooperative game whose elements are the entities of transmission, or vertices of the graph.

The concept of non-fast transmission of information (Mercik 2017) allows to analyse the changes in the so-called intellectual capital (IC) of various participants involved in the movement of intellectual capital between the participants of this process. We assume that the IC transmission process takes place at various levels and with varying intensity. We intend to present the situation associated with the transmission of IC naturally in the form of a planar graph in which there is the possibility of multiple connections between adjacent vertices. Moreover, we assume that this graph also allows the mapping of the logical structure of transmission, in which continuation may require an additional condition to be met (e.g. the need to implement two of the three connections entering the node - the so-called OR condition). Similarly, there may be conditions on the output of a given vertex e.g. only one transmission may start from it (i.e. an EX-OR condition). So designed graph is calibrated using relevant transmission units and will serve to form a matrix of distances between all vertices. Further, the distance matrix will allow the creation of a taxonomic dendrite needed to determine the likelihood of execution of specific coalitions of vertices. Using these probabilities we

© Springer International Publishing AG 2017
N.T. Nguyen et al. (Eds.): ICCCI 2017, Part II, LNAI 10449, pp. 13–21, 2017.
DOI: 10.1007/978-3-319-67077-5_2

can determine the Shapley values for cooperative games with pre-coalitions, which in turn will help to estimate the relative value of each node involved in the non-fast transmission of knowledge.

The article is set up as follows. First introduction. The next section outlines the way in which transmission of knowledge is modelled. The necessary preliminaries connected with the game theoretical language of modelling and ways of calculating power indices for a simple voting game together with taxonomy of object of a graph is presented in the following section. The next section presents assessment of a connection between vertices. This section describes a procedure to assessment of paths and importance of nodes in a given communication graph. After that, the next section presents transformation of a communication graph into taxonomy dendrite being an equivalent of it. The last section presents an example of simple graph with logical structure, its assessment, dendrite equivalent and evaluation of nodes in communication graph. Some conclusions and suggestions for future research are presented at the end of the paper.

2 Preliminaries

The model will use the following elements: a flat graph with a logical structure for the entry and exit of vertices and the possibility of multiple binary relationships between any two vertices; the definition of s-path, i.e. a path with the length s connecting any two vertices, the term coalition formed from the elements of the set of vertices[1] including a full coalition; and the Shapley value of payoffs for the coalition and the elements of the coalition in a cooperative game in which the players are vertices.

A graph G is an ordered pair $G = \langle N, U \rangle$ wherein each u edge corresponds to at least one pair of ordered vertices, $\langle x, y \rangle \in N \times N$ such that $\langle x, u, y \rangle \in N \times U \times N$. A graph G has no so-called loops, by assumption. Because we assume that all the vertices of the graph are involved in the transmission of intellectual capital, all possible paths in the graph must be analysed. If $a_{ij} > 0$ there is connection between every two nodes i and j and G represents the structure of existing relations by incidence matrix $A(G) = \left[a_{ij} \right]_{nxm}$, clearly.

For modelling transmission of intellectual capital we assume that it is enough to know the set of nodes with the logical structure of connections and nodes, weights of each node, w_i, and connection $p_{i,j}^r$. Following (Mercik 2017) the assessment, $t_{i,j}$, of a connection between two vertices equals

$$t_{i,j}^r = \alpha_{i,j} w_{i,j} p_{i,j}^r, \tag{1}$$

where: $\alpha_{i,j}$ – is initially equal to 1, $r = 1, 2, \ldots$ (representing multi connections between any two nodes).

---

[1] Because, naturally, not all paths can be implemented (this is usually not a complete graph) we are talking about the so-called pre-coalitions in the sense of (Owen 1977).

Definition. Path $s_{i,j}^r$ is a sequence of edges and vertices joining vertices i and j, $s_{i,j}^r = \{i, u_{ik}, \ldots, u_{lj}, j\}$ for $k, l \neq i, j$, $k \neq l$, $r = 1, 2, \ldots$. Each path can be interpreted as a specific permutation of the vertices from the set of vertices that are on this path. This fact is used in the process of determining the Shapley value for the cooperative game formed by those vertices.

Definition (Mercik 2016). Let i and j be two neighbouring vertices connected at least by one relationship $\{u_{i,j}^r, r = 1, 2, \ldots\}$. If their respective weights are w_i and w_j, $w_{ij} = w_i + w_j$. If i and j vertices are connected by at least one path, $\{s_{i,j}^r, r = 1, 2, \ldots\}$, the weight of the relationship equals for $k, l \in s_{ij}$, $k \neq i, l \neq j$, $w_{ij} = w_{ik} + \sum_{\substack{(k,l) \in s_{ij} \\ k \neq l}}$

$w_{kl} + w_{lj} - \sum_{\substack{k \in s_{ij} \\ k \neq i,j}} w_k$. Moreover, $w_{ij} \geq w_i + w_j$ and can be different, for each path

$s_{i,j}^r$. It is possible that for every $r = 1, 2, \ldots$ path $s_{i,j}^r$ has the same weight $w_{i,j}$.

Definition. For different q, weight w_{ij} meets the inequality $w_{ij} \geq q$. The path $s_{i,j}^r$ is called a q-path and denoted by $^q s_{i,j}^r$. Changing parameter q allows for example to analyse propagation of information among nodes.

For any non-zero value $a_{ij} \in A(G)$ by replacing a_{ij} by w_{ij} one can obtain a weighted incidence matrix $A_w(G)$.

Formally the cooperative game on the set of vertices N is described by a set of vertices and the characteristic function v. The coalition is any subset of vertices ($cardN = 2^N$), including the grand coalition N made up of all the vertices. The characteristic function of the game (N, v) is the real function v defined for the set of all coalitions, interpreted so that for the coalition T size $v(T)$ it is the amount that the coalition T is able to achieve on its own.

With a fixed ordering, the contribution of a player to the game (N, v) is what he brings to the coalition composed of all the players preceding him in this ordering. This division is called the Shapley value (Shapley 1953) of the game v and is denoted by $\varphi(v)$. The components of This vector's components are Shapley values of individual players (vertices) - their contributions in the division given by the Shapley value. Formally, for any game (N, v) the Shapley value of the game $\varphi(v)$ is given by the formula $\varphi(v) = (\varphi_1(v), \ldots, \varphi_N(v))$,

$$\varphi_i(v) = E_\Pi(v(H_{\pi,i}) - v(H_{\pi,i}\setminus\{i\})) = \sum_{\pi\in\Pi} \frac{v(H_{\pi,i}) - v(H_{\pi,i}\setminus\{i\})}{n!}, \qquad (2)$$

where Π is the set of all permutations of the set N. E is the expected value, and with a fixed permutation $\pi \in \Pi$ by $H_{\pi,i}$ we denote the set of all those players who, in this permutation, occur not later than the player i (i.e. $H_{\pi,i} = \pi^{-1}$ for $1, 2, \ldots, \pi(i)$).

In the definition of Shapley value, each coalition has the same probability of realization. But this is not the case if we consider the vertices in the communication graph as the players, i.e. the arithmetic mean (expected value) has to be replaced with the weighted average using probabilities describing the likelihood of realisation of a given coalition (path). For this purpose we use the weight $w_{i,j}$ of a path connecting the vertices i and j in such a way that the greater the length of the path, the smaller is the chance for the realization of this path. The relationship between the weight of the path and the probability is inversely proportional. Thus, the probability assigned to the coalition is estimated by the expression

$$\frac{w_{i,j}^{min}}{w_{i,j}}, \tag{3}$$

where $w_{i,j}^{min}$ means the smallest of weights $w_{i,j}$ calculated for a given graph.

3 Estimating the Significance of a Communication Node in the Graph

The proposed estimation of significance of the vertex is a result of the implementation of the following steps:

Step 1. Marking all connections in the graph by assigning (formula 1) respective values $t_{i,j}^{r}$ to all $i \neq j; i, j = 1, \ldots, N$.

Step 2. Transformation of the graph shown in the matrix $A_w(G)$ into a dendrite showing the shortest connections in terms of length of path $s_{i,j}^{r}$. We propose to use this method known as Wroclaw taxonomy.

Step 3. On the basis of the dendrite the values v should be determined for the selected coalitions of vertices.

Step 4. For each permutation of the set of vertices $\{1, 2, \ldots, N\}$ calculate the value $\varphi(v)$ and then, using the weights determined by (3) calculate the $\varphi_i(v)$ as a weighted average according to Eq. (2).

The resulting value in the fourth step will describe the importance of a given vertex of the graph of non-fast transmission of information. The value has all the good qualities of the Shapley value and can be easily interpreted.

Note that the calculations in step 4 have a considerable computational complexity, namely $O(n!)$. They can be reduced to the value $O(n2^{n-1})$ by replacing the analysis of all permutations with an analysis of all sets preceding a given vertex in permutations. It is also possible to use approximate algorithms, e.g. those specified in http://powerslave. val.utu.fi/.

4 Transformation of the Graph into Dendrite T

The Wroclaw Taxonomy Method (Florek et al. 1951) is based upon the idea of a spanning tree which is an undirected, acyclic and connected graph. The use of Wroclaw taxonomy guarantees that the connections (paths) in a dendrite achieved after te transformation of the communication graph are shortest (so-called Riker's postulate Riker 1962). Using this method we build a graph, whose nodes are all subsets of the set of the n analysed nodes – all the possible pre-coalitions. The edges of the graph are the distances between the respective vertices. The distances are found from the matrix D of distances between nodes (formula 1).

The set of the nodes of the obtained dendrite contains all possible pre-coalitions, whereas the set of edges contains the distances between them. The dendrite determines the shortest possible dendrite ordering (acyclic complete graph). Using this minimum ordering, we determine weights proportional to the probability of the formation of each coalition.

In order to generate partial orderings it is necessary to find among the coalitions the most probable coalition (a starting point of ordering). The choice of the most probable coalition is set arbitrarily. From among all coalitions we select the one that is composed of the smallest number of nodes (Riker 1962). If there is more than one such coalition, we choose the one that has the greatest value of w_i.

Assume that the selected most probable coalition is coalition K. Denote the corresponding node of the dendrite as W_k. The dendrite T has the property that for any of its nodes there will be a path connecting it with W_k. Thus, for any coalition P, it is possible to find the distance d between that coalition and coalition K. The distance is the sum of the lengths of all edges connecting W_p (the vertex corresponding to coalition P) with W_k. As the dendrite T determines the minimum ordering, the distance d between W_p and W_k is the minimum distance in terms of the similarity between the nodes (it is not the minimum distance in the ordinary sense).

For a given coalition P, the distance $d_{i,j}$ is related to the probability of the formation of that coalition. The greater the distance $d_{i,j}$, the smaller the probability of the formation of coalition P. The relation between the distance d and the probability of the formation of coalition P is inversely proportional. Coalition distance, d_{ij}, for $\{i,j\} \in E$, is defined as the length of the edge $\{i,j\}$ of the spanning tree. The dendrite T is therefore in general a nonlinear order over coalitions from the set of all coalitions. The sum of the lengths of edges of spanning tree T is called the length of the order and the best order is the one for which the length of the order is the least.

The described dendrite is therefore equivalent to communication graph. Its design ensures its uniqueness, and its character enables application of game theory approach and calculation of the power index for the games with precoalitions. Consequently, at a given size of, for example, intellectual capital contained in the individual vertices, we receive the assessment of their significance taking into account not only the size of the capital but also the complexity of the structure in which this intellectual capital moves.

5   Example

Analyse the example in Fig. 1. In this graph we have four nodes. We also assume that we are dealing with a directed graph, i.e. the graph with the lower number from a pair of connected vertices is the source-vertex. Let the weights of vertices, w_i, are respectively: 2, 2, 4, 3. Figure 1 presents these values in the upper part of the vertex.

Suppose also that the vertices of the graph have different logical conditions on inputs and outputs. Figure 1 the value in the lower part of the vertex on the left denotes the number of connections which must be completed before the signal goes to another vertex. Similarly, the value on the right side describes the number of connections

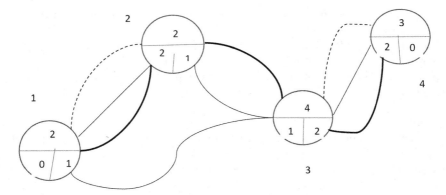

Fig. 1. Various connections (represented by the different kinds of arrows: the dashed line - type a; solid line - type b; the solid bold line - type c) between four objects with different logical structure of entry and output of each node.

simultaneously activated from a given vertex. E.g. the input value of 0 means that the given vertex is the source, and the input value of 2 means that the start of transmission from a given vertex requires the completion of two different transmissions leading to it.

Calculations of the significance of the vertex start with determining the weights of connections. For the graph in Fig. 1 we receive: $w_{1,2} = 4; w_{1,3} = 6; w_{2,3} = 6; w_{3,4} = 7$.

$t_{i,j}^{r}(1)$ is weight of direct connections between each pair of vertices, where different types of a, b or c of relationships is marked by $r = 1, 2, 3$. If there is a connection between two vertices, but it is not direct communication we calculate the value of the path connecting them. Of course, for some vertices there are mixed connections, i.e. direct and path connections ($\alpha_{i,j} = 1$).

Table 1. Distances between every pair of nodes from the example

	Nodes			
		1	2	3	4
Nodes	1	0	4	12	33
	2		0	12	33
	3			0	14

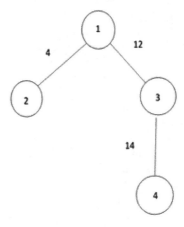

Fig. 2. Dendrite spinned over nodes from the example

Table 2. Permutations of grand coalition, the probability of their realization and contributions made by the individual elements (vertices) to the value of the grand coalition.

Coalition	v(1) = 2	v(2) = 2	v(3) = 4	v(4) = 3	Length of connection	Probability of coalition	Weighted values of elements in a given permutation of coalition N			
							{1}	{2}	{3}	{4}
1-2-3-4	2	10	0	35	34	0.882353	1.764706	8.823529	0	30.88235
1-2-4-3	2	10	35	0	48	0.625	1.25	6.25	21.875	0
1-3-2-4	2	0	0	43	59	0.508475	1.016949	0	0	21.86441
1-3-4-2	2	4	0	31	56	0.535714	1.071429	2.142857	0	16.60714
1-4-2-3	2	0	43	0	72	0.416667	0.833333	0	17.91667	0
1-4-3-2	2	4	31	0	56	0.535714	1.071429	2.142857	16.60714	0
2-1-3-4	10	2	0	35	30	1	10	2	0	35
2-1-4-3	10	2	35	0	44	0.681818	6.818182	1.363636	23.86364	0
2-3-1-4	0	2	0	43	54	0.555556	0	1.111111	0	23.88889
2-3-4-1	43	2	0	0	56	0.535714	23.03571	1.071429	0	0
2-4-1-3	0	2	43	0	68	0.441176	0	0.882353	18.97059	0
2-4-3-1	43	2	0	0	56	0.535714	23.03571	1.071429	0	0
3-1-2-4	10	0	4	41	46	0.652174	0	0	2.608696	26.73913
3-1-4-2	0	16	4	29	68	0.441176	0	7.058824	1.764706	12.79412
3-2-1-4	0	0	4	41	46	0.652174	0	0	2.608696	26.73913
3-2-4-1	41	0	4	0	72	0.416667	17.08333	0	1.666667	0
3-4-1-2	29	16	4	0	44	0.681818	19.77273	10.90909	2.727273	0
3-4-2-1	41	0	4	0	48	0.625	25.625	0	2.5	0
4-1-2-3	0	0	42	3	46	0.652174	0	0	27.3913	1.956522
4-1-3-2	0	15	30	3	54	0.555556	0	8.333333	16.66667	1.666667
4-2-1-3	0	0	42	3	46	0.652174	0	0	27.3913	1.956522
4-2-3-1	42	0	0	3	58	0.517241	21.72414	0	0	1.551724
4-3-1-2	30	15	0	3	30	1	30	15	0	3
4-3-2-1	42	0	0	3	34	0.882353	37.05882	0	0	2.647059
Shapley value for element of coalition N = {1.2.3.4}							0.324677	0.100063	0.270942	0.304318

Source: own calculations.

Assuming respectively that $p_{i,j}^r$, for a, b, c are 1, 2 and 3 one may find the following direct attributes: $t_{1,2}^1 = 4; t_{1,2}^2 = 8; t_{1,2}^3 = 12; t_{1,3}^1 = 12; t_{2,3}^1 = 12; t_{2,3}^2 = 18; t_{3,4}^1 = 14; t_{3,4}^2 = 21; t_{3,4}^3 = 7$.

Calculated parameters lead to matrix of taxonomy distances and taxonomy dendrite respectively (Table 1 and Fig. 2).

The achieved dendrite describing the taxonomical valuation allows to determine the pre-coalitions and the likelihood of their realization. As we can see, the following winning coalitions described by the value of the function v exist: $v(1, 2) = max\left(t_{1,2}^r\right) = 12$, $v(1, 3, 4) = max\left(t_{1,3,4}^r\right) = max\left(t_{1,3}^r\right) + max\left(t_{3,4}^r\right) = 12 + 21 = 33$, $v(1, 2, 3, 4) = 45$. The values thus obtained allow for the calculation of the contribution individual vertices make to all permutations of vertices. Distances in Table 1 also allow to determine the probability of a given permutation. Table 2 shows the calculated values.

From Table 2 one may evaluate elements of the graph shown in Fig. 1. As shown, the output proportion of knowledge between the nodes 0.18182; 0.18182; 0.36364; 0.27273 was adjusted by taking into account the structure and nature of connections into the proportion corresponding to the Shapley values: 0.324677; 0.100063; 0.270942; 0.304318. In our opinion, the latter proportions better correspond to the estimated role of each vertex in the flow of intellectual capital (knowledge) through this graph.

6 Conclusions

The proposed a priori evaluation of the elements of graph representing the transmission of intellectual capital via Shapley value may also be used for any other kind of information transmission, not only the slow one.

The very essential logical structure of all transmissions was up to now evidently omitted. This novel approach eliminate this gap.

In the future, the problem of summing such graphs into bigger unit should be analysed. It seems that connection between two graphs by single or multi connections looks relatively easy. The sum of two games where at least one player is common for both of them needs more attention. Some ideas how to solve the problem can be found in Malawski (2017).

References

Florek, K., Łukaszewicz, J., Perkal, J., Steinhaus, H., Zubrzycki, S.: The Wroclaw Taxonomy (in Polish). Przegląd Antropologiczny XVII (1951)

http://powerslave.val.utu.fi/. Accessed 27 Mar 2017

Malawski, M.: A note on positions and power of players in multicameral voting games. In: Springer Transactions on Computational Collective Intelligence, vol. 10450 (2017, forthcoming)

Mercik, J.: A power-graph analysis of non-fast information transmission. In: Nguyen, N.T., Tojo, S., Nguyen, L.M., Trawiński, B. (eds.) ACIIDS 2017. LNCS, vol. 10191, pp. 89–99. Springer, Cham (2017). doi:10.1007/978-3-319-54472-4_9

Mercik, J.: Formal a priori power analysis of elements of a communication graph. In: Nguyen, N.T., Trawiński, B., Fujita, H., Hong, T.-P. (eds.) ACIIDS 2016. LNCS, vol. 9621, pp. 410–419. Springer, Heidelberg (2016). doi:10.1007/978-3-662-49381-6_39

Myerson, R.B.: Graphs and cooperation in games. Math. Oper. Res. **2**, 225–229 (1977)

Owen, G.: Values of games with a priori unions. In: Henn, R., Moeschlin, O. (eds.) Mathematical Economy and Game Theory. Lecture Notes in Economics and Mathematical Systems, vol. 141, pp. 76–88. Springer, Heidelberg (1977). doi:10.1007/978-3-642-45494-3_7

Riker, W.H.: The Theory of Political Coalitions. Yale University Press, New Haven (1962)

Rosenthal, E.C.: Communication and its costs in graph-restricted games. Soc. Netw. **10**, 275–286 (1988)

Shapley, L.S.: A value for n-person games. In: Kuhn, H.W., Tucker, A.W. (eds.) Contributions to the Theory of Games, vol. 2, pp. 307–317 (1953). (Ann. Math. Stud. 28)

MDBR: Mobile Driving Behavior Recognition Using Smartphone Sensors

Dang-Nhac Lu[1(✉)], Thi-Thu-Trang Ngo[2], Hong-Quang Le[3],
Thi-Thu-Hien Tran[3], and Manh-Hai Nguyen[4]

[1] University of Engineering and Technology,
Vietnam National University in Hanoi, Hanoi, Vietnam
nhacld.dill@vnu.edu.vn
[2] Posts and Telecommunications Institute of Technology, Hanoi, Vietnam
trangnttl@ptit.edu.vn
[3] Academy of Journalism and Communication, Hanoi, Vietnam
{lehongquang, tranthithuhien}@ajc.edu.vn
[4] HoChiMinh National Academy of Politics,
Hanoi, Vietnam
nguyenmanhhai@hcma.vn

Abstract. The driving behavior is interested approach in human life service provider, special using various smartphone sensors. We proposed an efficient framework for recognizing driving behavior using smartphone sensors. It names Mobile Driving Behavior Recognition Systems (MDBRS). The system implement while users put and change their smartphones dynamic due to their trips. The synchronous Practice Swarm Optimization (PSO) is used to auto select suitable features extracted from sensor data. The online user activity is predict by classification algorithms via only accelerometer signal. Hence, the system recognizes online behavior base on training data set by Artificial Neural Network (ANN). It auto predicts abnormal behavior from seven activity such as stop, moving, acceleration, deceleration, turn left, turn right and U-turn. MDBRS experiment on walking, bicycle, motorbike, bus and car and announce safety or unsafe behavior by abnormal behaviors predicted. The system also allowing update data training set by user confirmation from feedback module and achieve higher results with 86.71% accuracy.

Keywords: Activity recognition · Online behavior recognition · Detecting behavior · PSO algorithm

1 Introduction

The mobile phone is indispensable device in modern life and there are a lot of app in this area special approaching human activity, behavior recognition using sensor signals [1]. The personal data analysis by sensor equipment is more cared about researchers with many utilizing [2]. The activity could be build up driving behavior, but the challenges of some behaviors are difficult and complex activity. Driving behavior system is meaningful with traffic participant and defined by any activity analysis technical base on values such as: distances, gaps [3], time, angle [4] and velocity [5].

© Springer International Publishing AG 2017
N.T. Nguyen et al. (Eds.): ICCCI 2017, Part II, LNAI 10449, pp. 22–31, 2017.
DOI: 10.1007/978-3-319-67077-5_3

Hence, smartphone users could be help from recognizing system when they are moving with their smartphone.

The accelerometer is the most commonly used sensor for reading motion signals and utilize for various application [6, 7]. However, it is a lot of noise, needed to combine with other signals [8] and using more resources in app. The smartphone sensors signal are time series data easy to collected but difficult to analysis. It depends on devices quality, environment conditional and sensitive applying model. So that, feature extraction technical is important with human activity recognition, driving behavior problem and have mathematical basic in analysis and applying [9, 10].

Vavouranakis et al. [11] proposed method to recognize driving behavior by smartphone sensor. However, the mobile is fixed in car and reoriented data by sensor fusion method, mobile coordinate system be reoriented by its coordinate system. The windowing technical utilize with 5 s on accelerometer data. Then, abnormal behavior are predicted by thresholds of x and y-axis accelerometer values. The 12 distinguish events about six safe and six unsafe behaviors deployed by their method.

Li et al. [4] developed system for detected dangerous driving behavior. It gathers accelerometer signal with ground truth position on taxi. The yaw angles is estimated by transformation matrix converter between vehicles coordinate and smartphone coordinate system and helping them to detect behavior on manual dangerous driving behavior set which define by accelerometer value threshold and 90% accuracy was received by their experiment.

Xu et al. [5] detected human behavior rules base on accelerometer with Fourier transform take 1–8 s point for analysis and calculated velocity from accelerometer.

The one of challenges with driving behavior recognition using accelerometer is signal quality in difference devices and complexity of abnormal behavior. Hence, our problems is detecting user activity while they are moving, then the sequence of activity corresponding to their trip is background for system predicts and announce abnormal behavior.

MDBRS implement by three phases: firstly, it collects label data, analysis and extract suitable features subset by PSO. Secondly, it predicts seven activity by some classification algorithms as Random Forest (RB), Naïve Bayes (NB), k-Nearest Neighbor (KNN) and Support Vector Machine (SVM). Finally, it predicts abnormal driving behavior by ANN. The system also use feedback technical aim to update training data and gains higher accuracy. The experiment on walking, bicycle, motorbike, bus, car and predicted safe and unsafe behavior with higher traditional method.

2 Mobile Driving Behavior Recognition System

Hereinafter, we show (MDBRS) in Fig. 1. MDBRS framework. The system composed from three main phases: firstly, the label data collected from each volunteer and features selection aim to predefine input value for predict activity. The sensors data is preprocessed by extracted into a set of representative features after that PSO will choosing the suitable subset feature for driving activity recognition. Secondly, the best features subset and data training is used for classifying and online recognition activity on smartphone. Finally, we suggest an abnormal driving behavior data set within k linear activity is an instance with label. Recently predicted activity is combined with k-1 previous aim to recognize behavior of users. The feedback module help system to

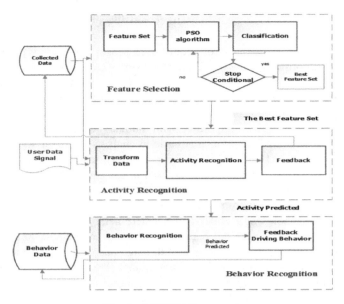

Fig. 1. MDBRS framework

update instance, which has attached label into training data set and gain higher accuracy on prediction.

2.1   Data Preprocessing

In the phase 1, signal data label obtain from smartphone on their pocket, handbag, or in hands, etc. while they are moving. Hence, the orientation and signal value is frequent changing. Our paper suggest an approach to preparing data for classifier by collected and transform data base one windowing technical and feature extraction. The features set is combined by time-base, power-base and frequency-base domain. It is a base for dynamic and suitable selected features system with complex activity. An approach to solve noisy accelerometer data is reoriented. We use accelerometer, gyroscope and magnetometer sensors to transform accelerometer data from the smartphone coordinate system to the Earth coordinate system [12]. Considering that, a(x, y, z) is data point in coordinate system then earth coordinate system of a' (x', y', z') by R matrix and it is computed by function below:

$$\begin{pmatrix} x' \\ y' \\ z' \end{pmatrix} = R \begin{pmatrix} x \\ y \\ z \end{pmatrix} \tag{1}$$

Following, directly analyze with amount raw sensor data is needed a lot of time or memory space. The windowing technical is usually choice in time series data then it also apply to calculate value of features from root mean square, sample correlation coefficient, cross-correlation, vertical and horizontal accelerometer energy of window signal, energy of M coefficient Fourier values, Signal magnitude area, average Energy of X, Y, Z axis, and the entropy of signal. These total 18 features apply for detection in system and is shown by Table 1. The suggestion feature set for MDBRS system.

Table 1. The suggestion feature set for MDBRS system

The features	Variable name
Root mean square (RMS)	X_{rms}
Correlation coefficient	$CorreCo_{xy}$, $CorreCo_{xz}$, $CorreCo_{yz}$
Cross-correlation	$Cross_{xy}$, $Cross_{xz}$, $Cross_{yz}$
Signal magnitude area	SMA
The accelerometer energy of windows	Ev, Eh
Statistical value: mean, variance, standard deviation	sM, sV, sSD
The energy of M coefficient Fourier	EM
Average energy of X,Y, Z axis	E_x, E_y, E_z
The entropy	H

In fact, any approaching chose several kind of feature suitable with data and problem in their field. The selected features is usually via experiment and no method agree to all problem. Hence, we suggest using PSO algorithm to select suitable features for improving prediction accuracy base on wrapper method. The PSO was introduced by Kennedy and Eberhart [13]. In PSO, each potential solution is corresponding to particle and assumption that, x = $[x_1, x_2, ..., x_D]$ with $X_i = (X_1, X_2, ..., X_D)$ are features at particle i^{th} and the x_k^i is particle position; the v_k^i is particle velocity; the p_k^i is the best individual particle position; p_k^g is the best swarm position; c_1, c_2 are cognitive and social parameters; r_1, r_2 is random numbers between 0 and 1. The PSO algorithm is express below:

1. **Initialize**
   a. Set constants k_{max}, c1, c2.
   b. Randomly initialize particle position $x_0^i \in D$ in R^n
      i=1:p
   c. Set k =1
2. **Optimize**
   a. Evaluate all fitness values f_k^i at X^i. The fitness
      f_k^i = 1/Acc. The Acc values is classification
      accuracy by set of feature of x_k^i)
   b. If $f_k^i \leq f_{best}^i$ then $f_{best}^i = f_k^i$, $p_k^i = x_k^i$
   c. If $f_k^i \leq f_{best}^g$ then $f_{best}^g = f_k^i$, $p_k^g = x_k^i$
   d. If stopping condition is satisfied then goto 3
   e. Update all particle velocities v_k^i Equation (2) for
      i =1,...,p
   f. Update all particle position x_k^i Equation (3) for i
      =1,...,p
   g. Increment k
   h. Go to (2.a)
3. **Terminate**. The best subset of features corresponding with best fitness values.

The position particle individual updates following:

$$x_{k+1}^i = x_k^i + v_{k+1}^i \tag{2}$$

With the velocity of particle individual updates following:

$$v_k^i = v_k^i + c_1 r_1 (p_k^i - x_k^i) + c_2 r_2 (g_k^i - x_k^i) \tag{3}$$

When phase 1 finished, the best of subset feature is chosen and utilizing to predict activity from online accelerometer signal sensors data.

2.2 The Online Training Model

The abnormal driving behaviors are usually assessed with complex activity. Some activity happen and repeat during all user trips with sort time and suddenly. There are any method detect abnormal driving behavior base on some measurement. It has good accuracy because fix position or know orientation and predefine conditionals of smartphone sensors. With dynamic changing position or some unpropitious condition when vehicles moving, the collected data is a lot of noise and difficult analysis. Thence, algorithm for detecting is more affect to predicted results.

There are two approach to classify data as offline and online training. The offline method in advance model, which compute on personal computer or server then client to send input and parameters of model. The online method implements computation, recognition on smartphone. The offline method has more propitious conditional and resources. However, it depends on linking and services. Nowadays, hardware and devices quality is more improved. So that, we use online training method on smartphone and assess by some appropriate algorithm such as RF, NB, SVM, KNN, which applied in researches in this field and have shown appropriate accuracy. The WEKA tool has used and integrated in MDBRS for classification and recognition. Experiment in this paper indicated that, RF is appropriate and higher accuracy.

2.3 Driving Behavior Recognition

The definition context safety or not depend on realities issue and opinions. Some activity repeating to reflect abnormal driving behavior. It is also belonging personality of user habit. Hence, MDBRS use k-series activity aim to predict abnormal driving behavior via accelerometer sensors signal. The system is also monitoring and predicting behavior at status of vehicle on real time. When systems has detected current driving activity a_c then combined with k-1 previous activity to instance consist k linear activity as $(a_{c-k-1}, \ldots, a_{c-1}, a_c)$. In fact, abnormal behaviors have realized after one or several complex activity, which are abnormal and repeating. Hence, series activity in instance is basic to predict them. The problems affect to prediction accuracy is value of k and personal habit.

The abnormal behavior training dataset is build up from series activity while they are driving on sleeping, drunk and frequency swing with high velocity. The clustering technical with k-means clustering algorithm is used to set of k linear activity into three class and the label of abnormal behavior instance number i is shown by $S_i(S_{i_1}, S_{i_2}, \ldots, S_{i_k}, l_i)$. This indicate that, abnormal behavior has built up from sequence basic activity. The feedback module also could be updated personal behavior instance by confirm method when they are performing. It shows in Fig. 3.

The ANN algorithm is applied by some research in human activity and behavior recognition system. The neurons will be received inputs from other neurons. The value of each input is determined by a weight associated with them. The sum of input weights computed and value output is according to its transfer function. It shows in Fig. 2 (Left).

The layer on ANN is sets of neurons and combine with another once. The neurons in a layer do not interconnect with each other, but interconnect with neurons in other layers. Neural networks can have one or more layers between input and output layers. They are hidden layers and expressed in Fig. 2 (Right).

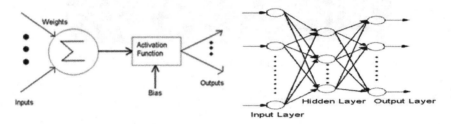

Fig. 2. Model of a neuron (Left) and basic neural network structure (Right)

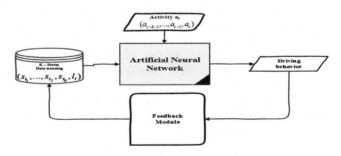

Fig. 3. Driving behavior recognition

2.4 Online Feedback Module

The training data set collects by any difference subjects with advanced supposed conditional. However, each user might have different in habits. Therefore, the prediction accuracy may be fall down when the system used by another or new users. Okeyo et al. [15] developed the idea to incrementally update the training data set by utilizing real-time feedbacks from users. As the system provides the activity, behaviors prediction, user can confirm the right of the result. The newly instance data labeled, which correcting from users is then added to the training data set. It is really mean-ingful with abnormal behavior, which happening and depend on user habit, complexity when user is moving. Specially, characteristic behavior of difference users will be recognize and update. The idea is express in Fig. 1 and some the interfaces of MDBRS system is shown in Fig. 4.

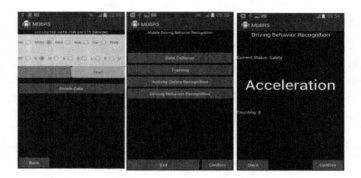

Fig. 4. The interfaces of MDBRS

3 Experiment and Results

3.1 Experiment Environment

We implemented MDBRS on the Android Operating System from 4.0 to 5.0 platform. The dataset with activity labels collected by 20 subjects when they are driving by walking, bicycle, motorbike, bus and car. They freely carry a Samsung galaxy S4, Quad-core 1.6 GHz Cortex-A15 processor, 2 GB of Ram, 2600 mAh battery, Android 4.2.2 Jelly Bean OS. The sevens activity recognition are {stop, moving, acceleration, deceleration, turn left, turn right and U-turn}. The Weka tool is used for deploying on our framework to predict the vehicle status. The classification was used such as Random Forest, KNN, Naive Bayes, SVM. In each case, the default setting is used by setting parameter of algorithms. We also used 10-fold cross validation for evaluating the accuracy of each classification algorithm.

3.2 Data Collection

The signal data is collected from three types of sensors as acceleration, gyroscope and magnetic sensor signal with 50 Hz frequency. These sensor returns x, y, and z coor-dinates values at point. The raw data stream is first cut out 1 s at the starting point, and 3 s at the end point, cause these periods time are usually redundant. Then, split into a window by 6 s size and the overlapping time is one second. We collected at least 200 samples for each activity from subjects, improving and contained meaningful habit characteristic of users.

The abnormal driving behavior collected by motorbike, walking then use the k-means clustering algorithm with k = 3 base on series activity of users, reflecting to three label of abnormal driving behavior such as sleeping driving, drunk and frequency swing by high velocity. After that, the training dataset with normal and abnormal behavior is used for recognizing driving behavior.

3.3 The Accuracy of Detection System Result

The Accuracy of Activity Recognition
The first scenario (S1), we deploy on about 3500 sample; the system will build up model for subjects to predict activity and abnormal behaviors base on two choosing feature method. The result when MDBRS deploy with traditional on18 features and using PSO to select features, it shows in Fig. 5.

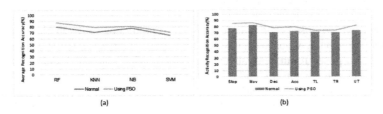

(a)   (b)

Fig. 5. Activity recognition accuracy of algorithms using PSO and traditional

In the Fig. 5(a) is shown the results between two method select features, the average of recognition accuracy when user uses PSO is always higher. With RF is max as 87.66% and SVM is min with 71%. It indicates that, the RF algorithm is suitable in MDBRS system.

In the Fig. 5(b) is shown the average activity recognition accuracy by four algorithms {RF, K-NN, NB,SVM} on normal method using 18 features and using PSO algorithm for optimization. Labels of activity in figure is short of them. It is reflected that, moving status is higher than other activity and deceleration, acceleration, turn left, turn right is difficult than.

The second scenario (S2), MDBRS predict activity with the best feature subset, which selected by PSO and Random Forest algorithm predict activity and the system use feedback service update training dataset. The results is shown in Fig. 6. It indicated

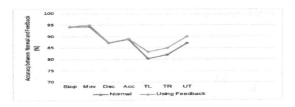

Fig. 6. Normal and feedback recognition accuracy

that, the feedback information has improved data training quality corresponding the higher accuracy. The accuracy of the abnormal activity are lower than normal, specially abnormal turn left and turn right activity is lowest.

The results of activity predicted is affected to accuracy of behavior recognition, if it is higher than then the users behaviors are more exact when MDBRS running.

The Accuracy of Driving Behavior Recognition

The first, we propose basic behavior training dataset contain 300 abnormal behavior instances aim to detect current abnormal behavior base on activity. Through experiments, we chose k = 5 with 5 linear activity are built for each instance. The users could be updated to training dataset by behavior with feedback information. The parameters value of ANN is default setting with behavior detecting accuracy is 83.12% and when using feedback up to 86.71%.

4 Conclusion and Future Works

In this paper, we proposed a flexible framework to predict current vehicle activity base on smartphone sensor, when user moving, dynamic changing position and direction. Besides, our proposed framework also using PSO to select suitable features. Following, MDBRS detect vehicle activity and this is basic for recognition driving behavior using ANN. It utilizes real-time feedbacks from users also use to increase the activity prediction accuracy. In the experiments, MDBRS can achieve on average 86.71% accuracy for predicted driving behavior. Furthermore, Random Forest classifier is a promising one for our framework. In the future, we are planning to further improve the current framework to increase accuracy and integrated in traffic simulation system for any traffic problems.

References

1. Reyes-Ortiz, J.-L., Oneto, L., Sama, A., Parra, X., Anguita, D.A.I.-O.: Transition-aware human activity recognition using smartphones. Neurocomput. Int. J. **171**, 754–767 (2016). (D.A.I.-O. Anguita: http://orcid.org/000-0002-4943-3021)
2. Krishnaswamy, S., Gaber, M.M., Sousa, P.A.C., Menasalvas, E.: MARS : A Personalised Mobile Activity Recognition System
3. Kumar, M.K., Prasad, V.K.: Driver behavior analysis and prediction models. 6(4), 3328–3333 (2015)
4. Li, F., Zhang, H., Che, H., Qiu, X.: Dangerous Driving Behavior Detection Using Smartphone Sensors, pp. 1902–1907 (2016)
5. Xu, H., Zhang, L., Zhai, W.: Detection of human movement behavior rules using three-axis acceleration sensor. In: Jin, D., Lin, S. (eds.) Advances in Multimedia, Software Engineering and Computing. Advances in Intelligent and Soft Computing, vol. 1, pp. 647–652. Springer, Heidelberg (2011). doi:10.1007/978-3-642-25989-0_103
6. Jain, A., Kanhangad, V.: Exploring orientation and accelerometer sensor data for personal authentication in smartphones using touchscreen gestures. Pattern Recognit. Lett. **68**, 351–360 (2015)

7. Kalra, N., Bansal, D.: Analyzing driver behavior using smartphone sensors: a survey. Int. J. Electron. Electr. Eng. **7**(7), 697–702 (2014)
8. Ferrer, S., Ruiz, T.: Travel behavior characterization using raw accelerometer data collected from smartphones. Procedia-Soc. Behav. Sci. **160**, 140–149 (2014)
9. Bayat, A., Pomplun, M., Tran, D.A.: A study on human activity recognition using accelerometer data from smartphones. Proc. Comput. Sci. **34**, 450–457 (2014)
10. Yu, J., Chen, Z., Zhu, Y., Chen, Y., Kong, L., Li, M.: Fine-grained abnormal driving behaviors detection and identification with smartphones. **1**(c), 1–14 (2016)
11. Vavouranakis, P., Panagiotakis, S., Mastorakis, G., Mavromoustakis, C.X., Batalla, J.M.: Recognizing driving behaviour using smartphones. In: Batalla, J.M., Mastorakis, G., Mavromoustakis, Constandinos X., Pallis, E. (eds.) Beyond the Internet of Things. IT, pp. 269–299. Springer, Cham (2017). doi:10.1007/978-3-319-50758-3_11
12. Lu, D.-N., Nguyen, T.-T., Ngo, T.-T., Nguyen, T.-H., Nguyen, H.-N.: Mobile online activity recognition system based on smartphone sensors. In: Akagi, M., Nguyen, T.-T., Vu, D.-T., Phung, T.-N., Huynh, V.-N. (eds.) ICTA 2016. AISC, vol. 538, pp. 357–366. Springer, Cham (2017). doi:10.1007/978-3-319-49073-1_39
13. Eberhart, R., Kennedy, J.: A new optimizer using particle swarm theory. In: Proceedings of the Sixth International Symposium on Micro Machine and Human Science, MHS 1995, pp. 39–43 (1995)
14. Liu, Z., Wu, M., Zhu, K., Zhang, L.: SenSafe: a smartphone-based traffic safety framework by sensing vehicle and pedestrian behaviors, vol. 2016 (2016)
15. Okeyo, G., Chen, L., Wang, H., Sterritt, R.: Dynamic sensor data segmentation for real-time knowledge-driven activity recognition. Pervasive Mob. Comput. **10**, 155–172 (2014)

Adaptive Motivation System Under Modular Reinforcement Learning for Agent Decision-Making Modeling of Biological Regulation

Amine Chohra$^{(\boxtimes)}$ and Kurosh Madani

Images, Signals, and Intelligent Systems Laboratory (LISSI/EA 3956),
Paris-East University (UPEC), Senart Institute of Technology,
Avenue Pierre Point, 77127 Lieusaint, France
{chohra,madani}@u-pec.fr

Abstract. In this paper, an adaptive motivation system under modular reinforcement learning is suggested for agent decision-making modeling of biological regulation. For this purpose, first, main concepts of drives, rewards, action selection under modular reinforcement learning as well as an adaptive priority process are developed. Second, experiments and results are presented and analyzed demonstrating the efficiency of the suggested concepts. Finally, a discussion is given in conclusion with regard to related works. The obtained results demonstrate how the suggested adaptive motivation system can be used by an agent learning (on-line) to select appropriate actions, during a navigation task from a starting position to a goal position (external goal), i.e., in each moving step, in order to reach an external goal as well as to satisfy internal goals (drives such as hunger, thirst, ...); predicting a promising result in future to demonstrate how the nature of the interaction (stimulation-drive, social-drive, ...) influences the agent behavior.

Keywords: Decision-making · Adaptive goal-directed behavior · Agent–environment interactions · Motivation · Modular reinforcement learning · Action selection

1 Introduction

According to Hull's theory [1], all behavior is motivated either by an organism's survival and reproductive needs giving rise to primary drives (such as hunger, thirst, sex, and the avoidance of pain) or by derivative drives that have acquired their motivational significance through learning [2].

In psychology, ethology, and computer science 'motivation' does not refer to a specific set of readily identified processes, though for practical purposes motivation can be discussed in terms of 'drives' and 'incentives', push and pull of behavior [3]. Among most influential theories of motivation in psychology is Hull's drive theory [1]. Hull's theory followed principles of physiological homeostasis that maintains bodily conditions in approximate equilibrium despite external perturbations. In effect, homeostasis is achieved by processes that trigger compensatory reactions when the

© Springer International Publishing AG 2017
N.T. Nguyen et al. (Eds.): ICCCI 2017, Part II, LNAI 10449, pp. 32–42, 2017.
DOI: 10.1007/978-3-319-67077-5_4

value of a critical physiological variable departs from the range required to keep the animal alive [4]. Many other theories have been influential in the design of motivational systems for artificial agents, as discussed in [5].

Thus, in artificial intelligence and computer science, to be autonomous, an agent requires an internal motivation system that appropriately values the actions available to it and generates its goals. The reinforcement learning [6], which is a learning and action selection paradigm, is of great importance in building such motivation systems from agent-environment interaction. In effect, an agent should have a number of internal reward functions built in, rather than relying on a single external reward function. Animals, for instance, rely on satiating hunger to learn certain tasks rather than using the single bit of reward from death, perhaps due to starvation but perhaps due to some other cause, as a learning reward [7]. Thus, for an agent, motivation system is the reward generating mechanism that expresses the agent's internal goals.

In this paper, an adaptive motivation system is suggested for agent decision-making modeling of biological regulation. In fact, this paper is a part of the developed research works in order to achieve a framework integrating drives, personality traits, and emotions for adaptive agent decision-making modeling of biological regulation and psychological mechanisms, discussed in [8], deduced from a perception-action cycle scheme inspired from [9]. For this, main concepts of drives, rewards, action selection under modular reinforcement learning as well as an adaptive priority process are developed in Sect. 2. Second, experiments and results are presented and analyzed in Sect. 3. Finally, a discussion is given in Sect. 4 with regard to related works.

2 Adaptive Motivation System (Modular Reinforcement Learning)

Hull's idea that reward is generated by drive reduction is commonly used to connect reinforcement learning to a motivational system [1, 2]. Drive reduction (consequence of the agent-environment interaction) is directly translated into a reward signal delivered to a reinforcement learning algorithm. Thus, in the following, an adaptive motivation system, under reinforcement learning, is developed as a reward generating mechanism that appropriately values the actions available to it and generates its goals.

2.1 Drives

In the suggested adaptive motivation system, internal goals are expressed by a number of n drives $d_1, d_2, ..., d_i, ..., d_n$ which could be related to hunger (Food), thirst (Water), ... Each drive d_i should have its Priority-Parameter P_{Pi} (depending on the environment) which determines its Drive-Priority D_{Pi} given its Satiation-Level S_{Li}. This is achieved as in [10] giving Drive-Priority $D_{Pi} \in [0, 1]$ by Eq. (1):

$$D_{Pi} = 1 - S_{Li}^{\tan\left(\frac{P_{Pi}\pi}{2}\right)}, \tag{1}$$

where $S_{Li} \in [0, 1]$ is the Satiation-Level, and $P_{Pi} \in [0, 1]$ is the Priority-Parameter.

This results in a curve of $D_{Pi} = f(S_{Li})$ for each P_{Pi} fixed with regard to the environment. Examples of such curves are given in Fig. 1: $D_{P1} = f(S_{L1})$ curve with a *low* $P_{P1} = 0.25$ and $D_{P2} = f(S_{L2})$ curve with a *high* $P_{P2} = 0.75$. Note that, for each $S_{Li} = 0$ the drive priority $D_{Pi} = 1$, and for each $S_{Li} = 1$ the drive priority $D_{Pi} = 0$.

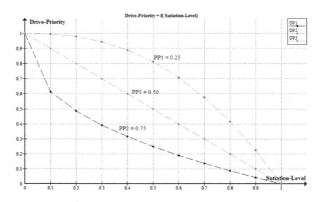

Fig. 1. $D_{P1} = f(S_{L1})$ with a *low* $P_{P1} = 0.25$ (curved downwards in black), $D_{P3} = f(S_{L3})$ with a $P_{P3} = 0.50$ (linear curve in grey), and $D_{P2} = f(S_{L2})$ with a *high* $P_{P2} = 0.75$ (curved upwards in grey).

Thus, Priority-Parameter P_{Pi} is used to affect the shape of a priority curve that determines Drive-Priority D_{Pi} given its Satiation Level S_{Li}. Also, Priority-Parameter allows agent to adjust its drive priorities without changing underlying drive process.

For instance, an agent having two primary drives (d_1 and d_2 related to hunger and thirst, respectively) in an environment where Food is abundant and Water is scarce. This results in a situation where drive d_1 related to hunger (Food) is easy to satiate, and where drive d_2 related to thirst (Water) is difficult to satiate. Consequently, this needs from agent to choose *low* priority parameter (e.g., $P_{P1} = 0.25$) for hunger (Food), and *high* priority parameter (e.g., $P_{P2} = 0.75$) for thirst (Water). In order to balance the satiations of the two drives, agent should be able to learn to value Water over Food in such environment.

2.2 Rewards

Ethology, comparative psychology, and neuroscience have shown that observable behavior is influenced by internal factors and by external factors [11]. In this paper, drive reduction (consequence of agent-environment interaction) is directly translated into a reward signal delivered to a reinforcement learning, inspired from Hullian learning [1, 10, 12]. In fact, Hull specified the hypothetical property supposed to be shared by all reinforcers that gave them the ability to reinforce behavior: they all reduced drive. When we are hungry, Hull argued, we are in a state of drive that activates us. Food, when eaten, reduces that hunger drive (the source of reinforcement:

reduction of drive, reward). Thus, many of the physical stimuli that serve as rewards (food, water, ...) can be conceived as drive reducers.

Behavior is finally a simple consequence of the level of drive at the moment and the ability of stimuli present to trigger stimulus-response habits (previously created by drive reduction). This simple relationship could be expressed as (Hull's postulate):

$$\text{Behavior strength} = D \times H, \tag{2}$$

where Behavior strength is actual behavioral response strength (from a stimulus), H is associative learned habit strength, and D is the level of drive at the moment.

From this, when an agent performs an action from t to t + 1, it results for drives d_i in a change in satiation levels from $S_{Li}(t)$ to $S_{Li}(t + 1)$. For each time step, reward $r_i(t + 1)$ generated by each drive d_i is obtained by multiplying the difference in satiation between time steps $S_{Li}(t + 1) - S_{Li}(t)$ by the drive priority D_{Pi}, as in Eq. (3):

$$r_i(t+1) = D_{Pi}(t)(S_{Li}(t+1) - S_{Li}(t)). \tag{3}$$

Note that, since $D_{Pi} \in [0, 1]$, $r_i(t + 1)$ is positive to reward if $(S_{Li}(t + 1) - S_{Li}(t)) > 0$, null if $(S_{Li}(t + 1) - S_{Li}(t)) = 0$, or negative to punish if $(S_{Li}(t + 1) - S_{Li}(t)) < 0$.

At this stage, in each different state internal goals d_i are coupled by the requirement that they share actions (same possible actions of the agent in each state), i.e., a common action space in a problem of multiple-goal reinforcement learning. Indeed, the underlying formalism for many reinforcement learning algorithms is the Markov decision process. The reward function R(s, a) defines the expected one-step payoff for taking action a in state s. Goal of reinforcement learning is to discover an optimal policy that maps states to actions so as to maximize discounted long term reward.

Thus, in multiple-goal reinforcement learning problem, each decision process (each internal goal d_i) has a distinct state space, but they share a common action space, and are required to execute the same action on each time step. Then, n decision processes implicitly define a larger composite decision process. Formally, the aim is to find optimal policy for this composite decision process [13] which could be defined as the policy that maximizes summed discounted reward across n decision processes, under the assumption that rewards are comparable [14]. Then, agent aims to maximize composite reward as in Eq. (4):

$$R(s, a) = \sum_{i=1}^{n} R_i(s, a). \tag{4}$$

2.3   Action Selection ('Greatest Mass' Sarsa(0)-Learning)

The action selection problem is to find approximate solutions to multiple-goal rein-forcement learning problem referenced to modular reinforcement learning problem (decomposition of a complex multi-goal problem into a collection of single-goal learning processes, modeled as Markov decision processes).

Several approaches have been developed in order to solve such multiple-goal problem as a modular reinforcement learning problem [13–16]. From these approaches and under the assumption that rewards are comparable which allows to assume the composite reward as a sum, 'Greatest Mass' Sarsa(0)-Learning approach performs better than 'Greatest Mass' Q-Learning, Top Q-Learning, and Negotiated W-Learning approaches for a certain class of composite Markov decision processes as proved in [13], even if the convergence was not proved.

'Greatest Mass' Sarsa(0)-Learning generates a composite module (an overall Q-value function) as a sum of the Q-values of the individual modules $Q_i(s, a)$, in Eq. (5):

$$Q(s, a) = \sum_{i=1}^{n} Q_i(s, a). \tag{5}$$

Sarsa(0)-Learning (on-line) algorithm, used in this paper, is developed from [6].

2.4 Adaptive Priority Process

A motivated agent has to adapt to and survive in a situated, dynamic environment. However, to reach this behavioral flexibility via reinforcement learning in changing environments may demand mechanisms that modulate the criterion of internal assessment to influence behavior in an adaptive manner [17]. Hullian drive-based architecture developed in [10] includes both expression of intended purpose of any motivated architecture, combined with learning of priorities by reinforcement.

Ashby proposes the definition that: a form of behavior is 'adaptive' if it maintains the essential variables within physiological limits [18]. The thesis that 'adaptation' means the maintenance of essential variables within physiological limits is thus seen to hold not only over the simpler activities but over the more complex activities.

Thus, instead of a simple priority adjustment heuristic developed in [10], the suggested adaptive priority process, in this paper, is based on a set of if-then rules inspired from the viability zone concept [18, 19] reducing or enhancing the motivation values in order to maintain the homeostasis of the internal variables. From this, variations of satiation levels S_{L1} and S_{L2} lead to modulate priority parameters P_{P1} and P_{P2} which consequently influence drive priorities D_{P1} and D_{P2} in order to maintain homeostasis of the internal variables, i.e., internal goals (primary drives) d_1 and d_2.

As satiation level $S_{Li} \in [0, 1]$, this range is divided in three intervals $N = [0, 0.3]$, $M = [0.3, 0.7]$, and $F = [0.7\ 1]$. Then, the developed set of if-then rules is as follows:

If $S_{L1} \in N$ and $S_{L2} \in N$, Then $P_{P1} = 0.50$ and $P_{P2} = 0.50$,
If $S_{L1} \in N$ and $S_{L2} \in M$, Then $P_{P1} = 0.75$ and $P_{P2} = 0.50$,
If $S_{L1} \in N$ and $S_{L2} \in F$, Then $P_{P1} = 0.99$ and $P_{P2} = 0.01$,
If $S_{L1} \in M$ and $S_{L2} \in N$, Then $P_{P1} = 0.50$ and $P_{P2} = 0.75$,
If $S_{L1} \in M$ and $S_{L2} \in M$, Then $P_{P1} = 0.50$ and $P_{P2} = 0.50$,
If $S_{L1} \in M$ and $S_{L2} \in F$, Then $P_{P1} = 0.99$ and $P_{P2} = 0.01$,
If $S_{L1} \in F$ and $S_{L2} \in N$, Then $P_{P1} = 0.01$ and $P_{P2} = 0.99$,
If $S_{L1} \in F$ and $S_{L2} \in M$, Then $P_{P1} = 0.01$ and $P_{P2} = 0.99$,
If $S_{L1} \in F$ and $S_{L2} \in F$, Then $P_{P1} = 0.50$ and $P_{P2} = 0.50$.

3   Experiments and Results

In this Section, an artificial environment is simulated considering the task of navigating from a Start state (S), i.e., from one state to another in each moving step, until to a Goal state (G) in a grid environment of size 10×10 states. From each state, four actions are possible (right, up, left, and down), as illustrated in Fig. 2(a). Also, from each state, the agent senses its surrounding environment within a small perceptual radius, as illustrated in Fig. 2(a). Note that this artificial environment is simulated in the goal of setting up a hard action selection problem [15].

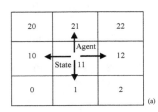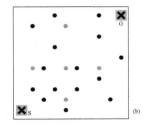

Fig. 2. (a) The agent with a small perceptual radius. (b) An environment with abundant Food (15 resources in black) and scarse Water (5 resources in grey), a Start (S), and a Goal (G).

Thus, in order to experiment the suggested adaptive motivation system, an agent having two primary drives d_1 (hunger) and d_2 (thirst) is simulated in an environment scattered with two types of resources: abundant Food (e.g., 15 resources) and with scarse Water (e.g., 5 resources), as illustrated in Fig. 2(b). Given its limited sensory information, agent needs to develop an adaptive balancing drives during the achievement of the task of navigation from S to G.

3.1   Experimental Setup

Initially, the agent start from the state S (at $t = 0$). Then, a moving step leads to: no consumption, a consumption from a Food resource, or a consumption from a Water resource, until the reaching of the goal G:

- the initial satiation levels are set to 0.50, i.e., $S_{L1} = 0.50$ and $S_{L2} = 0.50$,
- a satiation level penalty (-0.00015) is applied to both S_{L1} and S_{L2} after each moving step without consumption,
- a satiation level gain ($+0.08$) is applied to S_{L1} and a satiation level penalty (-0.00015) is applied to S_{L2}, after each moving step with a Food consumption,
- a satiation level gain ($+0.08$) is applied to S_{L2} and a satiation level penalty (-0.00015) is applied to S_{L1}, after each moving step with a Water consumption.

Concerning modular reinforcement learning, three learning cases are developed:

- learning case with the individual module related to Food noted $Q_F(s, a)$,
- learning case with the individual module related to Water noted $Q_W(s, a)$,
- learning case with the composite module noted $Q_{F+W}(s, a)$.

Sarsa(0)-Learning (on-line) parameters: α the constant step-size parameter ($0 < \alpha <= 1$) and γ the discount rate ($0 <= \gamma <= 1$) are set to $\alpha = 0.009$, $\gamma = 0.80$, ε-greedy $= 0.10$, and the number of episode $= 3500$ for each individual module.

3.2   Experiments and Results

In order to validate the suggested adaptive motivation system three experiments are developed. Each experiment is analyzed with regard to the results:

- the step number to achieve the navigation task from start S until to reach the goal G (to evaluate how much is the necessary number of step to satisfy the external goal),
- the number of Food consumption during this task (to evaluate the satisfaction level of the internal goal related to hunger),
- the number of Water consumption during this task (to evaluate the satisfaction level of the internal goal related to thirst),
- the evolution of satiation levels S_{L1} and S_{L2} are given in function of steps (to analyze and compare satiation levels of the two internal goals).

3.2.1   Experiment 1
In this experiment, same priority parameters (i.e., $P_{P1} = 0.50$ and $P_{P2} = 0.50$) are given for the two drives d_1 and d_2. The obtained results are given in Table 1 and Fig. 3.

Table 1. Experiment 1 results.

	Step number	Food consumption	Water consumption
$Q_F(s, a)$	72	17/15	0/5
$Q_W(s, a)$	59	7/15	1/5
$Q_{F+W}(s, a)$	48	7/15	0/5

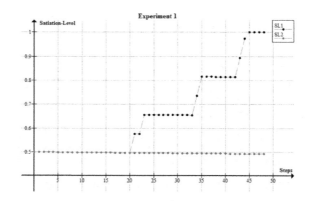

Fig. 3. Learning-case $Q_{F+W}(s, a)$: S_{L1} evolution (black), and S_{L2} evolution (grey), from Start state to Goal state, with same priority parameters $P_{P1} = 0.50$ and $P_{P2} = 0.50$.

These results demonstrate that the agent (same priority parameters) cannot balance the satiations of the two drives for the three learning cases. More, as the environment is with abundant Food and scarse Water, agent learned to value Food over Water what is the inverse that what it should do in such environment.

3.2.2   Experiment 2

In this experiment, different priority parameters (i.e., $P_{P1} = 0.25$ and $P_{P2} = 0.75$) are given for the two drives d_1 and d_2. The obtained results are given in Table 2 and Fig. 4. These results are slightly better than those of Experiment 1, for the three learning cases, the agent tried to balance the satiations of the two drives. Agent tried to learn to value Water over Food what is the appropriate direction of what it should do in such environment. Consequently, if agent navigate in a new environment, different priority parameters should be fixed again and an adaptive priority process is then necessary.

Table 2. Experiment 2 results.

	Step number	Food consumption	Water consumption
$Q_F(s, a)$	62	18/15	3/5
$Q_W(s, a)$	100	16/15	1/5
$Q_{F+W}(s, a)$	44	8/15	4/5

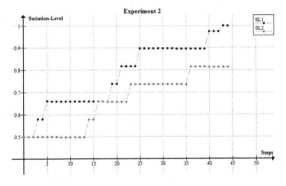

Fig. 4. Learning-case $Q_{F+W}(s, a)$: S_{L1} evolution (black), and S_{L2} evolution (grey), from Start state to Goal state, with different priority parameters $P_{P1} = 0.25$ and $P_{P2} = 0.75$.

3.2.3   Experiment 3

In this experiment, priority parameters are set from the suggested adaptive priority process (from initial $P_{P1} = 0.50$ and $P_{P2} = 0.50$ and then updated after each episode).

The obtained results are given in Table 3, Fig. 5, and an additional result is given in Fig. 6, in order to evaluate the evolution of the priority parameters P_{P1} and P_{P2}. It is clear that these results are better than those of Experiment 1 and Experiment 2, agent (adaptive priority process) learned to value Water over Food in such environment, particularly for the learning case $Q_{F+W}(s, a)$, allowing to balance the drives. Also, agent is able to learn in new environment (on-line-learning, adaptive priority process).

Table 3. Experiment 3 results.

	Step number	Food consumption	Water consumption
$Q_F(s, a)$	69	5/15	1/5
$Q_W(s, a)$	84	13/15	10/5
$Q_{F+W}(s, a)$	61	6/15	7/5

The evolution of priority parameters in Fig. 6 (related to Fig. 5) shows that from the Step 0 until Step 20, both S_{L1} and S_{L2} progresses from 0.5 to around 0.66, then S_{L2} surpasses first 0.7 at Step 21 which activates one of the rule of the adaptive priority process to set P_{P1} to 0.99 and P_{P2} to 0.01 (giving the priority to d_1 over d_2 in order to balance their satiation levels) until S_{L1} surpasses 0.7 at Step 45 which activates another rule to set P_{P1} to 0.5 and P_{P2} to 0.5.

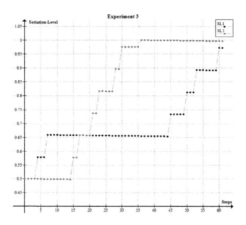

Fig. 5. Learning-case $Q_{F+W}(s, a)$: S_{L1} evolution (black), and S_{L2} evolution (grey), from Start state to Goal state, with the adaptive priority process.

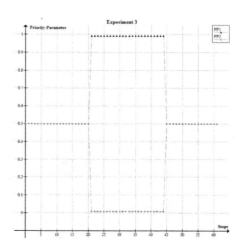

Fig. 6. Learning-case $Q_{F+W}(s, a)$: P_{P1} evolution (black), and P_{P2} evolution (grey).

4   Conclusion

In this paper, an adaptive motivation system under modular reinforcement learning is suggested for agent decision-making modeling of biological regulation. For this, main concepts of drives, rewards, action selection under modular reinforcement learning ('Greatest Mass' Sarsa(0)-Learning) as well as an adaptive priority process have been developed. Then, experiments and results are presented and analyzed. Obtained results demonstrate how the suggested adaptive motivation system can be used by an agent learning (on-line) to select appropriate actions, during a navigation task from a starting position to a goal position, i.e., in each moving step, in order to reach an external goal as well as to satisfy internal goals; predicting a promising result in future to demonstrate how the interaction nature (stimulation-drive, social-drive, …) influences the agent behavior.

With regard to the works developed in [10], two main differences are to be noted:

- two internal goals are considered in [10] while two internal goals and one external goal (the task goal) are considered in this paper,
- another important difference: a large perceptual radius is used in [10] while a small perceptual radius (a hard action selection problem) is used in this paper,
- finally, a simple priority adjustment heuristic is used in [10] while suggested adaptive priority process uses a set of rules inspired from viability zone [18, 19].

Another important point concerns the problem of reward incomparability (Sect. 2.2) and the problem of convergence (Sect. 2.3), authors in [14] highlight the problem of reward incomparability, leading them to the result that it is impossible to construct an ideal arbitration function without injecting external domain knowledge. In addition, the problem of convergence of 'Greatest Mass' Sarsa(0)-Learning can be overcome if a transition model of the external world environment is known [20].

Finally interesting alternatives for future works could be to investigate: the underlying mechanisms of hedonic value developed in [17] to enhance the adaptation for motivated agents, and the intrinsic and extrinsic motivation concepts [2].

References

1. Hull, C.L.: Principles of Behavior, An Introduction to Behavior Theory. D. Appleton-Century Co, New York (1943)
2. Singh, S., Lewis, R.L., Barto, A.G., Sorg, J.: Intrinsically motivated reinforcement learning: an evolutionary perspective. IEEE Trans. Auton. Ment. Dev. **2**(2), 70–82 (2010)
3. Halliday, T.: Motivation. In: Halliday, T.R., Slater, P.J.B. (eds.) Causes and Effects. Blackwell Scientific, Oxford (1983)
4. Cannon, W.B.: The Wisdom of the Body. W. W. Norton, New York (1932)
5. Savage, T.: Artificial motives: a review of motivation in artificial creatures. Connect. Sci. **12**, 211–277 (2000)
6. Sutton, R.S., Barto, A.G.: Reinforcement Learning: An Introduction. MIT Press, Cambridge (1998)

7. Brooks, R.A.: The role of learning in autonomous robots. In: Proceedings of the Fourth Annual Workshop on Computational Learning Theory (COLT 1991), pp. 05–10 (1991)
8. Chohra, A., Madani, K.: Biological regulation and psychological mechanisms models of adaptive decision-making behaviors: drives, emotions, and personality. In: Nguyen, N.-T., Manolopoulos, Y., Iliadis, L., Trawiński, B. (eds.) ICCCI 2016. LNCS, vol. 9875, pp. 412–422. Springer, Cham (2016). doi:10.1007/978-3-319-45243-2_38
9. Warren, W.H.: The dynamics of perception and action. Psychol. Rev. **113**(2), 358–389 (2006)
10. Konidaris, G., Barto, A.: An adaptive robot motivational system. In: Nolfi, S., Baldassarre, G., Calabretta, R., Hallam, J.C.T., Marocco, D., Meyer, J.-A., Miglino, O., Parisi, D. (eds.) SAB 2006. LNCS, vol. 4095, pp. 346–356. Springer, Heidelberg (2006). doi:10.1007/11840541_29
11. Breazeal, C., Brooks, R.A.: Robot emotion: a functional perspective. In: Fellous, J.-M., Arbib, M. (eds.) MIT Press, pp. 271–310 (2005)
12. Berridge, K.C.: Rewards learning: reinforcement, incentives, and expectations. Psychol. Learn. Motiv. **40**, 223–278 (2001). Academic Press
13. Sprague, N., Ballard, D.: Multiple-goal reinforcement learning with modular Sarsa(0). Technical report 798, The University of Rochester (2004)
14. Bhat, S., Isbell Jr., C.L., Mateas, M.: On the difficulty of modular reinforcement learning for real-world partial programming. In: American Association Artificial Intelligence, pp. 318–323 (2006)
15. Humphrys, M.: Action selection methods using reinforcement learning. Ph.D. dissertation, University of Cambridge, Technical report (1997)
16. Karlsson, J.: Learning to solve multiple goals. Ph.D. thesis, University of Rochester (1997)
17. Cos, I., Canamero, L., Hayes, G.M., Gillies, A.: Hedonic value: enhancing adaptation for motivated agents. Adapt. Behav. **21**(6), 465–483 (2013)
18. Ashby, W.R.: Design for a Brain. Chapman & Hall, London (1952)
19. Meyer, J.-A.: Artificial life and the animat approach to artificial intelligence. Artif. Intell. (1995). Academic Press
20. Kompella, V.R., Kazerounian, S., Schmidhuber, J.: An anti-hebbian learning rule to represent drive motivations for reinforcement learning. In: del Pobil, A.P., Chinellato, E., Martinez-Martin, E., Hallam, J., Cervera, E., Morales, A. (eds.) SAB 2014. LNCS, vol. 8575, pp. 176–187. Springer, Cham (2014). doi:10.1007/978-3-319-08864-8_17

Computational Swarm Intelligence

Simulated Annealing for Finding TSP Lower Bound

Łukasz Strąk[✉], Wojciech Wieczorek, and Arkadiusz Nowakowski

Institute of Computer Science, University of Silesia in Katowice,
Będzińska 39, 41-205 Sosnowiec, Poland
lukasz.strak@gmail.com,
{wojciech.wieczorek,arkadiusz.nowakowski}@us.edu.pl

Abstract. Held and Karp's theory has been proposed in the early 1970s in order to estimate an optimal tour length for the Travelling Salesman Problem. The *ascent* method, which is based on this theory, makes it possible to obtain a graph, which contains a large number of edges common with the optimal solution. In this article, we presents a new algorithm of simulated annealing for the same purpose. Our approach improves the quality of obtained results and makes it possible to receive a greater number of edges common with the optimal solution. The *ascent* method, suggested by Helsgaun, was applied for comparison since it is well documented, achieves good results and has an available implementation.

Keywords: Held-Karp · Simulated annealing · TSP

1 Introduction

Held and Karp published in 1970 and 1971 studies concerning an estimation of the optimal tour length for the TSP and they applied this theory in the optimization process for the salesman problem. Neither the Held-Karp theory, nor the *ascent* method guarantee that the obtained result will be a correct TSP solution. However, over a short period of time the algorithm promptly finds very good solutions - both in terms of the number of edges of the obtained graph common with the optimum of the problem, and the sum of the weights of this graph edges, in relation to the sum of weights of the optimal solution edges. The result is an estimated value $C \leq C^*$ (C is the estimated optimal solution, C^* is the optimal solution). The TSPLIB studies have shown that the algorithm very frequently gives lower bound less than 1% of optimum [1]. The probability analysis of the results obtained from the HK LO (Held-Karp Lower Bound) algorithm can be found in [2–4]. The work [5] describes analyses of special cases for 1-tree and its generalization on k-tree.

Despite the fact that the graph obtained from the *ascent* method might not have all 2-degree vertices, the technique has been applied in many optimization algorithms for the traveling salesman problem successfully. In these algorithms, the obtained graph is treated as a quasi-optimal solution, which is commonly

© Springer International Publishing AG 2017
N.T. Nguyen et al. (Eds.): ICCCI 2017, Part II, LNAI 10449, pp. 45–55, 2017.
DOI: 10.1007/978-3-319-67077-5_5

used within two contexts: evaluating the optimal tour length and the neighbour-hood function.

This paper presents using a random technique - Simulated Annealing for finding Held-Karp Lower Bound. Other known algorithms described in litera-ture are deterministic. Our approach is stochastic and can be started multiple times for the same problem instance. Existing technique describes in the domain literature have no ability to avoid local optimum, in contrast to the proposed solution.

The structure of this paper is as follows: Sect. 2 includes a description of basic concepts and algorithms connected with the Held and Karp's theory. Section 3 describes the *ascent* method and our algorithm. The last two sections present experiments and conclusions.

2   The Held-Karp Theory and Simulated Annealing

The basis for the Held-Karp Lower Bound algorithm is the concept of 1-tree (see the Definition 1 and Fig. 1).

Definition 1. *(1-tree) The 1-tree is a connected weighted graph that contains exactly one cycle.*

Fig. 1. Examples of three different 1-tree(s) from the same set of vertices (points) in the Cartesian plane.

The 1-tree graph is the relaxation of the tree - without acyclic constraints. Each correct solution for the Traveling Salesman Problem is a 1-tree (each vertex has degree 2). The Held and Karp's work presents an algorithm for generating a special 1-tree called min-1-tree. It is as follows: create a *minimum spanning tree T* for the $\mathcal{V} - \{v_s\}$ set. Then connect the vertex v_s with two vertices located within the *minimum spanning tree T* using the edges with the smallest weights. Figure 2 shows an exemplary 1-tree obtained from the Held-Karp algorithm.

Remark 1. The cost of min-1-tree is related to a choice of vertex v_s.

Held and Karp in their breakthrough work [6,7] have shown that a change of a edge weight by using formula (1) does not cause a change of edges that belong to the optimum solution, but it causes a change of edges that belong to min-1-tree. This observation will be used in our approach, which modifies the edge weights by maximizing the number verticies with degree 2 in a min-1-tree.

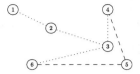

Fig. 2. Example of min-1-tree (dashed and dotted lines). Vertex 5 is selected as v_s and connected with two vertices in the tree (dashed edges). Dotted edges form a *minimal spanning tree*.

Let T^1 be a 1-tree and vertex $i \in T^1$. According to equation (1) all edges adjacency to i is increased by π_i (all others vertices move away from i by π_i). The sum of penalties is constant, if each vertex has the same degree. The sum of penalties is equal to: $\sum_{i=1}^{n} \pi_i$. This implies that the optimum of the TSP problem, which is min-1-tree (increased by vector π) is equal to $2\sum_{i=1}^{n} \pi_i$.

$$d_{ij} = c_{ij} + \pi_i + \pi_j, \tag{1}$$

where: i, j are vertices, c_{ij} is the original distance matrix, and π_i and π_j are the additional costs that cause a change of edges weights in a min-1-tree [6,7]. The sum of weights of the optimal solution changes according to formula (2).

$$\max_k w(\pi^k) = \sum_{(i,j)\in T^k} d_{ij}^k - 2\sum_{i=1}^{n} \pi_i^k \tag{2}$$

where: k is the iteration number, n is the number of vertices, T^k is a set of edges that belong to min-1-tree in an iteration k, which takes into account the weight of edges specified by formula (1). Formula (2) constitutes the optimization objective function. It consists in finding a transformation, $C \to D$ (i.e., C tends to D), given by the vector $\pi = (\pi_1, \pi_2, ..., \pi_n)$ that maximizes the lower bound Eq. (2).

2.1   The *ascent* Method

A suitable method for finding π that maximizes w is subgradient optimization called *ascent* [6,7]. A subgradient is a generalization of the gradient concept. It is an iterative method in which the maximum is approximated by stepwise changes of π. At each step π is changed in the direction of the subgradient, according to Eq. (3).

$$\pi^{k+1} = \pi^k + t^k \cdot v^k \tag{3}$$

where: k is the iteration number, v^k is a subgradient vector, and t^k is a positive scalar, called the step size. An exemplary subgradient optimization process is presented by Fig. 3.

It was proven that in a case when $t^k \to 0$ for $k \to \infty$ and $\sum t^k = \infty$ formula (3) converges to optimum [1,8]. In practice, however, this process has too slow convergence to be useful.

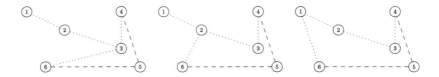

Fig. 3. An exemplary 1-tree in its optimization process.

2.2   Simulated Annealing

Within the method of solution evaluation, in accordance with formula (2), the longest operation is defining min-1-tree, which is built on the basis of *minimal spanning tree* algorithm of a $O(n \log n)$ complexity. The majority of randomized algorithms require a large number of solution space samplings, which in this case is unacceptable. It is one of the reasons why the Simulated Annealing algorithm was chosen. The drawback of the *ascent* algorithm, which we tried to avoid, is the lack of moving back possibility in a case in which the undertaken step does not yield the desired results. The iterative method, offered by Held and Karp is deterministic. It has both advantages and disadvantages as well. An undoubted advantage is the fact that the obtained result can be predicted easier. On the other hand, in case of bad results the next run of the algorithm is useless, since the same result will be obtained. Hence, the algorithm is unable to break through a possible impasse. Our solution is based on Simulated Annealing algorithm and has a feature of avoiding local optima. Functioning within the area of current solution neighbourhood and randomness are also the advantages of this algorithm. Our approach is presented as the Algorithm 1.

Algorithm 1. Simulated Annealing

```
 1: function Ascent(N, Temp, t⁰)
 2:     e = π⁰ = 0                                               ▷ Initial
 3:     for k = 0 → N do
 4:         e' = GetNeighbor(e, tᵏ)
 5:         Δ = Fitness(e') − Fitness(e)                         ▷ Maximize fitness
 6:         if Δ ≥ 0 then                                        ▷ Better solution
 7:             e = e'
 8:         else
 9:             if exp(−Δ/Temp) ≥ Uniform(0, 1) then            ▷ Accept worse solution with
      probability
10:                 e = e'
11:             end if
12:         end if
13:         Temp = Cooling(Temp)
14:         tᵏ⁺¹ = Cooling(tᵏ)                                   ▷ Decrease step size
15:     end for
16: end function
```

The parameters of the algorithm are: k is the iteration number, N - number of iterations, $Temp$ - temperature (changeable in time), t^0 - initial step of the algorithm. During every consecutive step of the algorithm, the neighborhood of the current position e is explored. In the event of finding a better solution, such solution is accepted as the current one. If the solution obtained from neighborhood is worse, in relation to the current one, such a solution is accepted with some probability. The higher the temperature is, the higher the probability. $Uniform$ function returns a random value from the $[0, 1)$ range of a uniform distribution. After each algorithm iteration the cooling process takes place - both of temperature and step as well. Algorithm 2 presents evaluation function applied in the solution.

Algorithm 2. Simulated Annealing

1: **function** FITNESS(π^k)
2:     $T' = MinSpanningTree(\pi^k)$
3:     $T = MinOneTree(T')$
4:     **return** $w(\pi^k) = L(T) - 2 \sum_{i=1}^{n} \pi_i^k$ ▷ See (2)
5: **end function**

The evaluation of the solution takes place after obtaining a *minimum spanning tree* and after creating min-1-tree on its basis. The evaluation is consistent with formula (2). The evaluated solution is built on the basis of the π^k vector. Algorithm 3 presents a scheme of creating a new solution on the basis of π^k and step t^k.

Algorithm 3. Simulated Annealing

1: **function** GETNEIGHBOR(e, t^k)
2:     **return** $e + Normal(t^k) \cdot (0.7v^k + 0.3v^{k-1})$
3: **end function**

A new vector π obtained from *GetNeighbor* function of the Algorithm 3 is drawn from normal distribution of a $\mu = 0.5t^k$ and standard deviation $\sigma = t^k/6$. Values v^k represent the degree of a vertex in an iteration k.

The subject matter literature is abundant in a great many various cooling schemes, including two most popular - geometric and linear. On the basis of many experiments, linear cooling was applied in the solution, which is a linear function dependent on the number of iterations and initial value. This type of cooling was applied with a view to decreasing the value of temperature $Temp$ and step t^k.

3   Experiments

This section presents the results of the tests of the solution presented in this study. The obtained results were compared with the implementation of the Held-Karp Lower Bound algorithm, prepared by Helsgaun, which contains a series of improvements in relation to the original version [1,9]. In order to standardize the computing time, both algorithms were implemented in C# language (Helsgaun source version is written in C++ language). The tests were conducted on an i7 965 computer, but only one core was used in order to minimize the influence of the operating system. Additionally, the algorithm has been run 30 times. The estimated total computing time, excluding the time needed for writing and testing algorithms, is approximately 602 h.

The Helsgaun algorithm is parameterless. In order to compare both algorithms the same number of iterations had to be applied. The value of these parameters was defined on the basis of the number of iterations obtained from the Helsgaun algorithm. The most time-consuming operation in both algorithms is the calculation of the value of the evaluation function, which is defined on the basis of the *minimum spanning tree* algorithm. With the same number of iterations the difference in the executions times is a result of using object-oriented programming approach in our implementation, which was used to improve the readability of the code. Table 1 contains the number of iterations used in all tests.

Table 1. The number of iterations obtained from the Helsgaun HK Lower Bound algorithm

Problem	Number of iterations	Problem	Number of iterations	Problem	Number of iterations	Problem	Number of iterations
bayg29	103	*gr120*	347	*gr202*	1602	*pcb442*	491
bays29	158	*pr124*	241	*tsp225*	280	*ali535*	3216
berlin52	186	*ch130*	286	*pr226*	1104	*gr666*	1687
eil76	99	*pr136*	583	*a280*	424	*pcb1173*	1264
kroA100	588	*pr152*	145	*pr299*	1935	*pr2392*	2390
eil101	165	*kroA200*	498	*pr439*	2333		

Instances from the TSPLIB were chosen randomly in relation to the size of the problems including the small (29–100 vertices), medium (100–600 vertices), and big instances of problems (600–2392 vertices).

3.1   Analysis of the Values of Parameters

The first stage of the tests was the analysis of the influence of the parameters on the results obtained from the algorithm. This analysis concerned two of three parameters, since the number of iterations of our approach is defined by the precise number of iterations of the parameterless Helsgaun algorithm (the algorithm has stopped after the certain number of iterations automatically). Figure 4 presents results obtained for exemplary instance of the problems. The examined ranges are $Temp \in [60, 270]$ and $t_0 \in [1, 19]$. Sampling in case of temperature was every 30 (60,90,...,270), for step every 2 (1,3,...,19). Every run of the algorithm was retaken 30 times, and the results were averaged out.

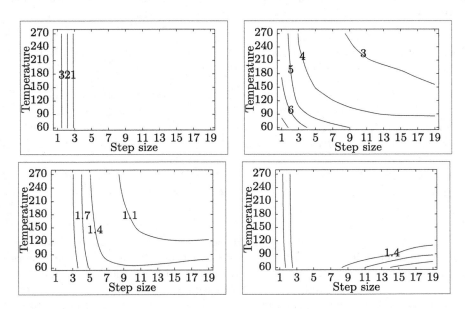

Fig. 4. The contour chart of the results obtained from sampling different values of $Temp$ parameters and t_0 for selected instances of problems from the TSPLIB. The presented problems are respectively *gr202, pr439, gr666, pr2392*

In all tested temperatures the algorithm yielded best results for the highest values of the parameter. In case of the second parameter t_0 (size of the step) higher values and allow us to obtain the best result (from 11 to 19), apart from several instances where value 1 proved to be the best. It is connected with the weights of edges (distances between vertices) for a particular instance of a problem. For small distances, a large value of the step is too big and the algorithm might surpass the optimum. For the *gr202* problem, in order to obtain the best result, it is sufficient to use an adequately large step - above 3. In case of the highest analysis of the *pr2392* problem, the tolerance for both parameters is very high. The best results can be achieved by selecting $Temp$ from the range of $[3, 19]$ and t_0 from the range $[120, 270]$. On the basis of the achieved results, we can construe that a high value of the $Temp$ parameter - 270 most often makes it possible to achieve the best results. It is much more difficult to adjust the parameter of the initial step of the algorithm. It is also important to emphasize that the achieved results were obtained on the basis of the number of iterations from Table 1.

3.2 The Comparison of Algorithms

In this subsection the simulated annealing algorithm for the Held and Karp theory and the implementation of this theory applied by Helsgaun were compared. The algorithm presented in this study was launched on the settings presented in Table 2, which were defined on the basis of the analysis from the previous subsection. Similarly as in the previous tests, every run of the algorithm was retaken 30 times, and the results were averaged out.

Table 2. The best settings of the simulated annealing algorithm for the HK Lower Bound, taking into account the number of iteration from Table 1

Problem	$Temp$	t_0	Problem	$Temp$	t_0	Problem	$Temp$	t_0	Problem	$Temp$	t_0
bayg29	270	11	gr120	270	11	gr202	270	11	pcb442	270	11
bays29	270	5	pr124	270	19	tsp225	270	5	ali535	270	19
berlin52	270	11	ch130	270	7	pr226	270	19	gr666	270	19
eil76	270	1	pr136	270	19	a280	270	1	pcb1173	270	11
kroA100	270	11	pr152	270	19	pr299	270	11	pr2392	270	11
eil101	270	1	kroA200	270	11	pr439	270	19			

Table 3 presents the results of the Helsgaun algorithm [1,9] tests and the solution presented in the study. Similarly as in the previous tests, the results are averaged out for 30 reruns. Apart from the average, the Mann-Whitney-Wilcoxon statistics test and a 5% significance were used in the tests. Every name of the problem includes a number of vertices (the last figures in the name). "HK" refers to the original HK Lower Bound algorithm, in turn "HKSA" to the solution proposed in this work. The length of the tour is calculated on the basis of the original unit of measure located in the problem file. "Mean difference" denotes an absolute difference in the obtained mean lengths of solutions from both algorithms.

A worse computing time, in case of the solution suggested in the work, stems from an object approach at the stage of implementing the algorithm. Both algorithms have the same number of iterations. The Simulated Annealing algorithm was statistically better 15 times. In turn, the Helsgaun algorithm 7 times. The first algorithm, in case when it was statistically better, significantly differed in terms of the difference in the mean tour length. The biggest difference was 1701.18 and it was the difference for the gr666. Despite the victories, the Helsgaun algorithm only slightly differed from the algorithm suggested in the article (the difference in mean tour distances was never bigger than 1). The reason why it proved statistically better is that it is a deterministic algorithm and its standard deviation equal zero (zero scatter of results). Both algorithms were equal in only one case - berlin52. It was connected with the fact that both algorithms achieved optimal solutions in every rerun.

Table 3. The results of the comparison of Helsgaun algorithm with the solution presented in the study. The column "statistically better" was filled on the basis of the Mann-Whitney-Wilcoxon test

Problem	Time [s]		Optimum	Mean distance		Statistical better	Mean difference
	HK	HKSA		HK	HKSA		
bayg29	0.01	0.01	1610	1607.78	**1607.83**	HKSA	0.05
bays29	0.01	0.01	2020	**2013.46**	2013.45	HK	0.01
berlin52	0.04	0.02	7542	7542	7542	-	0
eil76	0.03	0.04	538	536.92	**536.93**	HKSA	0.01
kroA100	0.33	0.37	21282	**20936.42**	20936.4	HK	0.02
eil101	0.1	0.1	629	**627.46**	627.22	HK	0.24
gr120	0.34	0.38	6942	6886.97	**6910.29**	HKSA	23.32
pr124	0.24	0.27	59030	54826.09	**55440.31**	HKSA	614.22
ch130	0.26	0.28	6110	6073.9	**6075.05**	HKSA	1.15
pr136	0.64	0.69	96772	94012.17	**94249.46**	HKSA	237.29
pr152	0.23	0.25	73682	62728.77	**63950.75**	HKSA	1221.98
kroA200	1.02	1.14	29368	29016.27	**29062.21**	HKSA	45.94
gr202	7.38	7.87	40160	39984.04	**40054.96**	HKSA	70.92
tsp225	0.76	0.86	3861	**3827.25**	3826.94	HK	0.31
pr226	3.1	3.4	80369	75469.48	**77056.72**	HKSA	1587.24
a280	1.63	1.43	2579	**2566**	2565.93	HK	0.07
pr299	7.46	8.08	48191	47244.01	**47379.92**	HKSA	135.91
pr439	29.86	39.47	107217	104279.66	**104977.01**	HKSA	697.35
pcb442	4.87	8.1	50778	50485.7	**50493.27**	HKSA	7.57
ali535	78.1	112.5	202339	200296.17	**200861.63**	HKSA	565.46
gr666	62.25	82.94	294358	290042.78	**291743.96**	HKSA	1701.18
pcb1173	112.57	96.4	56892	**56351.01**	56350.15	HK	0.86
pr2392	686.97	720.25	378032	**373489.22**	373489.03	HK	0.19

3.3 Similarity to the Optimal Tour

The aim of the previous experiments was to compare the obtained tour length with the optimal solution. Now, we are going to compare the number of tours' edges that also belong to the optimal solution. The results are presented in Table 4.

It can be noticed that the number of edges oscillates at the level above 80% and in some case equals even 90%. In the case of the *berlin52* problem the optimal tour is identical with the optimal solution.

Table 4. The number of common edges of the obtained tour with an optimal solution denominated in percentage. If the value equals 100%, the obtained tour is identical with the problem optimal solution.

Problem	Gap in tour length [%]		Tour edges in optimum [%]		Problem	Gap in tour length [%]		Tour edges in optimum [%]	
	HK	HKSA	HK	HKSA		HK	HKSA	HK	HKSA
bayg29	0.14	0.14	89.66	**91.95**	*gr202*	0.44	**0.26**	88.12	**93.25**
bays29	0.32	0.32	86.21	**88.97**	*tsp225*	**0.87**	0.88	**89.78**	89.24
berlin52	0.00	0.00	100.00	100.00	*pr226*	6.10	**4.12**	85.40	**85.90**
eil76	0.20	0.20	85.53	**87.89**	*a280*	**0.50**	0.51	85.36	85.36
kroA100	1.62	1.62	86.00	**87.40**	*pr299*	1.97	**1.68**	84.62	**84.83**
eil101	**0.24**	0.28	83.17	**88.68**	*pr439*	2.74	**2.09**	87.24	**88.90**
gr120	0.79	**0.46**	85.83	**87.25**	*pcb442*	0.58	**0.56**	83.94	**84.20**
pr124	7.12	**6.08**	**89.52**	88.84	*ali535*	1.01	**0.73**	82.24	**82.66**
ch130	0.59	**0.57**	85.38	**87.67**	*gr666*	1.47	**0.89**	85.89	**87.35**

4   Conclusion

On the basis of the obtained results we can claim that the Held-Karp theory implemented in the Simulated Annealing algorithm yields very good results. In the majority of cases it makes it possible to achieve better results. The biggest advantage of the algorithm is its randomness, which means that many instances of the algorithm can be run, e.g. parallel, and better results can be achieved. In turn, the Helsgaun algorithm is easy to slaunch, since it is parameterless. The solution suggested in the study has three parameters, however, the values of $Temp$ parameters and the number of iterations is easy to estimate, which was presented in the tests. Both algorithms approximate very well the optimal solution, both nominally - the length of the tour and in terms of the number of edges common with the optimal solution.

Good results of the randomized algorithm lead to a conclusion that different randomized algorithms are also capable of improving the results. In the future, we would like to focus on creating an algorithm that deterministically looks at the min-1-treeand try to improve the final solution.

References

1. Helsgaun, K.: Eur. J. Oper. Res. **126**, 106–130 (2000)
2. Goemans, M.X., Bertsimas, D.J.: Probabilistic analysis of the Held and Karp lower bound for the Euclidean traveling salesman problem. Math. Oper. Res. **16**(1), 72–89 (1991)
3. Valenzuela, C.L., Jones, A.J.: Estimating the Held-Karp lower bound for the geometric TSP. Eur. J. Oper. Res. **102**(1), 157–175 (1997)

4. Johnson, D.S., McGeoch, L.A., Rothberg, E.E.: Asymptotic experimental analysis for the Held-Karp traveling salesman bound. In: Proceedings of the Seventh Annual ACM-SIAM Symposium on Discrete Algorithms, SODA 1996, Philadelphia, PA, USA, pp. 341–350. Society for Industrial and Applied Mathematics (1996)
5. Westerlund, A., Göthe-Lundgren, M., Larsson, T.: A note on relatives to the Held and Karp 1-tree problem. Oper. Res. Lett. **34**(3), 275–282 (2006)
6. Karp, R.M., Held, M.: The traveling-salesman problem and minimum spanning trees. Oper. Res. **18**(6), 1138–1162 (1970)
7. Held, M., Karp, R.M.: The traveling-salesman problem and minimum spanning trees: Part II. Math. Program. **1**(1), 6–25 (1971)
8. Polyak, B.T.: A general method of solving extremum problems. Dokl. Akad. Nauk SSSR **174**(1), 33 (1967)
9. Helsgaun, K.: An effective implementation of k-opt moves for the Lin-Kernighan TSP heuristic. Technical report, Roskilde University (2006)

A Cellular Automaton Based System
for Traffic Analyses on the Roundabout

Krzysztof Małecki[1(✉)], Jarosław Wątróbski[1], and Waldemar Wolski[2]

[1] West Pomeranian University of Technology,
Zolnierska Str. 52, 71-210 Szczecin, Poland
{kmalecki,jwatrobski}@wi.zut.edu.pl
[2] University of Szczecin, Mickiewicza Str. 64, 71-101 Szczecin, Poland
wwolski@wneiz.pl

Abstract. The paper presents an analysis of the impact of road conditions, the distance between vehicles and the number of pedestrians on the roundabout capacity. The study was based on a developed cellular automaton (CA) model and the implemented simulation system. The developed CA model extends the basic traffic model with a braking mechanism. It also reflects the actual technical conditions of vehicles (acceleration and braking depending on the dimensions and functions of the vehicle, as well as the driving at a roundabout of different speeds that are appropriate for the size of the vehicle). The study was based on the example of a two-lane roundabout with four two-lane inlet roads.

Keywords: Capacity of roundabout · Cellular automaton (CA) · Roundabout traffic simulation · Weather conditions · Pedestrians · Distance between vehicles

1 Introduction

Traffic study is an important and current issue that can be implemented in a number of ways, such as traffic observations, road structure modifications, and computer simulations. The latter are made on the basis of specialized software. This paper proposes a micro-simulator of road traffic, specifically for roundabout. The main topic of the roundabout approach is the current trend, such as in Poland, the exchange of intersections for the roundabouts. Roundabouts solve many complications of traditional intersections with designated roads: main and sub-ordered. Roundabouts reduce the number of collision points at intersections, not only between vehicles but also between vehicles and pedestrians [1]. It translates directly into the possible number of accidents [2]. An important aspect of entering roundabouts is to force the driver to slow down when approaching the roundabout, which results both from the necessity of giving priority to the vehicles at the roundabout and the proper construction of the roundabout. Leaf and Preusser [3] have shown that vehicle speed has a significant impact on pedestrian injuries during accidents. Based on the research conducted in Sweden [4], the main factors affecting the safety of the roundabout are the number of roads at the roundabout, the diameter of the island, and the speed and intensity of traffic. The study authors argue that single-line roundabouts are the safest and building roundabouts with a large radius is associated with higher permissible speeds, which can increase the

© Springer International Publishing AG 2017
N.T. Nguyen et al. (Eds.): ICCCI 2017, Part II, LNAI 10449, pp. 56–65, 2017.
DOI: 10.1007/978-3-319-67077-5_6

number of accidents. On the other hand, a radius of less than 10 m usually forces the driver to turn the bend and drive almost straight without slowing down.

Much work has been done on the study and analysis of roundabout traffic [5–7], but they do not address the effectiveness of traffic (in the meaning of the capacity of roundabout) considering the different road conditions and distance between vehicles. Factors affecting the capacity of roundabouts are specified in [8].

The aim of this article is to investigate a roundabout capacity in relation to the weather conditions on the road, pedestrians and the distance between vehicles. To achieve the aim a cellular automata (CA) model was developed and a computer simulation was run, using software developed for the purposes of road traffic simulation.

2 Related Work

There are various simulation solutions related to traffic research. The available literature describes methods and algorithms [9, 10], large simulation software systems [11–13], small applications [14], hardware simulators [15, 16], hardware-software simulators [17], models [18, 19], traffic analyses [20] and traffic lights algorithms and simulations [21]. An example of a micro-simulator available through the website is the traffic simulator developed by Martin Treiber [22]. It allows analysis of traffic in the aspect of onramp, offramp, road works, uphill and routing. The user observes the situation on the road and has the ability to change certain parameters, e.g. the amount and the speed of vehicles for strictly defined cases.

Basic mathematical models of car traffic, based on cellular automata, are presented in [23–25] and then extended in [26–28]. The traffic rules for roundabouts are addressed in [29], single-lane roundabouts in [30–32] and multi-lane roundabouts in [33–35]. Pedestrian models are considered in [36, 37]. For the purposes of this publication, an original model for multi-lane roundabout was used, which applies to the current road traffic regulations and makes it possible to study the capacity of roundabout in various aspects, including the impact of dimensions of roundabout, weather conditions or distance between cars on its capacity. The research is aimed at making modelling simulations of road traffic more realistic.

3 Proposed Approach

The simulating environment presented in this paper is based on the author's CA model of the multi-lane roundabout with multi-lane inlet roads. This model is based on the widely discussed Nagel-Schreckenberg model (N-Sch) [9] and its modified version, Hartman's model [27]. Originally, the N-Sch model was developed to simulate traffic on highways, without distinguishing the type of vehicle. It was assumed that the length of a single cell of CA corresponds to 7.5 m of road, which is the average length of the car along with the surrounding space. Thus, the speed unit in the N-Sch model corresponds to a real speed of about 27 km/h. In order to truly reflect urban traffic, according to [27], vehicles were classified according to their length and number of occupied cells (Table 1).

Table 1. Vehicle classification according to their length. Source: own research.

Type of vehicle	Length of vehicle [m]	Length of vehicle [cells of CA]
Motorcycle	2.5	1
Car	5	2
Van	7.5	3
Minibus	10	4
Bus, commercial car	12.5	5
Truck	15	6

In the developed model, the length of a single cell reflects a 2.5 m of real road. This translates into the speed of moving vehicles presented in Table 2.

Table 2. Comparison of velocity in the N-Sch model and the developed model presented in this article. Source: own research.

Speed $\left[\frac{cells}{iteration}\right]$	Speed in N-Sch model $\left[\frac{km}{h}\right]$	Speed in presented model $\left[\frac{km}{h}\right]$
0	0	0
1	27	9
2	54	18
3	81	27
4	108	36
5	135	45
6	162	54

Stopping vehicle on the road is dependent on many factors [38, 39]: the type of surface (asphalt, concrete, etc.), the surface conditions (dry, wet, snowy, cleared of snow, sleet), the technical condition of the vehicle, driver experience and his psychophysical state and vehicle speed. Discussed earlier model N-Sch not recognizes such factors. It treats the inhibition process as a speed reduction by 1 or a speed reduction to the specified value. This paper focuses, among others, on the generalization of N-Sch model about braking coefficients.

Based on measurements of braking length for different speed and the state of the road surface [38], here is fixed a maximum speed of vehicles on dry, wet and snow-covered roads (Fig. 1). The article focuses primarily on speeds in city traffic.

Extrapolating values, it can determine the length of the deceleration road for the speeds appropriate for this model (Table 3).

According to Table 3, and by choosing the proportional length of braking on dry roads, here is proposed the maximum speed of cars and trucks on wet and snow-covered road (Table 4).

An important factor in braking is the adhesion of the tires to the road. It depends on the type of road surface and its condition. Best adhesion factor has a dry road, and the worst – icy road. Table 5 shows the coefficients of adhesion for different road surfaces.

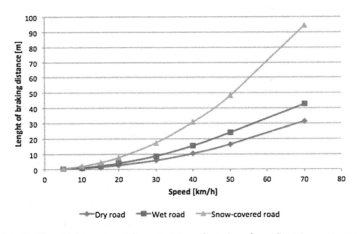

Fig. 1. Braking distance for different states of road surface. Source: own research.

Table 3. Extrapolation of road length of inhibition for the presented model. Source: own research.

Cell/iteration	km/h	Dry road [m]	Wet road [m]	Snow-covered road [m]
1	9	0.469	0.728	1.525
2	18	1.999	3.365	6.151
3	27	4.501	7.136	13.855
4	36	7.975	12.041	24.637
5	45	12.421	18.08	38.497

Table 4. The maximum speed of vehicles for different states of road surface. Source: own research.

Type of vehicle	Dry road	Wet road	Snow-covered road
Car	5	4	3
Truck	2	1	1

In the tire adhesion studies [38], the adhesion coefficients for various types of pavement (concrete, asphalt, basalt, etc.) and the road surface conditions (wet, dry, snowy, iced) have been determined. In this paper, for the purposes of simulation, the braking coefficients for the asphalt road are assumed, assuming that this is the most commonly used road surface in cities.

The coefficients described in Table 5 are used to simulate delayed braking. At factor 1 (dry asphalt), the vehicle is braking according to the N-Sch model; otherwise the value is multiplied by the braking factor and rounded down to total value. The resulting difference is memorized and added in the next iteration. A mechanism of braking is presented as:

Table 5. Braking factors for CA model. Source: own research.

Road surface conditions	Braking factors
Dry asphalt road	1
Wet asphalt road	0.6
Snow-covered asphalt road	0.3

$$v_{t+1} = \lfloor (v_t - k) * r + z_{t-1} \rfloor, \tag{1}$$

where: v_t – the current speed, k – the value by which the speed is reduced (according to the N-Sch), r – the braking factor, z_{t-1} – the value saved from previous braking.

The above mechanism is used to slowing down and stopping vehicles. It simulates the behaviour of tires on wet or snow-covered road surfaces. Thanks to this solution, the vehicle does not slow down like in the N-Sch model, but it decreases its speed gradually, and according to the researches presented and described earlier. In the case of wet or snow-covered road, braking can cause unexpected tire slippage and consequently may cause collisions.

Collisions are mostly a consequence of not adjusting speed to road conditions and keeping the distance between vehicles too low. Such events can cause road congestion. The presented solution includes a special mechanism controlling such situations. If two vehicles collide, as in the real situation, both vehicles are going to the side of the road. Usually, it takes some time. In the simulation environment, the collision detection mechanism, after a fixed period of time, removes both vehicles involved in the collision, simulating move to the side of the road.

Simulating the phenomenon of the collision it has been adapted to the overall concept of cellular automata. The small bumps caused by incorrectly selected speed and bad distances from the preceding vehicle were attempted. It is limited only to the collision of two vehicles going in the same direction. In the main menu of the application, the user has access to information about the number of collisions and slippages. It can also change the maximum speed and set the required distance between vehicles. These settings allow user to conduct research on adjusting speed and distance to prevailing weather conditions, to reduce the number of potential road collisions and thereby increase road capacity.

4   The System Developed for the Simulation

The application was developed in JavaScript and it can be operated in web browsers as well as by means of a console. Running the application by means of a console is possible via NodeJS runtime environment based on V8 engine in the Chrome browser. Running the application this way is considerably faster, as the program may operate without the graphics layer. The application makes use of many free tools enabling the programming works. Figure 2 shows the structure of the application together with applied tools.

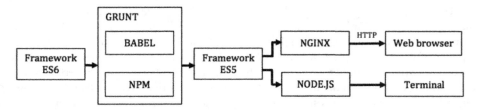

Fig. 2. Application structure. Source: own research.

Frequent problems with running the application are due to different runtime environments. The problems are solved by the combination of Vagrant and Ansible tools. Vagrant enables management of virtual machines, offering identical runtime conditions for applications. Ansible is used to ensure that the virtual machine is always equipped with any indispensable libraries. Upon starting, the program compares the current state of the machine with the expected one and carries out any necessary set-ups. Before that, it is necessary to prepare the configuration files that define the dependencies. The application is written in accordance with the latest standards of ECMAScript 6. To enable correct operation of the application in web browsers, the Babel transpiler was used in order to change the code into the one compliant with ECMAScript 5. The process of transpilation and providing the application to the www server was automated by means of the Grunt program.

5   Experimental Results

The main goal of this paper was to develop a model based on cellular automata, aimed at investigating the impact of the road surface condition, the distance between vehicles and the number of pedestrians on the roundabout capacity.

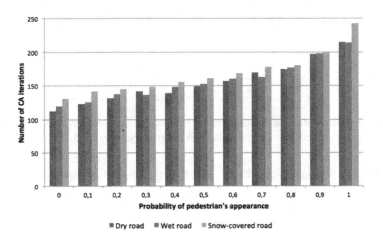

Fig. 3. The capacity of roundabout in terms of condition of road surface and the probability of pedestrian's appearance. Source: own research.

The experiment, the result of which is shown in Fig. 3, was related to the study of the impact of road surface condition on the capacity of roundabout (in terms of the number of vehicles to pass through the roundabout). As a result of the study, the number of iterations of the CA is observed. In fact, it can be recalculated for the time needed to drive a number of vehicles through the roundabout, under certain road conditions. The values in the figures show that both pedestrians and weather conditions affect traffic delays, it means they reduce the capacity of a road or roundabout.

The condition of the road surface is understood as the weather conditions causing changes in the surface state of the road, i.e. the road becomes wet or snow-covered. The road, for which conditions are treated as in classical N-Sch model, is a dry road.

The results shown in Figs. 4 and 5 are another result of experiments based on the developed CA model and the computer simulation environment. It has been verified if the change of vehicle speed will affect the number of possible slips on the road and thus will affect the number of collisions. Various maximum speeds were determined according to Table 4, all trials were conducted for wet road. Brake ratios were determined, the distance between vehicles (values 1–4 in Figs. 4, 5) and the probability of pedestrian entry on the road strips (values 0, 0.5, 1 in Figs. 4, 5) were changed. Figure 4 shows that the speed of vehicles and pedestrians affects the number of vehicles slipping and the number of collisions.

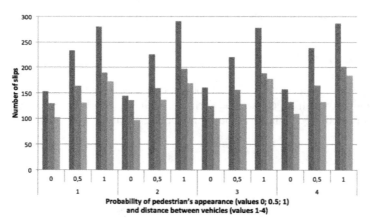

Fig. 4. The number of slips on the wet road in terms of distance between vehicles (values 1–4) and the probability of pedestrian's appearance on the road (values 0; 0.5; 1).

The distance between vehicles does not affect the number of slips, but it impacts on the number of collisions. While analysing Figs. 4 and 5, the greater the distance between vehicles is, the greater the safety of driving is. Figure 5 shows the number of collisions that decreases with the distance between vehicles.

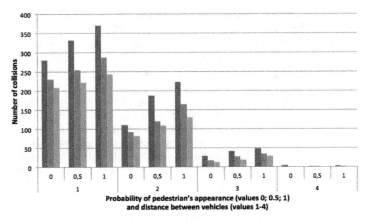

Fig. 5. The number of collisions on the wet road, in terms of distance between vehicles (values 1–4) and the probability of pedestrian's appearance on the road (values 0; 0.5; 1).

6 Conclusion

This paper focused on applying cellular automata in modelling a roundabout capacity in relation to different weather conditions on the road, distance between vehicles and the number of pedestrians. A model was developed and implemented in the form of a simulation system, which served to carry out the research study described herein.

Basic traffic models have been extended with a braking mechanism that takes into account the state of the road surface. Based on prior research, the maximum speeds and braking factors on various surfaces (dry, wet, snow-covered) were determined and, first time, were implemented in a simulator based on a CA. Studies on the impact of road weather conditions, number of pedestrians and the distance between vehicles on the capacity of roundabout were conducted. The results of the research are presented in figures. The introduction of the braking mechanism, for weather road conditions, was justified by the fact that the braking process was based on the basic traffic models. And now, computer simulation can be more realistic.

Further work will focus on extending the braking mechanism to other real situations, such as the driver response time (braking start delay), state of car brake, etc.

References

1. Webpage Created (2013). https://nextstl.com/2013/10/mythbusters-tackles-four-way-stop-v-roundabout-traffic-throughput/. Accessed Jan 2017
2. Transportation Research Board of the National Acad: National Cooperative Highway Research Program Report 572 - Roundabouts in the Unites States (2007)
3. Leaf, W.A., Preusser, D.F.: Literature review on vehicle travel speeds and pedestrian injuries, US Department of Transportation, National Highway Traffic Safety Administration (1999)

4. Brude, U., Larsson, J.: What roundabout design provides the highest possible safety? Nordic Road Transp. Res. **12**(2), 17–21 (2000)

5. Macioszek, E.: Relationship between vehicle stream in the circular roadway of a one-lane roundabout and traffic volume on the roundabout at peak hour. In: Mikulski, J. (ed.) TST 2014. CCIS, vol. 471, pp. 110–119. Springer, Heidelberg (2014). doi:10.1007/978-3-662-45317-9_12

6. Macioszek, E., Sierpiński, G., Czapkowski, L.: Problems and issues with running the cycle traffic through the roundabouts. In: Mikulski, J. (ed.) TST 2010. CCIS, vol. 104, pp. 107–114. Springer, Heidelberg (2010). doi:10.1007/978-3-642-16472-9_11

7. Macioszek, E.: Analysis of significance of differences between psychotechnical parameters for drivers at the entries to one-lane and turbo roundabouts in Poland. In: Sierpiński, G. (ed.) Intelligent Transport Systems and Travel Behaviour. AISC, vol. 505, pp. 149–161. Springer, Cham (2017). doi:10.1007/978-3-319-43991-4_13

8. Wang, R., Liu, M.: A realistic cellular automata model to simulate traffic flow at urban roundabouts. In: Sunderam, V.S., van Albada, G.D., Sloot, P.M.A., Dongarra, J.J. (eds.) ICCS 2005. LNCS, vol. 3515, pp. 420–427. Springer, Heidelberg (2005). doi:10.1007/11428848_56

9. Nagel, K., Schreckenberg, M.: A cellular automata model for freeway traffic. J. de Phys. I **2**, 2221–2229 (1992)

10. Chowdhury, D., Santen, L., Schadschneider, A.: Statistical physics of vehicular traffic and some related systems. Phys. Rep. **329**, 199–329 (2000)

11. Fellendorf, M.: VISSIM: a microscopic simulation tool to evaluate actuated signal control including bus priority. In: 64th Institute Transportation Engineers (ITE) Annual Meeting, Technical paper, Session 32, Dallas, TX, pp. 1–9 (1994)

12. Barcelo, J., Ferrer, J.L., Montero, L.: AIMSUN: Advanced Interactive Microscopic Simulator for Urban Networks, User's Manual, Departament d 'Estadística i Investigació Operativa, UPC (1997)

13. Krajzewicz, D., Erdmann, J., Behrisch, M., Bieker, L.: Recent development and applications of SUMO-simulation of urban mobility. Int. J. Adv. Syst. Meas. **5**, 128–138 (2012)

14. Popescu, M.C., Ranea, C., Grigoriu, M.: Solutions for traffic lights intersections control. In: Proceedings of the 10th WSEAS (2010)

15. Han, X., Sun, H.: The implementation of traffic signal light controlled by PLC. J. Chang. Inst. Opt. Fine Mech. **4**, 029 (2003)

16. Kołopieńczyk, M., Andrzejewski, G., Zając, W.: Block programming technique in traffic control. In: Mikulski, J. (ed.) TST 2013. CCIS, vol. 395, pp. 75–80. Springer, Heidelberg (2013). doi:10.1007/978-3-642-41647-7_10

17. Jaszczak, S., Małecki, K.: Hardware and software synthesis of exemplary crossroads in a modular programmable controller. Prz. Elektrotech. **89**(11), 121–124 (2013)

18. Sierpiński, G.: Theoretical model and activities to change the modal split of traffic. In: Mikulski, J. (ed.) TST 2012. CCIS, vol. 329, pp. 45–51. Springer, Heidelberg (2012). doi:10.1007/978-3-642-34050-5_6

19. Sierpiński, G.: Travel behaviour and alternative modes of transportation. In: Mikulski, J. (ed.) TST 2011. CCIS, vol. 239, pp. 86–93. Springer, Heidelberg (2011). doi:10.1007/978-3-642-24660-9_10

20. Karoń, G., Mikulski, J.: Transportation systems modelling as planning, organisation and management for solutions created with ITS. In: Mikulski, J. (ed.) TST 2011. CCIS, vol. 239, pp. 277–290. Springer, Heidelberg (2011). doi:10.1007/978-3-642-24660-9_32

21. Małecki, K., Pietruszka, P., Iwan, S.: Comparative analysis of selected algorithms in the process of optimization of traffic lights. In: Nguyen, N.T., Tojo, S., Nguyen, L.M., Trawiński, B. (eds.) ACIIDS 2017. LNCS, vol. 10192, pp. 497–506. Springer, Cham (2017). doi:10.1007/978-3-319-54430-4_48

22. Webpage: http://traffic-simulation.de. Accessed Dec 2016

23. Nagel, K., Wolf, D.E., Wagner, P., Simon, P.M.: Two-lane traffic rules for cellular automata: a systematic approach. Phys. Rev. E **58**(2), 1425–1437 (1998)

24. Biham, O., Middleton, A.A., Levine, D.: Phys. Rev. A **46**, 6124 (1992)

25. Chowdhury, D., Schadschneider, A.: Self-organization of traffic jams in cities: effects of stochastic dynamics and signal periods. Phys. Rev. E **59**, 1311–1314 (1999)

26. Małecki, K., Iwan, S.: Development of cellular automata for simulation of the crossroads model with a traffic detection system. In: Mikulski, J. (ed.) TST 2012. CCIS, vol. 329, pp. 276–283. Springer, Heidelberg (2012). doi:10.1007/978-3-642-34050-5_31

27. Hartman, D.: Head leading algorithm for urban traffic modelling. Positions **2**, 1 (2004)

28. Gwizdałła, T.M., Grzebielucha, S.: The traffic flow through different form of intersections. In: International Conference on Computer Information Systems and Industrial Management Applications (CISIM), pp. 299–304. IEEE (2010)

29. Belz, N.P., Aultman-Hall, L., Montague, J.: Influence of priority taking and abstaining at single-lane roundabouts using cellular automata. Transp. Res. Part C: Emerg. Technol. **69**, 134–149 (2016)

30. Wang, R., Ruskin, H.: Modeling traffic flow at a single-lane urban roundabout. Comput. Phys. Commun. **147**, 570–576 (2002)

31. Lakouari, N., Ez-Zahraouy, H., Benyoussef, A.: Traffic flow behaviour at a single lane roundabout as compared to traffic circle. Phys. Lett. Sect. A: Gen. At. Solid State Phys. **378** (43), 3169–3176 (2014)

32. Belz, N.P., Aultman-Hall, L., Lee, B.H.Y., Gårder, P.E.: An event-based framework for non-compliant driver behavior at single-lane roundabouts. Transp. Res. Rec.: J. Transp. Res. Board Nat. Acad. **2402**, 38–46 (2014). Washington, DC

33. Wagner, P., Nagel, K., Wolf, D.: Realistic multilane traffic rule for cellular automata. Phys. A **234**, 687–698 (1997)

34. Wang, R., Ruskin, H.J.: Modelling traffic flow at multi-lane urban roundabouts. Int. J. Mod. Phys. C **17**(5), 693–710 (2006)

35. Schroeder, B., Rouphail, N., Salamati, K., Bugg, Z.: Effect of pedestrian impedance on vehicular capacity at multilane roundabouts with consideration of crossing treatments. Transp. Res. Rec.: J. Transp. Res. Board Nat. Acad. **2312**(10), 14–24 (2012)

36. Was, J.: Cellular automata model of pedestrian dynamics for normal and evacuation conditions. In: Proceedings of the 5th International Conference on Intelligent Systems Design and Applications, ISDA 2005, pp. 154–159. IEEE (2005)

37. Wąs, J., Gudowski, B., Matuszyk, P.J.: New cellular automata model of pedestrian representation. In: El Yacoubi, S., Chopard, B., Bandini, S. (eds.) ACRI 2006. LNCS, vol. 4173, pp. 724–727. Springer, Heidelberg (2006). doi:10.1007/11861201_88

38. Webpage: http://prawko-torun.pl/droga-zatrzymania-a-czas-reakcji-kierowcy. Accessed Nov 2016

39. Bułka, D., Walczak, S., Wolak, S.: Braking process - legal and technical aspects in terms of simulation and analysis. In: Proceedings of the 3rd Conference on Rozwój techniki samochodowej a ubezpieczenia komunikacyjne (2006). (in Polish)

The Swarm-Like Update Scheme for Opinion Formation

Tomasz M. Gwizdałła[⊠]

Department of Solid State Physics, University of Łódź,
Pomorska 149/153, 90-236 Łódź, Poland
tomgwizd@uni.lodz.pl

Abstract. The question, how to describe the individual's position concerning some particular issue and especially the factors influencing its change is the topis of different studies for tens of years. The dynamics of opinions change is usually adopted from ideas related to the physical description of magnetism including especially some form of interaction between spins. In our paper we are going to propose the scheme based on formulation of popular global optimization mechanism - the Particle Swarm Optimization. We consider our proposition as some form of comeback to the roots, since PSO is based on the analysis of behavior of flocks of animals. We present the background of the model and some comparisons with earlier studied approaches.

1 Introduction

There is a long history of attempts having the mathematical description of human behavior as a goal. We can look back up till psychological works of French [1] or sociological ideas of Lewin [2]. In their papers they tried to study the phenomena occurring between different persons in groups and portray them in some analytical form. It seams that the crucial impulse in the progress of the simulation of social effects came from the analogy between the behavior of some agents equated to the members of groups and the dynamics of spins in magnetic materials. This approach enabled to create so-called voter model in 1973 [3]. It adapted the idea of Glauber dynamics [4] and finally proposed the probabilistic model for opinion change.

Since then a lot of models have been established. As the most famous we can point out: Galam's majority model [5], deterministic and one-dimensional Sznajd model [6] expanded later significantly by Stauffer and collaborators [7]. The main similarity between these approaches is that all of them can be considered within the frame of physical notions mentioned earlier. All of them follow the spin magnetism analogy. Certainly, there exist also another models. an interesting is e.g. the Hegselmann-Krause [8] model with dynamic topology.

Presented paper is the modification and enhancement of the approach shown in our earlier paper [9]. We studied there the influence of topology on the opinion formation in the Glauber-oriented model. As the best measurement of opinion we chose the results of elections which, however taking place relatively rarely, can

© Springer International Publishing AG 2017
N.T. Nguyen et al. (Eds.): ICCCI 2017, Part II, LNAI 10449, pp. 66–75, 2017.
DOI: 10.1007/978-3-319-67077-5_7

be, in our opinion, considered as the most spectacular and actual expression of people's preferences. The seminal plot for the analysis is shown in Fig. 1, which is the modification of the Fig. 2 from paper [9].

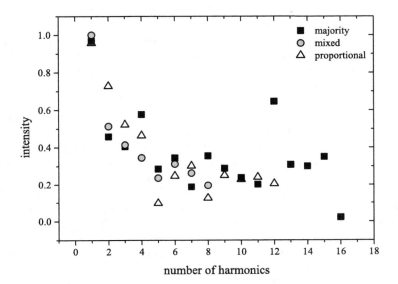

Fig. 1. The amplitudes of Fourier spectra for selected electoral systems. The data for majority system are determined on the basis of elections in USA and UK, those for the mixed one are obtained for Germany and the ones for proportional voting came from Denmark.

The idea is to consider the discrete Fourier spectrum of results instead of the results itself. That is why we selected several countries which have the particular voting systems and the data for these countries were collected for sufficiently long amount of time, in order to enable to determine the amplitudes of many harmonics. Finally we selected USA and UK (First Past The Post - FPTP majority system), Germany (mixed majority proportional system) and Denmark (pure proportional d'Hondt system). For all major parties, the results obtained in successive elections were treated with the Fourier transform and then all the values of coefficients for the same system were averaged. This method has certainly some merits, as well as drawbacks. Such a procedure can disclose at least generally some features characteristic for presented systems. For instance there is a visible significance of higher harmonics for FPTP voting. On the other hand it can be the objection that the error of every point is large and no general conclusions can be made based on it or that the number of elections and so the number of harmonics is to low to illustrate the dependence. Anyway, we hope that considering Fourier transforms we can better discuss the opinion formation properties than by analyzing the raw results.

The main effort in the paper is to reconsider the update technique for the system. The inspiration for our model is the Particle Swarm Optimization (PSO). We think that some basics of the PSO as the technique reproducing the swarm behavior of individuals can be also efficiently applied to model the opinion formation. It is widely known that PSO was proposed by Kennedy and Eberhart [10] as the technique where we try to obtain the global optimum by weighting the social influence of swarm and the independent position of particular individual. Since the opinions, as considered in researches, are mapped onto discrete space (like casting the vote in elections) we are especially interested in the discrete PSO (dPSO). Such a model, usually called binary PSO (bPSO), due to only two-state configurational space was originally proposed by the same authors [11].

Although the typical problems solved by PSO are described by the floating-point variables, also bPSO/dPSO are the points of interest of a lot of authors, obviously in the context of optimization. Among the interesting approaches we can enlist for example the discrete PSO shown in [12] used in operation research. Lee et al. [13] proposed the correction to bPSO moving it a little closer to genetic algorithm by using the mutation operator. Khanesar et al. [14] proposed the slight change in the understanding of velocity in binary environment Bansal and Deep [15] increased the role of previous state in the analysis of velocity and applied it to the classical knapsack problem. Also very recently the bPSO is an interesting topic and a lot of new approaches can be observed like non-parametric PSO [16], or analyses strongly related to some particular properties like bounding the velocity [17] or the influence of inertia [18]. Binary PSO was also applied to magnetic phenomena within the frame of well known, and also mentioned later in this paper, Ising model [19].

2   Model

The model studied in this paper can be interpreted rather as the agent model than based on the Cellular Automata, however a lot of features which has to presented is common for both approaches. Typically we have to define three properties: the set of states, the topology of system and the update procedure.

The opinion scheme is related to the so-called Nolan diagram. According to this idea we can distinguish different scopes and study the individual's opinion about particular problems separately. Originally, two opinions are taken into account: the one corresponding to social view and the one corresponding to economic view. The opinion can be presented on a two-dimensional plot with the intensity described with real numbers, typically from interval $[-1, 1]$. In our approach we limit this model and we allow the individual to have just two possibilities for every scope. It corresponds somehow to analyzing only the appropriate quadrant instead of full information about the position. It limits also the number of opinions so we have only 4 possibilities when considering the simplest, two-scope, model.

If we consider scopes independently, we can describe everyone of them with the magnetic-like variables. Due to this analogy we can call them spins and denote S_{ij}^t, where t is time step, i enumerates individuals and j - opinions. In our case $j \in \{1, 2\}$, $i \in \{1, 2, .., N\}$ (N is the number of individuals) and $S_{ij}^t \in \{-1, 1\}$.

In the presented paper we do not consider all geometries shown in the mentioned earlier one [9]. We concentrate on the most realistic, at least in our opinion, Barabasi-Albert (BA) model. This model of interpersonal connections is well known especially for several features. Among the most important are those ones that it reproduces correctly the scale-free structure recognized in societies and revealed in such environments like social networks, citation networks or internet. Its crucial property is the power-law character of the dependence describing the number of connections between the nodes-individuals, That is why we consider this approach to be better than every other geometry based on constant number of neighbors within the given n-dimensional system.

For the BA topology it is also easy to use both update methods which we want to compare in our paper. The first one is the Glauber update which we consider as comparative material. It is the most often used approach in opinion formation simulation since the first, mentioned earlier, paper written by Clifford [3]. It is based on the determination of opinion change probability (P) with the formulas typical for statistical physics.

$$P = \frac{1}{1 + \exp(\frac{-\Delta E}{k_B T})} = \frac{1}{1 + \exp(-\beta \Delta E)}. \tag{1}$$

In order to use the above formula we have to extend the analogy between opinion and spin into the notions of energy and temperature. For simplicity we can assume the Boltzmann constant $k_B = 1$ and temperature (T or $\beta = \frac{1}{k_B T}$) remains as a parameter of simulation.

The value of energy change (ΔE) can be easily calculated by assumption that we use the Ising-like model fo description of inter-personal interaction.

$$\Delta E = -J * \sum_{neighbors} \left(s_{i,updated} * s_{j,updated} - s_{i,old} * s_{j,old} \right). \tag{2}$$

According to this formula, the uniformity is preferred (the ferromagnetic state) but with increasing temperature this order vanishes. Certainly, the number of neighbors (colleagues) can significantly influence the value of ΔE thus the strong pressure can force some fast opinion changes. The results for this model was presented in [9] and the further remarks will be related to the mode called there a "two-mode" option.

Here we are going to introduce the model based on formulas well known in the procedure of Particle Swarm Optimization. The starting point is the formula describing the velocity of motion of solution in the search space.

$$v_{ij}(t) = \chi \Big(\omega v_{ij}(t-1) + c_1 * rnd() * (x_{global,j}(t) - x_{ij}(t))$$

$$+ c_2 * rnd() * (x_{ij}^{best}(t) - x_{ij}(t)) \Big) \qquad (3)$$

In the Eq. 3 there are four parameters: c_1 and c_2, learning factors, corresponding to the influence of previous global and local optimum respectively, ω - the inertia factor decreasing the former value of velocity and χ constricting the total change of velocity.

For the search performed in real coordinates the position of individual is updated using the equation.

$$x_{ij}(t+1) = x_{ij}(t) + v_{ij}(t) \qquad (4)$$

For the problem studied in the paper we use the binary PSO where velocity does not define the motion of individual but is remapped onto $[0,1]$ interval and the obtained number determines the probability of successive state. As the remapping function, usually the sigmoid function is used.

$$sig(v) = \frac{1}{1+e^{-v}} \qquad (5)$$

And the value returned by formula 5 is then used to calculate the next position.

$$x_{ij}(t+1) = \begin{cases} -1, & if \ rnd \geq sig(v_{ij}(t)) \\ 1, & if \ rnd < sig(v_{ij}(t)) \end{cases} \qquad (6)$$

The crucial idea of our proposition is to redefine formula 3 by introducing some values which are not related to the best solutions but to current average positions of population and the group of neighbors of given individual. We propose to use formulas shown by Eqs. 7 and 8

$$v_{ij}(t) = \omega_{inert} v_{ij}(t-1) + c_{all} * rnd() * (x_{all,j}(t) - x_{ij}(t))$$

$$+ c_{group} * rnd() * (x_{ij}^{group}(t) - x_{ij}(t)) \qquad (7)$$

The above equation almost directly reproduces the formula 3. The inertia factor remaines unchanged and the differences are emphasized with the change of notation. The first term, originally related to the best global solution, in our model is related to the average position of all individuals in the sample $x_{all,j}(t)$. Certainly, this pair of values is located somewhere inside the square $\{(1,1)(-1,1)(-1,-1)(1,-1)\}$. The coefficient responsible for the influence of whole population we call c_{all}. The second term corresponds to the influence of the closest environment. It is determined in the same way as the first one but the average $x_{ij}^{group}(t)$ is taken over all neighbors of individual in the BA network. The coefficient responsible for this feature we call c_n.

Finally we change the meaning of χ parameter. Instead of multiplying the velocity we change the value of probability after using formula 6. We take into consideration that most of people are conservative in their opinion. The word

"conservative" means here the reluctance to change their own position. Therefore, the probability of change is multiplied by some factor, called c_{self}.

$$prob_{change} = c_{self} * prob_{change} \qquad (8)$$

When compared with the Glauber model, our proposition can be described
by the greater number of parameters. In the further part of paper they will be
presented as tuple containing values of $(\omega_{inert}, c_{all}, c_n, c_{self})$.

We want also to remark once more that we consider both coordinates separately. The update can be performed in synchronous, as well as asynchronous
way. In the synchronous update, all averages are calculated between so-called
Monte-Carlo Sweeps. Performing on MC Sweep means that we do an update
of N individuals (usually not all of them but few of them several times) in the
arbitrary, sampled randomly order. In the presented simulations $N = 8019$ was
chosen in order to follow our earlier studies where the same number of individuals
was taken into account.

3   Results

The main effort of our approach is related to the construction of Fourier spectra
based on the both models presented in the previous section. The comparison is
shown in Fig. 2

The upper plot shows the data for selected parameters of Glauber update
while the lower plot is prepared for our PSO-like method. Among a lot of tuples
considered we select just three in order to rather show qualitative differences
than to concentrate on quantitative ones. All spectra are prepared by averaging
over 100 runs and over intensities calculated for every of four considered opinions.
The values are normalized to the value of first harmonic. Several properties can
be observed. With relatively small range of parameters we can reproduce the
same behavior as we did for Glauber model. It is also interesting that for spectra
characterized by linear dependence, the dispersion of points is significantly larger
than for those for Glauber update. Considering the single runs we can easily
select some of them when the higher harmonics have larger values.

In order to find differences between the approaches, we present Fig. 3 where
the average radii of differet groups are shown in the function of simulation phase.
The plots are prepared with the help of the earlier defined values. When calculating the average position of individuals in whole population $x_{all,j}(t)$ we determine
the average distance of all individuals from this point. This characteristic corresponds to the process of unifying the opinions. With one state prevailing, the
radius would go to 0 while for uniform distribution between opinions, the center would tend to the center of coordinate system and the radius would reach
its maximal value $\sqrt{2}$. We can observe that for PSO-like update for high temperatures, all opinions are represented uniformly and only differences between
particular runs come from some small oscillations. The similar picture is confirmed by the lower plot. When preparing, it we calculate the averages for all

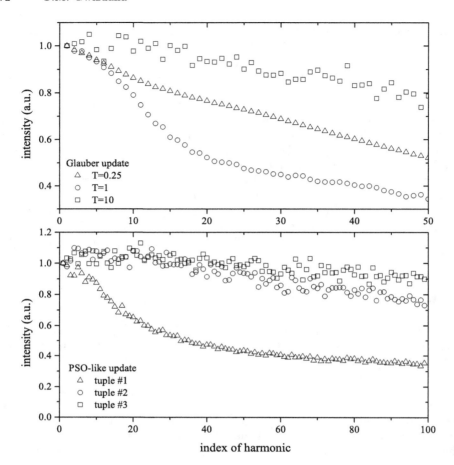

Fig. 2. The comparison of Fourier spectra for selected sets of values for Glauber and PSO-like updates. The plots for Glauber update are chosen among those from Fig. 5 of paper [9]. The tuples for PSO-like update are as follows: tuple#1 = (0.25,5,0.1,0.25); tuple#2 = (0.1,0.75,0.25,0.5); tuple#3 = (0.1,0.75,0.5,0.25).

neighbors of every individual $x_{ij}^{group}(t)$ and then the distances between the individual (always form set $\{\pm 1, \pm 1\}$) are averaged over whole sample. For PSO-like update there are only slight deviations below the $r \approx 1.7$ line which corresponds to the situation where half of point is in distance 2 and a quarters in distances 0 and $2 \times \sqrt{2}$.

We have to mention here that such dependence can be a result of applying the pure Barabasi-Albert network. The mechanism of network creation ensures the shape of degree curve to be of famous n^{-3} character but does not supports the creation of real groups.

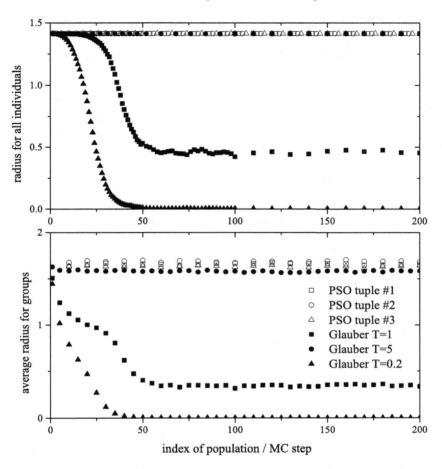

Fig. 3. The comparison of average radii from the center of whole population (upper plot) and the center of group of neighbors/colleagues (lower plot). The tuples for PSO-like update are as follows: tuple#1 = (0.1,0.750.95,0.25); tuple#2 = (0.1,0.5,0.5,0.5); tuple#3 = (0.1,2.0,0.75,0.5). The number of point in different intervals of plots are chosen in such a way to present clearly the different character of behavior of presented dependences during different phases of simulation.

In the Fig. 4 we show the dependence of number of opinion changes (expressed in percentage of population number) in the function of simulation phase. We can see that both methods have some advantages over the other one. The Glauber update gives us the larger range of results. On the other hand there exist the phase of reaching the stability (which is however not long) and for the PSO-like update we studied only limited model with limited parameter range. It might be pointed out that even very humble selection of simulation parameters can lead to significantly different characteristics, describing especially very important parameter in social simulations - the number of opinion changes.

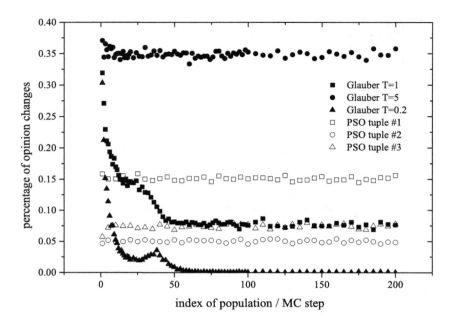

Fig. 4. The comparison of number of opinion changes. The tuples for PSO-like update are as follows: tuple#1 = (0.1,0.750.95,0.25); tuple#2 = (0.1,0.5,0.5,0.5); tuple#3 = (0.1,2.0,0.75,0.5).

4   Conclusions

Comparing the models presented in our paper we are going to show some possibilities related to our proposition. The crucial point of every opinion formation simulation is to ensure several properties. The first one is to control the number of opinion change. Here the features of both models are similar with the need, for PSO-like update, to study in more detail some special regions of parameter space.

The other feature is the possibility to individualize agents. In the Glauber model we can change just one parameter - the temperature. The PSO-like model gives us more possibilities by changing all four presented coefficients. We can easily create e.g. some nonconformists or opportunists by changing the c_{self} value from some distribution. We think also about the increase of number of parameters and make them dependent on time. It can correspond to some social pressure of e.g. media or to the reaction on some spectacular incidents.

References

1. French Jr., F.R.: A formal theory of social power. Psychol. Rev. **63**, 181–194 (1956)
2. Lewin, K.: Field Theory in Social Science: Selected Theoretical Papers. Harper & Brothers, New York (1951)

3. Clifford, P., Sudbury, A.: A model for spatial conflict. Biometrika **60**, 581–588 (1973)
4. Glauber, R.J.: Time-dependent statistics of the Ising model. J. Math. Phys. **4**, 294–307 (1963)
5. Galam, S.: Minority opinion spreading in random geometry. Eur. Phys. J. B **25**, 403–406 (2002)
6. Sznajd-Weron, K., Sznajd, J.: Opinion evolution in closed community. Int. J. Mod. Phys. C **11**, 1157 (2000)
7. Stauffer, D., Sousa, A.O., de Oliveira, S.: Generalization to square lattice of Sznajd sociophysics. Int. J. Mod. Phys. C **11**, 1239 (2000)
8. Hegselmann, R., Krause, U.: Opinion dynamics and bounded confidence: models, analysis and simulation. J. Artif. Soc. Soc. Simul. **5**, 1–24 (2002)
9. Gwizdałła, T.M.: The influence of cellular automaton topology on the opinion formation. In: Malyshkin, V. (ed.) PaCT 2015. LNCS, vol. 9251, pp. 179–190. Springer, Cham (2015). doi:10.1007/978-3-319-21909-7_17
10. Kennedy, J., Eberhart, R.: Particle swarm optimization. In: Grefenstette, J.J. (ed.) Proceedings of IEEE International Conference on Neural Networks, Perth, Australia, pp. 1942–1948. IEEE Service Center, Piscataway (1995)
11. Kennedy, J., Eberhart, R.C.: A discrete binary version of the particle swarm algorithm. In: 1997 IEEE International Conference on Systems, Man, and Cybernetics, 1997. Computational Cybernetics and Simulation, vol. 5, pp. 4104–4108 (1997)
12. Rameshkumar, K., Suresh, R.K., Mohanasundaram, K.M.: Discrete particle swarm optimization (DPSO) algorithm for permutation flowshop scheduling to minimize makespan. In: Wang, L., Chen, K., Ong, Y.S. (eds.) ICNC 2005. LNCS, vol. 3612, pp. 572–581. Springer, Heidelberg (2005). doi:10.1007/11539902_70
13. Lee, S., Soak, S., Oh, S., Pedrycz, W., Jeon, M.: Modified binary particle swarm optimization. Prog. Nat. Sci. **18**, 1161–1166 (2008)
14. Khanesar, M.A., Teshnehlab, M., Shoorehdeli, M.A.: A novel binary particle swarm optimization. In: 2007 Mediterranean Conference on Control Automation, pp. 1–6 (2007)
15. Bansal, J.C., Deep, K.: A modified binary particle swarm optimization for knapsack problems. Appl. Math. Comput. **218**, 11042–11061 (2012)
16. Beheshti, Z., Shamsuddin, S.M., Hasan, S.: Memetic binary particle swarm optimization for discrete optimization problems. Inf. Sci. **299**, 58–84 (2015)
17. Liu, J., Mei, Y., Li, X.: An analysis of the inertia weight parameter for binary particle swarm optimization. IEEE Trans. Evol. Comput. **20**, 666–681 (2016)
18. Gunasundari, S., Janakiraman, S., Meenambal, S.: Velocity bounded boolean particle swarm optimization for improved feature selection in liver and kidney disease diagnosis. Expert Syst. Appl. **56**, 28–47 (2016)
19. Gwizdałła, T.M.: Different versions of particle swarm optimization for magnetic problems. In: Proceedings of the 13th Conference on Genetic and Evolutionary Computation, GECCO 2011, pp. 5–6. ACM, New York (2011)

A Comparative Study of Different Variants of a Memetic Algorithm for ATSP

Krzysztof Szwarc$^{(\boxtimes)}$ and Urszula Boryczka

Institute of Computer Science, University of Silesia,
Bedzinska 39, 41-205 Sosnowiec, Poland
{krzysztof.szwarc,urszula.boryczka}@us.edu.pl
http://ii.us.edu.pl/

Abstract. In this paper we present a computational study of how different local search methods and the choice of an algorithm stage in which they are applied affect the performance of Memetic Algorithm (MA) solving Asymmetric Traveling Salesman Problem (ATSP). This study contains a comparison of quality of solutions obtained (both in terms of the value of the objective function and the performance time of the method) by sixteen variants of the Memetic Algorithm. Considerable amount of a given problem's instance and Wilcoxon Signed-Rank Test were used to ensure the impartiality of gained results.

Keywords: Memetic algorithm · Asymmetric Travelling Salesman Problem · Metaheuristics

1 Introduction

The Memetic Algorithm, also referred to as the hybrid Evolutionary Algorithm, was defined by Pablo Moscato in the publication released in 1989 [7]. Over 25 years on, the metaheuristic still finds numerous applications in solving optimization problems [3,6,10], while the freedom of choice of a local search method and a hybridization form have created a knowledge gap with regard to the accurate configuration of the metaheuristic. In order to determine a recommended approach to hybrid Evolutionary Algorithm design, the paper measures an effectiveness of the selected variants of the Memetic Algorithm - varying in terms of a local search method applied (Greedy Local Search, Hill Climbing, Tabu Search, and Simulated Annealing) and a hybridization form (after the generation of an initial population, before selection operation, after crossover and mutation operations as well as after the evolutionary phase).

The Travelling Salesman Problem (TSP) is a classical and well-researched optimization problem, formulated in 1930 by an Austrian mathematician Karl Menger [4], that involves finding the shortest Hamiltonian cycle in a complete weighted graph. The popularity of the problem stems from the fact that it models many problems encountered in business practice - its asymmetric variant (characterized with the possibility of varying weights of edges connecting the same

© Springer International Publishing AG 2017
N.T. Nguyen et al. (Eds.): ICCCI 2017, Part II, LNAI 10449, pp. 76–86, 2017.
DOI: 10.1007/978-3-319-67077-5_8

vertices), representing line infrastructure located in urban areas, has become the basis for many logistical problems (it was used, for example, in the process of planning the mobile collection of waste electrical and electronic equipment [8] and optimizing waste collection in Sweden [11]). The fact that the Travelling Salesman Problem belongs to the class of $\mathcal{NP}$-hard problems together with its utilitarian applications decided about its choice as a test-bed for different variants of Memetic Algorithm.

The paper comprises the following six parts: the introduction of the meta-heuristic approach as well as combinatorial optimization problem, the formulation of the Asymmetric Travelling Salesman Problem, the characteristics of the Memetic Algorithm, the presentation of research methodology, the discussion of the results obtained in the study, and the conclusion with the proposals for future work.

2 Problem Formulation

Based on the paper authored by Pataki [9], the following Travelling Salesman Problem formulation was adapted: for a directed graph $D = (N, A)$, with weighted arcs represented as c_{ij} (where $i, j \in \{1, 2, \ldots, n\}$), a route (a directed cycle comprising all n cities) of minimal length is sought. The asymmetric variant of the problem (ATSP) used in the study is characterized with the possibility of the occurrence of inequality $c_{ij} \neq c_{ji}$.

A decisive variable x_{ij} representing the edge between vertices i and j in the solution found:

$$x_{ij} = \begin{cases} 1 \text{ when the edge } (i, j) \text{ is part of the route contstructed,} \\ 0 \text{ otherwise.} \end{cases} \quad (1)$$

The objective function assuming the minimization of the weights of the edges making up the solution is as follows:

$$\sum_{i,j} c_{ij} x_{ij} \rightarrow min. \quad (2)$$

The constraints ensuring exactly one visit of the travelling salesman to every city are presented below:

$$\sum_{i} x_{ij} = 1 \quad \forall_{j \in \mathbb{N}}, \quad \sum_{j} x_{ij} = 1 \quad \forall_{i \in \mathbb{N}}. \quad (3)$$

Such a formulation of the TSP problem may result in the occurrence of solutions representing separate cycles instead of one cycle, so it is necessary to introduce additional constraints (MTZ):

$$u_1 = 1, \quad 2 \leq u_i \leq n, \quad u_i - u_j + 1 \leq n(1 - x_{ij}), \quad \forall_i \neq 1, \quad \forall_j \neq 1. \quad (4)$$

3   Memetic Algorithm

Pablo Moscato, inspired by Dawkins' [2] work, defined the nature of the Memetic Algorithm as the imitation of cultural evolution, assuming in practice the combination of the evolutionary algorithm with the local search heuristic used by particular individuals (interpreted as a learning procedure followed by individual representatives of a population). The general structure of the algorithm was presented in the following way: an initial population is created and its representatives perform local search in order to find a local optimum or improve a solution (until a stopping condition is satisfied) [7]. After this action is completed, interaction occurs between the individuals in the population presented by cooperation or competition. Competition can be compared to the process of selection, known from the genetic algorithm, while cooperation - to the crossover mechanism or analogous ways of creating new individuals.

The study assumed the possibility of hybridization in one of the four MA stages - after generating an initial population, before a selection operation, after crossover and mutation operations, and after the evolutionary stage. The pseudocode of the MA with the hybridization forms highlighted is presented in Algorithm 1.

Algorithm 1. Memetic Algorithm

1: P=InitializePopulation();
2: B=GetBestSolution(P);
3: P=LocalSearch(P, B); [Optional]
4: **while** !stop **do**
5:     P=LocalSearch(P, B); [Optional]
6:     P'=SelectParents(P);
7:     P'=Crossover(P');
8:     P'=Mutation(P');
9:     P'=LocalSearch(P', B); [Optional]
10:     P=BuildNextGeneration(P, P');
11:     B'=GetBestSolution(P);
12:     **if** $F(B') < F(B)$ **then**
13:         B=B';
14:     **end if**
15: **end while**
16: P=LocalSearch(P, B); [Optional]
17: return B;

Function *LocalSearch* performs the local search and the check whether the best solution in the population is designated with a value of the objective function more favorable that the result B' - if so, it replaces B.

4   Research Methodology

In order to evaluate the effectiveness of particular variants the metaheuristic on a representative sample, $10,000$ instances of the Asymmetric Travelling Salesman Problem were generated. They were described with the parameters presented in the Table 1. The use of a relatively small number of vertices and performed only one run of trial are justified with the necessity to limit the time needed to conduct the experiment (the verification of the results on a substantial sample was necessary to eliminate the impact of non-determinism of the algorithms on the research results).

Table 1. Parameters of test tasks

No.	Parameter	Value
1	Minimum number of cities	2
2	Maximum number of cities	20
3	Minimum value of the weights of edges	0.01
4	Maximum value of the weights of edges	20

It was assumed that the weight of the edges between vertex i and node i is always equal 0, while the other weights adapt pseudorandom values from the selected range (to an accuracy of two decimal places). The study was conducted on a computer with parameters given in Table 2.

Table 2. Parameters of the computer used to conduct the study

No.	Parameter	Value
1	Processor	Intel CORE i7 920 (4 cores, 2660 MHz do 2.93 GHz)
2	RAM	6 GB (DDR3, 1333 MHz)
3	Hard drive	512 GB SATA 7200 rev.
4	Operating system	Windows 7 Professional N Service Pack 1 64-bit

4.1   Algorithms Efficiency Evaluation

The evaluation of the effectiveness of particular variants of the MA was conducted based on the total execution time and the error relative to the best value in relation to the lowest result achieved in the study. It was assumed that the quality of the algorithm would be determined based on the following formula:

$$q = i_w \cdot w + i_c \cdot c, \tag{5}$$

where q is a coefficient of the quality of the algorithm, i_w is a significance of the value of the objective function, w is a normalized (min-max) mean surplus of the value of the objective function, i_c is a significance of time and c is a normalized (min-max) value of cummulative performance time.

Additionally, the formula (5) was equipped with the constraints (Eq. (6)) ensuring the total value of weights equaling 1 and it was determined that $i_w = 0.75$ and $i_c = 0.25$ (the values correspond with the estimated priority of particular parameters for utilitarian applications). It was assumed that the lower the value of coefficient q, the better the performance of the algorithm:

$$i_w + i_c = 1, \quad i_w, i_c \in \mathbb{R}, \quad i_w, i_c \geq 0, \quad i_w, i_c \leq 1. \tag{6}$$

The results achieved by subsequent pairs of variants of the MA (denoted further as $C1$ and $C2$) were checked for statistically significant differences using the non-parametric one-sided Wilcoxon signed-rank test. The level of significance was placed at 0.05 (p-values exceeding this level of significance reveal that there are no grounds to question the hypothesis assuming that the $C1$ method yields lower results compared to the results achieved with the $C2$ method).

4.2   General Configuration of Memetic Algorithm

The study used the rank based selection and succession with partial replacement (10% of the worst individuals in population P' are replaced by the same amount of the best individuals from P). The initial population was generated on a pseudorandom basis and individuals were coded in path representation. The stopping condition was defined as the performance of one hundred generations until the last change of the B solution.

4.3   Crossover and Mutation

The study assumed the likelihood of crossover and mutation operation at the level of 70% and 10% respectively (they were established based on empirical studies conducted). The choice of one of the implemented operators occurred according to the parameters presented in Table 3 (they were discussed in more detail in [5, 12]).

Table 3. Operators used in the study and the likelihood of their application

Crossover		Mutation	
Crossover operator	Likelihood of crossover operator application [%]	Mutation operator	Likelihood of mutation operator application [%]
CX	20	Displacement	25
OX	10	Exchange	25
PMX	20	Insertion	25
SCX	50	Simple Inversion	25

4.4 Local Search

In order to prevent premature convergence, it was assumed that local search operations were conducted for the solutions of the individuals with the likelihood of 10%. 80% of the results underwent Lamarckian evolution, while the remaining 20% corresponded with the Baldwinian effect (the former assumes that an individual chromosome is being replaced as a result of local search operation, while the latter ensures only the replacement of the fitness of the chromosome, but does not modify the chromosome; the methods were discussed in work [1]).

The neighborhood of the current path is defined as a set of paths varying from it with the location of two cities only, with the initial vertex determined. Based on general multiplication law, the number of neighboring solutions is $(n - 1) \cdot (n - 2)$, where n is the number of all cities including the initial city.

Local search is conducted with the following methods: Greedy Local Search (GLS), Hill Climbing (HC), Tabu Search (TS), and Simulated Annealing (SA). In contrast to GLS, HC finds the best solution within the neighborhood in each iteration, while GLS accepts the first solution, which is better than the current one. TS is equipped with the acceptance criterion, allowing for the performance of a forbidden move when it led to the improvement of the best known result, and with the list of candidates comprising half the neighborhood (modeling the symmetrical variant of the TSP. Irrespective of the type of a problem, the neighborhood constraint allowed for the elimination of cycles and, as a result, the achievement of noticeably better results). In the implemented algorithm of SA, the initial temperature t is determined with the use of the following formula:

$$t = \frac{-\Delta f}{ln(P_0)}, \tag{7}$$

where t is an initial temperature, Δf is a mean value of the deterioration in a solution and P_0 is a likelihood of moving to a worse solution in the first iteration.

The parameters used for particular local search algorithms are presented in Table 4. They were established based on empirical studies conducted.

Table 4. Parameters of the local search methods used in the study

Method	Parameter	Value
TS	Tabu period (iterations)	5
SA	Cooling rate (α)	0.98
SA	Length of epoch (L)	200
SA	Likelihood of moving to a worse solution in the first iteration (P_0)	0.3
GLS/HC/TS/SA	Maximum number of iterations without improving a solution	100

5   Experimental Results

In order to simplify the records of the results, the variants of the Memetic Algorithm were implemented, which is presented in Table 5. The results obtained in the course of the study are presented in Table 6. Based on the Wilcoxon test (Tables 7 and 8 for the surplus value of the objective function and the time needed to perform the methods respectively), no grounds were identified to reject the conclusions that were formulated based on the results. Additionally, the dependency between the execution time and the value of the objective function was noticed (there were many significant differences between the algorithms according to the results presented in Tables 7 and 8).

Table 5. The coding of the variants of the Memetic Algorithm

No.	Local search method	Local search stage	Code (C)
1	Greedy local search	After generating an initial population	$A1$
2	Hill climbing	After generating an initial population	$A2$
3	Tabu search	After generating an initial population	$A3$
4	Simulated annealing	After generating an initial population	$A4$
5	Greedy local search	Before selection operation	$A5$
6	Hill climbing	Before selection operation	$A6$
7	Tabu search	Before selection operation	$A7$
8	Simulated annealing	Before selection operation	$A8$
9	Greedy local search	After the crossover and mutation operations	$A9$
10	Hill climbing	After the crossover and mutation operations	$A10$
11	Tabu search	After the crossover and mutation operations	$A11$
12	Simulated annealing	After the crossover and mutation operations	$A12$
13	Greedy local search	After evolutionary stage	$A13$
14	Hill climbing	After evolutionary stage	$A14$
15	Tabu search	After evolutionary stage	$A15$
16	Simulated annealing	After evolutionary stage	$A16$

According to the results presented in Table 6, the best variant of the Memetic Algorithm in terms of the value of the objective function and the quality coefficient q is the Tabu Search metaheuristic applied before selection, while the worst is Greedy Local Search after the evolutionary stage. The longest time was characteristic of the methods with hybridization occurring in the evolutionary loop and using Simulated Annealing.

Table 6. The results of the study

Code (C)	Average relative error [%]	w	Total execution time [h]	c	q
$A1$	3.03	0.813	1.9	0.004	0.61
$A2$	2.93	0.785	2.07	0.006	0.59
$A3$	0.72	0.176	2.56	0.01	0.135
$A4$	1.37	0.355	4.26	0.025	0.273
$A5$	1.3	0.336	2.12	0.006	0.254
$A6$	1.09	0.278	2.51	0.01	0.211
$A7$	0.08	0.0	32.08	0.272	0.068
$A8$	0.16	0.022	114.08	1.0	0.267
$A9$	2	0.529	2.74	0.012	0.4
$A10$	1.75	0.46	2.71	0.012	0.348
$A11$	0.09	0.003	34.05	0.29	0.074
$A12$	0.19	0.03	112.75	0.988	0.27
$A13$	3.71	1.0	2.18	0.007	0.752
$A14$	3.7	0.997	1.41	0.0	0.748
$A15$	2.47	0.658	2.59	0.01	0.496
$A16$	1.75	0.46	4.89	0.031	0.353

Table 7. Wilcoxon test results for the value of the objective function

C2\C1	A1	A2	A3	A4	A5	A6	A7	A8	A9	A10	A11	A12	A13	A14	A15	A16
A1	ND	0.99	1.00	1.00	1.00	1.00	1.00	1.00	1.00	1.00	1.00	1.00	0.00	0.00	1.00	1.00
A2	0.01	ND	1.00	1.00	1.00	1.00	1.00	1.00	1.00	1.00	1.00	1.00	0.00	0.00	1.00	1.00
A3	0.00	0.00	ND	0.00	0.00	0.00	1.00	1.00	0.00	0.00	1.00	1.00	0.00	0.00	0.00	0.00
A4	0.00	0.00	1.00	ND	0.98	1.00	1.00	1.00	0.00	0.00	1.00	1.00	0.00	0.00	0.00	0.00
A5	0.00	0.00	1.00	0.02	ND	1.00	1.00	1.00	0.00	0.00	1.00	1.00	0.00	0.00	0.00	0.00
A6	0.00	0.00	1.00	0.00	0.00	ND	1.00	1.00	0.00	0.00	1.00	1.00	0.00	0.00	0.00	0.00
A7	0.00	0.00	0.00	0.00	0.00	0.00	ND	0.00	0.00	0.00	0.17	0.00	0.00	0.00	0.00	0.00
A8	0.00	0.00	0.00	0.00	0.00	0.00	1.00	ND	0.00	0.00	1.00	0.00	0.00	0.00	0.00	0.00
A9	0.00	0.00	1.00	1.00	1.00	1.00	1.00	1.00	ND	1.00	1.00	1.00	0.00	0.00	0.00	1.00
A10	0.00	0.00	1.00	1.00	1.00	1.00	1.00	1.00	0.00	ND	1.00	1.00	0.00	0.00	0.00	0.88
A11	0.00	0.00	0.00	0.00	0.00	0.00	0.83	0.00	0.00	0.00	ND	0.00	0.00	0.00	0.00	0.00
A12	0.00	0.00	0.00	0.00	0.00	0.00	1.00	1.00	0.00	0.00	1.00	ND	0.00	0.00	0.00	0.00
A13	1.00	1.00	1.00	1.00	1.00	1.00	1.00	1.00	1.00	1.00	1.00	1.00	ND	0.73	1.00	1.00
A14	1.00	1.00	1.00	1.00	1.00	1.00	1.00	1.00	1.00	1.00	1.00	1.00	0.27	ND	1.00	1.00
A15	0.00	0.00	1.00	1.00	1.00	1.00	1.00	1.00	1.00	1.00	1.00	1.00	0.00	0.00	ND	1.00
A16	0.00	0.00	1.00	1.00	1.00	1.00	1.00	1.00	0.00	0.12	1.00	1.00	0.00	0.00	0.00	ND

Table 8. Wilcoxon test results for the execution time

C2\C1	A1	A2	A3	A4	A5	A6	A7	A8	A9	A10	A11	A12	A13	A14	A15	A16
A1	ND	0.00	0.00	0.00	0.00	0.00	0.00	0.00	0.00	0.00	0.00	0.00	0.00	1.00	0.00	0.00
A2	1.00	ND	0.00	0.00	1.00	0.00	0.00	0.00	0.00	0.00	0.00	0.00	0.00	1.00	0.00	0.00
A3	1.00	1.00	ND	0.00	1.00	1.00	0.00	0.00	1.00	1.00	0.00	0.00	1.00	1.00	0.04	0.00
A4	1.00	1.00	1.00	ND	1.00	1.00	0.00	0.00	1.00	1.00	0.00	0.00	1.00	1.00	1.00	0.00
A5	1.00	0.00	0.00	0.00	ND	0.00	0.00	0.00	0.00	0.00	0.00	0.00	0.00	1.00	0.00	0.00
A6	1.00	1.00	0.00	0.00	1.00	ND	0.00	0.00	0.00	0.48	0.00	0.00	1.00	1.00	0.00	0.00
A7	1.00	1.00	1.00	1.00	1.00	1.00	ND	0.00	1.00	1.00	0.00	0.00	1.00	1.00	1.00	1.00
A8	1.00	1.00	1.00	1.00	1.00	1.00	1.00	ND	1.00	1.00	1.00	1.00	1.00	1.00	1.00	1.00
A9	1.00	1.00	0.00	0.00	1.00	1.00	0.00	0.00	ND	1.00	0.00	0.00	1.00	1.00	0.00	0.00
A10	1.00	1.00	0.00	0.00	1.00	0.52	0.00	0.00	0.00	ND	0.00	0.00	1.00	1.00	0.00	0.00
A11	1.00	1.00	1.00	1.00	1.00	1.00	1.00	0.00	1.00	1.00	ND	0.00	1.00	1.00	1.00	1.00
A12	1.00	1.00	1.00	1.00	1.00	1.00	1.00	0.00	1.00	1.00	1.00	ND	1.00	1.00	1.00	1.00
A13	1.00	1.00	0.00	0.00	1.00	0.00	0.00	0.00	0.00	0.00	0.00	0.00	ND	1.00	0.00	0.00
A14	0.00	0.00	0.00	0.00	0.00	0.00	0.00	0.00	0.00	0.00	0.00	0.00	0.00	ND	0.00	0.00
A15	1.00	1.00	0.96	0.00	1.00	1.00	0.00	0.00	1.00	1.00	0.00	0.00	1.00	1.00	ND	0.00
A16	1.00	1.00	1.00	1.00	1.00	1.00	0.00	0.00	1.00	1.00	0.00	0.00	1.00	1.00	1.00	ND

Table 9. Aggregate results taking into account the local search algorithm

Local search algorithm	Mean value of the coefficient q
Greedy local search	0.5
Hill climbing	0.47
Tabu search	0.19
Simulated annealing	0.29

Table 10. Aggregate results taking into account the form of hybridization

Form of hybridization	Mean value of the coefficient q
After the generation of an initial population	0.4
Before selection operation	0.2
After crossover and mutation operations	0.27
After the evolutionary phase	0.59

The results of the study were further analyzed, determining the mean value q, while taking into account the local search algorithm (Table 9) and the form of hybridization (Table 10). Based on the results, it was stated that the best results were achieved with Tabu Search ($q = 0.19$) and when hybridization was placed before selection ($q = 0.2$). The worst results were obtained for the methods involving Greedy Local Search ($q = 0.5$) and hybridization occurring after the evolutionary stage ($q = 0, 59$).

6 Conclusions and Future Work

The study led to the formulation of the following conclusions: Memetic Algorithm design should involve the thorough analysis of type and sequence (hybridization form) of local search occurrence. The best value of the quality coefficient q can be attributed to the methods with hybridization before selection, whereas the worst - to the algorithms with local search occurring after the evolutionary stage. The worst results for the quality coefficient q are achieved by the methods using Greedy Local Search, while the best - Tabu Search algorithms. The longest computation time is characteristic of the variants of the Memetic Algorithms with hybridization occurring after the crossover and mutation operations, whereas the shortest - the methods of local search placed after the generation of the initial population. Based on the values determined for the quality coefficient q, it is recommended to use the Memetic Algorithm with Tabu Search occurring before selection (additionally, it is the best variant in term of the value of the objective function). It is recommended to use MA with TS occuring after generating an initial population in the situations when computational time is limited. The least recommended variant of the Memetic Algorithm is the method using Greedy Local Search after the evolutionary stage.

Future work will involve further research into the effectiveness of the particular variants of the Memetic Algorithm with regard to other optimization problems and the instances of the Asymmetric Travelling Salesman Problem with a larger number of vertices. It is planned to analyze the convergence of algorithms and computational cost measured in fitness function calls. Additionally, we would like to compare the obtained results with the solutions found by local search methods and MA variants using only the crossover/mutation loop.

References

1. Castillo, P., Arenas, M., Castellano, J., Merelo, J., Prieto, A., Rivas, V., Romero, G.: Lamarckian Evolution and the Baldwin Effect in Evolutionary Neural Networks (2006). http://www.arxiv.org/PS_cache/cs/pdf/0603/0603004v1.pdf
2. Dawkins, R.: The Selfish Gene. Oxford University Press, Oxford (1976)
3. Dib, O., Manier, M., Caminada, A.: Memetic algorithm for computing shortest paths in multimodal transportation networks. Transp. Res. Procedia **10**, 745–755 (2015)
4. Held, M., Hoffman, A., Johnson, E., Wolfe, P.: Aspects of the traveling salesman problem. IBM J. Res. Dev. **28**(4), 476–486 (1984)
5. Larranaga, P., Kuijpers, C.M.H., Murga, R.H., Inza, I., Dizdarevic, S.: Genetic algorithms for the travelling salesman problem: a review of representations and operators. Artif. Intell. Rev. **13**(2), 129–170 (1999)

6. Lau, H., Agussurja, L., Cheng, S., Pang Jin, T.: A multi-objective memetic algorithm for vehicle resource allocation in sustainable transportation planning. In: International Joint Conference on Artificial Intelligence (IJCAI 2013), Beijing, China, 3–9 August 2013, pp. 2833–2839 (2013)
7. Moscato, P.: On Evolution, Search, Optimization, Genetic Algorithms and Martial Arts - Towards Memetic Algorithms (1989)
8. Mrowczynska, B., Nowakowski, P.: Optymalizacja tras przejazdu przy zbiorce zuzytego sprzetu elektrycznego i elektronicznego dla zadanych lokalizacji punktow zbiorki. Czasopismo Logistyka 2, 593–604 (2015)
9. Pataki, G.: The bad and the good-and-ugly: formulations for the traveling salesman problem. Technical report CORC 2000–1 (2000)
10. Shaikh, M., Panchal, M.: Solving asymmetric travelling salesman problem using memetic algorithm. Int. J. Emerg. Technol. Adv. Eng. 2(11), 634–639 (2012)
11. Syberfeldt, A., Rogstrom, J., Geertsen, A.: Simulation-based optimization of a real-world travelling salesman problem using an evolutionary algorithm with a repair function. Int. J. Artif. Intell. Expert Syst. (IJAE) 6(3), 27–39 (2015)
12. Zakir, A.: Genetic algorithm for the traveling salesman problem using sequential constructive crossover operator. Int. J. Biometrics Bioinform. (IJBB) 3(6), 96–105 (2010)

Improving ACO Convergence with Parallel Tempering

Rafał Skinderowicz[✉]

Institute of Computer Science, Silesia University, Sosnowiec, Poland
`rafal.skinderowicz@us.edu.pl`

Abstract. Parallel Tempering (PT) is an efficient Monte Carlo simulation method known from statistical physics. We present a novel PT-based Ant Colony Optimization algorithm (PTACO) in which multiple replicas of the Ant Colony System enhanced with a temperature parameter (ACST) are executed in parallel. Based on computational experiments on a set of TSP and ATSP instances we show that the PTACO converges (in terms of solutions quality) significantly faster than the ACS and is competitive to the state-of-the-art Ant Colony Extended algorithm.

Keywords: Ant Colony System · Parallel Tempering · Simulated Annealing · Travelling salesman problem

1 Introduction

Ant Colony Optimization (ACO) algorithms, including the ACS and MAX-MIN Ant System (MMAS), belong to a class of nature-inspired metaheuristics which were successfully applied to solve many difficult optimization problems. Typically, they do not guarantee to find an optimum within a (very) limited computational time available in real-world scenarios. However, they generate approximate solutions of a good quality and are a viable alternative to exact, time-consuming, methods [5]. The quality of solutions generated by the ACO can be seen as a function of the computation time, i.e. if more time is spent on computations we can expect to find a solution of a better quality (algorithm converges). Numerous ideas on improving the ACO convergence were proposed in the literature, including changes to the inner structure of the algorithms, combining them with efficient, problem-related local search (LS) and other metaheuristics [6]. In this article, we present how the convergence of the ACO could be improved using the ideas borrowed from the PT – a proven method in statistical physics [7].

The main contribution of this paper is the introduction of the novel ACO algorithm based on the PT method (also known as the replica exchange Markov chain Monte Carlo, MCMC). Results of preliminary computational experiments conducted on a set of TSP and ATSP instances from the well-known TSPLIB repository are presented. The results show that the efficiency of the PTACO is comparable to the state-of-the-art ACO algorithms, including the MMAS and Ant Colony Extended (ACE).

© Springer International Publishing AG 2017
N.T. Nguyen et al. (Eds.): ICCCI 2017, Part II, LNAI 10449, pp. 87–96, 2017.
DOI: 10.1007/978-3-319-67077-5_9

The structure of the article is as follows. Section 2 contains a concise review of the related work in the literature. In Sect. 3 the ACS is briefly described. Section 4 contains a brief characteristic of the PT. The proposed PTACO algorithm is presented in Sect. 5, while the results of the computational experiments are shown in Sect. 6. A summary is presented in Sect. 7.

2   Related Work

Initial applications of the PT concerned statistical physics, but it was later successfully applied in biology, engineering, and material science [7]. PT was also applied to solve more general optimization problems, including image analysis, Bayesian data analysis, and risk analysis. Little work was done on applying the PT in the field of combinatorial optimization. In our interest is especially a work of Wang et al. who applied the PT to the TSP and showed that the resulting performance was better than that of the Simulated Annealing (SA) [15]. Li et al. proposed a hybrid of the PT and SA and showed that it performed slightly better than the PT for the problem of continuous function optimization and the TSP [11].

To the best of our knowledge, no work on applying the PT to any of the ACO algorithms was done. Somewhat related is the research on improving the ACO convergence with the SA. Most often the SA plays the role of an LS used to improve the quality of the solutions generated by the ants. Successful applications concerned the parallel-machine scheduling problem [2], course timetabling problem [1] and vehicle routing problem [3], among others. Zhu et al. proposed a hybrid of the Ant System (AS) and SA for solving the Quadratic Assignment Problem [16]. In the algorithm, the decision which solutions are used to update a pheromone matrix is based on the Metropolis criterion of the SA. Some similarities can be found in the work of Citrolo and Mauri [4] who developed a hybrid of the Monte Carlo and ACO for the Protein Structure Prediction. An idea how to improve the convergence of the ACS and a problem-specific LS was demonstrated in our work on solving the Sequential Ordering Problem [14].

3   Ant Colony System

The Ant Colony System is one of the most successful algorithms from the ACO family which, in turn, belongs to nature-inspired metaheuristics [6]. In the ACS, a number m of artificial ants iteratively construct solutions to the optimization problem tackled. Usually the problem is modeled using a complete, weighted graph $G = (V, A)$, where V is a set of nodes and $A = \{(i,j)|(i,j) \in V, i \neq j\}$ is a set of edges connecting the nodes. In the context of the TSP, the nodes correspond to the cities that have to be visited by a salesman who travels between them using roads represented by the edges of the graph G. If the cost (distance) in both directions is the same, the problem is *symmetric* and the graph G is undirected. Otherwise, the edges are directed and the problem becomes *asymmetric* (ATSP). The solution to the TSP is the shortest cycle containing each city exactly once.

In every iteration of the ACS, each ant builds a complete solution to the problem starting, typically, from a randomly chosen node. In the following steps, it extends its partial solution by choosing one of the unvisited nodes. This choice is based on a so-called *pseudo-random proportional rule* [5]. This rule defines the probability of being selected for each of the unvisited nodes based on the *artificial pheromone trails* and additional knowledge of the problem (distances between the cities in the case of the TSP). The pheromone trails are typically represented in the form of a real-valued matrix. In the beginning, the matrix is initialized with the same, small values. Later, the values are updated according to the *local* and *global* pheromone update rules. The former involves decreasing the pheromone values of the edges as they are traversed by the ants. The latter is performed after the ants have completed their solutions. It involves increasing the pheromone levels on the trails corresponding to the components of the best solution found so far (also called global best). The aim of the global pheromone update is to focus the construction process on the regions of the problem solution space containing the current global best solution (exploitation). On the other hand, the local pheromone rule aims at increasing the exploration of unvisited regions of the solution space. For the details, please refer to [5].

4   Parallel Tempering

Parallel tempering, also called replica exchange MCMC, is an MCMC method developed for efficiently sampling probability distributions with a complex structure [7]. The main idea of the PT is to run multiple replicas (instances) of an MC simulation, each at a different temperature. The high-temperature instances are able to sample large volumes of a solution search space (phase space in physics), whereas the low-temperature instances are able to precisely sample local areas of the solution space. A good sampling of the solution search space is obtained by allowing the instances at different temperatures to exchange configurations with probability described by the Metropolis criterion. In other words, the instances at high temperatures "feed" the low-temperature instances which act as local optimizers. It is worth noting that the temperature of each MC simulation remains constant. Thus it differs from the SA in which the temperature is gradually lowered.

The Parallel Tempering requires the setting of temperature levels for the parallel MC simulations. Kone and Kofke suggested that the optimal choice of temperatures should be that 23% of possible transitions between simulations are accepted [10].

5   Parallel Tempering Based ACO

In the proposed PTACO algorithm a number M of the ACS-based replicas are executed in parallel. The replicas differ from the ACS in the following way. In the

ACS the search process is guided by the pheromone memory which is reinforced iteratively by increasing the pheromone levels on the trails which correspond to the components of the best solution found so far (global best solution). This process leads to a good overall convergence, but it is also strongly exploitation-oriented and prone to getting trapped in local optima [6]. On the contrary, each of the ACS replicas in the PTACO updates pheromone based on the current *active solution* which may not necessarily be the current global or even iteration best solution. Specifically, each of the solutions produced by the ants becomes *a candidate* for replacing the current active solution. The decision whether to accept the solution is taken according to the Metropolis criterion known from the SA. It is based on the difference in qualities (route lengths in the case of the TSP) of the candidate and active solutions, and the *temperature* parameter T. The temperature was added to the ACS so that not only better but also worse quality solutions could replace the active solution. This was intended to increase the probability of escaping local minima. It is worth noting, that if the $T = 0$ only the better quality solutions are accepted and the process becomes equivalent to that of the *unmodified* ACS. For convenience, we will henceforth denote the described *ACS with temperature* as the ACST.

```
1  for  i ← 0 to M − 1 do
2  |    Tᵢ ← calculate_temperature(i)
3  |    replicaᵢ ← new instance of the ACST with Tᵢ
4  end
5  while Not reached stopping criterion do
6  |    for i ← 0 to M − 1 do in parallel
7  |    |    Run replicaᵢ for N_swap iterations
8  |    end
9  |    G ← min(Cost(global_best_replica(0)),…,Cost(global_best_replica(M−1)))
10 |    j ← U{0, M − 2}          // Randomly select two adjacent temp. levels
11 |    ΔE ← (Cost(active_solution_replica(j)) − Cost(active_solution_replica(j+1)))/G
12 |    Δβ ← 1/Tⱼ − 1/T_{j+1}
13 |    if U(0, 1) < min(1, exp(ΔβΔE)) then
14 |    |    Set the temp. of replicaⱼ and replica_{j+1} to T_{j+1} and Tⱼ, respectively
15 |    |    swap(replicaⱼ, replica_{j+1})          // Keeps replicas ordered by temp.
16 |    end
17 end
```

Fig. 1. Parallel tempering based ACO (PTACO).

The pseudocode of the PTACO algorithm is shown in Fig. 1. The PTACO starts with creating M replicas (instances) of the ACST algorithm (lines 1–4) each with its own temperature level T_i, $i \in \{0, 1, \ldots, M\}$. The temperature levels are strictly increasing, i.e. $T_0 < T_1 < \ldots < T_{M-1}$. Next, the replicas are allowed to execute N_{swap} iterations each (lines 6–8) after which a trial to swap the replica's configuration at T_j (j is chosen randomly, line 10) with the

```
 1  for i ← 1 to #iterations do
 2  │   for j ← 1 to #ants do
 3  │   │   route_Ant(j)[1] ← U{1, #nodes}          // Start from random position
 4  │   end
 5  │   for k = 2 to #nodes do                      // Build complete solutions
 6  │   │   for j ← 1 to #ants do
 7  │   │   │   route_Ant(j)[k] ← select_next_node(route_Ant(j))
 8  │   │   │   local_pheromone_update(route_Ant(j)[k − 1], route_Ant(j)[k])
 9  │   │   end
10  │   end
11  │   for j ← 1 to #ants do            // Local update on the closing edges
12  │   │   local_pheromone_update(route_Ant(j)[#nodes], route_Ant(j)[1])
13  │   end
14  │   local_best ← select_best(route_Ant(1), route_Ant(2), . . . , route_Ant(#ants))
15  │   if Cost(local_best) < Cost(global_best) then
16  │   │   global_best ← local_best
17  │   end
18  │   for j ← 1 to #ants do
19  │   │   Δ ← (Cost(route_Ant(j)) − Cost(active_solution))/Cost(global_best)
20  │   │   if Δ < 0 then
21  │   │   │   active_solution ← route_Ant(j)      // Always accept a better move
22  │   │   else if U(0, 1) < e^{−Δ/T} then
23  │   │   │   active_solution ← route_Ant(j)      // Accept a worse move
24  │   │   end
25  │   end
26  │   global_pheromone_update(active_solution)
27  end
```

Fig. 2. ACS with temperature (ACST) used in the PTACO.

replica's configuration at T_{j+1} is considered (lines 9–16). Instead of exchanging the internal states of the replicas we can simply swap their temperature levels (line 14). This swap creates a *random walk in the temperature* space. When a replica drifts to a higher temperature, it can overcome "energy barriers" in the solution search space. After it returns to a low-temperature level, it moves locally in a small region [7].

The decision whether to do the swap between T_j and T_{j+1} levels is based on the difference between the costs (qualities) ΔE of the replicas' current *active solutions* (line 11) and the difference between reciprocals of T_j and T_{j+1} (line 12). It is an adaptation of the PT formula presented in [15]. The resulting probability is compared with a random variable taken uniformly from the range $(0, 1)$ (line 13). It is worth noting that if the cost of the active solution of the (upper temperature) replica T_{j+1} is lower than the respective cost of the (lower temperature) replica T_j the probability equals 1. The difference between the costs in the calculation of ΔE is divided by the cost of the best so far solution found by all the replicas. It is a *normalization step* intended to keep the calculations independent of the edge weights of the problem instance.

The temperature levels are calculated according to:

$$T_i = \frac{-0.01}{\ln(p_{\mathrm{L}} + (2^i - 1)(p_{\mathrm{H}} - p_{\mathrm{L}})/(2^{M-1} - 1))}, \tag{1}$$

where $i \in \{0, 1, \ldots, M - 1\}$; p_{L} and p_{H} are parameters denoting the lowest and highest probability of accepting a move for which $\Delta E = 0.01$, respectively. Please note, that $\Delta E = 0.01$ if the relative difference between the replicas' active solutions is 1% – a value chosen arbitrarily to allow exchanges only if the differences are relatively small. The denominator in Eq. 1 is responsible for "spacing" the temperature levels so that the probability of accepting a worse move (with negative ΔE) at the level T_{i+1} is twice the probability at the level T_i.

Pseudocode of the ACST is shown in Fig. 2. The main part of the algorithm (lines 1–17) is identical to that of the ACS. The only differences concern selecting the current active solution (lines 18–25) and the global pheromone update (line 26) which is performed using the current active solution and not the global best solution as in the ACS. The decision whether to replace the current solution with a candidate solution is taken based on the difference Δ in costs of the candidate and active solutions (line 19). This difference is normalized by dividing it by the cost of the global best solution. A better solution is always accepted (lines 20–21), while the worse solution is accepted with the probability, calculated similarly to the SA, based on Δ and T (temperature) values.

6   Experiments

A series of computational experiments[1] was conducted to investigate the behavior of the PTACO and assess its efficiency. The main motivation behind the experiments was to show the viability of the proposed idea (PTACO), therefore we have focused on a few relatively small TSP and ATSP instances. A total of 12 problem instances was selected from the well-known TSPLIB repository, including 9 symmetric and 3 asymmetric ones, namely: *st70*, *pr76*, *gr96*, *kroA100*, *lin105*, *gr137*, *pr152*, *u159*, *d198*, *ftv70*, *kro124p* and *ftv170*.

The PTACO requires setting a number of parameters, some of which are required by the ACST replicas. The parameters that were not subject of the investigation were set as follows based on recommendations in the literature and: number of ants, $m = 10$; $\beta = 2$; $\psi = 0.01$ and $\rho = 0.1$, local and global pheromone evaporation ratio, respectively; $q_0 = \frac{n-20}{n}$, where n is the size of the problem; size of the candidate lists, $cl = 25$; $p_{\mathrm{L}} = 0.1\%$ and $p_{\mathrm{H}} = 50\%$ (see Eq. 1).

The stopping criterion was set that exactly $n \cdot 10^4$ (n being the size of the problem) solutions were constructed and evaluated by the ants in all replicas in total. For this reason, the higher the number of replicas M, the shorter the runtime of each replica. The computations were repeated 30 times for each configuration of the parameter values and the problem instance.

---

[1] Source code is available at https://github.com/RSkinderowicz/PTACO.

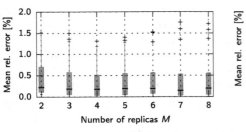
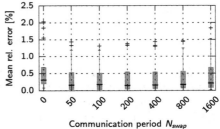

(a) Mean rel. solution error vs the number of replicas M (parallel runs)

(b) Mean rel. solution error vs the communication period

Fig. 3. Boxplots of the mean solution error (relative to optima) of the PTACO algorithm for the 12 TSP and ATSP instances considered.

The aim of the first part of the experiments was to investigate how the number of replicas M influence the quality of the solutions generated by the PTACO. Figure 3a shows a boxplot of the mean solution error obtained by the PTACO vs the number of replicas, $M \in \{2, 3, \ldots 8\}$, used. The communication period N_{swap} was set to 100, i.e. a single exchange of the temperature levels between a pair of replicas was allowed every 100 iterations. As can be seen, the differences between the results for a different number of replicas are small what can be explained by a relatively big computational budget and the small sizes of the TSP/ATSP instances considered. Nevertheless, too few ($M = 2$) or too many ($M \geq 6$) replicas worsened results relative to $M \in \{3, 4, 5\}$. Overall, the best results were obtained for $M = 4$ and this setting was used in the next stages of the experiments.

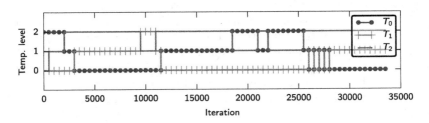

Fig. 4. An example random "walk" of $M = 3$ replicas between the temperature levels in the PTACO for the *kro124p* ATSP instance ($N_{\text{swap}} = 500$).

Apart from the number of replicas, the communication period N_{swap} is another important parameter of the PTACO, as the replicas are allowed to swap (exchange) the temperature levels every N_{swap} steps. A plot of an example "walk" in the temperature space of the replicas is shown in Fig. 4. As can be seen, the replicas change temperature levels from higher to lower and vice versa even multiple times during the algorithm execution. Several values of N_{swap} were considered, namely: 0, 50, 100, 200, 400, 800, 1600, i.e. they were increased exponentially. For $N_{\mathrm{swap}} = 0$ the communication between replicas was not allowed. Figure 3b shows a boxplot of the mean solution error obtained for the considered N_{swap} values. As can be seen, except for $N_{\mathrm{swap}} = 0$ and $N_{\mathrm{swap}} = 1600$, the differences between results were small, with pointing to $N_{\mathrm{swap}} = 100$ as the best choice. This confirms that the communication between the replicas resulting in the exchange of the temperature levels is beneficial for the convergence, however, the PTACO is not very sensitive to the communication frequency.

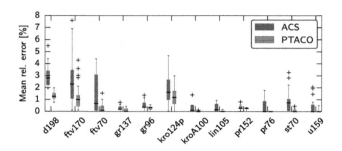

Fig. 5. Boxplot of the mean relative error for the PTACO and ACS algorithms for the TSP and ATSP instances considered.

Based on the presented findings, in the final part of the experiments, we compared the proposed PTACO with the ACS algorithm on which replicas in the PTACO are based. Most of the ACS and ACST parameters had the same values as the ACST replicas in the PTACO, except for $q_0 = 0.9$ and the local pheromone evaporation ratio $\psi = 0.1$. These values were taken after [8]. The computational budget was the same as for the PTACO to make the comparison fair. The PTACO was run with $M = 4$ and $N_{\mathrm{swap}} = 100$. Figure 5 presents a boxplot of the mean solution error obtained by the algorithms for the TSP and ATSP instances considered. As can be seen, there is a clear advantage of the PTACO over the ACS in almost every case. In fact, the highest mean relative error of 1.31% was obtained by the PTACO for the *d198* TSP instance, while the same value for the ACS was 2.93%. This advantage was further confirmed by the *two-sided non-parametric Wilcoxon rank-sum test*. Only for the *u159* instance, there was no statistically significant difference between the results of the PTACO and ACS (at the significance level $\alpha = 0.05$), in the other cases, the results of the PACO were significantly better.

The good performance of the PTACO was also confirmed by comparing the results with the results of the MMAS and Ant Colony Extended, which are the top-performing ACO algorithms [8]. As can be seen in Table 1 in 8 out of 12 cases the PTACO obtained the lowest mean solution value, while in the remaining 4 cases the best results belonged to the ACE.

Table 1. Comparison of the mean solution values for the ACS, MMAS, ACE and PTACO algorithms. The ATSP instances were marked with "*". All algorithms were allowed to construct $n \cdot 10^4$ solutions, where n is the size of the problem.

Instance	Optimum	ACS	MMAS [8]	ACE [8]	PTACO	PTACO (best)
ftv70*	1950	1981.80	1972.52	1968.16	**1954.73**	1950
st70	675	680.60	677.34	676.42	**675.57**	675
pr76	108159	108680.50	108391	108251	**108159.00**	108159
gr96	55209	55496.93	55440.2	55428.4	**55367.00**	55210
kroA100	21282	21347.60	21303.9	21298.6	**21285.97**	21282
lin105	14379	14421.50	14382.1	14385.5	**14380.57**	14379
kro124p*	36230	36908.43	36844.3	**36460.8**	36660.03	36230
gr137	69853	70024.40	70099.2	**69863.5**	69937.63	69853
pr152	73682	73938.43	73877.5	**73766.8**	73860.00	73682
u159	42080	42219.63	42135.3	42199.8	**42096.53**	42080
ftv170*	2755	2824.30	2821.44	2824.08	**2787.77**	2758
d198	15780	16242.50	15969.1	**15813.3**	15987.17	15901

7   Summary

Based on the presented computational study, we can conclude that the proposed PTACO algorithm is a viable alternative to the existing ACO algorithms. The results obtained for the set of 12 TSP and ATSP instances are competitive with the state-of-the-art ACO variants [8]. It is worth adding that the PTACO could be easily parallelized without affecting its convergence. It stems from the fact that, except for the communication events, there is no dependency between the parallel replicas of the ACST at different temperature levels. Our previous work showed that the ACS can be efficiently parallelized using modern GPUs, thus a similar approach should also work for the PTACO [13].

The main disadvantage of the proposed PTACO algorithm is the necessity to set values of the additional PT-related parameters, including the number of replicas, communication period and temperature levels. Fortunately, the presented computational study suggests that the algorithm is not very sensitive to the parameters settings which should facilitate practical applications.

Finally, the presented work is more a *proof of a concept* than a thorough study of the application of PT to the problem of improving convergence of the ACO. One possible direction for future work could involve automatic adjustment of the temperature levels in the PTACO following the ideas presented in [9]. Another interesting idea is to check if all the parallel instances of the ACST at different temperature levels indeed require the same computation time. It is possible that more time should be spent on exploring the problem solution space at lower temperatures as those at higher temperatures "move" more quickly [7]. Finally, the PTACO could be compared to the ACS with a Restart Procedure which involves running multiple replicas (instances) of the ACS simultaneously [12].

Acknowledgments. This research was supported in part by PL-Grid Infrastructure.

References

1. Ayob, M.B., Jaradat, G.M.: Hybrid ant colony systems for course timetabling problems. In: Proceedings of the 2nd Conference on Data Mining and Optimization, DMO 2009, Universiti Kebangsaan Malaysia, 27–28 October 2009, pp. 120–126. IEEE (2009)
2. Behnamian, J., Zandieh, M., Ghomi, S.F.: Parallel-machine scheduling problems with sequence-dependent setup times using an aco, SA and VNS hybrid algorithm. Expert Syst. Appl. **36**(6), 9637–9644 (2009)
3. Chen, C.-H., Ting, C.-J.: A hybrid ant colony system for vehicle routing problem with time windows. J. East. Asia Soc. Transp. Stud. **6**, 2822–2836 (2005)
4. Citrolo, A.G., Mauri, G.: A hybrid Monte Carlo ant colony optimization approach for protein structure prediction in the HP model. In: Graudenzi, A., Caravagna, G., Mauri, G., Antoniotti, M. (eds.) Proceedings of Wivace 2013 - Italian Workshop on Artificial Life and Evolutionary Computation, EPTCS, Milan, Italy, 1–2 July 2013, vol. 130, pp. 61–69 (2013)
5. Dorigo, M., Stützle, T.: Ant Colony Optimization. MIT Press, Cambridge (2004)
6. Dorigo, M., Stützle, T.: Ant colony optimization: overview and recent advances. In: Gendreau, M., Potvin, J.Y. (eds.) Handbook of Metaheuristics. International Series in Operations Research & Management Science, vol. 146, pp. 227–263. Springer, Boston (2010). doi:10.1007/978-1-4419-1665-5_8
7. Earl, D.J., Deem, M.W.: Parallel tempering: theory, applications, and new perspectives. Phys. Chem. Chem. Phys. **7**(23), 3910–3916 (2005)
8. Escario, J.B., Jimenez, J.F., Giron-Sierra, J.M.: Ant colony extended: experiments on the travelling salesman problem. Expert Syst. Appl. **42**(1), 390–410 (2015)
9. Katzgraber, H.G., Trebst, S., Huse, D.A., Troyer, M.: Feedback-optimized parallel tempering Monte Carlo. J. Stat. Mech.: Theory Exp. **2006**(03), P03018 (2006)
10. Kone, A., Kofke, D.A.: Selection of temperature intervals for parallel-tempering simulations. J. Chem. Phys. **122**(20), 206101 (2005)
11. Li, Y., Protopopescu, V.A., Arnold, N., Zhang, X., Gorin, A.: Hybrid parallel tempering and simulated annealing method. Appl. Math. Comput. **212**(1), 216–228 (2009)
12. Skinderowicz, R.: Ant colony system with a restart procedure for TSP. In: Nguyen, N.-T., Manolopoulos, Y., Iliadis, L., Trawiński, B. (eds.) ICCCI 2016. LNCS, vol. 9876, pp. 91–101. Springer, Cham (2016). doi:10.1007/978-3-319-45246-3_9
13. Skinderowicz, R.: The GPU-based parallel ant colony system. J. Parallel Distrib. Comput. **98**, 48–60 (2016)
14. Skinderowicz, R.: An improved ant colony system for the sequential ordering problem. Comput. Oper. Res. **86**, 1–17 (2017)
15. Wang, C., Hyman, J.D., Percus, A., Caflisch, R.: Parallel tempering for the traveling salesman problem. Int. J. Mod. Phys. C **20**(04), 539–556 (2009)
16. Zhu, J., Rui, T., Liao, M., Zhang, J.: Simulated annealing ant colony algorithm based on backfire method for QAP. In: Yu, X., Lienhart, R., Zha, Z., Liu, Y., Satoh, S. (eds.) The 4th International Conference on Internet Multimedia Computing and Service, ICIMCS 2012, Wuhan, China, 9–11 September 2012, pp. 100–105. ACM (2012)

Modeling Skiers' Dynamics and Behaviors

Dariusz Pałka and Jarosław Wąs[(✉)]

Department of Applied Computer Science, Faculty of Electrical Engineering,
Automatics, IT and Biomedical Engineering, AGH University of Science and
Technology, Mickiewicza 30, 30-059 Krakow, Poland
{dpalka,jarek}@agh.edu.pl

Abstract. The paper presents a proposal of an adaptation of our model
of the skier's dynamics and behaviors for different groups of downhill
skiers: beginner, intermediate and advanced. First, we propose parametrization of our model for different groups, and next we test certain characteristics of motion for different populations of skiers. We have found
that by introducing diverse populations with different behavioral patterns, we obtain a much more realistic simulation results, for instance,
a more accurate distribution of skiers on ski-slopes and more realistic
interactions between them.

Keywords: Skiers' modeling · Social force model · Granular flow · Intelligent particles

1 Introduction

The behavior of downhill skiers on ski slopes is a good example of a self-
organizing system. Particular skiers, following some rules which include collision
avoidance, ski slope boarders avoidance, speed control and many others, can be
simulated as a system of interacting intelligent particles. In our previous paper
we proposed a model of skiers' movement rules and behavior [8], where each skier
can be understood as an intelligent agent. The proposed approach is based on
Social Force Model (SFM) [4] as well as social distances related to the configuration of other skiers and the topology of a simulated ski slope including obstacles,
slope inclination, etc. In our model [8] we proposed an application of different
social forces like: way-point forces, skiers' repulsion forces - using a concept of
skier's social ellipses, obstacle repulsion forces, and slope edge repulsion inspired
by Biot - Savart law.

In this paper we would like to analyse how different characteristics of particular skiers affect the behavior of the whole system. Such analyses are important
in maintaining safety and developing downhill skiers infrastructure. In order to
address the issue, we have generated a set of tests with different populations of
skiers displaying different skiing skills.

© Springer International Publishing AG 2017
N.T. Nguyen et al. (Eds.): ICCCI 2017, Part II, LNAI 10449, pp. 97–106, 2017.
DOI: 10.1007/978-3-319-67077-5_10

2   Related Works

The idea of intelligent particles as the main component of complex, collective systems modeling crowd dynamics is well-known in literature [1–3, 10]. In 1995 Helbing and Molnár proposed the model of crowd dynamics simulation (Social) based on molecular dynamics, where consecutive positions of pedestrians are calculated as a superposition of different forces, such as: force of pursuit of the goal, obstacles repulsion or interactions between pedestrians. The Social Force Model was the main inspiration for our model of skiers' dynamics [8].

Skiers' dynamics has been the subject of research for many years. In his extensive book, Howe presented principles of skiing mechanics [6], while different aspects of physics of skiing were discussed in a book by Lind and Sanders [9]. Jentschura and Fahrbach [7] took into account physical aspects of the carving style. In [11] Schmitt and Muser analysed parametric aspects of skiing, such as reaction times and stopping performance, while Shealy et al. [12] measured distributions of skiers' velocities.

Holleczek and Tröster in [5] presented a particle-based model dedicated for ski slope traffic. The whole system was simulated thanks to physical and social forces.

3   Simulation Model

Similarly to the convention adopted in [8], in this study a skier is represented as an autonomous object characterized by mass, parameters of his skis and the level of his skiing skills.

In this model the movement of each autonomous intelligent particle (representing a skier) is the result of the following three components:

- Physical forces – that is forces that act on a skier moving down the slope and performing turns.
- Social forces – that is forces resulting from individual motivations and causing particular behaviours of intelligent particles (for example, a need to avoid a collision with obstacles or other skiers). We use the Social Force Model adapted to downhill, recreational skiing.
- Additional social behaviours – such as slowing down when exceeding the skier's safe speed or braking before the end of the slope.

In the study we tested how parameters describing the style of skiing and the level of skiing skills influenced safety and the capacity of the slope.

In order to check it, we introduced three profiles of skiers - beginner, intermediate and advanced - which differ by simulation parameters, and next we analysed the characteristics of the slope following the procedure used in [8] for each of these profiles separately.

3.1 Simulation Parameters for Particular Profiles of Skiers

Social Forces. The neighbor repulsion forces.

The model of neighbor repulsion forces between skiers used in the study is based on the model of social forces proposed in [8,13]. In this model the social area of objects (skiers) is an ellipse with a skier on the semi-major axis. If a skier moves with the maximum possible speed, he is in a focal point, while if his speed equals 0, he is in the centre of the ellipse[1]. The semi-major axis of the ellipse overlaps the direction of the velocity vector between two skiers ("observer" and "intruder").

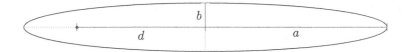

Fig. 1. The skier's social ellipse.

The values describing the social area presented in Fig. 1 are as follows:

$$b = r_0 \tag{1}$$

$$a = b + s_v v \tag{2}$$

$$d = \frac{v}{v_{max}} \sqrt{a^2 - b^2} \tag{3}$$

where:

- r_0 - is the radius of the social area for the skier with $v = 0$
- v - is the relative speed of the skier
- s_v - is the scale factor for v
- v_{max} - is the maximal possible speed of the skier (the world record)

For particular profiles of skiers we introduced differences in parameters r_0 and s_v. For the profile 'beginner', the values of r_0 and s_v are the highest, and for the profile 'advanced' – the lowest. It reflects the fact that beginner skiers try to maintain a greater distance from other persons on the slope, because their maneuvering skills are not good yet, and getting closer to other skiers may end in a collision.

In the study we used a linear model for social distance forces presented in [8,13].

Slope Edge Repulsion Force

The role of this force is to keep the skier along a slope route. The value of the force is the higher, the closer to the border of the slope a skier is. As suggested

---

[1] In this case, on the basis of Eq. 2, the eccentricity equals 0 and the ellipse is a circle.

in [8], this force is calculated in a way that is consistent with Biot-Sawart law (analogously to the way in which the Lorentz force is calculated). The force formula is:

$$F_{edges}(r_\alpha, v_\alpha) = A_1 v_\alpha \times \int_{C_L + C_R} \frac{dl \times r'}{|r'|^3} \qquad (4)$$

where:

- v_α - is the velocity of a skier
- dl - is a vector whose magnitude is the length of the differential element of the slope edge
- r' - is a displacement vector from the slope element dl to the point at which the force value is computed (the position of a skier)
- C_L and C_R - are respectively left and right slope edges in the skier's field of view
- A_1 - is the scaling constant, denoting the skier's desire to avoid crossing the edge of the slope

Because beginner skiers try to move away from the middle part (the axis) of the slope and keep far from the edge of the slope to avoid going beyond it (due to their weak maneuvering skills), the value of A_1 assumed for them is the highest, which results in the greatest repulsion force from the edge of the slope. However, for advanced skiers, who have no problems with maneuvering near the edge of the slope, the smallest value of A_1 was assumed.

All types of social forces act only when their causes (e.g. another skier, a part of the edge of the slope, or an obstacle) are placed in the skier's field of view. For particular profiles of skiers we assumed different angles of view. The more advanced the skier is, the bigger the angle is. The most advanced skiers react to a potential danger (e.g. other skiers) from a greater area, while less advanced skiers focus only on a narrow fragment of the slope in front of their current direction of movement. For each of these profiles two separate angles of vision were introduced: for static objects (such as the edge of the slope) and for moving objects (such as other skiers in motion).

Social Behaviors. The model proposed in [8] and used in this study includes three kinds of social behaviour:

- Slowing down when exceeding the skier safe speed – this behaviour is presented as Algorithm 1.

  In this simulation V_{safe} is randomly generated with uniform distribution within a range depending on the skier's skiing level. The range for V_{safe} has the lowest values for beginner skiers and the highest values for advanced skiers. In the study we have also assumed that speed ranges for particular profiles of skiers are disjoint.

– Braking before the end of the slope – makes skiers perform a braking maneuver before reaching the bottom station of the ski lift. This behaviour has been presented as Algorithm 2.

For particular profiles of skiers different values of V_{danger} and d_{end} have been assumed - the more advanced skiing skills, the higher the speed V_{danger} of reaching the end of the slope and the lower the distance d_{end} to the bottom station when the skier starts to brake are possible.

Algorithm 1. Skier slowing down

---

if $speed > V_{safe}$ **and** $skierNotTurning$ **then**
  $startTurning()$;
end if

Algorithm 2. Skier brake

---

if $speed > V_{danger}$ **and** $r_{\alpha End} < d_{end}$ **then**
  $brake()$;
end if

Physical Forces. Because physical forces are the forces which act on physical objects, they do not depend on the skier's profile. The model of physical forces used in this study is the same as the model presented in [8]. A detailed description acting on the skier on the slope can also be found in [7,9].

4   The Results of the Simulation

4.1   Configuration of Performed Simulations

As in our previous study [8], here we also chose for our simulation ski route number 1 in Jaworzyna Krynicka ski resort in Poland. The photograph of the route is presented in Fig. 2, and this route is the most popular in the ski station. The route is varied and scenic, and is the longest one in the ski station, which is why it is selected by numerous skiers. The first part of the route is relatively easy, while the second part is rather difficult and precipitous. The ski station has several chair lifts and one cable car (gondola).

A Set of the Model Parameters for Particular Profiles of Skiers. In order to build characteristics of different skiers, on the basis of our observations we propose to take into consideration certain key parameters presented in Table 1. First of all, skiers have different margins of safety, thus, we have proposed different values of safe speeds, perception of dangerous distance before the bottom station of cable cars, as well as deceleration before the bottom station. Another crucial issue are interaction patterns between other skiers and the environment, thus, we have proposed to differentiate radii and major axis scale

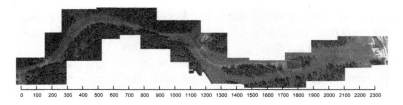

0 100 200 300 400 500 600 700 800 900 1000 1100 1200 1300 1400 1500 1600 1700 1800 1900 2000 2100 2200 2300

Fig. 2. Route number 1 in Jaworzyna Krynicka ski resort (Source [8])

factors of the social areas and slope edge repulsion forces for the proposed profiles of skiers. Other very important issues which should be taken into account are the angles of view for both static obstacles and for other skiers. On the basis of our empirical analysis and calibration, we have proposed parametrization of the mentioned crucial parameters: for instance, the radius of the social area and the social area major axis scale factor are connected with velocity ranges, while deceleration before the bottom station and dangerous speed before the bottom station are connected with the empirical observations.

According to parametrization presented in Table 1, we propose three characteristic profiles of skiers. We have identified behavioral patterns for beginner, intermediate, and advanced skiers (experts).

Table 1. Model parameters for particular profiles of skiers

Parameter	Symbol	Beginner	Intermediate	Advanced
The radius of the social area for speed $= 0$	r_0	10 m	8 m	6 m
The social area major axis scale factor	s_v	700	600	500
The slope edge repulsion force scale factor	A_1	1/4	1/5	1/6
The range of the skier's safe speed	V_{safe}	[5,10)	[10, 15)	[15,20)
Dangerous distance to the bottom station	d_{end}	80 m	60 m	40 m
Dangerous speed before the bottom station	V_{danger}	$3\frac{m}{s}$	$4\frac{m}{s}$	$5\frac{m}{s}$
Deceleration before the bottom station	a_{danger}	$3\frac{m}{s^2}$	$5\frac{m}{s^2}$	$6\frac{m}{s^2}$
The angle of view for static obstacles	φ_{static}	120°	130°	140°
The angle of view for other skiers	φ_{skiers}	100°	140°	180°

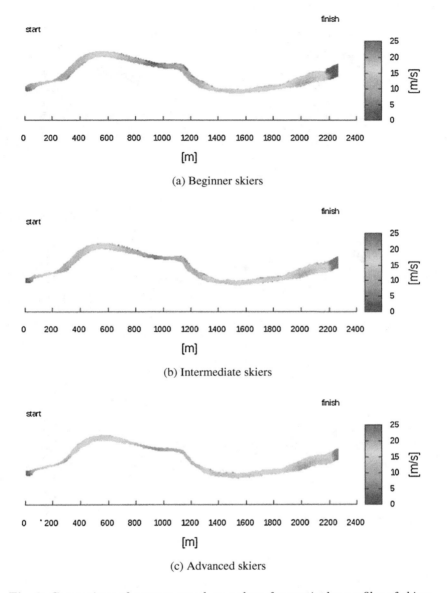

(a) Beginner skiers

(b) Intermediate skiers

(c) Advanced skiers

Fig. 3. Comparison of average speed on a slope for particular profiles of skiers.

4.2   The Results Obtained for Particular Profiles of Skiers

Taking into account speed characteristics (Fig. 3), speed is closely correlated with slope inclination. Of course, skiers move faster on steep sections of slopes, however, one can observe that each group follows specific safety rules: lower range of velocities for beginners and intermediate skiers. One can also observe more

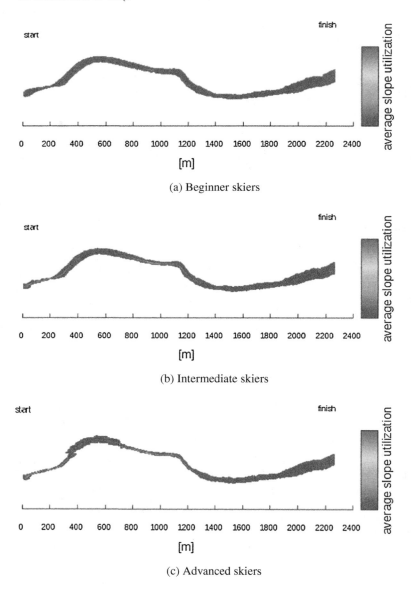

(a) Beginner skiers

(b) Intermediate skiers

(c) Advanced skiers

Fig. 4. Comparison of average slope utilization for particular profiles of skiers.

"economic" style of skiining in more experienced groups of skiers who are able to maintain more potential energy and convert it into kinetic energy on flatter sections of slopes. This is clearly visible in velocity characteristics (compare Fig. 3(a–c)).

A very interesting outcome is connected with the finding that near the boarder lines velocities are significantly lower in comparison to velocities in the center of routes, however, this is characteristic only for beginner and intermediate skiers. In the same situation advanced skiers have similar velocities as in the whole width of the route. This result is consistent with actual observations, where advanced skiers are able to keep safe speed independently on their position on a ski slope, although on very steep sections of slopes they avoid "deep corners" (Fig. 3(c)), as this would distort a smooth ride. We believe that the pursuit of fluency and smoothness of the skier's movement is an important factor in obtaining successive degrees of advancement (intermediate, advanced). An advanced skier is usually not focused only on performing basic manoeuvres and is able to observe the surroundings with a wider view-angle. The final macroscopic effects are smooth trajectories and homogeneous speed fields.

According to the slope utilization plot, it can be confirmed that advanced skiers tend to ride following a smooth, optimal trajectory (Fig. 4). The more advanced a skier is, the smoother such a trajectory is. It should be emphasized that in the group of beginners there are only few sections on the simulated ski slope, where such lines are visible Fig. 4(a). From a technical point of view, advanced skiers are able to utilize the whole slope, however, they often prefer to find optimized and relatively smoothed trajectories. It should be noted that the more experienced the skier is, the more optimized his trajectory is.

5   Conclusions

Over the last years downhill skiing has become more and more popular pastime, and one can observe rapid development of ski areas. We believe that computational methods, including modeling and simulations of skiers motion and behaviors, can be helpful in maintaining them safe. In the paper, on the basis of our model presented in [8], we have introduced specific motion characteristics of different types of skiers according to their skiing skills, namely beginner, intermediate and advanced skiers. We have proposed profiles of skiers including their: acceptable velocity, the margin of safety including slope edge repulsion, the social area size, the angle of view, and others. All proposed parameters are presented in Table 1.

We have performed a set of simulations for the real ski slope (Fig. 2) on the basis of such profiles, and it has been confirmed that taking into consideration different profiles of skiers improves the realism of the simulation. Advanced skiers are able to maintain similar speeds whether they are situated on the boarder line or in the center of the ski route, however, they tend to find optimized trajectories. Intermediate skiers are not able to maintain smooth and optimized trajectories as easily as advanced skiers, although they are able to maintain higher velocities and move more smoothly than beginners. Beginners have the lowest acceptable velocity, and they are not yet able to maintain smooth and optimized trajectories. We believe that our study is the next, substantial step in creating accurate models of skiers' behavior. In our further studies, we are going to collect real trajectories of skiers from mobile devices in order to build more and more credible simulations of skiers' dynamics.

References

1. Fridman, N., Kaminka, G.A.: Modeling pedestrian crowd behavior based on a cognitive model of social comparison theory. Comput. Math. Organ. Theory **16**(4), 348–372 (2010)
2. Fu, L., Song, W., Lv, W., Liu, X., Lo, S.: Multi-grid simulation of counter flow pedestrian dynamics with emotion propagation. Simul. Model. Pract. Theory **60**, 1–14 (2016)
3. Georgoudas, I.G., Koltsidas, G., Sirakoulis, G.C., Andreadis, I.T.: A cellular automaton model for crowd evacuation and its auto-defined obstacle avoidance attribute. In: Bandini, S., Manzoni, S., Umeo, H., Vizzari, G. (eds.) ACRI 2010. LNCS, vol. 6350, pp. 455–464. Springer, Heidelberg (2010). doi:10.1007/978-3-642-15979-4_48
4. Helbing, D., Molnár, P.: Social force model for pedestrian dynamics. Phys. Rev. E **51**(5), 4282–4286 (1995)
5. Holleczek, T., Tröster, G.: Particle-based model for skiing traffic. Phys. Rev. E **85**(5), 056101 (2012). http://pre.aps.org/abstract/PRE/v85/i5/e056101
6. Howe, J.: Skiing Mechanics. Poudre Press, Fort Collins (1983)
7. Jentschura, U.D., Fahrbach, F.: Physics of skiing: the ideal carving equation and its applications. Can. J. Phys. **82**(4), 249–261 (2004)
8. Korecki, T., Pałka, D., Wąs, J.: Adaptation of social force model for simulation of downhill skiing. J. Comput. Sci., March 2016. http://dx.doi.org/10.1016/j.jocs.2016.02.006
9. Lind, D.A., Sanders, S.P.: The Physics of Skiing. Springer, New York (2004). doi:10.1007/978-1-4757-4345-6
10. Nishinari, K., Kirchner, A., Namazi, A., Schadschneider, A.: Extended floor field CA model for evacuation dynamics. IEICE Trans. **87D**, 726–732 (2008)
11. Schmitt, K.U., Muser, M.: Investigating reaction times and stopping performance of skiers and snowboarders. Eur. J. Sport Sci. **14(Suppl. 1)**, S165–S170 (2014). PMID: 24444201
12. Shealy, J., Ettlinger, C., Johnson, R.: How fast do winter sports participants travel on alpine slopes? J. ASTM Int. **2**, 1–8 (2005)
13. Wąs, J., Gudowski, B., Matuszyk, P.J.: Social distances model of pedestrian dynamics. In: El Yacoubi, S., Chopard, B., Bandini, S. (eds.) ACRI 2006. LNCS, vol. 4173, pp. 492–501. Springer, Heidelberg (2006). doi:10.1007/11861201_57

Genetic Algorithm as Optimization Tool for Differential Cryptanalysis of DES6

Kamil Dworak[1,2]($\boxtimes$) and Urszula Boryczka[1]

[1] University of Silesia, Sosnowiec, Poland
{kamil.dworak,urszula.boryczka}@us.edu.pl
[2] Future Processing, Gliwice, Poland
kdworak@future-processing.com

Abstract. This article presents a new differential attack on the Data Encryption Standard (DES) reduced to 6 rounds, with the usage of the genetic algorithm (GA). The objective of the proposed attack is to indicate the last encryption subkey, used in the sixth cipher round, which makes it possible to define 48 from 56 primary key bits. The remaining 8 bits may be guessed by executing a brute-force attack. An additional heuristic negation operator was introduced to improve local search of proposed algorithm named NGA. The algorithm is based on the basic techniques of differential cryptanalysis. The results of the proposed NGA attack were compared with the simple genetic algorithm (SGA) and the simulated annealing (SA) attacks.

Keywords: Differential cryptanalysis · Genetic algorithm · DES · Cryptography · Simulated annealing

1 Introduction

Cryptography can be found in many IT systems. Contemporary ciphers transform plaintext into ciphertext with the usage of the generalized substitution-permutation S-P networks [1]. Many transformation rounds and long encryption keys made the cryptanalysis processes even more difficult [2]. Differential cryptanalysis was invented in 1990 by Biham and Shamir [3]. Even today, next to linear cryptanalysis, it is still one of the most popular attack techniques against many block ciphers [4].

Differential cryptanalysis can be insufficient. Despite the deliberately proposed attack, it is still a long-term process. To break the cryptographic algorithm is equivalent to solve the NP-difficult problem [1]. Any types of evolutionary computation methods, such as evolutionary algorithms (EA) are becoming more and more popular. Such algorithms are dedicated to many optimization problems, including these related to computer security. In the recent years, many papers on optimization of cryptanalytic processes, with the usage of different evolutionary techniques, such as EA and GA [5–8], Memetic Algorithms (MA) [9,10], Particle Swarm Optimization (PSO) [11,12] or SA [13], have been published. Usage of

© Springer International Publishing AG 2017
N.T. Nguyen et al. (Eds.): ICCCI 2017, Part II, LNAI 10449, pp. 107–116, 2017.
DOI: 10.1007/978-3-319-67077-5_11

the evolutionary computation techniques is becoming more and more popular; however, there is still a number of problems that needs to be examined.

The suggested attack is based on the usage of GA. These algorithms operate on a solutions set called a population [14,15] - in this case on a set of 48-bit subkeys. Each individual is subjected to evaluation, which determines the level of its usefulness, by the usage of the fitness function F_f. GA is inspired by nature; one can encounter operations, such as reproduction, mutation or fight for survival [14,15]. In order to achieve better GA functional quality, an additional heuristic negation operator was introduced.

During the first run, the suggested attack attempts to find 30 in 48 bits of the last K_6 subkey. The next run, for a different symmetric input difference, makes it possible to guess the subsequent 12 bits, and the remaining 6 bits can be guessed by executing a brute-force attack. At this phase, reversal of the key schedule process is possible, owing to that 48 in 56 bits of a primary key are known. Re-execution of a brute-force attack enables recreation of an original decryption key in a sensible period of time [1].

The next section presents the DES encryption algorithm. In third section the basics of a differential cryptanalysis of the examined cipher are discussed. The next section describes the suggested genetic attack. Fifth section includes the results of particular experiments, including the comparison of the selected algorithms. The conclusions and plans for the future can be found in the last section.

2   Data Encryption Standard

DES was developed in 1975 by IBM on the basis of the Lucifer cipher [1]. In 1977 it was accepted as a worldwide ISO standard [4]. The algorithm itself is described as a Data Encryption Algorithm DEA, however, in many papers it is referred to as DES. This cipher was the first one to be commercially used [16]. DES consists of a dozen of cycles, called rounds, which are responsible for intermingling of the partial cryptograms [1]. For several years, DES resisted any types of cryptanalytic attacks. The first more serious attack, by the usage of differential cryptanalysis [3].

DES was designed in such a way that an avalanche effect is present since the initiation of the algorithm [4]. The change of the one input bit cause the change of at least half of the output bits. Moreover, the state of each output bit depends on each input bit [1]. Authors decided to use reduced version of the encryption algorithm, named as $DES6$, which work as original DES.

The presented $DES6$ cipher is a 6-round encryption algorithm, transforming 64-bit plaintext blocks into 64-bit ciphertext block by using a 64-bit K encryption key [2,16]. Initially, the key is reduced to 56 bits by the deletion of the every 8th bit, used to verify encryption key correctness [4]. Then, K is subjected to disintegration into a set of six 48-bit subkeys, which are dedicated to each round of the algorithm, $K1, ..., K6$ - the detailed description of a key schedule can be found in [1,2,4,16,17]. Figure 1 presents 6-round DES algorithm.

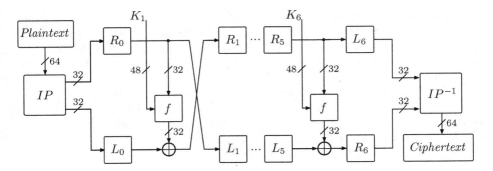

Fig. 1. *DES6* encryption algorithm

After the generation of subkeys, the encryption process can be initiated. Fragment of the plaintext is subjected to initial permutation IP. The generated block is divided into two 32-bit parts, right R and left L. Subsequently, six identical cycles are performed, where the right part R_{i-1} is transferred to round function f along with a K_i subkey. The generated block is subjected to exclusive disjunction with the left part L_{i-1}, thereby creating a new right part R_i. The new left L_i part corresponds to the right R_{i-1} side from the previous round.

After the completion of all rounds, the left L_6 and the right R_6 parts are connected into one 64-bit block of data, which is transferred for a bit reversal permutation IP^{-1}. The result of the particular bits transposition is a 64-bit block of ciphertext.

2.1 Round Function f

Each round of the algorithm use a pseudorandom function f, used to process data, presented in Fig. 2. On input, a 32-bit data portion is given, which is transferred to an expansion function E. The objective of this permutation is the equalization of length of a given block to make it the same size as a subkey, through duplication of the chosen bits. By allowing one bit to influence two substitutions, an avalanche effect is increased [4]. The obtained string is xored with the subkey bits, then it is divided into eight 6-bit $B1, ..., B8$ blocks.

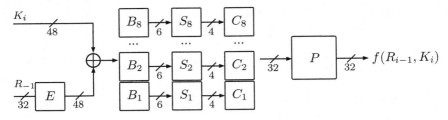

Fig. 2. Round function f of the *DES* encryption algorithm

Each of the B_j blocks is transferred to special substitution matrices called S-boxes S_j. These permutations are used for the data compression - they transform 6-bit input into 4-bit output. S_j consist of integer numbers ranging from 0 to 15, noted in matrices of the size of 16 columns and 4 rows. The first and the last bit of 6-bit B_j string indicates the row number. The remaining four bits indicate the number of a column, from which the chosen value will be returned [4,16].

S_j are the only nonlinear element of the *DES* cipher. Changing one bit in the input stream may lead to the alteration of all output bits. Each transformation of these hinders the cryptanalysis of the entire cipher. Information concerning the S_j project criteria may be found in [18].

In the end of the f function, the generated strings are concatenated with one other into one 32-bit block of data, which is then transferred to permutation P. Its objective is to copy each input bit into exactly one output bit, without duplication, nor omission of any of them [4].

3   Differential Cryptanalysis of the *DES6* Cipher

The proposed algorithm is based on the chosen plaintext attack. A cryptanalyst has access to the cipher, which allows him to select pairs of plaintexts, which differ in a specified way, and to analyze generated ciphertexts. In the case of *DES* algorithm, the above mentioned difference is calculated on the basis of the simple operation of exclusive disjunction, which may be written as $P' = P \oplus P^*$, where P and P^* are paired plaintexts. The pairs may be generated randomly, however, the P' difference has to be compatible with a specified value. Thereafter, one observes how the difference of a given pair changes over the course of the consecutive cipher rounds, until ciphertexts are generated. All cryptographs are generated with the same K key. By using the difference between the texts in the consecutive rounds for the bigger number of pairs, different probabilities are assigned, which can suggest the correctness of some subkeys [3,4].

Every block cipher is characterized by some sort of nonlinearity. In the case of the *DES* algorithm it is derived from the inside of f function, more precisely from the S_j [3]. It is impossible to find any formula, which would allow us to predict the function value for the next argument [3,4,17]. As it was previously mentioned, each of the differences is characterized by some probability, which determines how often the f function will return the expected value [3]. These differences will be called characteristics Ω. Assuming that $E = E(R_{i-1})$, input symmetric difference B' may be calculated on the basis of an expression:

$$B' = \overset{8}{\underset{j=1}{||}} B_j \oplus B_j^* = \overset{8}{\underset{j=1}{||}} (E_j(R_{i-1}) \oplus K_i) \oplus (E_j(R_{i-1}^*) \oplus K_i) = \overset{8}{\underset{j=1}{||}} E_j \oplus E_j^*. \quad (1)$$

On the basis of this expression, one may conclude, that used subkey has not any impact for B' difference. When the value of each B_j' is known, a set of all organized pairs of (B_j, B_j^*) is indicated for input symmetric difference, in accordance with the suggestion described in [17]:

$$\Delta(B_j') = \{(B_j, B_j \oplus B_j') : B_j \in (\mathbb{Z}_2)^6\}. \quad (2)$$

While calculating the output symmetric difference $C'_j = S_j(B_j) \oplus S_j(B^*_j)$, for each 4-bit pair, a schedule of all possible inputs for all possible outputs is calculated using formula presented in [17]:

$$IN_j(B'_j, C'_j) = \{B_j \in (\mathbb{Z}_2)^6 : S_j(B_j) \oplus S_j(B_j \oplus B'_j) = C'_j\}. \qquad (3)$$

In most cases the distribution will be steady. Cryptanalyst's job is to find schedules of the biggest unsteadiness. On the basis of the formula 3, an additional test set may be calculated by the usage of the following expression, formulated in [17]:

$$test_j(E_j, E^*_j, C'_j) = \{B_j \oplus E_j : B_j \in IN_j(E'_j, C'_j)\}. \qquad (4)$$

If in the set $test_j$ a number of elements is equal to the power of IN_j set, then $test_j$ must include K_{ij} bits [17].

4   Proposed *NGA* Attack

From the point of view of the cryptanalysis, IP and IP^{-1} permutation may be omitted. The attack begins with the selection of two most probable 3-round characteristics Ω^1_P and Ω^2_P, presented on Fig. 3.

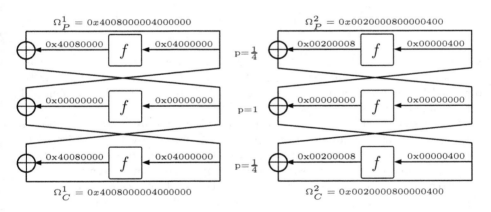

Fig. 3. Two 3-round characteristics Ω^1_P and Ω^2_P

Probability of each characteristic Ω is calculated through product of probabilities in the specified cipher rounds - in both cases it equals $P_\Omega = 1/16$. In the 4th round of S-box cipher S_2, S_5, S_6, S_7, S_8 for Ω^1_P and S_1, S_2, S_4, S_5, S_6 for Ω^2_P take the input B'_j difference and return the output C'_j difference, which equals zero. It creates the basis for the estimation of $I_1 = \{2, 5, 6, 7, 8\}$ for Ω^1_P and $I_2 = \{1, 2, 4, 5, 6\}$ for Ω^2_P sets.

The attack begins with the generation of the specified amount of plaintext pairs and their cryptographs, which symmetric difference is equal to consecutively proposed by Ω^1_P and Ω^2_P. Further algorithm description will pertain to

Ω_P^1 marked with Ω and I_1 set marked with I. Algorithm initiation for Ω_2^P requires only a different symmetric difference and I_2 set.

Not every generated pair, despite appropriate difference, is correct. Filtration is performed during this phase. For each pair a set of $test_j$ is determined, then its numerical amount is examined. If:

$$\bigwedge_{j \in I} |test_j| > 0 \tag{5}$$

then the pair may suggest the correct subkey bits, otherwise it is omitted. During the filtration phase, approximately 60% of all generated pairs are rejected.

The basis of the formulated attack is to guess the K_6 subkey. Owing to the knowledge of C' and R_5 part difference, equal to L_6, one can analyze different subkeys by comparing S-boxes output with C'. The execution of brute-force attack would require checking 2^{30} subkeys, that is approximately 1 073 741 824 combinations. In this case, GA can be used as a great optimization tool.

Each individual is represented by one 30-bit K_j subkey. Initial population consists of N randomly generated chromosomes. During the repeating algorithm iterations, under the influence of genetic operators, individuals evolve, improving the quality of their adaptation. The fitness function was defined as follows:

$$F_f = \sum_{i=0}^{n} L - H \sum_{j \in I} (S_j(B_j) \oplus S_j(B_j^*)), P^{-1}(R_6' \oplus L_3')), \tag{6}$$

where H stands for Hamming distance, L is a subkey length, P_Ω is the characteristic probability. The estimation of L_3' value is possible, and R_6' may be obtained by analyzing a pair of ciphertexts. F_f counts the number of identical bits between S-boxes result and C' differences.

In the suggested attack the tournament selection was used. It randomly chooses subset of individuals, from among these only one will be subjected to the crossover process. The remaining individuals are returned to the population, afterwards this process is repeated in order to find the second parent.

Descendant chromosomes are generated as the result of the crossover process. Point of parents bisection is chosen randomly from 1 to 30. Subkeys are intersected according to the chosen point, and then they exchange their genetic material. Newly created individuals may be subject to mutation operator, which objective is to randomly swap two chosen bits of the chromosome.

Newly generated individuals are a subjected to an additional heuristic negation operator. Entails negating each bit of the chromosome, with a certain probability P_n, and remembering the most favourable variant. The heuristic negation operator is activated last.

The execution of the above mentioned attack for Ω_P^1 and I_1 allows us to guess 30 in 48 subkey K_6 bits. Repetition of this attack for Ω_P^2 and I_2 will enable the obtainment of 12 additional bits. Complete K_6 subkey will lack only 6 bits of S-box S_3, which may be easily guessed by executing brute-force attack. Knowing the entire subkey, the disintegration of the K key process may be reversed; due to this 48 in 56 primary key bits will be found. The remaining 8 bits may be guessed by the re-execution of the brute-force attack.

The *SGA* attack is the same as the *NGA* one described above, except the additional heuristic negation operator. It used same crossover, mutation and selection operators. It was decided to skip describing *SGA* attack.

Table 1. F_f values for *SGA*, *NGA* and *SA* attacks

ID	Ω_P^1							Ω_P^2						
	Min	Med	Avg	Max	Std. dev.	F_b	t	Min	Med	Avg	Max	Std. dev.	F_b	t
SGA														
1	1039	1101	1106.6	1234	48.5	29	0	1229	1287	1303.4	1427	54.8	28	0
2	1179	1236	1244.7	1339	39.1	24	0	1124	1193	1209.8	1348	54.0	28	0
3	1275	1361	1367.4	1473	52.1	29	0	1226	1299	1311.8	1428	55.8	27	0
4	1301	1368	1371.6	1452	37.2	27	0	1043	1092	1101.2	1199	38.8	29	0
5	1391	1464	1470.1	1591	44.6	26	0	1192	1249	1262.5	1386	48.8	27	0
6	1326	1399	1423.0	1576	68.2	29	0	1215	1274	1283.1	1394	45.0	25	0
7	1241	1319	1321.7	1405	39.6	28	0	**1171**	**1249**	**1249.6**	**1348**	**47.9**	**30**	**71**
8	1150	1237	1233.0	1298	33.6	25	0	1293	1376	1386.2	1529	61.9	28	0
9	1095	1169	1172.2	1277	37.0	26	0	1224	1291	1293.6	1398	45.5	27	0
10	**1579**	**1681**	**1703.5**	**1911**	**83.5**	**30**	**64**	1261	1337	1353.1	1497	63.1	26	0
NGA														
1	**1125**	**1177**	**1178.3**	**1241**	**30.7**	**30**	**56**	**1284**	**1348**	**1356.2**	**1445**	**37.8**	**30**	**51**
2	1225	1284	1286.6	1344	31.2	27	0	**1232**	**1271**	**1278.4**	**1384**	**27.9**	**30**	**73**
3	**1368**	**1448**	**1443.7**	**1479**	**24.8**	**30**	**92**	**1282**	**1387**	**1382.4**	**1433**	**45.8**	**30**	**53**
4	**1352**	**1424**	**1425.1**	**1479**	**32.6**	**30**	**33**	**1090**	**1167**	**1162.9**	**1214**	**35.4**	**30**	**14**
5	**1471**	**1531**	**1531.0**	**1606**	**30.5**	**30**	**50**	**1263**	**1345**	**1349.3**	**1404**	**35.9**	**30**	**23**
6	**1399**	**1501**	**1502.4**	**1589**	**39.5**	**30**	**17**	1314	1350	1348.8	1394	18.2	28	0
7	**1310**	**1355**	**1358.0**	**1421**	**26.9**	**30**	**23**	**1239**	**1317**	**1309.0**	**1348**	**25.9**	**30**	**52**
8	**1239**	**1292**	**1291.0**	**1323**	**20.6**	**30**	**39**	**1363**	**1448**	**1452.6**	**1539**	**39.4**	**30**	**10**
9	1171	1219	1223.4	1283	29.0	28	0	**1270**	**1346**	**1352.5**	**1430**	**42.0**	**30**	**19**
10	**1665**	**1793**	**1803.6**	**1911**	**45.5**	**30**	**46**	**1344**	**1421**	**1423.7**	**1524**	**49.3**	**30**	**21**
SA														
1	1004	1075	1058.7	1143	20.6	22	0	1196	1221	1247.9	1322	22.4	20	0
2	1148	1186	1202.3	1269	20.8	19	0	1102	1145	1154.0	1211	19.8	26	0
3	1255	1289	1308.2	1352	20.8	19	0	1202	1229	1245.2	1311	23.0	25	0
4	1287	1326	1325.8	1387	18.4	15	0	1026	1057	1062.1	1119	18.1	23	0
5	1375	1430	1426.5	1473	18.1	16	0	1172	1217	1224.5	1313	25.1	24	0
6	1303	1326	1347.5	1434	23.8	19	0	1175	1225	1237.7	1300	20.2	19	0
7	1228	1295	1275.3	1327	20.3	17	0	1148	1196	1197.7	1247	18.0	18	0
8	1142	1243	1194.5	1249	20.0	18	0	1276	1296	1317.4	1387	20.7	24	0
9	1101	1115	1140.0	1181	17.5	15	0	1180	1226	1237.5	1289	20.7	22	0
10	1547	1547	1615.7	1722	28.2	18	0	1249	1279	1290.6	1338	18.0	22	0

4.1 Simulated Annealing Attack

At the beginning the initial temperature T_0 and the minimum temperature T_{MIN} are determined. In each algorithm's iteration, T temperature is reduced by a $n \in [0,1]$ constant and a new K_6' subkey is generated, through the change of one random bit into the opposite one. If the new individual proves to be better than his predecessor, he substitutes him. Additionally, a probability function is introduced, which enables the acceptance of the solution, in case it is worse:

$$Probability = \exp \left(\frac{F_f(K_6') - F_f(K_6)}{k \cdot T} \right), \tag{7}$$

where k is the Boltzmann constant. This process is repeated until the minimum temperature T_{MIN} is reached.

5 Experimental Results

All algorithms were implemented using the C# programming language and were executed on the same computer equipped with an Intel i5 processor clocked at 3.2 GHz. For SGA and NGA attacks the maximum allowed number of iterations was set to 100. Population in both algorithms consisted of 70 chromosomes. The tournament selection used 5 individuals to find each parent. The crossover probability P_c was set to 0.8, the mutation probability P_m to 0.02. The additional heuristic negation operator, used in NGA named as P_n, was set to 0.25.

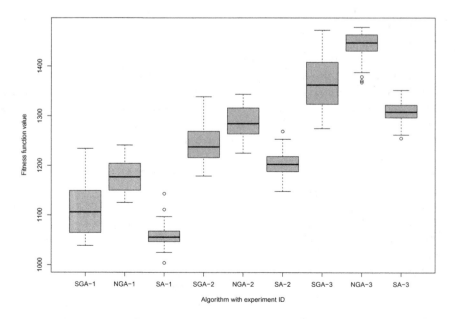

Fig. 4. Cryptanalysis results of *DES6* for *SGA*, *NGA* and *SA* attacks

The population leader was saved and was moved to the offspring population in every generation. For SA attack the starting temperature T_0 was set to 1000, the minimum temperature T_{MIN} to 0.001. The temperature T was decreased in each iteration by *coolingRate* set to 0.99. For each algorithm 160 pairs of plaintexts where generated randomly, all of them satisfy condition defined in expression 5. Also the number of correctly found bits have been counted for every experiment. All encryption keys were generated randomly.

Table 1 presents results of all algorithms. It contains statistical information about F_f in last iteration t of each algorithm and number of matching bits F_b.

Many times SGA algorithm found about 27–29 bits of correct encryption subkey. Unfortunately in cryptanalysis all bits of subkeys are required. Simple GA found valid subkey only 2 times. NGA attack find perfect subkey almost every time. 85% of tested subkeys where successfully broken. SA did not find any valid solution. In first experiment for Ω_P^1 it found 22 bits of subkey. All experiments were selected randomly. Figure 4 presents a results for first three experiments for each attack. Comparing all presented algorithms and their results can be easily notice, the proposed NGA algorithm was the most effective one.

6   Conclusions and Future Work

This paper presents genetic differential cryptanalysis directed to $DES6$ encryption algorithm. Attack was enriched by GA with an additional heuristic negation operator. The proposed algorithm was compared with simple GA (SGA) and simulated annealing (SA) attacks. The NGA attack turned out to be the best from all algorithms, it broke valid parts of K_6 in 85% of cases.

The presented attack should be tested on more complex variants of DES cipher such as $DES8$, $DES12$ or original DES algorithms. Also any adaptive techniques should be used to modify GA parameters during execution. Other metaheuristic alternatives like Particle Swarm Optimization (PSO) or Memetic Algorithms (MA) can provide interesting results.

References

1. Pieprzyk, J., Hardjono, T., Seberry, J.: Fundamentals of Computer Security. CRC Press, Inc., Boca Raton (2003)
2. Stallings, W.: Cryptography and Network Security: Principles and Practice, 5th edn. Pearson, New York (2011)
3. Biham, E., Shamir, A.: Differential cryptanalysis of DES-like cryptosystems. J. Cryptol. 4(1), 3–72 (1991)
4. Schneier, B.: Applied Cryptography: Protocols, Algorithms, and Source Code in C. Wiley, Hoboken (1996)
5. Dworak, K., Boryczka, U.: Differential cryptanalysis of FEAL4 using evolutionary algorithm. In: Nguyen, N.-T., Manolopoulos, Y., Iliadis, L., Trawiński, B. (eds.) ICCCI 2016. LNCS, vol. 9876, pp. 102–112. Springer, Cham (2016). doi:10.1007/978-3-319-45246-3_10

6. Song, J., Zhang, H., Meng, Q., Wang, Z.: Cryptanalysis of four-round DES based on genetic algorithm. In: Proceedings of IEEE International Conference on Wireless Communication, Network and Mobile Computing, pp. 2326–2329. IEEE (2007)
7. Huseim, H.M.H., Bayoumi, B.I., Holail, F.S., Hasan, B.E.M., El-Mageed, M.Z.A.: A genetic algorithm for cryptanalysis of DES-8. Int. J. Netw. Secur. 5, 213–219 (2007)
8. Tadros, T., Hegazy, A.E.F., Badr, A.: Genetic algorithm for DES cryptanalysis genetic algorithm for DES cryptanalysis. Int. J. Comput. Sci. Netw. Secur. 10, 5–11 (2010)
9. Dworak, K., Nalepa, J., Boryczka, U., Kawulok, M.: Cryptanalysis of SDES using genetic and memetic algorithms. In: Król, D., Madeyski, L., Nguyen, N.T. (eds.) Recent Developments in Intelligent Information and Database Systems. SCI, vol. 642, pp. 3–14. Springer, Cham (2016). doi:10.1007/978-3-319-31277-4_1
10. Garg, P., Varshney, S., Bhardwaj, M.: Cryptanalysis of simplified data encryption standard using genetic algorithm. Am. J. Netw. Commun. 4, 32–36 (2015)
11. Abd-Elmonim, W.G., Ghali, N.I., Hassanien, A.E., Abraham, A.: Known-plaintext attack of DES-16 using particle swarm optimization. In: Third World Congress on Nature and Biologically Inspired Computing, vol. 9330, pp. 12–16. IEEE (2011)
12. Jadon, S.S., Sharma, H., Kumar, E., Bansal, J.C.: Application of binary particle swarm optimization in cryptanalysis of DES. In: Deep, K., Nagar, A., Pant, M., Bansal, J.C. (eds.) Proceedings of the International Conference on SocProS 2011. AINSC, vol. 130, pp. 1061–1071. Springer, India (2012). doi:10.1007/978-81-322-0487-9_97
13. Nalini, N., Raghavendra, R.G.: Cryptanalysis of block ciphers via improved simulated annealing technique. In: Information Technology, ICIT, vol. 130, pp. 182–185. IEEE (2007)
14. Michalewicz, Z.: Genetic Algorithms + Data Structures = Evolution Programs, 3rd edn. Springer, London (1996). doi:10.1007/978-3-662-03315-9
15. Goldberg, D.E.: Genetic Algorithms in Search, Optimization and Machine Learning. Addison-Wesley Longman Publishing, Boston (1989)
16. Menezes, A.J., Oorschot, P.C., Vanstone, S.A.: Handbook of Applied Cryptography. CRC Press, Boca Raton (1997)
17. Stinson, D.R.: Cryptography: Theory and Practice. CRC Press, Inc., Boca Raton (1995)
18. O'Connor, L.J.: An analysis of product ciphers based on the properties of Boolean functions. Ph.D. thesis, University of Waterloo, Waterloo (1992)

Machine Learning in Medicine and Biometrics

Edge Real-Time Medical Data Segmentation for IoT Devices with Computational and Memory Constrains

Marcin Bernas$^{(\boxtimes)}$, Bartłomiej Płaczek, and Alicja Sapek

Institute of Computer Science, University of Silesia,
Bedzinska 39, 41-200 Sosnowiec, Poland
marcin.bernas@gmail.com, placzek.bartlomiej@gmail.com

Abstract. The Internet of Things (IoT) becomes very important tool for data gathering and management in many environments. The majority of dedicated solutions register data only at time of events, while in case of medical data full records for long time periods are usually needed. The precision of acquired data and the amount of data sent by sensor-equipped IoT devices has vital impact on lifetime of these devices. In case of solutions, where multiple sensors are available for single device with limited computation power and memory, the complex compression or transformation methods cannot be applied - especially in case of nano device injected to a body. Thus this paper is focused on linear complexity segmentation algorithms that can be used by the resource-limited devices. The state-of-art data segmentation methods are analysed and adapted for simple IoT devices. Two segmentation algorithms are proposed and tested on a real-world dataset collected from a prototype of the IoT device.

Keywords: Internet of Things · Data segmentation · Edge mining · Medical data

1 Introduction

The Internet of Things (IoT) [1] became one of major medium of gathering information via network of networks, where large number of devices with sensors is providing information. This information is used to enhance already existing services or allowing to provide new ones. This headway was brought by capabilities that the IoT adds to the society [2]. The IoT area is widely investigated by scientists. An overview of the IoT technology can be found in [2]. Another research was conducted to enable self organisation of the IoT sensor-equipped devices [3]. A lot of effort was put to define communication standards between the IoT devices. Sheng et al. [4] have presented a protocol stack developed specifically for the IoT domain by the Internet Engineering Task Force (IETF). Another research group was focused strictly on security issues in IoT [5]. In case of devices that are monitoring a medical activity, there is a need to send registered data

© Springer International Publishing AG 2017
N.T. Nguyen et al. (Eds.): ICCCI 2017, Part II, LNAI 10449, pp. 119–128, 2017.
DOI: 10.1007/978-3-319-67077-5_12

continuously or only in case of a defined event. This paper is focused on continuous data collection, which then could be used for historical data analysis or data mining [6]. Sending raw data can consume significant energy resources of the IoT device, thus several approaches have been introduced in the related literature, which are focused on data mining techniques for the IoT [7]. The term of edge mining was defined in IoT paradigm to discus data analysis for IoT devices [8]. In contrast to the traditional data mining, edge mining takes place on the wireless, battery-powered and smart sensing devices that sit at the edge points of the IoT. The fundamental research concerning data mining was made by Keogh et al. [9]. They were followed by pattern recognition research as motifs or symbolic aggregate approximation (SAX) [10] to develop a codebook [11]. The compressed dissimilarity measure (CDM) was proposed [12] to find motifs in a given time series represented as SAX elements. There are several algorithms that use in-network aggregation techniques [13]. The related works show that solutions based on discrete cosine transform (DCT) [14], lightweight temporal compression (LTC) [15], and enumeration of motifs (EMMA) [16] can give good result in applications for wireless sensor networks. The segmentation problem of time series was widely researched and many high level representations of time series have been proposed [17], e.g., Fourier transforms, wave-lets, symbolic mappings, piecewise linear representation (PLR) or motifs. The classical approach was extended [18] by fuzzy queries [19], weighted queries and multi-resolution queries. The existing edge mining techniques do not take under consideration the IoT devices with simple construction and low resources (especially nano sensors [20]). Thus, this research is focused on the segmentation methods that are characterized by linear computational complexity, reduced memory usage and do not require complex calculations. A modification of the lightweight algorithm named RAZOR, for IoT [21] was proposed in this paper along with SAX simplification [10]. Several modifications of the segmentation algorithms are presented that have linear complexity and could be applied for simple IoT devices. The paper is organized as follows. Details of the considered IoT model as well as the proposed algorithms are discussed in Sect. 2. Results of experiments are presented in Sect. 3. Finally, conclusions are given in Sect. 4.

2   Proposed Model

Data readings collected from sensors by the IoT devices can be represented as time series (TS). The data collected over a long time period allows finding relations and patterns that describe the registered process [22] (e.g. changes of temperature or pressure). TS can be defined as ordered set X of real values x_i registered in equal time periods: $X = \{x_1, x_2, \ldots, x_i, \ldots, x_m\}$. The sensors data can be collected in an indefinitely long period, while memory and processing power of IoT devices are always limited, thus the data has to be processed as n-size subsequences. The subsequence of TS $X = \{x_1, x_2, \ldots, x_m\}$ of length k is defined as:

$$X_{n,k} = \{x_n, x_{n+1}, \ldots, x_{n+k-1}\}, \text{where}: 1 \leq n \leq m - k + 1, k \leq m. \quad (1)$$

The subsequences of TS can be obtained by using one of following segmentation approaches [13]: sliding windows, top-down, or bottom-up strategy. The top-down and bottom-up strategies require processing of data sequences that have significant length. In case of small IoT device, the available memory can be insufficient for execution of these strategies. Therefore, the sliding window segmentation strategy was used in the proposed method with linear complexity. This strategy can be used in any IoT device to reduce the amount of transmitted data. Based on the introductory research [9,18], different state-of-the-art representations of data subsequence (D) were selected for this study, i.e.: mean, median, PLR, wavelets transform and Fourier transform. The data collection system considered in this study is based on IoT principles and application of subsequence representation (D). It requires a device with access to internet via wireless communication. The IoT based system was illustrated in Fig. 1.

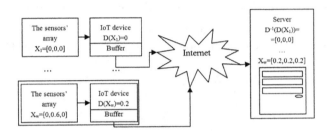

Fig. 1. Overview of the IoT implementation

To reduce the number of transmissions two approaches were considered that utilize fixed size data subsequence and variable size data subsequence. In the first case k values from sensors are collected and then its representation is created. Thus, the data sent to the server are defined as:

$$V_c = \{D(X_{i,k}, k) : i = 0, 1, ..., \lfloor m/k \rfloor - 1\} \tag{2}$$

where D is a subsequence representation function and m denotes size of TS.

The server obtains the representations of subsequences and restores TS values by using an opposite operation (D^{-1} function). This approach do not burden the IoT device, however the precision of the restored data cannot be assured. The second approach assumes that each subsequence can have different size. The size of subsequence (k) is calculated each time based on representation error and takes into account the buffer size (k_{max}). In this case the subsequence is represented as a pair: $D(X_{(i,k)}), k$, where k is a size of the subsequence and i is an index of the first element in the sequence. The size k for every subsequence is calculated by using Eq. 3.

$$k = \max\{x : ER(X_{i,x}) < \beta \wedge ER(X_{i,x+1}) > \beta) \wedge x \leq k_{max}\},$$
$$ER(X_{i,x}) = RMSE(D^{-1}(D(X_{i,x})), X_{i,x}), \tag{3}$$

where: $x \in N$, i is the beginning of subsequence, k_{max} is maximal subsequence size, and β is a defined maximal root mean square error (RMSE).

In the second approach (Eq. 3) the device, for each iteration, checks if error requirement is met and there is still sufficient free memory available. At each step the RMSE value is calculated. Thus, this approach is more computationally complex than the previous one with fixed subsequence size assumption. Nevertheless, the variable subsequence size approach guarantees the constant error rate of incoming data at the cost of various numbers of messages. This is especially important in case of medical data anomalies (related with given disease) that otherwise could be lost in the process. All these approaches are based on D and D^{-1} representation functions that could be applied for simple IoT sensors. The next subsections present the functions that were selected by taking into account their computational complexity and memory requirements.

2.1 Representation Functions with Linear Complexity

The two simplest description methods use basic statistic estimators: mean and median value. In this case the functions D and D^{-1} are defined in Eqs. 4 and 5.

$$
\begin{aligned}
D_{mean}(X_{i,k}, k) &= \frac{\sum_{j=0}^{k-1} x_{i+j}}{k} = \bar{x}, \\
D_{mean}^{-1}(\bar{x}, k) &= \hat{X} = \{\hat{x}_j = \bar{x} : j = 1, ..., k\},
\end{aligned} \tag{4}
$$

$$
\begin{aligned}
D_{median}(sort(X_{i,k}), k) &= x_{i+\lceil k/2 \rceil} = \bar{x}, \\
D_{median}^{-1}(\bar{x}, k) &= \hat{X} = \{\hat{x}_j = \bar{x} : j = 1, ..., k\},
\end{aligned} \tag{5}
$$

where: $sort$ is the function that sorts data in ascending order, $\lceil \rceil$ - ceiling function.

Other methods define the linear approximation of a given subsequence. Two the most common implementation of linear approximation were used: piecewise linear representation (PLR) and regression method. As a result, each subsequence is represented by pair (x_1, x_2). The advantage of regression is a more precise fit of the function to the data points at the cost of rapid changes of TS on the borders of the subsequence that can be interpreted as sudden events. PLR is the most commonly used method of linear approximation. It describes TS as a series of lines. This method is often used for fast search through the TS and for the fuzzy queries. There are several algorithms available for finding the linear representation of TS. Due to computation constrains the simplest representation was used, where the first and last element of a sequence is selected. Thus D and D^{-1} functions are defined as follows:

$$
\begin{aligned}
D_{PLR}(X_{i,k}, k) &= (x_i, x_{i+k-1}), \\
D_{PLR}^{-1}(x_i, x_{i+k-1}, k) &= \hat{X} = \{x_i + (j-1)\frac{x_{i+k-1} - x_i}{k-1} : j = 1, ..., k\},
\end{aligned} \tag{6}
$$

$$
\begin{aligned}
D_{reg}(X_{i,k}, k) &= (r(x_i), r(x_{i+k-1})), \\
D_{reg}^{-1}(r(x_i), r(x_{i+k-1}), k) &= D_{PLR}^{-1}(r(x_i), r(x_{i+k-1}), k),
\end{aligned} \tag{7}
$$

where: r denotes the liner regression function implemented in accordance with the definition given in [23].

2.2 SAX

The more sophisticated approaches assume that each subsequence $X_{i,k}$ is described by symbols [9]. In this research we assume that three symbols can be used: constant (corresponding to average), increase or decrease, and peak. The symbols are presented in Fig. 2.

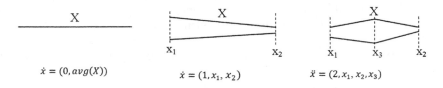

$$\dot{x} = (0, avg(X)) \qquad \dot{x} = (1, x_1, x_2) \qquad \ddot{x} = (2, x_1, x_2, x_3)$$

Fig. 2. Symbols representing subsequence

The first two symbols are using representation of average and PLR, while the third representation considers two lines. The symbols were defined as follows:

$$
\begin{aligned}
D_{SAX,0}(X_{i,k}, k) &= (\bar{x}, k), \bar{x} = D_{mean}(X_{i,k}, k), \\
D_{SAX,0}^{-1}(\bar{x}, k) &= D_{mean}^{-1}(\bar{x}, k), \\
D_{SAX,1}(X_{i,k}, k) &= (\dot{x}, k), \dot{x} = D_{PLR}(X_{i,k}, k), \\
D_{SAX,1}^{-1}(\dot{x}, k) &= D_{PLR}^{-1}(x_i, x_{i+k-1}, k), \\
D_{SAX,2}(X_{i,k}, k) &= (\ddot{x}, k), \ddot{x} = (x_i, x_{i+\lfloor k/2 \rfloor}, x_{i+k-1}),
\end{aligned}
\tag{8}
$$

$$
D_{SAX,2}^{-1}(\ddot{x}, k) =
\begin{cases}
x_i + j \frac{x_{i+\lfloor k/2 \rfloor} - x_i}{\lfloor k/2 \rfloor} & : j = 0, ..., \lfloor k/2 \rfloor, \\
x_{i+\lfloor k/2 \rfloor} + j \frac{x_{i+k-1} - x_{i+\lfloor k/2 \rfloor}}{\lceil k/2 \rceil - 1} & : j = 1, ..., \lceil k/2 \rceil - 1.
\end{cases}
\tag{9}
$$

The symbol for a data subsequence is selected based on the following algorithm (Algorithm 1):

1. Obtain next k samples as $X_{i,k}$ subsequence,
2. Define table tab $= \infty, \infty, \infty$,
3. If $(x_i < x_{\lceil (i+k)/2 \rceil}$ and $x_{i+k-1} < x_{\lceil (i+k)/2 \rceil})$ or $(x_i > x_{\lceil (i+k)/2 \rceil}$ and $x_{i+k-1} > x_{\lceil (i+k)/2 \rceil}$ then tab[3] = RMSE$(D_{SAX,3}^{-1}(D_{SAX,3}(X_{i,k}, k)), X_{i,k})$ else
   For $j = 1$ to 2 do tab[j] = RMSE$(D_{SAX,j}^{-1}(D_{SAX,j}(X_{i,k}, k)), X_{i,k})$,
4. Select the best describing symbol as $s = \arg\min(\text{tab}[j])$,
5. Generate the symbol s as best describing the $X_{i,k}$ sequence.

In the proposed algorithm (Algorithm 1) the symbols are build based on $X_{i,k}$ subsequence. The third symbol is considered, if the middle value is larger or smaller than the values of the first and the last element in the sequence (lines 2–3). Finally, the best fitted symbol is selected (lines 4–5). The symbol is then sent to the buffer with its descriptive parameters.

2.3 Motif

The last considered approach is based on motif. This approach is especially effective, when the TS has a periodic character. Several motif implementations were proposed in the literature [18]. Taking under consideration the linear computational complexity requirement the following implementation based on RAZOR framework [21] was proposed (Algorithm 2):

1. Define number of motifs stored as p parameter,
2. Create p vectors of size k initialized with zeros and store them as table $M[i], i = 1...p$,
3. Set RMSE values between motifs: Initialize matrix $R[p,p]$ with value ∞, set most similar motif as: $minM = 0$, $m_1 = 1$, $m_2 = 2$,
4. Obtain the next k samples as $X_{i,k}$ subsequence,
5. Calculate average value as $\bar{x} = D_{mean}(X_{i,k}, k)$,
6. $X'_{i,k} = \{x'_{i+j} = x_{i+j} - \bar{x} : j = 0..., k-1, x_{i+j} \in X_{i,k}\}$
7. For $x = 1$ to p do $RM = \text{RMSE}(X'_{i,k}, M[x])$
8. If $\min(RM) <= minM$ then s $= \arg\min_x(RM)$; store subsequence as motif s of average $\bar{x}$,
9. If $\min(RM) > minM$ then add $X_{i,k}$ as m_1 motif to transfer buffer; store subsequence as motif s of average $\bar{x}$, $M[p] = X_{i,k}$;
   For $x = 1$ to p do If $m_1 <> x$ then $R[x, m_1] = \text{RMSE}(M[m_1], M[x])$; find closest motifs as:
   $(m_1, m_2) = \arg\min_{m_1, m_2}(R[m_1, m_2])$; $minM = R[m_1, m_2]$,
10. Go to step 4.

Algorithm 2 is initialized with motifs represented by vectors with zero elements (lines 1–3). The motifs are equal to each other, therefore their RMSE value is 0. In the first loop motifs 1 and 2 are selected as the closest. The lines 4–7 define the main loop operations for each subsequence of length k. In case of motif, the objective is to compare the shape and not values directly, thus the average value is subtracted from the subsequence. Then the similarity between the subsequence and a motif is evaluated by means of RMSE. If the most similar motif is found (line 8) the subsequence is represented by this motif and the motif is send to the buffer (line 8). However, if the defined motifs are more similar to each other than to the analysed subsequence, this subsequence is treated as a new motif. In this case the shape of motif is send to a server and the sequence is stored as the new motif (line 9).

In case, when each subsequence can have various size (parameter k), the initial motifs have length of k_{max} (line 2), but only their first k elements are processed (lines 4–10) for next subsequence $X_{i,k}$. Other approaches to the subsequence representation problem include Fourier or wavelet transformations. However, the considered platform was not able to compute such transforms in real-time thus these approaches were not considered in this study.

3   Experiments

The proposed algorithms were implemented in three identical IoT devices. The devices were presented in Fig. 3.

Fig. 3. The implementation of IoT device: (a) back, (b) front

The device is based on Arduino platform with 8-bit microcontroller. Each device can support up to 8 one wire sensors collecting medical data. For the experiment purposes, four DS18B20 temperature sensors were used to collect patient or environment data. The data are stored in memory for processing. The data from transmission buffer of size n are transmitted to an IoT server by using the Wi-Fi interface. In this solution ESP8266 module was used with implemented TCP/IP stack. Using HTTP protocol the data are transmitted to the server via URL request frame. The buffer is considered as full if $n = 50$ bytes of data are ready for sending. The device sends the segmented frames and real data for comparison purposes at the same time. To verify the proposed approaches, the sum of frames transmitted as well as the data loss was measured. The size of subsequence k is the parameter of all proposed approaches, however due to the device memory limitations k_{max} was set to 25. For motif method the number of motifs ($p = 6$) was determined based on preliminary experiments. The selected set-tings ensure good performance of the motif method in initial stage, when changes of motifs are frequent as well as in second phase, when the set of motifs is stable. Moreover, for $p = 6$ memory requirements of the motif method are met by the selected hardware architecture. Then all functions were investigated using fixed size of the subsequence. The results are presented in Fig. 4.

The results firmly show that the proposed SAX representation for small subsequence gives lower error at the cost of slightly higher data transmission. The overhead is caused by symbol selection, which in case of 3 values is additional 2 bits to send. However, for sake of simplicity, the byte was reserved as symbol representation. Nevertheless, the registered SAX implementation overhead was 6% to PLR and 12% in case of average/mean value. The SAX RMSE value is smaller to PLR and mean methods by 19% and the proportion changes with length of

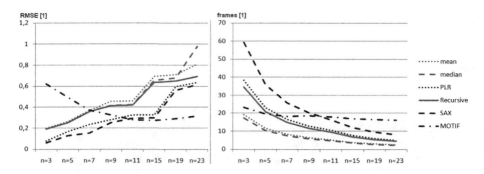

Fig. 4. The average results of the algorithms: (a) RMSE, (b) transmissions number

sequence, thus for longer subseries the SAX and motif implementation is preferred. According to obtained results, the construction of motifs for subseries
smaller than 9 data points is not preferred. The short motifs are not containing
useful information about the TS and give unreliable results. According to the
obtained results, the subseries above 9 data points contain enough information
to be able to provide reliable results. The research considering the change of
window length was conducted according to Eq. 3. The results for the analysed
methods are presented in Fig. 5 for β as a parameter defining the maximal RMSE
value.

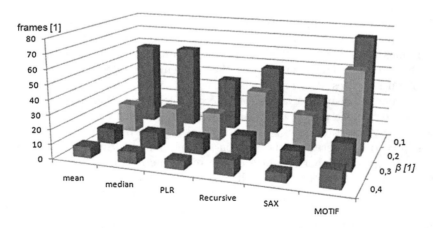

Fig. 5. Result obtained for various RMSE error (β parameter)

Similarly, as in case of fixed size frames the proposed SAX approach proved
to be superior in case of all RMSE threshold, with exception of $\beta = 0.1$. On
the other hand the proposed motif approach is not giving good results. The
motif algorithm does not scale well with size of subsequence. In case of too short

or to long subsequences the motifs changes very often and thus the number of transmissions increases. The change of motif size force to create new patterns and in simplest approach increases the number of messages sent. Nevertheless, the changing pattern caused in case of medical data to be forwarded without change, and thus reduce error rate of atypical data.

4   Conclusion

The proposed algorithms can be used to reduce number of medical data transmissions by simple IoT devices with low computing power and memory size. Six methods were considered in this study. Two modifications of SAX and MOTIF algorithms were introduced. The results show that for a small subsequence, with up to 13 data elements, the proposed SAX algorithm ensures the lowest RMSE value. If the number of data elements is above 13 then the motifs approach allows the error rate to be kept at a low level with slightly higher number of transmissions. If segments of variable length are considered then the simple motif implementation is inferior to the SAX method. Results of this research show that it is possible to implement the segmentation algorithms in IoT medical devices, even with very limited memory and low computing power. In future, the algorithms will be tested on more than one type of devices. The granular computing paradigm will be used for data representation. Another topic for future research covers investigation of spatial and temporal correlations between data collected from IoT medical devices.

References

1. Charith, P., Chi, H., Srimal, J.: The emerging Internet of Things marketplace from an industrial perspective: a survey. IEEE Trans. Emerg. Top. Comput. **3**(4), 34–42 (2015)
2. Atzori, L., Iera, A., Morabito, G.: The Internet of Things: a survey. Comput. Netw. **54**(15), 2787–2805 (2010)
3. Athreya, A., Tague, P.: Network self-organization in the Internet of Things Networking and Control (IoT-NC). In: IEEE International Workshop, pp. 25–33 (2013)
4. Sheng, Z., Yang, S., Yu, Y., Vasilakos, A., Mccann, J., Leung, K.: A survey on the IETF protocol suite for the Internet of Things: standards, challenges, and opportunities. IEEE Wirel. Commun. **20**(6), 91–98 (2013)
5. Ning, H., Liu, H., Yang, L.: Cyberentity security in the Internet of Things. Computer **46**(4), 46–53 (2013)
6. Wesołowski, T.E., Porwik, P., Doroz, R.: Electronic health record security based on ensemble classification of keystroke dynamics. Appl. Artif. Intell. **30**(6), 521–540 (2016)
7. Tsai, C., Lai, C., Chiang, M., Yang, L.: Data mining for Internet of Things: a survey. Commun. Surv. Tutor. **99**, 1–21 (2013)
8. Gaura, E., Brusey, J., Allen, M., Wilkins, R., Goldsmith, D., Rednic, R.: Edge mining the Internet of Things. IEEE Sens. J. **13**(10), 3816–3825 (2013)
9. Keogh, E., Chakrabarti, K., Pazzani, M., et al.: Knowl. Inf. Syst. **3**, 263 (2001). doi:10.1007/PL00011669

10. Bishop, C.: Pattern Recognition and Machine Learning. Springer, Heidelberg (2006). doi:10.1007/978-1-4615-7566-5
11. Murakami, T., Asai, K., Yamazaki, E.: Vector quantiser of video signals. Electron. Lett. **18**, 1005–1006 (1981)
12. Lin, J., Keogh, E., Lonardi, S., Chiu, B.: A symbolic representation of time series, with implications for streaming algorithms. In: Proceedings of the ACM SIGMOD, San Diego, CA, USA, pp. 9–12 (2003)
13. Ji, S., Xue, Y., Carin, L.: Bayesian compressive sensing. IEEE Trans. Signal Process. **56**, 2346–2356 (2008)
14. Bello, J.: Measuring structural similarity in music. IEEE Trans. Audio Speech Lang. Process **19**, 2013–2025 (2011)
15. Schoellhammer, T., Greenstein, B., Osterweil, E., Wimbrow, M., Estrin, D.: Lightweight temporal compression of microclimate datasets. In: Proceedings of the 29th Annual IEEE International Conference on Local Computer Networks, Tampa, FL, USA, 16–18 November 2004, pp. 516–524 (2004)
16. Lin, J., Keogh, E., Lonardi, S., Patel, P.: Finding motifs in time series. In: Proceedings of the 2nd Workshop on Temporal Data Mining, Edmonton, Canada, 23–26 July 2002, pp. 53–68 (2002)
17. Agrawal, R., Faloutsos, C., Swami, A.: Efficient similarity search in sequence databases. In: Lomet, D.B. (ed.) FODO 1993. LNCS, vol. 730, pp. 69–84. Springer, Heidelberg (1993). doi:10.1007/3-540-57301-1_5
18. Hunter, J., McIntosh, N.: Knowledge-based event detection in complex time series data. In: Horn, W., Shahar, Y., Lindberg, G., Andreassen, S., Wyatt, J. (eds.) AIMDM 1999. LNCS, vol. 1620, pp. 271–280. Springer, Heidelberg (1999). doi:10.1007/3-540-48720-4_30
19. Kudłacik, P., Porwik, P., Wesołowski, T.: Fuzzy approach for intrusion detection based on user's commands. Soft Comput. **20**(7), 2705–2719 (2016)
20. Balasubramaniam, S., Kangasharju, J.: Realizing the internet of nano things: challenges, solutions, and applications. Soft. Comput. **46**(2), 62–68 (2013)
21. Danieletto, M., Bui, N., Zorzi, M.: A compression and classification solution for the Internet of Things. Sensors **14**(1), 68–94 (2014). doi:10.3390/s140100068
22. Bernas, M., Płaczek, B.: Period-aware local modelling and data selection for time series prediction. Expert Syst. Appl. **59**, 60–77 (2016)
23. Preacher, K., Curran, P., Bauer, D.: Computational tools for probing interactions in multiple linear regression, multilevel modeling, and latent curve analysis. J. Educ. Behav. Stat. **31**(4), 437–448 (2006)

A Privacy Preserving and Safety-Aware Semi-supervised Model for Dissecting Cancer Samples

P.S. Deepthi[1,2(✉)] and Sabu M. Thampi[1]

[1] Indian Institute of Information Technology and Management-Kerala,
Trivandrum 695581, Kerala, India
deepthisath@gmail.com
[2] University of Kerala, Trivandrum 695034, India

Abstract. Research in cancer genomics has proliferated with the advent of microarray technologies. These technologies facilitate monitoring of thousands of genes in parallel, thus providing insight into disease subtypes and gene functions. Gene expression data obtained from microarray chips are typified by few samples and a large number of genes. Supervised classifiers such as support vector machines (SVM) have been deployed for prediction task. However, insufficient labeled data have resulted in a paradigm shift to semi-supervised learning, in particular, transductive SVM (TSVM). Analysis of gene expression data using TSVM revealed that the performance of the model degenerates in the presence of unlabeled data. We address this issue by using a representative sampling strategy which ensures safety of the classifier even in the presence of unlabeled data. We also address the issue of privacy violation when classifier is shipped to other medical institutes for analysis of shared data. We propose a safety aware and privacy preserving TSVM for classifying cancer subtypes. Performance of TSVM with SVM and accuracy loss of the proposed TSVM are also analyzed.

Keywords: Gene expression · Transductive support vector machine · Safety · Privacy

1 Introduction

Bioinformatics technologies like microarrays help to increase knowledge on human genomic information and improve biomedical research. Gene expression profiling using microarrays has been found efficient in the prognosis of different cancer subtypes [10]. Data mining techniques such as clustering [5,10] and classification [7], [14] have been employed for disease outcome prediction and class discovery from gene expression datasets. A major challenge associated with these data is the "curse of dimensionality" attributed to the number of genes. Another issue associated with gene expression classification is the high cost of microarray technology which limits the availability of sufficient samples to make accurate

© Springer International Publishing AG 2017
N.T. Nguyen et al. (Eds.): ICCCI 2017, Part II, LNAI 10449, pp. 129–138, 2017.
DOI: 10.1007/978-3-319-67077-5_13

prediction [4]. Hence it would be useful if medical institutes can collaborate and carry out analysis on joint data. But these data contain sensitive information like gene markers, disease etc., which may reveal an individual's identity. Thus privacy is a major concern while mining such data.

Supervised learning techniques like SVM have been deployed to predict disease subtypes from gene expression data. However, the classifier built by SVM consists of some intact samples of the training data and it naturally violates privacy due to the presence of these instances. Shipping the classifier to other medical institutes or data mining sites will expose the content of some individuals present in the training set. As result, there is violation of patient privacy during the release of the classifier.

Considering the above problems, we propose a safety and privacy aware classification system for sample prediction. We improve the performance of TSVM using a representative sampling strategy to choose the most representative samples from unlabeled data. Further, the model undergoes a privacy-preserving transformation in which the support vectors in the training data are transformed. This ensures privacy and facilitates shipping of the classifier to other users or medical institutes without revealing the support vectors.

The paper advances with related works in Sect. 2. The methods and datasets used are detailed in Sect. 3. The experimental details and discussion are given in Sect. 4 and conclusions in Sect. 5.

2   Related Works

There have been several attempts to uncover disease subtypes from gene expression data using supervised and unsupervised techniques. While supervised methods require sufficient labeled data, unsupervised methods require informative gene selection, which is a difficult task with unlabeled data. The privacy issues related to sharing of patient information have led to privacy-preserving research in microarray analysis. In [2], the authors propose a distributed recursive feature elimination (RFE)-SVM for preserving privacy. Informative gene selection using RFE-SVM is carried out using secure multi-party computation involving miners at local sites. The two parties at local sites further send the encrypted score of RFE-SVM to a central data miner. In another work, privacy-preserving calculation of Fisher score for informative gene selection has been explored [6]. It follows a semi-honest model and uses ElGamal cryptographic system for computing scores. A PCA-based gene clustering is proposed in [11]. In this method, secure multiparty computation is used for calculating eigen vectors and transmitted to a trusted central site.

Traditional supervised classifiers are designed to work with abundant labeled data. However, a large number of samples that do not have clinical follow-up are disposed. This has resulted in the investigation of semi-supervised techniques for disease prognosis [1, 3, 15]. In [13], the authors have combined TSVM with gene selection to make accurate prediction using unlabeled data. But in many cases, incorporating unlabeled data degenerates the performance of semi-supervised learning techniques. This necessitates the need for safety-aware semi-supervised learning which ensures similar performance as that of supervised learning. Privacy is also a major concern while mining such data. In this work, we use a safety and privacy preserving TSVM model to predict cancer subtypes. A representative sampling strategy is devised to choose most representative samples from unlabeled data in order to improve the performance of TSVM. This model then undergoes a privacy-preserving transformation in which the support vectors in the training data are transformed. This learned classifier can be shared among other medical institutes without privacy violation.

3 Proposed Work

The proposed system selects prospective gene markers and subsequently incorporates safety and privacy into TSVM technique. Figure 1 shows the methodology of the system. Semi-supervised Silhouette filter is used to select important genes from the tissue samples. The partitioned data undergoes the safety-aware TSVM algorithm in which the basic SVM classification is used to iteratively predict and select the transductive samples. After each prediction, the representative sampling algorithm is used to find the most representative samples for updating the training set. The process is repeated for a number of times to get a more accurate classification. The final classifier model is saved and undergoes privacy-preserving transformation. The classifier performance using the transformed decision function is measured using true positive rate and compared to that using original decision function. The saved classifier is reloaded, and the test data that is not considered for the building of the classifier is used for prediction. The accuracy of the prediction is calculated.

3.1 Supervised Silhouette Filter

Supervised simplified Silhouette filter [8] selects features from an induced partition of the feature set considering the correlation between the features and class. The simplified Silhouette calculates the distance between features and medoids and a correlation measure called symmetric uncertainty which is a normalized version of information gain. The correlation measure, symmetric uncertainty (SU) is defined using information gain as

$$SU(X_j, X_j) = 2\left[IG(X_i|X_j)/H(X_i) + H(X_j)\right] \tag{1}$$

where IG is the information gain and H is the entropy The dissimilarity between two features is measured as

$$SU_{dis} = \frac{1 - SU(X_i, X_j) + |SU(X_i, C) - SU(X_j, C)|}{2} \tag{2}$$

A value of 0 indicates that features are completely correlated, and 1 means that they are independent. The medoid is chosen as the feature that is correlated with other features of the cluster and the class

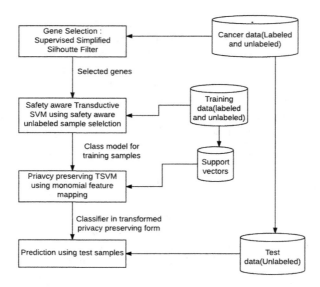

Fig. 1. Overall methodology

$$\eta_r = argmax \left\{ 1/2 \left[\sum_{X_j \epsilon C_r} SU(X_i, X_j)/(|C_r| - 1) + SU(X_i, C) \right] \right\} \quad (3)$$

where η_r is the medoid of the cluster.

$$argmax_{X_i \epsilon C_r} \left\{ (1 - SU(X_i, \eta_r) + SU(X_i, C))/2 \right\} \quad (4)$$

3.2 Proposed Safety-Aware and Privacy Preserving TSVM

TSVM is a semi-supervised version of SVM that takes into account unlabeled data for the learning process. We describe the transductive procedure used in incorporating unlabeled data into SVM. First, the available working set W(0) is used for training SVM. A representative sampling algorithm is used to choose best transductive samples on both sides of the hyperplane that separates the class, the positive and negative candidate sets are then identified. Thus, in the first iteration, a transductive set $T^{(i)}t = T^+ \bigcup T^-$ is created. Let $A^{(0)}t = \phi$. Further, $T^{(1)}t$ and the initial working set are combined for retraining the classifier, and the process is repeated. $A^{(1)}t$ and $T^{(2)}t$ are intersected to form the second resultant transductive set $D^{(2)}t$. Thus, the samples common between the first and the second set are contained in the resultant set. The first set is then taken off from the initial working set, while the second resultant set is incorporated into it. The training set thus obtained helps to obtain a more reliable discriminant hyperplane. Those samples whose labels have remained consistent throughout the process are retained in the final transductive set.

To make the training process safe, representative sampling is used during each iteration, as given in Algorithm 1. The algorithm finds several low-density separators among which those with large diversity are retained. Hence, a two-stage approach is deployed - first, search for several large-margin low-density separators and then, choose representative ones. A local search is made to identify different candidate large-margin low-density separators and k-means algorithm is applied to choose the representative ones. Algorithm 1 describes the steps involved. This algorithm ensures that the accuracy achieved will be at least that with only labeled data, and the performance degeneration due to incorporation of unlabeled samples can be mitigated. In order to ensure privacy, the classifier model used for making prediction is transformed to support vector independent decision function. We extend the monomial feature mapping proposed in [12] to make the TSVM model a privacy-preserving one. The classifier comprises of the kernel parameter g, the bias term b, support vectors $(SV1, y1); \dots; (SV_m, y_m)$ and their corresponding supports. Original decision function is as follows:

$$f(X) = \sum_{i=1}^{m'} \alpha_i y_i exp(-g \|SV_i - x\|^2) + b \tag{5}$$

Data: Set of labelled data x_i, y_i, set of unlabeled data $\hat{y}$
Result: Best candidate separator $\hat{y}_t^{best}$
begin
    Randomly sample N number of $\hat{y}$
    for *n=1 to N* **do**
        while *not converged* **do**
            Fix $\hat{y}n$, solve w_n, b_n using SVM solver
            Fix w_n, b_n
            Perform sorting to update $\hat{y}n$ w.r.t TSVM's objective function
        end
    end
    Apply k-means clustering algorithm for S where k=T.
    Select $\hat{y}$ with the minimum objective value within each cluster
end

Algorithm 1. Proposed Representative Sampling Algorithm

The transformed decision function is of the form:

$$\sum_{i=1}^{m'} c_i + \sum_{d=1}^{n} \phi_d(x) \left((2g)^d/d! \sum_{i=1}^{m'} c_i \phi_d(SV_i) \right) \tag{6}$$

$$= w_0 + \sum_{d=1}^{\infty} \phi_d(x) \cdot w_d \tag{7}$$

where

$$c_i = \alpha_i y_i exp(-g \|SV_i\|^2) \tag{8}$$

for $i = 1$ to m'

$$f(x) = exp(-g \|x\|^2) \sum_{i=1}^{m'} c_i exp(2gSV_i.x) + b \tag{9}$$

On substitution, the decision function becomes

$$f(x) = exp(-g \|x\|^2)(w_0 + \sum_{d=1}^{\infty} \Phi_d(x).w_d) + b \tag{10}$$

In this transformed decision function, support vectors have been avoided, and instead linear combination of monomial features is given. Hence, the new decision function ensures privacy by hiding support vectors, thus facilitating shipping of the classifier model to other medical institutes for joint analysis of data.

3.3 Experimental Setup

The publicly available datasets leukemia, small round blood cell tumors (SRBCT), diffuse large B-cell lymphomas (DLBCL)and mixed-lineage leukemias (MLL) are used for experiments [9]. The particulars of the datasets are given in Table 1. TSVM classification is experimented using the data sets leukemia, DLBCL, MLL, and SRBCT. The training data set is selected as 20%, 30% and 40% for each data set, for each run. The basic normalization algorithm is used to scale the data between the range $[-1, +1]$. The Gaussian kernel function is used for conducting experiments. The experiments were carried out in different stages. First, we carried out experiments with SVM on selected genes and compared the accuracy with TSVM. In the second stage, we employed the proposed

Table 1. Datasets

Dataset	#Subtypes
Leukemia (72 samples, 5147 genes)	Acute lymphoblastic leukemia (ALL) - 47 samples
	Acute myeloid leukemia (AML) - 25 samples.
SRBCT (83 samples, 2308 genes)	Ewings sarcoma (EWS) - 29 samples
	Neuroblastoma (NB) - 18 samples
	Burkitt's lymphoma (BL) - 11 samples
	Rhabdomyo-sarcoma (RMS) - 25 samples
MLL (72 samples, 12533 genes)	ALL - 24 samples
	MLL - 20 samples
	AML - 28 samples
DLBCL (77 samples, 7070 genes)	Diffuse large B-cell lymphomas (DLBCL) - 58 samples
	Follicular lymphoma (FL) - 19 samples

safety algorithm and investigated the performance improvement over TSVM and SVM. In the final stage, we incorporated privacy into the safety-aware TSVM and carried out the experiments.

4 Results and Discussion

The proposed classifier was implemented in Java and experiments were carried out on a PC with Intel Core i3-4010U Processor, 500 GB hard disk, and 4 GB RAM. Semi-supervised Silhouette filter has been used to identify gene markers that can dissect the tissue sample classes. It helps to eliminate redundant features and selects the relevant ones that can discriminate the classes using a medoid-based approach. The details of these gene markers are listed in Table 2.

Table 2. Gene markers

Dataset	Gene no.	Gene id
Leukemia	1336	M23197_at
	1375	M27891_at
	3545	X95735_at
	4969	M31523_at
SRBCT	187	296448
	1389	770394
	1434	784257
	1954	814260
MLL	11	31317_r_at
	2543	36239_at
	7310	41225_at
DLBCL	447	D55716_at
	913	HG4074-HT4344_at
	2598	X16983_at t
	4135	U06863_at

The average accuracy and standard deviation of the datasets for SVM, TSVM, and privacy-preserving TSVM are given in Table 3. During the experiments, we found that there is a performance degeneration of TSVM over DLBCL data with an accuracy of only 62%. This is due to the presence of unlabeled data and wrong selection of transductive samples. Hence, we applied safety algorithm and accuracy has been increased to 76% which is better than that with SVM. The representative sampling algorithm selects discriminant samples iteratively, thus improving the accuracy of prediction.

Figure 2(a) depicts the maximum accuracy achieved through SVM and TSVM. For all datasets except DLBCL, TSVM outperforms SVM. Figure 2(b) gives the accuracy of SVM, TSVM and privacy preserving TSVM.

Table 3. Results

Dataset	Test set	Training set (%)	SVM	TSVM	Privacy + TSVM
Leukemia	36	20	62.32 ± 0.45	90.75 ± 0.72	81.53 ± 0.26
		30	62.53 ± 0.68	90.80 ± 0.42	82.32 ± 0.68
		40	63.12 ± 0.74	90.92 ± 0.38	82.53 ± 0.72
SRBCT	41	20	72.12 ± 0.46	94.65 ± 0.72	84.34 ± 0.66
		30	72.62 ± 0.74	94.23 ± 0.72	84.12 ± 0.82
		40	72.76 ± 0.88	94.75 ± 0.92	84.53 ± 0.98
MLL	36	20	70.04 ± 0.68	92.15 ± 0.72	82.34 ± 0.62
		30	70.36 ± 0.72	92.75 ± 0.94	82.53 ± 0.76
		40	70.64 ± 0.86	92.88 ± 0.72	82.86 ± 0.86
DLBCL	38	20	68.23 ± 0.24	62.75 ± 0.72	60.53 ± 0.66
		30	68.64 ± 0.56	62.88 ± 0.96	60.68 ± 0.72
		40	68.88 ± 0.96	62.92 ± 0.84	60.78 ± 0.80

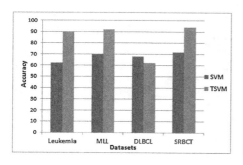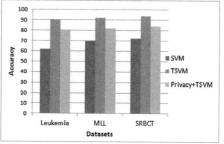

Fig. 2. (a) Comparison of SVM and TSVM (b) comparison of SVM, TSVM and privacy preserving TSVM

Figure 3(a) displays the accuracy of SVM, TSVM, and safety TSVM on DLBCL dataset. The accuracy of DLBCL has been improved to 76%. Figure 3(b) gives the results of safety SVM and privacy-preserving safety TSVM. Due to the transformation, there is a slight drop in accuracy, still yielding better results than SVM and TSVM.

Figure 4(a) shows the comparison of TSVM with privacy-preserving safety TSVM. Figure 4(b) depicts comparison of SVM, safety SVM and privacy-preserving safety TSVM. It is evident that the incorporation of privacy does not result in much performance degradation.

From the experiments, we found that the presence of unlabeled data results in wrong sample selection, thus causing a drop in accuracy of the TSVM classifier. The proposed iterative sampling strategy helps to select more suitable samples, and hence accuracy has been increased. The privacy-preserving transformation changes the decision function into another form by hiding support vectors. But

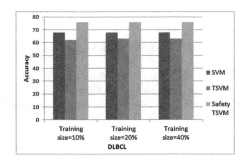
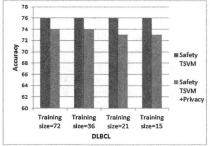

Fig. 3. (a) Accuracy of SVM, TSVM, safety SVM on DLBCL (b) safety SVM and privacy preserving safety SVM

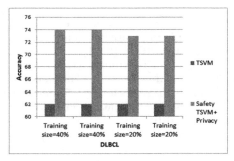
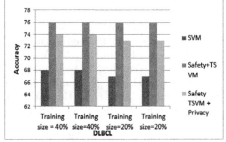

Fig. 4. (a) TSVM and privacy preserving TSVM on DLBCL (b) SVM, safety SVM and privacy preserving safety SVM on DLBCL

this transformation does not affect the performance of the classifier much and we attained accuracy comparable with that of the original decision function of TSVM and SVM.

5 Conclusions

An iterative transductive sample selection based on the representative sampling algorithm and privacy preserving classifier for dissecting cancer subtypes from gene expression data is proposed. Experimental results indicate that the proposed semi-supervised model ensures guaranteed accuracy compared to TSVM and SVM. The privacy violation in the case of support vectors in the classifier is solved using a transformation without much performance degradation. The privacy-preserving classifier ensures data to be shared among medical institutes for joint analytics without affecting patient privacy.

Acknowledgments. The research was financially supported by Department of Information Technology, Government of Kerala and the facilities were provided by Indian Institute of Information Technology and Management - Kerala.

References

1. Bair, E., Tibshirani, R.: Semi-supervised methods to predict patient survival from gene expression data. PLoS Biol. **2**, E108 (2004)
2. Camara, F., Samb, M.L., Ndiaye, S., Slimani, Y.: Privacy preserving RFE-SVM for distributed gene selection. Int. J. Comput. Sci. Issues 154–159 (2012)
3. Deepthi, P.S., Thampi, S.M.: Predicting cancer subtypes from microarray data using semi-supervised fuzzy C-means algorithm. J. Intell. Fuzzy Syst. **32**(4), 2797–2805 (2017)
4. Ein-Dor, L., Zuk, O., Domany, E.: Thousands of samples are needed to generate a robust gene list for predicting outcome in cancer. Proc. Natl. Acad. Sci. **103**(15), 5923–5928 (2006)
5. Eisen, M.B., Spellman, P.T., Brown, P.O., Botstein, D.: Cluster analysis and display of genome-wide expression patterns. Natl. Acad. Sci. **95**(25), 14863–14868 (1998)
6. Guo, S., Zhong, S., Zhang, A.: Privacy preserving calculation of fisher criterion score for informative gene selection. In: 2014 IEEE International Conference on Bioinformatics and Bioengineering (BIBE), pp. 90–96. IEEE (2014)
7. Haferlach, T., Kohlmann, A., Wieczorek, L., Basso, G., Te Kronnie, G., Bn, M.C., De Vos, J., Hernndez, J.M., Hofmann, W.K., Mills, K.I., Gilkes, A.: Clinical utility of microarray-based gene expression profiling in the diagnosis and subclassifcation of leukemia: report from the International Microarray Innovations in Leukemia Study Group. J. Clin. Oncol. **28**(15), 2529–2537 (2010)
8. Hruschka, E.R., Covoes, T.F.: Feature selection for cluster analysis: an approach based on the simplified Silhouette criterion. In: 2005 International Conference on Intelligent Agents, Web Technologies and Internet Commerce, pp. 32–38 (2005)
9. http://orange.biolab.si.datasets.psp. Accessed 6 June 2014
10. Jiang, D., Tang, C., Zhang, A.: Cluster analysis for gene expression data: a survey. IEEE Tran. Knowl. Data Eng. **16**, 1370–1386 (2004)
11. Li, X.: Privacy preserving clustering for distributed homogeneous gene expression data sets. In: Innovations in Data Methodologies and Computational Algorithms for Medical Applications, pp. 184–207. IGI Global (2012)
12. Lin, K.P., Chen, M.S.: On the design and analysis of the privacy-preserving SVM classifier. IEEE Trans. Knowl. Data Eng. **23**(11), 1704–1717 (2011)
13. Maulik, U., Mukhopadhyay, A., Chakraborty, D.: Gene-expression-based cancer subtypes prediction through feature selection and transductive SVM. IEEE Trans. Biomed. Eng. **60**(4), 1111–1117 (2013)
14. Salazar, R., Roepman, P., Capella, G., Moreno, V., Simon, I., Dreezen, C., Lopez-Doriga, A., Santos, C., Marijnen, C., Westerga, J., Bruin, S.: Gene expression signature to improve prognosis prediction of stage II and III colorectal cancer. J. Clin. Oncol. **29**, 17–24 (2010)
15. Zhang, X., Guan, N., Jia, Z., Qiu, X., Luo, Z.: Semi-supervised projective nonnegative matrix factorization for cancer classification. PloS ONE **10**(9), e0138814 (2015)

Decision Fusion Methods in a Dispersed Decision System - A Comparison on Medical Data

Małgorzata Przybyła-Kasperek[(✉)], Agnieszka Nowak-Brzezińska,
and Roman Simiński

Institute of Computer Science, University of Silesia,
Będzińska 39, 41-200 Sosnowiec, Poland
{malgorzata.przybyla-kasperek,agnieszka.nowak,roman.siminski}@us.edu.pl

Abstract. A dispersed decision-making system that use knowledge accumulated in separate knowledge bases is considered in this paper. The system has complex and dynamic structure. In previous papers, different fusion methods were applied in this system. The novelty of this paper is to examine an approach in which many different fusion methods from the rank or from the measurement level are used simultaneously in one decision-making process. The obtained results were compared with the results obtained when fusion methods are used individually. In addition, the results were compared with an approach in which a dispersed system is not used, but the classifications generated based on each local base are directly aggregated using fusion methods. The conclusions indicated the legitimacy of the use of a dispersed system.

Keywords: Dispersed system · Fusion methods · Combining classifiers

1 Introduction

Methods of inference were created in order to be used when we have a single knowledge base and when we want to make a decision. Examples of inference systems can be found in the papers [7,13]. Over time, a more complex problem arose that involved the simultaneous use of various methods of inference in order to improve the efficiency of inference. Moreover, the approach of dividing a single knowledge base with respect to features or objects and then inferring based on these smaller bases in order to achieve greater efficiency was considered [1,3]. When ensembles of classifiers are considered, different fusion methods are used in order to combine the predictions of the base classifiers. There are different types of fusion methods. Some of them belong to the rank level or to the measurement level [2,4–6].

In this paper a set of classifiers is considered. It is assumed that dispersed knowledge is given in advance - data dispersion is not a part of system's work. The form of local knowledge bases may be different, there are no constraints imposed on the sets of attributes or the sets of objects. This issue has been considered by one of the authors in earlier papers. In the paper [8], a dispersed

© Springer International Publishing AG 2017
N.T. Nguyen et al. (Eds.): ICCCI 2017, Part II, LNAI 10449, pp. 139–149, 2017.
DOI: 10.1007/978-3-319-67077-5_14

system with a dynamic structure has been proposed. This system reflects well the relationships between classifiers in clusters. In this paper this system is used. The use of various fusion methods in this system was analyzed in the papers [9–11]. However, these methods were applied separately. In this article an approach in which many methods are used simultaneously in order to generate decisions is considered. Such approach may improve average and poor results. Moreover, the decisions that are confirmed by a group of fusion methods can be considered as more reliable. An important aspect of this paper is also the comparison of these results with the results obtained without the use of the dispersed system. Decisions are made by using fusion methods directly to vectors that were generated based on local knowledge bases.

The paper is organized as follows. Section 2, briefly describes the dispersed decision-making system. Section 3 section describes fusion methods and an approach that is used in this paper. The Sec. 4 shows a description and the results of experiments carried out using two medical data sets from the UCI Repository: the Lymphography and the Primary Tumor data set are used. Finally, a short summary is presented in Sect. 5.

2   Description of Dispersed System with Negotiations

A system that was proposed in the article [8] is used in this paper. Below only short description of the system is included. As was mentioned earlier, knowledge in a dispersed form is provided to the system, i.e. knowledge accumulated in a set of local decision tables. Local tables that are considered must have the same decision attribute. A resource agent ag manages one local decision table $D_{ag} = (U_{ag}, A_{ag}, d_{ag})$. Agents who make decisions for a test object in a similar way are connected into clusters. At first, classification that is made by each agent is stored in a vector of probabilities and then in a vector of ranks. The vectors are defined using the modified m_1-nearest neighbors classifier. Some relations between agents are defined based on these vectors. Then, initial clusters are defined. These clusters contain agents in friendship relation. At the end, the negotiation process is realized in order to define final clusters.

For each cluster a superordinate agent is defined. This agent is called a synthesis agent as. The synthesis agent has access to aggregated knowledge, which was created by the combination of decision tables of resource agents belonging to one cluster. Based on each aggregated decision table a vector of probabilities or a vector of ranks is generated. This is accomplished by using the modified m_3-nearest neighbors classifier. Global decisions are generated using these vectors and fusion methods.

3   Selected Fusion Methods and Simultaneous Use of These Methods

Fusion methods that are known from the literature can be divided into three groups - the abstract level, the rank level and the measurement level. Assigning a method to one of the above groups depends on the result generated by classifiers used in this method. In the first group, each classifier generates only one decision class. In the second group, classifiers generate vectors of ranks associated to decision classes according to the probability that the decision class is correct. In the third group, classifiers generate vectors of probabilities. In this paper methods from the rank level and from the measurement level are considered. These methods were used separately in earlier articles. In the paper [9], two methods from the rank level - the Borda count method and the highest rank method - were used. In the paper [11], two methods from the rank level - the intersection method and the union method - were considered and two methods from the measurement level - the product rule and the weighted average method - were used. In the paper [10], seven methods from the measurement level - the maximum rule, the minimum rule, the median rule, the sum rule, the probabilistic product method, the method that is based on the theory of evidence and the method that is based on decision templates - were used. Based on the experiments, it was found that the probabilistic product method, the method that is based on the theory of evidence and the method that is based on decision templates did not provide satisfactory results. Therefore, in this paper these methods will not be used.

Based on the comparisons contained in the articles mentioned above, it could not be said that one method always gives better results than other methods. For different data sets different methods were appropriate. Therefore, without the previous analysis of all methods, the best method for a given set of data can not be determined. In this paper, the fusion methods are used simultaneously to generate a common decision. This is implemented as follows. First, each fusion method generates decisions. Then the final decisions are made by aggregating the results obtained by individual methods with using the simple voting method. In this way, the decisions will be confirmed by several fusion methods and probably the impact of the worst methods will be reduced. It is expected to receive quite good results without prior analysis of all available fusion methods. In addition, in this study an approach is used in which the dispersed system was not applied. In this approach, based on each local decision table a vector of probabilities or a vector of ranks was generated by using the modified m_3-nearest neighbors classifier. Then a group of fusion methods was applied in order to make decisions. Such an approach is considered to compare the results obtained using the dispersed system and to test the need for application of this complex system. Below a brief description of the methods that are used and an example of their application are given.

In the methods from the rank level, it is assumed that, based on each aggregated decision table, a vector of ranks $[r_{j,1}(x), \ldots, r_{j,c}(x)]$, j-th cluster and c is the number of the decision classes, is generated. The following methods are used.

The Borda Count

In this method for each decision class the sum of the number of classes ranked below it is calculated. Then the decisions that have the maximum value of the Borda count are selected.

The Intersection Method

In this method, at first the training process is realized. This stage aims to establish threshold value for each synthesis agent. For this purpose, the resource agents classify the training objects. Then, the lowest rank ever given by the resource agent to any true class is determined. The threshold for the synthesis agent is equal to the minimum of the thresholds for the resource agents belonging to the subordinate cluster. For the test object and for each synthesis agent a set of decisions that ranks are higher or equal to the threshold is defined. Then the intersection of these sets is calculated.

The Highest Rank Method

In this method, for each decision class the value $\min_j r_{j,i}(x)$ is calculated. Then decisions that have the minimum of these values are selected.

The Union Method

In this method, at first the training process is realized. The threshold value for resource agents is determined using the min-max procedure, which is described in detail in the paper [12]. The threshold for the synthesis agent is equal to the minimum of the thresholds for the resource agents belonging to the subordinate cluster. For the test object and for each synthesis agent a set of decisions that ranks are higher or equal to the threshold is defined. Then the union of these sets is calculated.

The following methods from the measurement level are used. It is assumed that, based on each aggregated decision table, a vector of probabilities $[\mu_{j,1}(x), \ldots, \mu_{j,c}(x)]$, j-th cluster and c is the number of the decision classes, is generated.

The Maximum Rule, the Minimum Rule and the Median Rule

In these three methods, for each decision class the maximum or the minimum or the median value of the probability values assigned to the class by each synthesis agent is determined. Then the decisions that have the maximum of these values are selected.

The Sum Rule and the Product Rule

In these two methods, for each decision class the sum or the product of the probability values assigned to the class by each synthesis agent is determined. Then the decisions that have the maximum of these values are selected.

The Weighted Average Method

In this method, at first the training process is realized. In this step, the error rate for each resource agent is determined based on the training set. Then the error rate for each synthesis agent e_{as_j} is calculated as the average of the error rates of the resource agents belonging to the subordinate cluster. Then, for each decision class v_i, the value is calculated according to the formula $\sum_{j=1}^{L} \omega_j \mu_{j,i}(x)$, where $\omega_j \propto \frac{1-e_{as_j}}{e_{as_j}}$. Finally, the decisions that have the maximum of these values are selected.

As was mentioned earlier, in this paper the fusion methods are used simultaneously. It was assumed that the methods from one level - rank or measurement - are used simultaneously. Thus, in the case of the rank level, four methods are implemented and the final decisions are made using the simple voting method. In the case of the measurement level, the situation is analogous, except that six methods from this level are used simultaneously. Let us consider the following example.

Example 1. Consider a dispersed decision-making system in which there are four resource agents $\{ag_1, ag_2, ag_3, ag_4\}$ and two synthesis agents were generated for the test object x. The agent as_1 is superordinate to the cluster $\{ag_1, ag_2, ag_3\}$ and the agent as_2 is superordinate to the cluster $\{ag_1, ag_3, ag_4\}$. The results of the predictions for the synthesis agents are equal to: for as_1 – the rank level $[1, 2, 3, 2]$, the measurement level $[0.55, 0.5, 0.3, 0.5]$; for as_2 – the rank level $[2, 1, 2, 3]$, the measurement level $[0.45, 0.5, 0.45, 0.2]$.

In the Borda count method the sum of the number of classes that were ranked below class: v_1 is equal to $(4-2) + (4-2) = 4$; v_2 is equal to $(4-1) + (4-1) = 6$; v_3 is equal to $(4 - 3) + (4 - 2) = 3$; v_4 is equal to $(4 - 2) + (4 - 3) = 3$. Thus, the final decision is v_2. In the intersection method at first, the resource agents classify the training objects. Let us assume that there are 5 training objects and ranks given to their correct classes are given in Table 1. Then the threshold values for the synthesis agents are calculated. The threshold values are equal respectively: 2 for as_1 and 2 for as_2. Then the intersection of the sets of decisions meeting the thresholds is calculated, i.e. $\{v_1, v_2, v_4\} \cap \{v_1, v_2, v_3\} = \{v_1, v_2\}$. In the highest rank method, for each decision class the highest rank is assigned. It is equal respectively: 1 for v_1, 1 for v_2, 2 for v_3 and 2 for v_4. Thus, the set of decisions is equal to $\{v_1, v_2\}$. In the union method, at first, the threshold value for each synthesis agent is calculated. The max-min procedure starts with the matrix given in Table 1. The minimum rank of each row is selected first. Only the minimum value is left in the rows of the new matrix, while the remaining cells are set to 0 (Table 2). The maximum value from the column is the resource agent threshold. Then the threshold values for the synthesis agents are calculated. It is equal to 1 for both agents. Then the union of the sets of decisions meeting the thresholds is calculated, i.e. $\{v_1\} \cup \{v_2\} = \{v_1, v_2\}$. Realizations of the maximum rule, the minimum rule, the median rule, the sum rule and the product rule are shown in Table 3. In the weighted average method the error rates of the resource agents are calculated first. As can be seen from Table 1 this is equal respectively: 0.2 for ag_1, 0.4 for ag_2, 0.2 for ag_3 and 0.6 for ag_4. Then, the error rates of the synthesis agents are calculated: 0.267 for as_1 and 0.333 for as_2. The weights of the synthesis agents are calculated: $\omega_1 = 0.579$, $\omega_2 = 0.421$. Finally, for each decision class, the value of the weighted average is calculated: 0.508 for v_1, 0.5 for v_2, 0.363 for v_3 and 0.374 for v_4. Thus, the set of decisions is equal to $\{v_1\}$.

Considering an approach in which all methods from the rank level are used simultaneously, decision v_2 is made (we have three votes for v_1 and four votes for v_2). Let us note that although most methods gave ambiguous answers, the set of methods made an unequivocal decision. Considering an approach in which all

Table 1. Ranks given to the correct classes of the training objects

Training object	ag_1	ag_2	ag_3	ag_4
x_1	1	1	1	1
x_2	1	2	2	2
x_3	1	1	1	1
x_4	1	1	1	2
x_5	3	2	1	3
Max	3	2	2	3

Table 2. Max-min procedure

Training object	ag_1	ag_2	ag_3	ag_4
x_1	1	1	1	1
x_2	1	0	0	0
x_3	1	1	1	1
x_4	1	1	1	0
x_5	0	0	1	0
Max	1	1	1	1

Table 3. Results for the maximum rule, the minimum rule, the median rule, the sum rule and the product rule

Method	v_1	v_2	v_3	v_4	Decisions made
Maximum rule	0.55	0.5	0.45	0.5	$\{v_1\}$
Minimum rule	0.45	0.5	0.3	0.2	$\{v_2\}$
Median rule	0.5	0.5	0.375	0.35	$\{v_1, v_2\}$
Sum rule	1	1	0.75	0.7	$\{v_1, v_2\}$
Product rule	0.2475	0.25	0.135	0.1	$\{v_2\}$

methods from the measurement level are used simultaneously, decisions v_1 and v_2 are made (we have four votes for v_1 and four votes for v_2). This situation is unfortunate because the set of methods did not generate an unequivocal answer, although most methods generated one decision.

As can be seen different results can be generated by a group of fusion methods. Therefore, it should be checked experimentally what results will be received.

4   Experimental Study

In the experiments, two medical data sets were used: the Lymphography data set and the Primary Tumor data set. These data sets are available in the UCI

Table 4. Approach of using several fusion methods in one decision-making process. Designations: $1 - WSD_{Ag1}^{dyn}$, $2 - WSD_{Ag2}^{dyn}$, $3 - WSD_{Ag3}^{dyn}$, $4 - WSD_{Ag4}^{dyn}$, $5 - WSD_{Ag5}^{dyn}$

System	Methods from the rank level		Methods from the measurement level	
	$m_1/p/m_2/m_3/u$	$e/e_{ONE}/\overline{d}_{WSD_{Ag}^{dyn}}/t$	$m_1/p/m_2/m_3$	$e/e_{ONE}/\overline{d}_{WSD_{Ag}^{dyn}}/t$
Lymphography data set				
1	2/0.05/1/2/0.01	0.068/0.227/1.159/0.01	2/0.05/1/2	0.136/0.136/1/0.01
2	1/0.05/2/8/0.01	0.091/0.295/1.205/0.01	2/0.05/7/6	0.136/0.136/1/0.01
3	17/0.05/3/9/0.02	0.114/0.182/1.068/0.01	15/0.05/5/8	0.114/0.136/1.023/0.01
4	13/0.05/4/5/0.03	0.136/0.273/1.159/0.01	17/0.05/6/10	0.136/0.159/1.023/0.02
5	10/0.05/3/1/0.01	0.182/0.500/1.318/0.13	1/0.05/9/9	0.205/0.250/1.045/0.15
Primary Tumor data set				
1	1/0.15/3/1/0.03	0.373/0.784/2.627/0.02	1/0.15/8/1	0.382/0.794/2.627/0.03
2	2/0.15/1/2/0.01	0.324/0.824/3.245/0.03	2/0.1/3/1	0.333/0.794/3.196/0.02
3	4/0.05/1/5/0.03	0.353/0.873/3.843/0.07	2/0.05/1/1	0.333/0.882/3.980/0.03
4	3/0.15/1/2/0.01	0.324/0.892/3.912/0.08	2/0.2/1/3	0.324/0.892/3.922/0.06
5	1/0.05/3/2/0.01	0.304/0.912/4.010/1.10	1/0.05/3/2	0.304/0.912/4.010/1.07

repository (archive.ics.uci.edu/ml/). Each data set was divided into a training set and a test set. The characteristics are as follows: Lymphography: # The training set - 104; # The test set - 44; # Conditional - 18; # Decision - 4; Primary Tumor: # The training set - 237; # The test set - 102; # Conditional - 17; # Decision - 22. The training sets were dispersed. Different divisions were considered - with 3, 5, 7, 9 and 11 resource agents (decision tables). The following designations are used for these systems: WSD_{Ag1}^{dyn} - 3 resource agents; WSD_{Ag2}^{dyn} - 5 resource agents; WSD_{Ag3}^{dyn} - 7 resource agents; WSD_{Ag4}^{dyn} - 9 resource agents; WSD_{Ag5}^{dyn} - 11 resource agents. For more details on data preparation, see [11]. Measures to determine the quality of classification are as follows: *estimator of classification error e* - if the correct decision of an object belongs to the set of global decisions generated by the system, then the object is properly classified; *estimator of classification ambiguity error* e_{ONE} - if only one, correct decision was generated to an object, then the object is properly classified; *the average size of the global decisions sets* $\overline{d}_{WSD_{Ag}^{dyn}}$ generated for a test set. There are some parameters in a dispersed system. For a detailed description of these parameters, see [8]. Below only designations and optimization process has been described. The following designations were used for parameters: m_1, p - parameters which are used in the process of generating system's structure; m_2 - parameter which is used in the process of generating aggregated decision tables; m_3 - parameter which is used in the process of generating vectors of ranks or vectors of probabilities. The main results presented in this paper concern the simultaneous use of fusion methods from the rank or the measurement level in a dispersed decision system. In this approach, the parameters were optimized as follows. Tests for different parameter values were carried out: $m_1, m_2, m_3 \in \{1, \ldots, 20\}$,

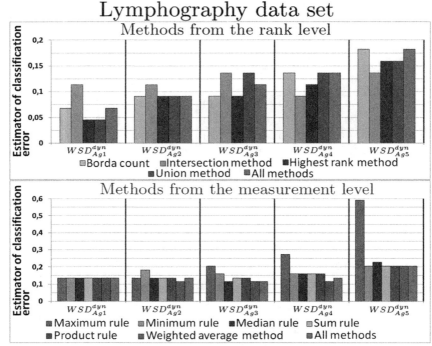

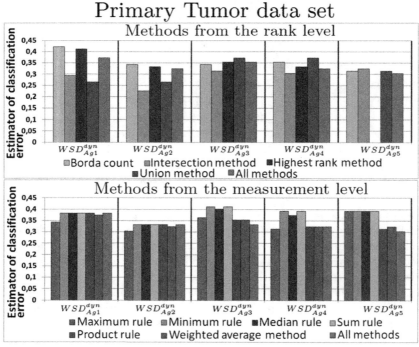

Fig. 1. Comparison of the results

and $p \in \{0.05, 0.1, 0.15, 0.2\}$. From all of the obtained results, one was selected that guaranteed a minimum value of estimator of classification error (e), while maintaining the smallest possible value of the average size of the global decisions sets $(\overline{d}_{WSD_{Ag}^{dyn}})$. In tables presented below the best results, obtained for optimal values of the parameters, are given. In the Table 4 the following information is given: the name of dispersed decision-making system (System); the selected, optimal parameter values (Parameters); the three measures discussed earlier e, e_{ONE}, $\overline{d}_{WSD_{Ag}^{dyn}}$; the time t needed to analyze a test set expressed in minutes.

Table 5. Approach of using several fusion methods in one decision-making process – without a dispersed decision-making system

System	Methods from the rank level		Methods from the measurement level	
	m_3/u	$e/e_{ONE}/\overline{d}_{WSD_{Ag}^{dyn}}/t$	m_3	$e/e_{ONE}/\overline{d}_{WSD_{Ag}^{dyn}}/t$
Lymphography data set				
WSD_{Ag1}^{dyn}	1/0.01	0.091/0.295/1.205/0.01	10	0.136/0.136/1/0.01
WSD_{Ag2}^{dyn}	1/0.01	0.136/0.227/1.091/0.01	3	0.182/0.182/1/0.01
WSD_{Ag3}^{dyn}	9/0.02	0.159/0.273/1.136/0.01	5	0.159/0.182/1.045/0.01
WSD_{Ag4}^{dyn}	1/0.02	0.159/0.341/1.182/0.01	1	0.159/0.318/1.159/0.01
WSD_{Ag5}^{dyn}	1/0.02	0.205/0.523/1.341/0.01	1	0.227/0.477/1.250/0.01
Primary Tumor data set				
WSD_{Ag1}^{dyn}	1/0.02	0.412/0.824/2.667/0.01	1	0.402/0.784/2.588/0.01
WSD_{Ag2}^{dyn}	1/0.01	0.353/0.824/3.206/0.01	1	0.324/0.804/3.235/0.01
WSD_{Ag3}^{dyn}	1/0.02	0.373/0.902/3.902/0.01	1	0.353/0.882/3.873/0.01
WSD_{Ag4}^{dyn}	1/0.01	0.333/0.892/3.853/0.01	1	0.333/0.892/3.843/0.01
WSD_{Ag5}^{dyn}	1/0.04	0.284/0.902/4.510/0.01	1	0.294/0.912/4.520/0.01

As can be seen, in most cases, for the fusion methods from the rank level better results were generated than for the fusion methods from the measurement level. A comparison of the results for individual fusion methods (results from the papers [9–11]) and the results obtained when several methods are used simultaneously is presented in Fig. 1. As can be observed, an approach in which several methods are used simultaneously in one decision-making process does not generate the best results. However, it guarantees a certain averaging of the results of individual fusion methods and an improvement of the worst results. Therefore, the application of this approach is a good option in a situation where we can not test all possible fusion methods and choose the best method for a given set of data. The use of multiple methods at the same time ensures that we get better results than for fusion method, which will not be suitable for a given set of data.

The proposed approach of using several fusion methods simultaneously in a dispersed system has been compared with an approach in which dispersed system is not used. In this case, only one parameter m_3 is used to generate vectors of ranks or vectors of probabilities. The vectors generated based on local decision tables are directly aggregated using fusion methods. Table 5 shows the results for the considered sets of data. As can be seen in almost every case, the approach without the use of a dispersed system produces worse results.

5   Conclusions

In this article, an approach in which several fusion methods are used simultaneously in one decision-making process was considered. Dispersed medical data were used in the experiments: the Lymphography data set and the Primary Tumor data set. The results obtained using this approach were compared with the results obtained when only one fusion method is used. It has been noted that this approach improves the worst results of methods when they are used separately. It has also been verified that without using a dispersed system this approach generates much worse results.

References

1. Bazan, J.G.: Hierarchical classifiers for complex spatio-temporal concepts. In: Peters, J.F., Skowron, A., Rybiński, H. (eds.) Transactions on Rough Sets IX: Journal Subline. LNCS, vol. 5390, pp. 474–750. Springer, Heidelberg (2008). doi:10. 1007/978-3-540-89876-4_26
2. Gatnar, E.: Multiple-Model Approach to Classification and Regression. PWN, Warsaw (2008). (in Polish)
3. Gągolewski, M.: Data fusion: theory, methods, and applications. Institute of Computer Science Polish Academy of Sciences (2015)
4. Kittler, J., Hatef, M., Duin, R.P.W., Matas, J.: On combining classifiers. IEEE Trans. Pattern Anal. Mach. Intell. **20**(3), 226–239 (1998)
5. Kuncheva, L., Bezdek, J.C., Duin, R.P.W.: Decision templates for multiple classifier fusion: an experimental comparison. Pattern Recogn. **34**(2), 299–314 (2001)
6. Kuncheva, L.: Combining Pattern Classifiers Methods and Algorithms. Wiley, Hoboken (2004)
7. Nowak-Brzezińska, A.: Mining rule-based knowledge bases inspired by rough set theory. Fundamenta Informaticae **148**(1–2), 35–50 (2016)
8. Przybyła-Kasperek, M., Wakulicz-Deja, A.: A dispersed decision-making system - the use of negotiations during the dynamic generation of a systems structure. Inf. Sci. **288**, 194–219 (2014)
9. Przybyła-Kasperek, M.: Comparison of four methods of combining classifiers on the basis of dispersed medical data. In: Czarnowski, I., Caballero, A.M., Howlett, R.J., Jain, L.C. (eds.) IDT 2016. SIST, vol. 57, pp. 3–13. Springer, Cham (2016). doi:10.1007/978-3-319-39627-9_1
10. Przybyła-Kasperek, M.: Dispersed decision-making system with selected fusion methods from the measurement level - case study with medical data. Ann. Comput. Sci. Inf. Syst. **8**, 129–136 (2016)

11. Przybyła-Kasperek, M., Nowak-Brzezińska, A.: Intersection method, union method, product rule and weighted average method in a dispersed decision-making system - a comparative study on medical data. In: Nguyen, N.-T., Manolopoulos, Y., Iliadis, L., Trawiński, B. (eds.) ICCCI 2016. LNCS, vol. 9876, pp. 451–461. Springer, Cham (2016). doi:10.1007/978-3-319-45246-3_43
12. Przybyła-Kasperek, M., Wakulicz-Deja, A.: Comparison of fusion methods from the abstract level and the rank level in a dispersed decision-making system. Int. J. General Systems **46**(4), 386–413 (2017)
13. Simiński, R.: Multivariate approach to modularization of the rule knowledge bases. In: Gruca, A., Brachman, A., Kozielski, S., Czachórski, T. (eds.) MMI 4. AISC, vol. 391, pp. 473–483. Springer, Cham (2016). doi:10.1007/978-3-319-23437-3_40

Knowledge Exploration in Medical Rule-Based Knowledge Bases

Agnieszka Nowak-Brzezińska[(✉)], Tomasz Rybotycki, Roman Simiński,
and Małgorzata Przybyła-Kasperek

University of Silesia, ul. Bankowa 12, 40-007 Katowice, Poland
{agnieszka.nowak,roman.siminski,malgorzata.przybyla-kasperek}@us.edu.pl

Abstract. This paper introduces the methodology of domain knowledge exploration in so called rule-based knowledge bases from the medical perspective, but it could easily by transformed into any other domain. The article presents the description of the *CluVis* software with rules clustering and visualization implementation. The rules are clustered by using hierarchical clustering algorithm and the resulting groups are visualized using the tree maps method. The aim of the paper is to present how to explore the knowledge hidden in rule-based knowledge bases. Experiments include the analysis of the influence of different clustering parameters on the representation of knowledge bases.

1   Introduction

The aim of the paper is to present how to explore the knowledge hidden in rule-based knowledge bases (KBs), in the context of a medical domain. Rules (in the form of *if ... then* chains) are a very specific type of data. They can be given apriori by a domain expert or generated automatically from a dataset using an algorithm. If they are generated from a dataset, with a large number of objects from a given domain, they are usually short (a small number of premises on its left hand side), what requires their careful comparison. One of the way to manage a large set of rules in a given KB is their clustering. A crucial step in every clustering algorithm is finding a pair of the most similar rules. In this reasearch, rules are divided into a number of groups based on similar premises. A knowledge based system (KBS) produces an answer to a given input searching within the whole set of rules. To avoid it (to minimize the number of rules that need to be processed) only representatives of groups could be compared with the set of input data and the most relevant group of rules could be selected for an exhaustive searching. It is really crucial to find and describe all the factors which influence clustering results and inference efficiency as it would help in designing or partitioning of KBs in order to maximize KBS's effectiveness because only then we may speak about the optimal exploration of the given KB. Besides the similarity and clustering methods, there are also other parameters that influence on the sucess of the KB's exploration: the number of rules as well as the number of attributes. That is why the main goal of the article is to present the analysis

© Springer International Publishing AG 2017
N.T. Nguyen et al. (Eds.): ICCCI 2017, Part II, LNAI 10449, pp. 150–160, 2017.
DOI: 10.1007/978-3-319-67077-5_15

of the influence of KB' size on the clusters' representation described for example by a biggest representative's length or the number of unclustered rules, which may be treated as outliers.

As similarity measures play an important role in finding pairs of rules (or further, in the clustering process/forming the groups of rules) and deciding about the order of clustering of rules (or groups of rules), to identify a suitable similarity measure, it becomes necessary to compare the results of using different measures and choose the one which produces the optimal results. For this reason, the next section describes the motivation behind using a hierarchical type of the clustering algorithm. It includes the description of a rule-based form of knowledge representation. The rest of the paper is organized as follows. In Sect. 2.3 the description of the cluster analysis algorithm, used for rules clustering, is described. This section focuses on both similarity measures used for comparing the rules to one another and hierarchical clustering algorithms. Section 4 contains the results of the experiments. The article concludes with the summary of the proposed approach and the results of the experiments (Sect. 5).

2   Knowledge Base Creating Process

As it has been mentioned in the previous section, rules in a given KB can be given apriori by a domain expert (as a way of their domain knowledge representation in form of *if - then* chains) or generated automatically from a dataset using a dedicated algorithm. In this research, we assume that the rules are achieved through the execution of one of many possible algorithms (based on the rough set theory) designated for rules extraction from data.

2.1   Inducing the Rules from the Original Data

There are many existing algorithms for generating rules (so called *rule induction* methods) like $LEM1$, $LEM2$, and AQ [8]. Usually, the original datasets have got a form of a so-called decision table. $LEM2$ (Learning from Examples Module version 2) algorithm is an example of a global rule induction algorithm and the option of $RSES$ (Rough Set Exploration System) software. It is most frequently used since (in most cases) it gives better results. Original data, stored, for example, in the form of decision table, is used to generate rules from it. Running the $LEM2$ algorithm for such data the decision rules being obtained.

As an example lets us take a dataset used for contact lenses fitting, which contains 24 instances, described by 4 nominal attributes[1] and a decision attribute with 3 classes[2]. The piece of the original dataset is as follows:

---

[1] *Age of the patient*: (1) young, (2) pre-presbyopic, (3) presbyopic, *spectacle prescription*: (1) myope, (2) hypermetrope, *astigmatic*: (1) no, (2) yes and *tear production rate*: (1) reduced, (2) normal.

[2] 1: *hard contact lenses*, 2: *soft contact lenses* and 3:*no contact lenses*. Class distribution is following: 1: 4, 2: 5 and 3: 15.

```
1  1  1  1  1  3
2  1  1  1  2  2
...
24 3  2  2  2  3
```

Using the *RSES* system with the *LEM2* algorithm implementation the *KB* with 5 rules has been achieved. The source file of the *KB* is as follows:

```
RULE_SET lenses
ATTRIBUTES 5
 age symbolic
...
 contact-lenses symbolic
DECISION_VALUES 3
none
soft
hard
RULES 5
(tear-prod-rate=reduced)=>(contact-lenses=none[12]) 12
(astigmatism=no)&(tear-prod-rate=normal)&(spectacle-prescrip=
...
(spectacle-prescrip=myope)&(astigmatism=no)&(tear-prod-rate=normal)
&(age=young)=>(contact-lenses=soft[1]) 1
```

The rule: `(tear-prod-rate=reduced)=>(contact-lenses=none[12])` 12 should be read as: *if (tear-prod-rate=reduced)* **then** *(contact-lenses=none)* which is covered by 12 of instances in the original dataset (50% of instances cover this rule).

2.2 Managing the Set of Rules

Natural way of knowledge representation makes rules easily understood by experts and knowledge engineers as well as people not involved in the expert system building. The knowledge represented by the rules should be encountered and, if possible, should describe all possible cases that can be met during the inference process. In other words, there should not be a situation, in which, for some input knowledge there are no rules to be activated, as it woould mean that there is no new knowledge explored from this particular *KB* for a given input data. The domain experts need to have a proper tool to manage the rules properly. They need to be able to easily obtain the information about all uncharted areas in an explored domain, in order to complete it as soon as possible. Many papers show the results of clustering a large set of data [6] but rarely for such a specific type of data like rule-based knowledge representation. From a knowledge engineer point of view, it is important to come up with a tool which helps to manage the consistency and completeness of the created *KB*. Decision support systems (*DSS*), which are usually based on rule-based *KBs*, use rules to extract new knowledge - this process is called the inference process. Instead of searching every rule one by one, it is possible to find the most relevant group of rules (the representative of such group of rules matches the given information in the best possible way) and reduce the time necessary to give an answer to a user of such a system. In [7] the authors propose modularization of large KBs using methods for grouping the rules. The results obtained in the authors's previous research [5] show that in an optimal case, it was possible to find a given rule when only a few percent of the whole KB has been searched. As long as so much depends on the quality of representatives for groups of rules, it is necessary to choose the best possible clustering algoritm the one which creates optimal descriptions for groups of rules.

2.3   Algorithms for Rules Clustering and Visualization

In this paper, the authors present a hierarchical clustering algorithm which is based on the connection between the two nearest clusters. We start by setting every object as a separate cluster. In each step, the two most similar objects are merged and a new cluster is created with a proper representative for it. After a specified stop condition is reached[3], the clustering process for the rules (or their groups) is finished.

The pseudocode of the hierarchical clustering algorithm - namely Classic AHC (agglomerative hierarchical clustering) algorithm [3] - is presented as Pseudocode 1.

Pseudocode 1. *AHC* algorithm for rules.
Input: stop condition *sc*, ungrouped set of objects *s*
Output: grouped tree-like structure of objects

1. Place each object *o* from *s* into a separate cluster.
2. Build a similarity matrix *M* that consists of similarity values for each pair of clusters.
3. Using *M* find the most similar pair of clusters and merge them into one.
4. Update *M*.
5. **IF** *sc* was met end the procedure.
6. **ELSE REPEAT** from step 3.
7. **RETURN** the resultant structure.

Step 2 is the most important. A similarity matrix *M* is created based on the selected similarity measure. A pair of the two most similar rules (or groups of rules) are merged. Two parameters are given by a user: the similarity measure and the clustering method. Eventually, both of them result in achieving different clusterings. For this reason, during the experiments, the hierarchical algorithm has processed each possible combination of these parameters, repeatedly changing the number of groups[4]. Having a set of attributes A and their values V, rules'

---

[3] There are many possible ways to define the stop condition. For example, it can reach the specified number of groups, or reach the moment when the highest similarity is under a minimal required threshold (which means that the groups of rules are now more differential than similar to one another).

[4] In this task clustering is stopped when given number of clusters is generated.

premises and conclusions are built using pairs (a_i, v_i), where $a_i \in A, v_i \in V_a$. In a vector of pairs (a_i, v_i), the i-th position denotes the value of the i-th attribute of a rule. Some similarity measures use the frequency (f) of a given pair (a_i, v_i), which is equal to 0 if a pair (a_i, v_i) is not included in any of the rules from a given KB (if $x \notin KB$ then $f(x) = 0$). Having made the notations clear, it will be much easier to understand the definition of the analyzed measures. The authors have studied the following five similarity measures: Simple Matching Coefficient (SMC), the Jaccard Index (based on SMC) sometimes also called the weighted similarity coefficient [5], the Gower measure (widely known in the literature) and two measures inspired by the retrieval information systems: the Occurence frequency measure (OF) and so-called inverse occurence frequency measure (IOF) [2]. They all are included in the Table 1.

Table 1. Intra-cluster similarity measures

Measure	The method of calculating
SMC	$s_{SMC}(r_j, r_k) = 1$ **if** $r_j = r_k$ **else** 0.
Jaccard	$s_{Jaccard}(r_j, r_k) = \frac{1}{n}$ **if** $r_j = r_k$ **else** 0, n -# of attributes.
Gower	$s_{Gower}^i(r_j, r_k) = s_{jki}^a$
OF	$s_{OF}(r_j, r_k) = \frac{1}{1 + \log \frac{N}{f(r_j)} \cdot \log \frac{N}{f(r_k)}}$ if $r_j \neq r_k$ else 1
IOF	$s_{OF}(r_j, r_k) = \frac{1}{1 + \log(f(r_j)) \cdot \log(f(r_k))}$ if $r_j \neq r_k$ else 1

^{a} For numeric type of the attribute i the value s_{jki} is calculated as $1 - \frac{|r_j - r_k|}{range(i)}$, where $range(i)$ is the range of the values of i-th attribute. If i is categorical and $r_{ji} = r_{ki}$ then $s_{jki} = 1$ else it is equal to 0.

SMC (Simple Matching Coefficient)[5], as the simplest measure, calculates the number of attributes that match in the two rules. Unfortunately, it tends to favour longer rules thus it is better to use the $Jaccard$ measure, which is similar to SMC. However, it is more advanced as it also divides the result by the number of attributes of both objects so longer rules are not favoured any more. The authors also have used the inverse occurrence frequency (IOF) and the occurrence frequency (OF) measures[6]. The last analyzed measure is the $Gower$ similarity coefficient[7]. Besides the similarity measures used to find the most similar pair of rules (or groups of rules) it is also important to find an optimal

---

[5] If both compared objects have the same attribute and this attribute has the same value for both objects then add 1 to a given similarity measure. If otherwise, do nothing.

[6] IOF measure assigns a lower similarity to mismatches on more frequent values while the OF measure gives opposite weighting for mismatches when compared to the IOF measure, i.e., mismatches on less frequent values are assigned a lower similarity and mismatches on more frequent values are assigned a higher similarity.

[7] The most complex of the all used inter-cluster similarity measures as it handles numerical attributes and symbolic attributes differently.

clustering method. The most popular are known as Single Link (SL), Complete Link (CoL), Average Link (AL) and Centroid Link (CL) [3]. *SL* describes similarity of the clusters based on their most similar objects while *CoL* returns a similarity of a single pair of rules as similarity of two clusters (clusters are only as similar as their two most distinct rules). *AL* is described as a mean similarity of all rules within the examined clusters while *CL* considers only a similarity between two virtual objects called centroids of clusters. Usually, the centroids are described as a geometrical center of the cluster, however, in case of rules, defining a geometrical center of a cluster is a non-trivial task, thus representative of this cluster is used instead.

Both: a proper visualization of clustering results and the symbolic description of it has to be proposed. There are many methods of generating representatives. In this paper, a representative is considered to be an average rule of the cluster, basing on cluster's content. Its creation algorithm can be described as follows. With the following input data: a cluster C, and its threshold value t [%], withing the cluster's attributes set A, find such attributes that can be found in t rules and put them in set A'. Then for each attribute $a \in A'$ check if a is symbolic or numerical. If it is symbolic then calculate a modal value, if numerical - an average value and return the attribute-value pairs from A' as the cluster's representative. The representative created this way contains the attributes that are the most important for the whole cluster (they are common enough). In this manner the examined group is well described by the minimal number of variables[8]. It should be possible to characterize the clusters using a small number of variables (the number of attributes attained by this method strongly depends on the selected threshold value). As the visualization of rules' clusters two treemap algorithms have been used - a rectangular treemap and a circular treemap. The differences between the two are very distinct for each of them is based on different geometrical shapes and has different methods of deploying them.

3   CluVis

CluVis (Clustering Visualizer) is an application designed to group sets of rules generated by RSES [1] and visualizing them using selected treemap methods. It is first application capable of working on raw KBs as generated from RSES. It has been successfully used in previous research to group and visualize medical KBs generated from artificial data sets available on [4] as well as one generated from real medical data [5].

3.1   Organizing Knowledge Bases in CluVis

Automatically created KBs have a chance to contain some undesired (redundant) rules. It is essential to maintain simplicity of KBs thus some methods of

---

[8] However, the authors see the necessity to analyze more methods for creating clusters' representatives and their influence on the resultant structure efficiency.

eliminating these rules is required. As `CluVis` is used to transform unorganized *KB*s into the organized ones (presented as responsive visualization) it's a perfect tool for this task. When the rules are imported into the `CluVis` software, having selected the clustering parameters grouping can be performed. The importing phase is a process which consists of checking the format of the input file (it is scanned to see whether it has a proper format). In the next step, the rules are transformed from lines of text into hashmaps that are basically vectors[9], which are stored as singular clusters. Then, a similarity matrix is built using a similarity measure selected during the first phase. Then, the two most similar clusters are joined and the similarity matrix is updated. During the merging of two smaller clusters, their representative is generated. `CluVis` ends clustering after a given (as a clustering parameter) number of clusters is reached and the visualization can be generated.

3.2 Features of the CluVis Software

CluVis implements the basic *AHC* algorithm and the two treemap visualization algorithms described in Sect. 2.3. Some of its other features are an ability to generate screen shots of visualizations and reports of current grouping or chose cluster, dive deeper into the hierarchy (responsive visualization) and calculate a clustering quality measure. Along its main functionalities (many of which can be seen in Fig. 1), `CluVis` is capable of generating reports of grouping (to `txt` or special `xml` files) which contain detailed information about each obtained cluster and about clustering in general: index of an experiment, the name of the *KB* file, the number of attributes, objects (rules), nodes in the created hierarchy, created clusters as well as the information about the algorithm used to visualize the groups of rules, the similarity measure and clustering method. The most important information about the created structure (also stored in such report) is as follows: the biggest cluster's size (*BCS*), the biggest representative's length (*BRL*), the biggest representative's size (*BRS*), the number of ungrouped rules (*U*) and many other data. An example of the cluster representative (which contains 4 descriptors)[10] may look like:

```
(spectacle-prescrip=myope)&(age=young)&(astigmatism=no)&
(tear-prod-rate=normal)=>(contact-lenses=soft)
```

Graphical representation of clustering can be performed in two ways - with a fully hierarchical view or without it. Visualization is responsive which means that selecting a cluster and clicking it with the right mouse button will generate a new visualization representing the internal structure of the selected cluster.

---

[9] The *i*-th variable value is accessed by its name in map, not by its index.

[10] In this poarticular case, the authors have used the contact lenses dataset, Gower's similarity measure and *SL* clustering method. The representative presented here is the description of the clusters *J*5 which contains 5 elements and the size of its representative is equal to 4.

Possible visualization of the created rules' structure, generated by the *CluVis* software, is presented in Fig. 2. It is possible to save the generated visualization to a `png` file.

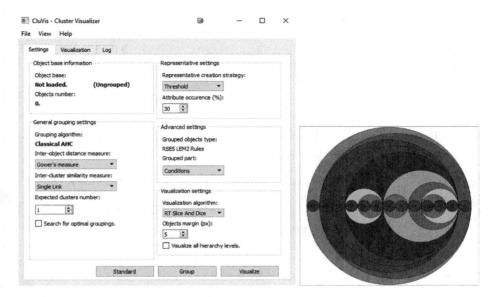

Fig. 1. CluVis's GUI.   **Fig. 2.** Circular visualization

`CluVis` is an open source application written in C++11 using QT graphic libraries. It is available in English and Polish and its source code can be downloaded from https://github.com/Tomev/CluVis.

4   Experiments

In this section, an experimental evaluation of 5 similarity measures on 7 different KBs [4] is presented.

All the details of analyzed datasets are included in Tables 2 and 3[11]. The optimal structure of KBs with rules clusters should contain the well separated

---

[11] The meaning of the columns in Table 3 is as follows: U - number of singular clusters in the resultant structure of grouping, BRS - a biggest representative's size - number of descriptors used to describe the longest representative, ARS - an average representative's size - an average number of descriptors used to describe cluster's representatives, $wARS$ - a weighted average representative's size (Attributes) - a quotient of an average number of descriptors used to describe cluster's representative in a given data set and the number of attributes in this dataset, BRL - a biggest representative's length - the number of descriptors in a biggest cluster's representative and BCS - a biggest cluster's size - number of rules in the cluster that contains the most of them. *Clusters* is the number of the created clusters of rules while *Nodes* is the number of nodes in the dendrogram representing the resultant structure.

Table 2. Basic characteristics for analyzed knowlegde bases

	Clusters	Attributes	Rules	Nodes
Arythmia	$12, 5 \pm 2, 5 \ (10 - 15)$	280	154	$295, 5 \pm 2, 5 \ (293 - 298)$
Audiology	$6, 9 \pm 3, 0 \ (4 - 10)$	70	42	$77, 2 \pm 3, 0 \ (74 - 80)$
Autos	$7, 8 \pm 2, 4 \ (4 - 10)$	26	60	$112, 2 \pm 2, 4 \ (110 - 116)$
Balance	$19 \pm 9, 1 \ (10 - 28)$	5	290	$560 \pm 9, 1 \ (550 - 560)$
Breast cancer	$11 \pm 1, 0 \ (10 - 12)$	10	130	$240 \pm 1, 0 \ (240 - 240)$
Diab	$29 \pm 19 \ (10 - 48)$	9	480	$940 \pm 19 \ (920 - 960)$
Diabetes	$29, 5 \pm 19, 7 \ (10 - 49)$	9	490	$950, 5 \pm 19, 7 \ (931 - 970)$

Table 3. Analysis of the rules' clusters structure and representatives

	BCS	BRL	U	BRS	ARS	wARS
Arythmia	$111, 5 \pm 42, 0$	$147, 4 \pm 1, 9$	$5, 8 \pm 4, 7$	$152, 1 \pm 4, 9$	$134, 0 \pm 11, 1$	$2, 1 \pm 0, 2$
Audiology	$30, 0 \pm 7, 9$	$66, 8 \pm 0, 5$	$3, 6 \pm 2, 7$	$67, 0 \pm 0, 5$	$49, 8 \pm 9, 1$	$1, 5 \pm 0, 32$
Autos	$37, 9 \pm 14, 3$	$10, 7 \pm 0, 6$	$3, 8 \pm 3, 0$	$11, 6 \pm 1, 8$	$8, 5 \pm 1, 7$	$3, 2 \pm 0, 6$
Balance	170 ± 95	$4, 0 \pm 0, 01$	$6, 9 \pm 9, 0$	$4, 0 \pm 0, 01$	$3, 6 \pm 0, 43$	$1, 4 \pm 0, 15$
Breast cancer	76 ± 32	$9, 0 \pm 0, 01$	$5, 1 \pm 3, 4$	$9, 0 \pm 0, 01$	$7, 1 \pm 1, 1$	$1, 4 \pm 0, 22$
Diab	320 ± 130	$4, 9 \pm 0, 46$	11 ± 13	$5, 4 \pm 0, 49$	$3, 4 \pm 0, 77$	$2, 8 \pm 0, 63$
Diabetes	$336, 6 \pm 135, 2$	$4, 9 \pm 0, 3$	$12, 5 \pm 14, 4$	$5, 5 \pm 0, 6$	$3, 3 \pm 0, 8$	$2, 8 \pm 0, 7$

groups of rules, and the number of such groups should not be too high. Moreover, the number of ungrouped rules should be minimal. Experiments were based on the following procedure. For 5 different similarity measure and 4 clustering methods (described in Sect. 2.3), changing the number of clusters, different clustering for 7 analysed KBs were achieved. The goal of this research is to check if there are any, statistically significant, correlations between the size of the KB measured by *Attributes*, *Clusters*, *Rules*, *Nodes* and the parameters analyzed here in this research such as: BCS, BRS, BRL, U, ARS and $wARS$. The results of this research are included in Table 4, where the symbol $*$ means that a given correlation value is statistically significant ($p < 0, 05$).

It can be seen in Table 4 that there is a statistically significant correlation between the number of attributes in the KB (*Attributes*) and the length of the biggest representatives (the more attributes we have the longer the representative's length). Another, quite obviuos, positive correlation is the one between the number of rules and the number of ungrouped rules (the more rules to cluster, the higher the number of rules that are dissimilar to the others and impossible to be merged with them). The results of the experiment also show the correlations that are not statistically significant. Such a correlation has to be determined for two pairs of variables: the number of ungrouped rules (U) and the number of attributes (*Attribute*) as well as a weighted average representative's size ($wARS$) and the number of attributes.

Table 4. Data gathered during the experiments.

	Attributes	Rules	Nodes	Clusters
BCS	-0.2297^*	0.8181^*	0.8255^*	0.3252^*
BRL	0.9779^*	-0.4026^*	-0.4024^*	-0.2523^*
U	$-0.0965, p = 0.107$	0.3475^*	0.3302^*	0.6342^*
BRS	0.9801^*	-0.3966^*	-0.3966^*	-0.2459^*
ARS	0.9837^*	-0.3705^*	-0.3704^*	-0.2300^*
wARS	$-0.0791, p = 0.187$	0.3222^*	0.3220^*	0.2024^*
Clusters	-0.2012^*	0.6269^*	0.6015^*	-

5   Conclusion

The article presents the description of the $CluVis$ software with rules clustering
and visualization implementation. The rules are clustered by using hierarchical
clustering algorithm and the resulting groups are visualized using the tree maps
method. The aim of the paper is to present how to explore the knowledge hidden
in rule-based KBs. Experiments include the analysis of the influence of different
clustering parameters as well as characteristic of a given KB on its representation. The results reveal that the number of outliers (unclustered rules) as well
as for example the length of the representatives of rules clusters or the size of
the biggest cluster depends strongily of the clustering parameteres and the size
of the input data (number of rules and attributes). The authors plan to extend
the research on the bigger KBs and more clustering parameters to analyze.

References

1. Bazan, J.G., Szczuka, M.S., Wróblewski, J.: A new version of rough set exploration system. In: Alpigini, J.J., Peters, J.F., Skowron, A., Zhong, N. (eds.) RSCTC
   2002. LNCS, vol. 2475, pp. 397–404. Springer, Heidelberg (2002). doi:10.1007/
   3-540-45813-1_52
2. Boriah, S., Chandola, V., Kumar, V.: Similarity measures for categorical data: a
   comparative evaluation. In: Proceedings of the 8th SIAM International Conference
   on Data Mining, pp. 243–254 (2008)
3. Dubes, R., Jain, A.K.: Clustering techniques: the user's dilemma. Pattern Recognit.
   8(4), 247–260 (1976)
4. Lichman M. UCI Machine Learning Repository, University of California (2013).
   http://archive.ics.uci.edu/ml
5. Nowak-Brzezińska, A.: Mining rule-based knowledge bases inspired by rough set
   theory. Fundamenta Informaticae **148**, 35–50 (2016). doi:10.3233/FI-2016-1421. IOS
   Press
6. Przybyła-Kasperek, M., Wakulicz-Deja, A.: The strength of coalition in a dispersed
   decision support system with negotiations. Eur. J. Oper. Res. **252**, 947–968 (2016)

7. Simiński, R.: Multivariate approach to modularization of the rule knowledge bases. In: Gruca, A., Brachman, A., Kozielski, S., Czachórski, T. (eds.) Man–Machine Interactions 4. AISC, vol. 391, pp. 473–483. Springer, Cham (2016). doi:10.1007/978-3-319-23437-3_40
8. Grzymala-Busse, J.W.: A new version of the rule induction system LERS. Fundamenta Informaticae **31**, 27–39 (1997)

Computer User Verification Based on Typing Habits and Finger-Knuckle Analysis

Hossein Safaverdi, Tomasz Emanuel Wesolowski$^{(\boxtimes)}$, Rafal Doroz,
Krzysztof Wrobel$^{(\boxtimes)}$, and Piotr Porwik

Institute of Computer Science, University of Silesia in Katowice, Katowice, Poland
{hossein.safaverdi,tomasz.wesolowski,rafal.doroz,krzysztof.wrobel,
piotr.porwik}@us.edu.pl

Abstract. The paper presents preliminary research conducted to assess the potential of biometric methods fusion for continuous user verification. In this article a novel computer user identity verification method based on keystroke dynamics and knuckle images analysis is introduced. In the proposed solution the user verification is performed by means of classification. The introduced approach was tested experimentally using a database which comprises of keystroke dynamics data and knuckle images. The results indicate that the introduced methods fusion performs better than the single biometric approaches.

Keywords: Biometrics · Verification · Keystroke dynamics · Finger knuckle

1 Introduction

Protecting electronically stored sensitive data is one of the most important tasks. Collections of our personal data or projects implemented by business are valuable assets to thieves specialized in cybercrime [13,14]. In the domain of electronic health records security the amount of individuals affected by security breaches, where sensitive personal data were stolen, increased a hundred times between the years 2014 and 2015 [20]. The criminals use modern techniques to perform various kinds of cyber attacks [15] that can be made as intrusions outside of the computer system but still a big part of intrusions consists of insider attacks [16]. Therefore the crucial task of computer security systems is to detect such intrusions and protect the data and resources from an unauthorized access. For this reason, the use of biometric methods in security systems is justified because these methods are highly effective in verifying the identity of computer system users [1,5,18].

In most cases a single biometric feature is used based on either behavioral (habits of using a computer mouse or typing on a keyboard) or physical characteristics (finger print, eye, knuckle imaging or face detection) [7,9,10]. In order to increase a protection level of computer system resources the novel security model introduced in this paper combines the analysis of user's typing habits with

© Springer International Publishing AG 2017
N.T. Nguyen et al. (Eds.): ICCCI 2017, Part II, LNAI 10449, pp. 161–170, 2017.
DOI: 10.1007/978-3-319-67077-5_16

finger knuckle imaging method [11,12,17], where keyboard events and knuckle image registration is performed using a dedicated software and device especially designed for this purpose [5,20].

The research presented in this paper was aimed at assessing the potential of the biometric methods fusion dedicated for Host-based Intrusion Detection Systems (HIDS). The long-term goal of the scientific project is to develop a computer user verification method working in real time and based on the fusion of typing habits and finger-knuckle pattern analysis [21,22]. However a method of analyzing the finger-knuckle patterns on the fly has not been implemented yet because of its high demand for resources. Before the decision to invest such resources was made it was necessary to verify if it is worth investing them and if the fusion is potentially interesting. This paper presents the preliminary results of an intrusion detection model based on the introduced novel approach.

2   The Proposed Approach for User Verification

The proposed user verification model is based on the fusion of keystroke dynamics and finger knuckle images analysis. The presented approach consists of two phases. At first user profiling is performed where a legitimate user activity (while working with a keyboard) is recorded and this user's finger knuckle images are acquired. Based on the retrieved data a user's profile is being established. The profile of a legitimate user can be used in the second phase for verification of a user that is working with a computer system. The proposed model of a user verification is presented in Fig. 1. After a user being verified starts working a keystroke verification is performed continuously. Based on the profiles of legitimate users and a recorded activity of an at the moment active user a decision if the recorded activity belongs to a legitimate user or an intruder is established. If the keystroke verification fails (a possibility of intrusion is detected), an additional verification is made by means of knuckle pattern recognition unit after taking a picture of the user's knuckles. After a successful verification an access is granted and the verification procedure continues with the keystroke based verification. If the verification has not been successful, an access to computer resources is denied and the security breach alert is generated.

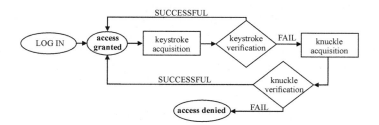

Fig. 1. The introduced user verification model

3   User Verification Based on Typing Habits Analysis

In order to allow that a computer user activity could be verified continuously all keyboard events generated by the user are recorded in the background, while a user is performing everyday tasks. In the proposed method a profile of a user is constituted of time dependencies that occur between the key events. The profiling is performed continuously in the background, it is practically transparent for a user, what is the important advantage of the proposed method and for this reason it can be used in HIDS that analyze the logs with registered user activity in real time to detect an unauthorized access. The keyboard events are captured on the fly. A data concerning a single event consists of a *prefix* describing the type of an event (key up or key down), followed by the time-stamp t of this event and an identifier id of a key that generated the event. Such a recorded raw data of a single j-th keyboard event can be presented in a form of a vector $\mathbf{e}_j = [prefix_j, t_j, id_j]$. Activity data analysis is carried out separately for each user identified in the system by a user identifier uid. All vectors $\mathbf{e}_j$ of the same user constitute this user's activity dataset. A total number of vectors $\mathbf{e}_j$ of a user is limited by a user activity recording time.

User's activity dataset in this form does not provide directly information on how a user interacts with a computer system through a keyboard so it is difficult to interpret. Therefore it is necessary to process recorded data by extracting time dependencies between keyboard events (generated while the user was working) in order to obtain characteristics of a user in a form of a profile. For the purpose of the presented research a profiling method was used that was indicated by the literature sources [18–20] as very efficient. As a result of the profiling for each user identified by uid a profile Θ^{uid} was obtained consisting of feature vectors $\Theta^{uid} = \{\mathbf{F}_1^{uid}, \mathbf{F}_2^{uid}, \dots, \mathbf{F}_z^{uid}\}$ characterizing the activity of this user in a computer system. Based on the obtained profiles and an activity of a current user using a computer system at the moment the verification is performed.

User verification consists in assigning a user to one of two possible classes: a legitimate user or an intruder. A classifier Υ maps the vector $\mathbf{F}$ of a given user to a class label c_j, where $j \in \{1, 2\}$:

$$\Upsilon(\mathbf{F}) \rightarrow c_j \in C. \tag{1}$$

In the proposed approach, the classifiers Υ returns a probability $\hat{p}_i(c_j|\mathbf{F})$, $j \in \{1, 2\}$ that a given object $\mathbf{F}$ belongs to a class c_j.

Based on the research presented in [4,20] the keystroke verification system was built using ensembles of classifiers EC_a. Each of the ensembles EC_a consists of four single classifiers: $\Upsilon^{(1)}$, $\Upsilon^{(2)}$, $\Upsilon^{(3)}$ and $\Upsilon^{(4)}$. The structure of a single ensemble is shown in Fig. 2. The study described in [4,20] indicates three ensembles as an optimal selection EC_a, $a = 1, \dots, 3$ for a verification system based on keystroke dynamics. The ensembles of classifiers EC_a work simultaneously and each of them is trained using a separate training set TS_a. At the input of each node Λ_a (Fig. 2) as a result of the use of classifiers $\Upsilon^{(i)}$ the following data matrix is introduced:

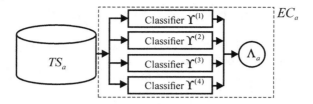

Fig. 2. The structure of a single ensemble of classifiers EC_a

$$\text{Input}\Lambda_a(\mathbf{F}) = \begin{bmatrix} \hat{p}_1(c_1|\mathbf{F}) \ \hat{p}_1(c_2|\mathbf{F}) \\ \hat{p}_2(c_1|\mathbf{F}) \ \hat{p}_2(c_2|\mathbf{F}) \\ \hat{p}_3(c_1|\mathbf{F}) \ \hat{p}_3(c_2|\mathbf{F}) \\ \hat{p}_4(c_1|\mathbf{F}) \ \hat{p}_4(c_2|\mathbf{F}) \end{bmatrix}. \tag{2}$$

Following the classification, each ensemble of classifiers EC_a, $a = 1, \ldots, 3$ generates a local decision Λ_a according to the soft voting formula 3:

$$\Lambda_a(\mathbf{F}) = \arg\max_{c_j \in C} \sum_{i=1}^{4} \hat{p}_i(c_j|\mathbf{F}), \, a = 1, \ldots, 3. \tag{3}$$

Next the class labels designated by (3) are being converted to the numerical values according to the rules (4).

$$\Gamma_a(\mathbf{F}) = \begin{cases} -1 \text{ if } \Lambda_a(\mathbf{F}) = c_1 \\ +1 \text{ if } \Lambda_a(\mathbf{F}) = c_2 \end{cases}. \tag{4}$$

Based on the converted local decisions $\Gamma_a(\mathbf{F})$ designated in each of the ensemble of classifiers EC_a a value of LS is determined as follows:

$$LS(\mathbf{F}) = \sum_{a=1}^{3} \Gamma_a(\mathbf{F}) \in L, a = 1, \ldots, 3. \tag{5}$$

The value of $LS(\mathbf{F})$ allows to determine the decision of keystroke based verification. The value greater than a threshold ϑ indicates that an activity of a verified user is compatible with a legitimate user's profile, the verified user is allowed to keep working and the process of keystroke verification is repeated continuously. Otherwise the user must proceed to knuckle verification phase. The influence of the threshold ϑ value on the keystroke verification accuracy is presented in the experiments section.

4　Finger Knuckle Based Verification

By means of a special device which consists of a box with a camera and three LED-lights inside it, the task of collecting the knuckle images has been done. The LED-lights are used to illuminate the inside space of the box to obtain better quality knuckle images and store them in our database. Next stage is to analyze the finger knuckle images and detect the furrows from them. To detect the furrows, we converted the images to gray scale then we utilized the Frangi filter (FRF). This filter allows to make dark objects visible on a light background and vice versa [3,8]. In our case, dark objects are the furrows, located on the knuckles. The values obtained after FRF are used as threshold to extract the furrows form images. After thresholding some noise could be misinterpreted as elements of the furrows and hinder the further analysis. Therefore by means of erosion and closing they are removed from the image. The image obtained in this way is proceeded to the next step which is skeletonization.

The detailed description of the acquisition process has been given in our previous paper [5].

4.1 Matching Process

During our investigation and analysis of the knuckle images we were confronted with the problem that the location and size of furrows of the same user may be different in subsequent images collected from this person. This problem was caused by high flexibility of a human skin and has been explained in [6]. One solution to minimize this disadvantage is to match the furrows before they are compared. For matching we applied the *Shape Contexts* (SC) and *Thin Plate Spline* (TPS). Figure 3 presents knuckle images before and after matching.

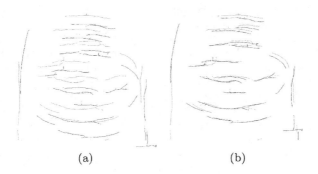

(a)   (b)

Fig. 3. Two overlapping knuckle images (marked by black and red color): (a) without matching, (b) with matching (Color figure online)

4.2 Points Extraction

One of the essential stages of our method is to convert the visible knuckle furrows into a chain of points. In order to extract the furrows from the skeletonized image we have to mark the end points and bifurcation points in the image. These points are marked by counting the number of black pixels in the vicinity 3×3 of the pixel being analyzed. This task has been accomplished by the following formula:

$$J(x,y) = \sum_{a=-1}^{1} \sum_{b=-1}^{1} I(x+a, y+b), \ x \in [2, \ldots, W-1], y \in [2, \ldots, H-1], \quad (6)$$

where (x, y) are the coordinates of analyzed point, W and H are the width and height of the image I, respectively.

Based on the calculated value of $J(x, y)$ and the formula (7), for each black pixel $p(x, y)$ we can assign a label p^T (end of the furrow) or p^B (bifurcation on the furrow):

$$p(x, y) = \begin{cases} p^T(x, y) \text{ if } J(x, y) = 2 \\ p^B(x, y) \text{ if } J(x, y) > 3 \end{cases} . \quad (7)$$

After labeling the end and bifurcation points the procedure of extracting the chains of points from the image begins. This task is done by the use of Algorithm 1:

Algorithm 1. Procedure of extracting the chains of points from the knuckle image.

Data: Thinned knuckle image; List of the image points with labels
$$P = \{p^{T_1}, p^{T_2}, \ldots, p^{B_1}, p^{B_2}, \ldots\}$$
Result: Set of chains of the points $C = \{c_1, \ldots\}$, where each chain c_i has
starting and ending points with any type label (p^T or p^B)

1 $C = \varnothing$;
2 $i = 1$;
3 **foreach** *labeled point $p \in P$ of the set P* **do**
4 　　add coordinates of the point $p \in P$ to the chain c_i;
5 　　**do**
6 　　　　move the analyzed point from the point p to the neighboring black pixel
　　　　　　p^* which does not belong to any chain from the list C;
7 　　　　add point p^* to the chain c_i;
8 　　　　set analyzed point $p = p^*$;
9 　　**while** *the analyzed point $p \notin P$*;
10 　　add chain c_i to set C;
11 　　$i = i + 1$;

4.3   Similarity Between the Knuckle Images

When an unknown person claims identity and provides a knuckle image to be verified, the system will go through the database and look for this person's reference (stored) knuckle images and compare them sequentially with the knuckle image provided at the moment. The similarity between the selected reference knuckle image I^R and the image being verified I^V is determined by finding the matching points that are in the same relative position in both images. After the analysis of all points in the reference image the obtained similarity values are averaged. The following formula is used to determine the similarity between the images I^R and I^V:

$$sim(I^R, I^V) = \frac{1}{m} \sum_{i=1}^{m} \min\left(d(c_i^R, c_1^V), d(c_i^R, c_2^V), \ldots, d(c_i^R, c_n^V)\right), \qquad (8)$$

where:
$C^R = \{c_1^R, c_2^R, \ldots, c_m^R\}$ – the set of chains describing the reference image I^R,
$C^V = \{c_1^V, c_2^V, \ldots, c_n^V\}$ – the set of chains describing the verified image I^V.

In our case $sim(I^R; I^V) = 0$ means that the images being compared are identical and as $sim(I^R; I^V)$ value increases the similarity decreases. In order to calculate the distance value between two chains of points being compared, the following formula is used:

$$d(c_i^R, c_j^V) = \frac{1}{r} \sum_{k=1}^{r} \min\left(euc(p_k^R, p_1^V), euc(p_k^R, p_2^V), \ldots, euc(p_k^R, p_s^V)\right), \qquad (9)$$

where r and s are the numbers of points in the chain c_i^R and c_j^V, respectively, while $euc(p_k^R, p_s^V)$ is the Euclidean distance between the point p_k^R in the reference chain c_i^R and the point p_s^V in the chain being verified c_j^V.

5   Fusion of the Methods

The system proposed in this work involves the fusion of two verification methods: keystroke and knuckle image verification. The ultimate decision of the user verification depends on both methods.

In the first step of the fusion the keystroke verification is carried out. If the result of the verification is positive, then the access to the system is granted, otherwise the second step of verification is introduced. In this step, the user being verified must provide a knuckle image. The knuckle image is compared with w reference images I^{Ri} from the database, where $i = 1, \ldots, w$. The general rule of the proposed fusion based verification system is presented below:

$$\text{decision} = \begin{cases} \text{access granted} & if & LS > \vartheta' \\ \text{access granted} & if & (LS > \vartheta'') \text{ and } (SIM > \lfloor \frac{w}{2} \rfloor) \\ \text{access denied} & otherwise \end{cases}, \qquad (10)$$

where:
$$SIM = \sum_{i=1}^{w} \mathbb{1}(sim(I^{Ri}, I^V) \leq \lambda), \; i = 1, \ldots, w. \qquad (11)$$

The $\mathbb{1}()$ is a indicator function defined as follows:

$$\mathbb{1}(sim(I^{Ri}, I^V) \leq \lambda) = \begin{cases} 1 & if & sim(I^{Ri}, I^V) \leq \lambda \\ 0 & & otherwise \end{cases}. \qquad (12)$$

Values of parameters ϑ', ϑ'' and λ have been determined in the experiments section.

6 Experimental Results

In this section, the results of experimental investigations examining the behavior of the proposed verification method have been presented. The experiments have been carried out using a database that consists of 4000 vectors $\mathbf{F}$ and 150 knuckle images acquired from 30 persons.

The classification module for a keystroke dynamics based verification consists of three ensembles of classifiers and four single classifiers in an ensemble: C4.5, Bayesian Network, Random Forest, Support Vector Machine. The used classifiers were chosen considering their high accuracy confirmed in [20]. During the first experiment optimal values of parameters ϑ', ϑ'' and λ were determined. These parameters are used in decision rule (10). The values of the mentioned parameters were determined by use of grid search procedure from the following sets $\vartheta', \vartheta'' \in \{1, 2, 3\}$, $\lambda \in \{10, 20, 30, \ldots, 100\}$.

In order to provide better statistical accuracy the results obtained by our system have been verified in the 10-fold cross-validation tests, so that the average values of evaluation metrics for all trials were calculated. Table 1 shows the seven best obtained results and the values of the parameters ϑ', ϑ'' and λ for which these results were obtained.

Table 1. The seven best results obtained in experiments together with the values of parameters used in experiments.

FAR [%]	FRR [%]	AER [%]	ACC [%]	ϑ'	ϑ''	λ
0.67 ± 0.11	2.16 ± 0.25	1.41 ± 0.22	98.96 ± 1.63	3	2	60
1.06 ± 0.24	2.67 ± 0.37	1.86 ± 0.26	98.57 ± 1.14	3	2	50
1.26 ± 0.19	3.53 ± 0.22	2.39 ± 0.23	98.06 ± 1.33	3	2	70
1.24 ± 0.21	3.65 ± 0.26	2.44 ± 0.28	97.95 ± 2.14	3	1	50
1.31 ± 0.30	3.54 ± 0.26	2.42 ± 0.25	97.90 ± 1.74	3	2	80
1.43 ± 0.25	3.48 ± 0.34	2.45 ± 0.29	97.89 ± 1.32	3	1	40
1.37 ± 0.33	3.36 ± 0.26	2.36 ± 0.22	97.86 ± 1.42	3	1	60

By analyzing Table 1 we can see that the optimal values of the parameters allowing to obtain the best result are equal to $\tau' = 3$, $\tau'' = 2$ and $\lambda = 60$.

The efficiency of the proposed fusion system has been compared with the efficiencies obtained by each of the biometrical verification methods separately. Therefore, the optimal values of parameters ϑ and λ have been determined once again, but this time the efficiency of each verification method (keystroke verification and knuckle verification) has been assessed independently. The results obtained by the fusion-based method and additionally by each individual verification method used in the fusion approach are presented in Table 2.

By analyzing Table 2, we can notice that the fusion of two methods allows to obtain better efficiency in classification than using only one of these methods separately.

Table 2. The comparison of the performance of various verification methods.

Method	FAR [%]	FRR [%]	AER [%]	ACC [%]
Keystroke	5.59 ± 1.16	7.94 ± 1.46	6.76 ± 1.72	97.83 ± 3.28
Knuckle	4.03 ± 0.21	7.76 ± 0.95	5.89 ± 1.23	95.24 ± 6.87
Keystroke + Knuckle	0.67 ± 0.11	2.16 ± 0.25	1.41 ± 0.22	98.96 ± 1.63

7   Conclusions

This work presents some elementary experiments of two biometric features, keystroke dynamics and finger knuckle prints analysis as well as the fusion of these two features. From the experiments we can infer that combining these two biometric features gives us more promising results than using only one feature. Therefore in the future we will continue this research and expand our experiments to work on a real time finger knuckle analysis method by monitoring the finger knuckles while the user types on a keyboard. Of course this task is so complicated to accomplish due to the constant movement of fingers and different positions of hands over a keyboard. Likewise, depending on how often we take a picture of the knuckles while users are performing everyday tasks this could cause some computer system performance problems. To overcome the mentioned problems the next research will focus on optimizing the method by finding a localization of the finger knuckles and on developing the fusion-based computer user continuous verification method.

References

1. Banerjee, S.P., Woodard, D.L.: Biometric authentication and identification using keystroke dynamics: a survey. J. Pattern Recogn. Res. **7**, 116–139 (2012)
2. Belongie, S., Malik, J., Puzicha, J.: Shape matching and object recognition using shape contexts. IEEE Trans. Pattern Anal. Mach. Intell. **24**, 509–522 (2002)
3. Ng, C.-C., Yap, M.H., Costen, N., Li, B.: Automatic wrinkle detection using hybrid hessian filter. In: Cremers, D., Reid, I., Saito, H., Yang, M.-H. (eds.) ACCV 2014. LNCS, vol. 9005, pp. 609–622. Springer, Cham (2015). doi:10.1007/978-3-319-16811-1_40
4. Doroz, R., Porwik, P., Safaverdi, H.: The new multilayer ensemble classifier for verifying users based on keystroke dynamics. In: Núñez, M., Nguyen, N.T., Camacho, D., Trawiński, B. (eds.) ICCCI 2015. LNCS (LNAI), vol. 9330, pp. 598–605. Springer, Cham (2015). doi:10.1007/978-3-319-24306-1_58
5. Doroz, R., et al.: A new personal verification technique using finger-knuckle imaging. In: Nguyen, N.-T., Manolopoulos, Y., Iliadis, L., Trawiński, B. (eds.) ICCCI 2016. LNCS (LNAI), vol. 9876, pp. 515–524. Springer, Cham (2016). doi:10.1007/978-3-319-45246-3_49
6. Fager, M., Morris, K.: Quantifying the limits of fingerprint variability. Forensic Sci. Int. **254**, 87–99 (2015)
7. Ferrer, M.A., Travieso, C.M., Alonso, J.B.: Using hand knuckle texture for biometric identifications. IEEE Aerosp. Electron. Syst. Mag. **21**(6), 23–27 (2006)

8. Iwahori, Y., Hattori, A., Adachi, Y., Bhuyan, M.K., Woodham, R.J., Kasugai, K.: Automatic detection of polyp using Hessian Filter and HOG features. Procedia Comput. Sci. **60**(1), 730–739 (2015)

9. Kasprowski, P.: The impact of temporal proximity between samples on eye movement biometric identification. In: Saeed, K., Chaki, R., Cortesi, A., Wierzchoń, S. (eds.) CISIM 2013. LNCS, vol. 8104, pp. 77–87. Springer, Heidelberg (2013). doi:10.1007/978-3-642-40925-7_8

10. Koprowski, R., Teper, S.J., Weglarz, B., Wylegała, E., Krejca, M., Wróbel, Z.: Fully automatic algorithm for the analysis of vessels in the angiographic image of the eye fundus. Biomed. Eng. Online **11**, 35 (2012)

11. Kumar, A., Ravikanth, C.: Personal authentication using finger knuckle surface. IEEE Trans. Inf. Forensics Secur. **4**(1), 98–110 (2009)

12. Morales, A., Travieso, C.M., Ferrer, M.A., Alonso, J.B.: Improved finger-knuckle-print authentication based on orientation enhancement. Electron. Lett. **47**(6), 380–382 (2011)

13. Porwik, P., Doroz, R.: Self-adaptive biometric classifier working on the reduced dataset. In: Polycarpou, M., Carvalho, A.C.P.L.F., Pan, J.-S., Woźniak, M., Quintian, H., Corchado, E. (eds.) HAIS 2014. LNCS (LNAI), vol. 8480, pp. 377–388. Springer, Cham (2014). doi:10.1007/978-3-319-07617-1_34

14. Porwik, P., Doroz, R., Wrobel, K.: A new signature similarity measure. In: Proceedings of the 2009 World Congress on Nature and Biologically Inspired Computing, NABIC 2009, pp. 1022–1027 (2009)

15. Raiyn, J.: A survey of cyber attack detection strategies. Int. J. Secur. Appl. **8**(1), 247–256 (2014)

16. Salem, M.B., Hershkop, S., Stolfo, S.J.: A survey of insider attack detection research. Adv. Inf. Secur. **39**, 69–90 (2008)

17. Usha, K., Ezhilarasan, M.: Finger knuckle biometrics - a review. Comput. Electr. Eng. **45**, 249–259 (2015)

18. Wesołowski, T.E., Porwik, P.: Keystroke data classification for computer user profiling and verification. In: Núñez, M., Nguyen, N.T., Camacho, D., Trawiński, B. (eds.) ICCCI 2015. LNCS, vol. 9330, pp. 588–597. Springer, Cham (2015). doi:10.1007/978-3-319-24306-1_57

19. Wesołowski, T.E., Porwik, P.: Computer user profiling based on keystroke analysis. In: Chaki, R., Cortesi, A., Saeed, K., Chaki, N. (eds.) Advanced Computing and Systems for Security. AISC, vol. 395, pp. 3–13. Springer, New Delhi (2016). doi:10.1007/978-81-322-2650-5_1

20. Wesolowski, T.E., Porwik, P., Doroz, R.: Electronic health record security based on ensemble classification of keystroke dynamics. Appl. Artif. Intell. **30**, 521–540 (2016)

21. Woodard, D.L., Flynn, P.J.: Finger surface as a biometric identifier. Comput. Vis. Image Underst. **100**(3), 357–384 (2005)

22. Xiong, M., Yang, W., Sun, C.: Finger-knuckle-print recognition using LGBP. In: Liu, D., Zhang, H., Polycarpou, M., Alippi, C., He, H. (eds.) ISNN 2011. LNCS, vol. 6676, pp. 270–277. Springer, Heidelberg (2011). doi:10.1007/978-3-642-21090-7_32

ANN and GMDH Algorithms in QSAR Analyses of Reactivation Potency for Acetylcholinesterase Inhibited by VX Warfare Agent

Rafael Dolezal[1,2,3(✉)], Jiri Krenek[1], Veronika Racakova[1],
Natalie Karaskova[2], Nadezhda V. Maltsevskaya[4],
Michaela Melikova[2], Karel Kolar[2], Jan Trejbal[1], and Kamil Kuca[1,2,3]

[1] Center for Basic and Applied Research,
Faculty of Informatics and Management, University of Hradec Kralove,
Rokitanskeho 62, 50003 Hradec Kralove, Czech Republic
rafael.dolezal@uhk.cz
[2] Department of Chemistry, Faculty of Sciences, University of Hradec Kralove,
Rokitanskeho 62, 50003 Hradec Kralove, Czech Republic
[3] Biomedical Research Center, University Hospital Hradec Kralove,
Hradec Kralove, Czech Republic
[4] State Budget Professional Education Institution,
Kosygina st. 17(3), Moscow 119334, Russia

Abstract. Successful development of novel drugs requires a close cooperation of experimental subjects, such as chemistry and biology, with theoretical disciplines in order to confidently design new chemical structures eliciting the desired therapeutic effects. Herein, especially quantitative structure-activity relationships (QSAR) as correlation models may elucidate which molecular features are significantly associated with enhancing a specific biological activity. In the present study, QSAR analyses of 30 pyridinium aldoxime reactivators for VX-inhibited rat acetylcholinesterase (AChE) were performed using the group method of data handling (GMDH) approach. The self-organizing polynomial networks based on GMDH were compared with multilayer perceptron networks (MPN) trained by 10 different algorithms. The QSAR models developed by GMHD and MPN were critically evaluated and proposed for further utilization in drug development.

Keywords: QSAR · ANN · Group method of data handling · Reactivators · AChE

1 Introduction

The current therapy of organophosphate (OP) poisoning uses aldoxime-based antidotes to reactivate the inhibited acetylcholinesterase (AChE) through releasing its serine from the complex with OP (e.g. pralidoxime, trimedoxime, obidoxime, HI-6) [1–3]. Although the present aldoxime reactivators may be applied in the OP poisoning treatment, their potency to recover AChE inhibited by different OP is limited and, thus,

© Springer International Publishing AG 2017
N.T. Nguyen et al. (Eds.): ICCCI 2017, Part II, LNAI 10449, pp. 171–181, 2017.
DOI: 10.1007/978-3-319-67077-5_17

further pharmaceutical investigation in this field is broadly encouraged. In this respect, much attention in various drug researches has lately been turned to exploration of the relationships between the structure and activity (SAR) in order to elucidate important physico-chemical properties required for the optimal therapeutic effect [4, 5]. SAR methodology as well as its quantitative alternative (QSAR) have been still more employed in drug discovery, utilizing benefits offered by advanced computational chemistry, bioinformatics, artificial intelligence and machine learning (e.g. multiple linear regression (MLR), partial least square regression (PLS), principal component analysis (PCA), support vector machines (SVM), random forests (RF), k-nearest neighbor, etc.). By means of these computer-aided drug design (CADD) methods, crucial features in chemical compounds can be revealed and translated into a mathematical relationship with the biological activity. If properly validated, the resulting QSAR models can be used for *in silico* predictions of improved drugs, which as such considerably save the time and costs vainly spent on experiments with randomly or non-rationally chosen compounds.

In current research of OP-inhibited AChE reactivators, several computational approaches such as flexible molecular docking, molecular dynamics, hybridized quantum chemistry/molecular mechanics simulations, ligand based 3D QSAR analyses have been applied to reveal SAR & QSAR models and to predict novel improved reactivators [1]. All of these studies contribute to the complex research of AChE reactivators although they rely rather on linear models correlating the reactivation potency and various molecular descriptors.

In the present article, we focus on the application of artificial neural networks (ANN) in developing a predictive QSAR model for 30 pyridinium aldoxime AChE reactivators. Special attention is devoted to implementation and research of the Group Method of Data Handling theory (GMDH), which may bring about considerable improvements in building reliable QSAR models through capability to solve extremely complex nonlinear systems. In simple terms, our objective is to reveal a significant correlation between a matrix of molecular descriptors and a column vector of experimentally determined reactivation potencies for VX-inhibited rat AChE. First, a set of 30 pyridinium aldoxime reactivators was modeled in HyperChem 8, then the optimized molecular models were submitted to Dragon 6, which provided 4885 molecular descriptors for each compound. Eventually, the data input containing the molecular descriptors and the reactivation potencies were data-mined in Matlab using standard and in-house codes as well. To find the best ANN QSAR model, we studied the performance of GMDH algorithm in designing and training multilayer network and compared it with the properties of classical perceptron network trained by 10 different optimization algorithms. The best obtained ANN QSAR model is briefly interpreted and its potential for design of novel pyridinium aldoxime antidotes discussed in this paper.

2   Group Method of Data Handling (GMDH)

In general, GMDH is a type of auto-regulative algorithm which approximates complex and non-linear systems through inductive modeling with extending polynomial series. This class of early deep learning methods has been ascribed to Ivakhnenko, who firstly developed and applied it within supervised analyses of multidimensional systems to predict their behavior as an interpolation problem [6]. The GMDH method can be simply described as a heuristic self-organization, heading towards a mathematical model with optimal complexity. The objectives of GMDH are to identify the most important patterns or hidden interdependencies in the input in order to predict the system response under different conditions. The solution by GMDH is searched for through a sorting-out procedure within the space G of polynomial functions g (1):

$$\tilde{g} = \arg\min_{g \subset G} \varphi(g), \quad \varphi(g) = f\left(M, S, n^2, T, F\right), \tag{1}$$

where φ is an external criterion of model g quality, M is number of variable sets, S is model complexity, n^2 is noise dispersion, T is number of data sample transformation, F is a type of reference function. The GMDH algorithm consequently evaluates possible models using mostly polynomial reference function F. Based on the Ivakhnenko's approach, every function y_n can be represented as an infinite Volterra-Kolmogorov-Gabor (VKG) polynomial (2):

$$\hat{y}_n = a_0 + \sum_{i=1}^{M} a_i x_i + \sum_{i=1}^{M}\sum_{j=1}^{M} a_{M+i+j} x_i x_j + \sum_{i=1}^{M}\sum_{j=1}^{M}\sum_{k=1}^{M} a_{M+M^2+i+j+k} x_i x_j x_k \ldots, \tag{2}$$

where $X(x_1, x_2, \ldots, x_M)$ is the independent variable input vector, $A(a_1, a_2, \ldots, a_R)$, $R \subset \aleph$ is the vector of weights. Interestingly, the GMDH algorithm automatically constructs a feed-forward network and determines the number of layers, the number of neurons and the effective input variables [7]. The unknown weights are calculated by the least square method, minimizing the mean square error function E (3) and applying a two-variable second-degree polynomial version of VKG (4):

$$E = \frac{\sum_{i=1}^{M}(y_i - G_i O)^2}{M}, \, y_i = f(x_{i1}, x_{i2}, x_{i3}, \ldots, x_{iM}), \tag{3}$$

$$G(x_i, x_j) = a_0 + a_1 x_i + a_2 x_j + a_3 x_i^2 + a_4 x_j^2 + a_5 x_i x_j. \tag{4}$$

So far, dozens of GMDH algorithm modifications have been reported in the literature differing in the type of supported input data, architectonic principles and implementation of various parameters. The basic combinatorial GMDH (COMBI) algorithm evaluates gradually complicated models by controlling the prediction on separated training and test data (e.g. with variation and complexity criteria). As a measure of network complexity, Akaike's Information Criterion (AIC) can be used and optimized within the training process (4):

$$AIC = n \log(E) + 2(N+1) + C, \tag{5}$$

where n is the number of observations in the training or test data set, N is the total number of neurons, C is a constant, and E the mean square error.

It is noteworthy that users need not define layers and neurons in this case because GMDH organizes the network topology on its own. The first layer of COMBI GMDH based network processes information in each single input variable, the second layer evaluates contributions of all pairs of the input variables, and, analogically, larger variable combinations are assessed in the next layers (Fig. 1). Depending on the external criteria for training termination, this exhaustive search is able to find a smooth function to fit the input data while avoiding over-fitting. Generally, GMDH neural networks are constructed in several steps: (i) splitting the input into training and test data sets; (ii) gradual generating variable combinations for neurons in a layer; (iii) training and calculation of the partial optimum criteria CR; (iv) selecting beneficial variables; (v) critical decision on forming a next layer. During the formation of GMDH network, neurons which do not contribute to improving the learning procedure can be pruned.

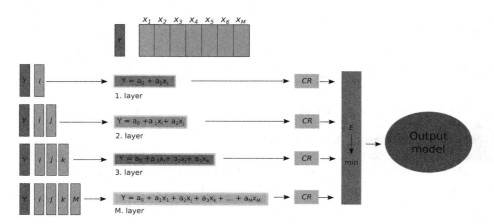

Fig. 1. Scheme of COMBI GMDH algorithm for a network model. Each layer is controlled by CR criteria, which takes into account the complexity and predictive performance of the model.

Besides COMBI algorithm, a number of novel variants of GDHM have been developed with improved capacities to solve specific problems [e.g. Multilayered Iteration (MIA), Objective System Analysis (OSA), Two-level algorithm (ARIMAD), Objective Computer Clusterization (OCC), Pointing Finger (PF), Analogues Complexing (AC), Neuro-fuzzy with Particle Swarm Optimization (NF-PSO)]. GMHD algorithms have been utilized in preparation of optimal educational tests, predicting explosive power, calculating properties of inorganic nano fluids, in cancer diagnosis, etc. [7]. For instance, Nikolaev and Iba developed an approach for growing higher-order networks implementing the multilayer GMDH algorithm and polynomial harmonic functions to predict time series [8]. Abdel-Aal Re described an abductive neuronet with GMDH based learning for complete feature ranking in data on various disease markers [9]. These few examples the demonstrate important position of GMDH among artificial intelligence methods and, thus, recommend its application to QSAR studies.

3   Problem Definition

QSAR analyses belong among very important tools in the current research and development of novel drugs due to their ability to reveal mathematical relationship between molecular structure and the elicited biological effect. Knowledge of this casual nexus in quantitative terms enables to predict *a priori* the biological activity of unseen compounds and to reveal molecular descriptors crucially influencing it. In fact, rational design of novel drugs would be considerably facilitated if medicinal chemists knew which chemical structure to synthesize and administer to patients in order to successfully treat particular diseases. However, revealing complex dependencies between the chemical structure and the biological activity is not straightforward and requires considerable effort of computational chemists, biologists and bioinformaticians.

In the present work, we focused on computational investigation of 30 chemical compounds belonging to the group of pyridinium aldoximes. These substances were synthesized in our laboratories and experimentally tested *in vitro* for the reactivation potency on rat VX-AChE. The set of the compounds exhibits a broad range of biological activities, which can be conveniently exploited to develop a QSAR model for *in silico* screening of potential reactivators. In order to develop a suitable QSAR model, we evaluated several artificial neural network models: (1) a three layer perceptron model with moderate and relative high number of neurons, (2) a self-regulated GMDH neuronet. The first mentioned neural nets were designed classically as feed-forward systems and trained with 10 different training algorithms. The robustness of the solutions was estimated from 20 repetitions of the learning process with random resampling of the input data into training and test sets. The results of the first approach were compared with outcomes provided by GMDH neuronet, which was also subjected to 20 repetions with resampling of the input data.

4   Proposed Solution

4.1   Input Data Preparation

Chemical structures of 30 pyridinium aldoximes and their reactivation potencies towards VX-inhibited rat AChE (*r*AChE) were obtained from Biomedical Research Center (University Hospital Hradec Kralove, Czech Republic). In the present work, we utilized these chemical structures as input data and generated computational chemistry models for them in HyperChem 8.0 program using molecular dynamics and geometry optimization techniques. As a trade-off between the accuracy and the speed, semi-empirical method PM3 was chosen to prepare consistent chemical models for subsequent calculation of the molecular descriptors. The resulting molecular models along with the reactivation potencies induced by 1.0 mM reactivator concentrations were imported into Dragon 6 to prepare a suitable input csv file for statistical analysis in Matlab.

In the present study, we calculated 4885 molecular descriptors for each of the 30 pyridinium aldoximes. The obtained molecular descriptors can be classified into 29 different groups (e.g. topological descriptors, 2D matrix-based descriptors, 3D matrix-based descriptors, WHIM descriptors, GETAWAY descriptors, charge descriptors, molecular properties, drug-like indices, etc.) enabling proper characterization of the studied compounds. However, some molecular descriptors assumed constant or zero values (e.g. number of triple bonds in the molecule) and had to be removed from descriptor matrix X. In total, 2741 molecular descriptors as independent variables and the reactivation potency as the supervisory signal y could be utilized in building QSAR models.

4.2   Designing ANN and GMDH for QSAR Analyses

In the QSAR analyses, we used Matlab 8.6 (2015b) program suite along with Parallel Computing and Neural Toolboxes to speed up the calculations employing a machine with 16 logical CPU cores (2x Intel Sandy Bridge E5-2470, 2.3 GHz, 96 GB RAM) and one NVIDIA Kepler K20 card (2496 shaders, 5 GB VRAM). For the calculations, several M-scripts were developed to remove constant vectors and NaN entries in the descriptor matrix, and to perform analyses of the data by ANN with 10 different training algorithms: (Broyden–Fletcher–Goldfarb–Shanno quasi-Newton backpropagation – BFG, conjugate gradient backpropagation with Powell-Beale restarts – CGB, conjugate gradient backpropagation with Fletcher-Reeves updates – CGF, conjugate gradient backpropagation with Polak-Ribiere updates – CGP, gradient descent with adaptive learning rate backpropagation – GDA, Gradient descent with momentum backpropagation – GDM, gradient descent with momentum and adaptive learning rate backpropagation – GDX, resilient propagation – RP, scaled conjugate gradient backpropagation – SCG, one-step secant backpropagation – OSS). Each of the optimization approaches was repeated 20 times with random input data resampling into the training, test and validation sets. A simplified code developed for testing the performance of ANN on the data is given below [10]:

```
import_data (X, Y);
clean (X, Y, [const, NaN]);
process_input (mapminmax);
net = feedforward (user_defined_topology(3 layers, neurons);
algorithms = {BFG,CGB,CGF,CGP,GDA,GDM,GDX,RP,SCG,OSS};
for i=1:size(algorithms, 2)
   for j=1:20
      net.trainFcn = algorithms{i}
      net_configured = configure (net, X', Y');
      net_configured = init(net_configured, GPU);
      [net_trained,tr] = train(net_configured, X', Y', useGPU);
      Y_pred = (net_trained (X', useGPU))';
      correlation (Y_pred, Y);
      evaluate_performance (tr);
   end
end
```

The above-mentioned schematic M-script performed learning of ANN and prediction of the reactivation potencies employing 10 different training algorithms and a three-layer feed-forward perceptron network. In the first variant, 30 and 15 neurons were assigned to the input and to the hidden layer, respectively. The performance of such ANN was compared with analogical networks containing 2741 and 1370 neurons in the input and the hidden layer, respectively. This configuration reflects the size of X matrix. In both the input and hidden layers, hyperbolic tangent sigmoid as the transfer function was implemented. The output of ANN was realized by one neuron with pureline transfer function. For the learning process, 70% of available cases were utilized, while the remaining 30% of cases were equally divided into test and validation sets. In order to estimate the robustness of the solutions, each training process was repeated 20 times and the resulting performance spectrum was statistically analyzed. Within the training process, the mean square error (E) was used as the objective function of the performance. The training procedure was terminated after 6 consecutive increases of E in the validation set, selecting that iteration which provided the lowest error. Gradient threshold, learning rate, and other parameters were used in the default mode.

For implementation of the GMHD network in QSAR analyses, an M-script was developed using the mean square error E as the objective function and two-variable second-degree polynomials G for defining network layers. Our code was inspired by the works of S.M.K. Heris (www.yarpiz.com). First steps in the procedure are importing the data, their cleaning and random splitting into the training and test sets in the ratio 85:15. The training algorithm based on multiple linear regression of poly-nomial coefficients a_i was set to use maximally 5 layers with maximally 30 neurons each (4). The main part of the GMDH algorithm is schematically illustrated below:

```
import_data (X, Y);
clean (X, Y, [const, NaN]);
params = {max_layers, max_neurons, alpha, ratio};
function gmdh = GMDH(params, X, Y)
    [X1, Y2, X2, Y2] = split (X, Y, ratio);
    Layers = cell(max_layers, 1);
    Z1 = X1;
    Z2 = X2;
    for l = 1: max_layers
        L = get_polynom_layer(Z1, Y1, Z2, Y2);
        if l>1
            if L(1).RMSE2 > Layers{l-1}(1).RMSE2
                break;
            end
        end
        if numel(L) > max_neurons
            L = L(1:max_neurons);
        end
        Z1 = reshape([L.Y1hat],nTrainData,[])';
        Z2 = reshape([L.Y2hat],nTestData,[])';
    end
Yhat = use_GMDH(gmdh, X);
evaluate (Y, Yhat);
end
```

In the first iteration, the function `get_polynom_layer` uses the training data set to define a second order polynomial G involving all pairwise molecular descriptor combinations. The output of the first iteration `L.Y1hat`/`L.Y2hat` is considered as the first network layer result and it is next processed as an input in the second iteration. Importantly, within the calculation of G, each pairwise G_{ij} is considered as a neuron, which can be removed from the network. Thus, only neurons predicting the output signal with relatively low E for the test set are kept in the network. If a new layer brings about higher E for the test set than the previous layer, then the layer is not added and the growing process of the network is terminated.

5   Results and Discussion

Investigations of the reactivation potencies of 30 pyridinium aldoximes towards VX-AChE and the molecular descriptors with a 3-layer perceptron networks provided different results depending on the size of the network and the used training algorithm. All the training algorithms exhibited significant instability after random resampling of the training, validation and test sets. The best predictions according to the highest Pearson's R were given by CGF, CGB, and CGP algorithms, which belong to conjugate gradient backpropagation group.

QSAR analyses by ANN showed that the model predictivity slightly worsens if the number of neurons is increased comparably to the size of the input data. As displayed in Fig. 2, the 30-15-1 topology exhibited overall a better performance than the 2741-1371-1 topology, which is obvious especially from R's for the test sets. The optimal predictivity across all the data sets was provided by the 30-15-1 network trained with CGF algorithm: $\overline{R} = \mathbf{0.58} \pm \mathbf{0.14}$. Analogical trend can be observed in MSE values for the validation sets. The lowest MSEs were again provided by CGF and CGB algorithms (Table 1). With the 30-15-1 topology network, CGF algorithm provided the best QSAR model ($\overline{MSE} = \mathbf{37.4} \pm \mathbf{28.8}$). In GMHD algorithm, neural networks were built automatically with limitation to deploy maximally 30 neurons in 5 layers. After 20 repetitions with data resampling, the GMHD methodology provided QSAR models characterized by $\overline{R} = \mathbf{0.56} \pm \mathbf{0.20}$ for all the data and $\overline{MSE} = \mathbf{140.1} \pm \mathbf{78.2}$ for the test data. Although the topology of the polynomial net is governed in the GMDH algorithm through eliminating instable neurons (e.g. by CR

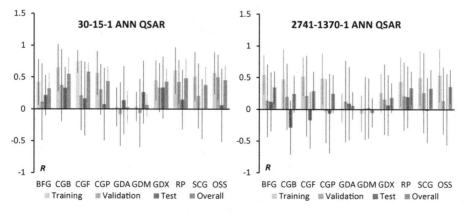

Fig. 2. 30-15-1 and 2741-1370-1 ANN QSAR models optimized by 10 different algorithms. Models are described by Pearson R correlation coefficients and standard sample deviation bars.

Table 1. Mean *MSE's* and the sample standard deviations *s* after 20 training repetitions

TA	BFG	CGB	CGF	CGP	GDA	GDM	GDX	RP	SCG	OSS
Performance in validation set – 30-15-1 ANN QSAR										
$\overline{MSE}$	182.7	44.4	**37.4**	81.4	6e + 30	4e + 32	282.9	101.4	73.2	96.0
$s\left(\overline{MSE}\right)$	104.7	56.3	**28.8**	84.7	1e + 31	1e + 23	98.5	137.0	58.8	80.9
Performance in validation set – 2741-1370-1 ANN QSAR										
$\overline{MSE}$	528.2	**63.3**	81.8	99.5	2e + 50	6e + 43	1930.5	4e + 05	72.5	147.1
$s\left(\overline{MSE}\right)$	989.2	**74.7**	76.1	97.1	8e + 50	3e + 44	2670.9	8e + 05	70.7	153.3

criteria), the procedure seems to be more sensitive to novel data, probably due to the effect of higher order terms in its definition.

Comparing the results obtained by the 3-layer perceptrons and the GMDH neuronets, we observed no significant benefits from application of the polynomial fitting scheme on the studied data. In terms of lower MSE, CGF and CGB algorithms provided rather better results than the GMDH. On the other hand, the present input data are undoubtedly burdened with noise and inaccuracy, which justifies some uncertainty in the prediction by any supervised learning method. However, in the case of the GMDH, we discovered some considerable discontinuities in the prediction of reactivation potencies for the test set (Fig. 3). Due to these irregular outlying estimates, MSE may dramatically increase. While the perceptron networks with sigmoid transfer functions can stay more or less stable, the GMDH polynomial networks seem to diverge more when processing specific vector of molecular descriptors. Although the studied compounds belong to one chemical domain, 30 cases along with 2471 variables represent a typical biomedical challenge for artificial intelligence to develop a reliable QSAR model from limited data.

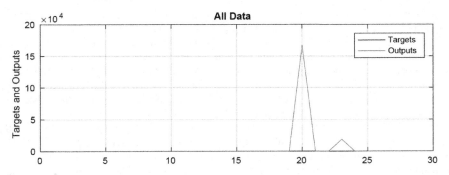

Fig. 3. An example of irregular over-estimations of the reactivation potencies (y-axis) provided by the GMDH algorithm for compound 20 and 23.

6 Conclusion

Even though the inductive GMDH algorithm exhibits powerful capability in learning non-linear problems, it is very sensitive to data-noise and tends to create over-complex polynomials in case of simple systems. In the present work, we revealed that the GMDH polynomial networks are able to build QSAR models for 30 pyridinium aldoxime reactivators of VX-AChE being practically comparable with those obtained by 3-layer perceptron networks. Nonetheless, the GMDH algorithm did not exceed the performance of the perceptron networks populated with relatively few neurons and trained by CGF or CGB algorithms. Moreover, a detailed investigation of the predictions by the GMHD algorithm within the test sets revealed significant over-estimations in some compounds, which probably resulted from sharp local optimization of the objective function by polynomials. Therefore, it will be useful for further QSAR analyses of the pyridinium aldoximes to focus on such supervised learning methods, which do not suppose highly complex interdependencies with sharp functional boundaries.

Acknowledgements. This work was supported by the project "Smart Solutions for Ubiquitous Computing Environments" FIM UHK, Czech Republic (under ID: UHK-FIM-SP-2017-2102). This work was also supported by long-term development plan of UHHK, by the IT4Innovations Centre of Excellence project (CZ.1.05/1.1.00/02.0070), and Czech Ministry of Education, Youth and Sports project (LM2011033).

References

1. Dolezal, R., Korabecny, J., Malinak, D., Honegr, J., Musilek, K., Kuca, K.: Ligand-based 3D QSAR analysis of reactivation potency of mono- and bis-pyridinium aldoximes toward VX-inhibited rat acetylcholinesterase. J. Mol. Graph. Model. **56**, 113–129 (2015)
2. Jokanovic, M., Prostran, M.: Pyridinium oximes as cholinesterase reactivators. Structure-activity relationship and efficacy in the treatment of poisoning with organophosphorus compounds. Curr. Med. Chem. **16**, 2177–2188 (2009)
3. Gorecki, L., Korabecny, J., Musilek, K., Malinak, D., Nepovimova, E., Dolezal, R., Jun, D., Soukup, O., Kuca, K.: SAR study to find optimal cholinesterase reactivator against organophosphorous nerve agents and pesticides. Arch. Toxicol. **90**, 2831–2859 (2016)
4. Waisser, K., Dolezal, R., Palat, K., Cizmarik, J., Kaustova, J.: QSAR study of antimycobacterial activity of quaternary ammonium salts of piperidinylethyl esters of alkoxysubstituted phenylcarbamic acids. Folia Microbiol. **51**, 21–24 (2006)
5. Patel, H.M., Noolvi, M.N., Sharma, P., Jaiswal, V., Bansal, S., Lohan, S., Kumar, S.S., Abbot, V., Dhiman, S., Bhardwaj, V.: Quantitative structure–activity relationship (QSAR) studies as strategic approach in drug discovery. Med. Chem. Res. **23**, 4991–5007 (2014)
6. Ivakhnenko, A.G.: Heuristic self-organization in problems of engineering cybernetics. Automatica **6**, 207–219 (1970)
7. Ebtehaj, I., Bonakdari, H., Zaji, A.H., Azimi, H., Khoshbin, F.: GMDH-type neural network approach for modeling the discharge coefficient of rectangular sharp-crested side weirs. Eng. Sci. Technol. Int. J. **18**, 746–757 (2015)
8. Nikolaev, N.Y., Iba, H.: Polynomial harmonic GMDH learning networks for time series modeling. Neural. Netw. **16**, 1527–1540 (2003)

9. Abdel-Aal, R.E.: GMDH-based feature ranking and selection for improved classification of medical data. J. Biomed. Inform. **38**, 456–468 (2005)
10. Dolezal, R., Trejbal, J., Mesicek, J., Milanov, A., Racakova, V., Krenek, J.: Designing QSAR models for promising TLR4 agonists isolated from *Euodia asteridula* by artificial neural networks enhanced by optimal brain surgeon. In: Nguyen, N.-T., Manolopoulos, Y., Iliadis, L., Trawiński, B. (eds.) ICCCI 2016. LNCS, vol. 9876, pp. 271–281. Springer, Cham (2016). doi:10.1007/978-3-319-45246-3_26

Multiregional Segmentation Modeling in Medical Ultrasonography: Extraction, Modeling and Quantification of Skin Layers and Hypertrophic Scars

Iveta Bryjova[1], Jan Kubicek[1(✉)], Kristyna Molnarova[1], Lukas Peter[1],
Marek Penhaker[1], and Kamil Kuca[2]

[1] VSB–Technical University of Ostrava,
FEECS, K450, 17. listopadu 15, 708 33 Ostrava-Poruba, Czech Republic
{iveta.bryjova,jan.kubicek,kristyna.molnarova.st,
lukas.peter,marek.penhaker}@vsb.cz
[2] Faculty of Informatics and Management, University of Hradec Kralove,
Rokitanskeho 62, 500 03 Hradec Králové, Czech Republic
kamil.kuca@uhk.cz

Abstract. In the clinical practice of the burns treatment, an autonomous modeling of the burns morphological structure is important for a correct diagnosis. Unfortunately, the geometrical parameters of burns and skin layers are subjectively estimated. This approach leads to the inaccurate assessment depending on the experience of an individual physician. In our research, we propose the analysis of multiregional segmentation method which is able to differentiate individual skin layers in the ultrasound image records. The segmentation method is represented by the mathematical model of skin layers while other structures are suppressed. Skin layers are consequently approximated by their skeleton with target of the layers distance measurement. The main applicable output of our research is the clinical SW *SkinessMeter 1.0.0* serving for an autonomous modeling and quantification of the skin layers.

Keywords: Multiregional segmentation · Burns · Skin layers · Ultrasound · Hypertrophic scars

1 Introduction

The hypertrophic scars after the skin burns represent the most frequent complication of burn trauma, especially on the deep level of burns, when it leads to the destruction of skin layers: epidermis and dermis. The hypertrophic scars are observable in the scale of the primary wound, and they are going out over niveau of the surrounding skin, they are painful, stiff, itchy, at the beginning rouge, and can spread up to the form of the scarred contracture with necessity of operating treatment [1].

For the clinical evaluation of the burn scars and their therapeutic response, several scales are routinely used: Patient and Observer Scar Assessment Scale, Visual Analog Scale, Manchester Scar Scale, but the Vancouver scale is the most frequently used. This

© Springer International Publishing AG 2017
N.T. Nguyen et al. (Eds.): ICCCI 2017, Part II, LNAI 10449, pp. 182–192, 2017.
DOI: 10.1007/978-3-319-67077-5_18

scale is based on the pigmentation assessment, height, compliance (pliability) and scar vascularization. The scars evaluation based on those scales is inherently subjective – depends on the experience of physician. One from the objective forms of the burns assessment and its time evolution is the kutometry using viscoelastic skin parameters on the base of the non-invasive suction method, and the ultrasound skin measurement of the skin thickness for the scar contractures thickness [1, 2, 11–14].

In our paper, we present a method of the ultrasound measurement of the individual skin layers with the aim of the geometrical parameters objective assessment from the ultrasound image records. The ultrasound skin diagnostic represent non-invasive approach allowing for a comparison of the tissue thickness in the affected area with the physiologic tissue. The ultrasound waves go through from one tissue structure to another, a part of the wave reflects itself, and it is detected by the sensor and expressed as an electric impulse. Time, needed for the reflection and return to the convertor is represented by the functional distance among individual tissue interfaces [3, 9–12].

The ultrasonography is a very sensitive method allowing for the scar tissue localization, scar thickness assessment and determination of the scar tissue scale in the context with the physiological tissue. The scar contracture thickness determination is an objective tool not only for the skin layers measurement, but also for the scar level assessment. The ultrasound measurement is more sensitive and specific method than subjective measurement methods (the ultrasound accuracy is approximately 0.1 cm) [4–8].

2 Modelling and Extraction of Skin Layers

The skin thickness measurement can be done by several ways. The most frequent method is ultrasonography utilizing wave frequency and attenuation. Based on the ultrasound theory, higher frequency we use, detailed structure we obtain. In an area of the dermatology, the probes with frequencies at least 20 MHz, but better with 30 MHz are commonly used. Skin layers are clinically evaluated by naked eyes. This approach can represent a source of significant subjective error, therefore an automatic segmentation allowing for a precise skin layers extraction and modelling is substantially important [15].

The key factor for burns modelling is the features extraction characterizing skin layers. Individual skin layers have different geometrical features therefore, it would be complicated to search any features precisely characterizing skin layers. Second important feature is the manifestation of the skin layers from the ultrasound image records. Fortunately, skin layers have significantly different brightness features in comparison with other imagined tissues in the ultrasound records. This important aspect is utilized in the multiregional segmentation model.

Segmentation model is aimed on the autonomous modelling of the individual skin layers from the ultrasound image records. We utilize the fact that the ultrasound histogram can be separated into a finite number of the segmentation classes in the dependence of feature's pixels lying inside of a respective class. By this way, each morphological structure (biological tissue) is characterized by a single region. Taking advantage this fact, the ultrasound image is separated into N isolated regions which classifies individual tissues based on its brightness features. In the result we obtain the mathematical model allowing for extraction of the skin layers in the form of the binary image. This model is consequently utilized for the distance measurement of the skin layers. The proposed algorithm for the skin layers modelling is depicted on the Fig. 1.

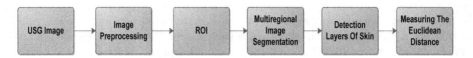

Fig. 1. Flow chart for skin layers modeling

2.1 Native Ultrasound Records

For testing of the segmentation procedure, we utilize data from the ultrasound device *Mindray M7*. We use the highest frequency 11 MHz in the superficialis mode. There is a simple rule that the higher ultrasound frequency we use, the higher spatial resolution, and lower penetration depth of the ultrasound wave we achieve. In the ideal case we would use the probe frequency 20–30 MHz. Nevertheless, we utilize only frequency 11 MHz because we only deal with the skin layers (epidermis and dermis). Their sublayers are omitted for the purpose of our work (Fig. 2).

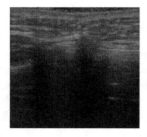

Fig. 2. Native data measured by the *Mindray M7* in the middle abdomen part when lesion is present

2.2 Image Preprocessing

Image preprocessing ensures a better visibility and the recognition of an individual skin layers. For this purpose, we are using the contrast transformation. Unfortunately, native ultrasound images are acquired in relatively lower resolution (800 × 600 px). Lower resolution is usually accompanied with worse contrast of the individual skin components leading to worse segmentation results. Therefore, we apply the contrast transformation ensuring a better skin layers image features (Fig. 3).

Optional alternative which is offered by our SW *SkinessMeter 1.0.0* is Region of Interest (RoI) extraction. It is valuable tool in the cases where we need to focus only on the particular part of the ultrasound image (Fig. 4). This procedure is linked with the problem of a lower number of RoI pixels. If we extract a RoI therefore, this image part

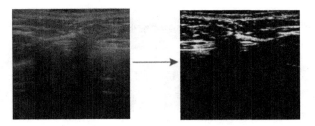

Fig. 3. Native ultrasound data from the *Mindray M7* (left) and contrast enhancement (right)

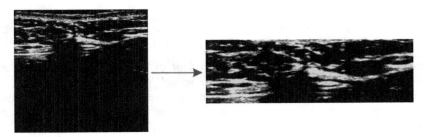

Fig. 4. Native image with contrast enhancement (left) and RoI extraction with cubic transformation (right)

is represented by a partition of pixels from the native image. This fact leads to the worse quality and recognition of the skin layers. In the case of the RoI selection, we simultaneously apply the cubic interpolation increasing spatial resolution in the RoI.

2.3 Multiregional Image Segmentation Procedure

The segmentation procedure represents a complex process which is composed from multilevel segmentation classifying pixels into predefined number (N) classes. As it was aforementioned, the analyzed ultrasound images have a lower resolution therefore, the segmentation model is connected with presence of the image noise which cannot be eliminated in the segmentation procedure. For this reason, the multiregional segmentation procedure is applied with the consequent morphological operations (filling holes and erosion). Those operations carries out the image smoothing in order to achieve an accurate model of the skin layers while the image noise is as suppressed as possible. The process of the segmentation is depicted on the Fig. 5.

2.3.1 Multiregional Otsu Segmentation Model

The multiregional segmentation method represents a sequence of the regions separated by the vector of the thresholdings. For this purpose, we have to solve the optimization problem of an optimal thresholding selection reliably approximating individual regions with regards original structures from the native ultrasound records. We suppose that the

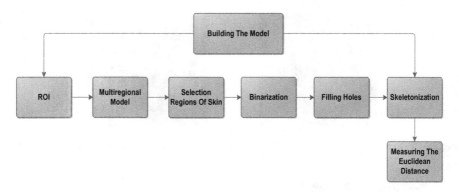

Fig. 5. Flow chart of the complex segmentation procedure

skin layers can be described by the feature vector of within-class variance (1) and between class variance (2) in the segmentation model.

If we analyze the skin layers manifestation therefore, it is evident that skin layers are represented by the brighter spectrum of higher intensities. Furthermore, this brightness spectrum does not exhibit any significant oscillations of the brightness scale. It means that the variance of such values should be minimized. This represents condition for (1).

We suppose that the brightness spectrum of the skin layers with comparison of other adjacent tissues should be significantly different. In this regards, if we computed variance between the skin layers and adjacent tissues therefore, we would obtain higher values of variance. This fact leads to the end that between class variance (2) should be maximized. Other words speaking, we are searching the optimal thresholding sequence minimizing parameter (1), and in the same time maximizing parameter (2).

In terms of computing efficiency is more effective to set the threshold value based on the between class variance, where the threshold value between the highest-variance threshold is determined as an ideal. For the calculation of between class variance is necessary to first calculate the weight and the average intensity values.

The pixels of the image have different shades of gray, those are labeled as L interval $[0,1,...,L]$. The number of pixels in given gray level i is signed as n_i. Whole number of pixels is given: $N = n_0 + n_1 + ... + n_L$. When we divide pixels into two groups 1 and 0 (0 represents the image background, 1 represents the image foreground), according to the level thresholding.

The weight W and the average intensity values μ are calculated by following equations:

$$W_0 = \sum_{i=1}^{k} \frac{n_i}{N} \qquad (1)$$

$$W_1 = \sum_{i=k+1}^{L} \frac{n_i}{N} \qquad (2)$$

$$\mu_0 = \sum_{i=1}^{k} \frac{n_i * i}{N_k} \tag{3}$$

$$\mu_1 = \sum_{i=k+1}^{L} \frac{n_i * i}{N - N_k} \tag{4}$$

Between class variance σ^2 is consequently given by equation:

$$\sigma^2 = W_0 * W_1 * (\mu_0 - \mu_1)^2 \tag{5}$$

Between class variance is calculated for every possible thresholding level and level with the highest value of between class variance is determined as the most ideal specified image is segmented according to it.

2.3.2 Design of Multiregional Segmentation Model

Segmentation method with only one thresholding is not suitable for processing of medical images. In this area, it is much better to use improved Otsu method with multiple level thresholding, which increases sensitivity of the skin layers detection for an image based on the histogram distribution into equally large areas. Specific thresholding level is used for each area. The analyzed image is consequently segmented according to all thresholding levels.

Individual pixels with different shade levels are labeled as L with range: $[0,1,\ldots,L]$. Number of thresholding levels is given as p. The size of one segmented area is given by equation:

$$a = \frac{L}{p} \tag{6}$$

Between class variance σ^2 is calculated similarly as Otsu method:

$$\sigma^2 = W_0 * W_1 * (\mu_0 - \mu_1)^2 \tag{7}$$

The number of divided regions of the histogram is equal to the number of thresholding levels p. The optimal thresholding levels for individual areas are given:

$$P_p = max_p(\sigma^2) \tag{8}$$

For the validity of the above formulas is necessary that the number of pixels in different shades of gray L is equal to 256 * j. Number of the thresholding levels p must belongs in the set: $[2*j,4*j,8*j]$, where j belongs to the set: $[1,2,\ldots,\infty]$. According to the practical results, it is recommended to use set: $[1,2,\ldots,8]$ [7].

The multiregional segmentation algorithm is described by following diagram (Fig. 6):

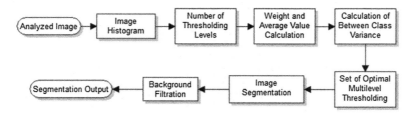

Fig. 6. Flow chart of the multiregional segmentation method

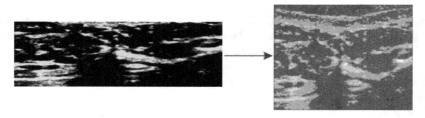

Fig. 7. RoI extraction with contrast transformation (left) and multiregional segmentation model containing 9 regions (right)

3 Testing and Evaluation of Segmentation Results

3.1 Testing of Multiregional Segmentation

We process the monochromatic ultrasound images (256 shade levels). Judging by the empirical testing, the optimal alternative appears itself 8 thresholding levels. It means that the multiregional model uses 9 regions. For this combination, the parameters (1) and (2) are calculated 256 times (for each intensity level). In the result, we obtain the segmentation output recognizing 9 areas (Fig. 7) where the skin layers are indicated by yellow levels. The multiregional segmentation is represented by the index matrix containing values 1–9 (9 segmentation classes). It means that the each class is represented by an unique number. Consequently, all the classes representing image background can be easily eliminated by a selection of a respective class number from the index matrix. The multiregional segmentation also offers the artificial color coding. This procedure shows result in the color scale in comparison with the native image where the individual structures are represented by shade levels. On the base of the experiments, we found out that the last three segmentation classes represent skin layers. Based on this fact, other classes can be reduced from index matrix (Fig. 8).

3.2 Morphological Adjustment of Segmentation Model

Last part of the segmentation procedure is an application of the morphological erosion. The segmentation outputs (Fig. 8) is accompanied by the noise which is hardly eliminated from the segmentation model. This noise is caused by a lower resolution and

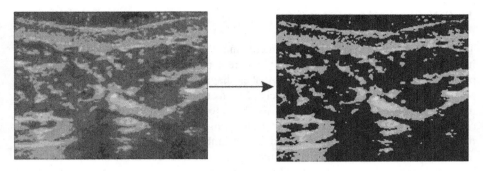

Fig. 8. Multiregional segmentation model containing 9 regions (left) and model of the skin layers (right)

worse homogeneity of the skin structures. The noise is manifested by the tiny clusters of pixels surrounding the main skin layers. The morphological erosion is able to eliminate such pixel clusters by setting a local probe with appropriate geometrical structure. In this regard we have to realize that the noise is directionally independent (horizontal and vertical orientation). Therefore, we are using the morphologic matrix probe (10×10 pixels) eliminating this noise in the both directions (Fig. 9).

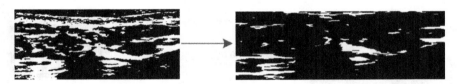

Fig. 9. Binarization of segmentation model (left) and morphological erosion with the morphological probe 10×10 (right)

Clinically important applicable output is the skin layers measurement based on the multilevel segmentation model. In order to achieve this task, it is necessary to approximate of the individual skin layers by its skeleton (Fig. 10). The morphological skeletonization simplifies the skin layers in the form of the single lines. Skeleton allows for precise localization of the skin layers for the measurement of their distance. The distance between individual layers is computed by the Euclidean distance.

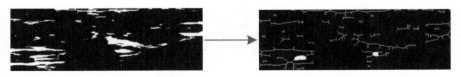

Fig. 10. Final skin layer model (left) and the morphological skeletonization (right)

4 SW *SkinessMeter 1.0.0*

The visual ultrasound image assessment and the measurement with the using of implemented markers are subjectively inherent and depend on the experience of a physician. A clinical expert sporadic makes diagnosis in comparison with radiologic physician making the diagnosis routinely. This fact often leads to the ambiguous results may significantly influence the diagnosis. Therefore, we have developed, with clinical cooperation, the SW *SkinessMeter 1.0.0* allowing for unified and objective measurement. By this way, a mistake caused by an ambiguous diagnostic representation is minimized. In the same time, an accuracy of a selected treatment is improved. It is important, especially for economically demanding therapy as it is the laser therapy of scars.

The SW *SkinessMeter 1.0.0* allows for two main functions for the burn skin assessment. The ultrasound images standardly contain besides skin layers also other tissues. Therefore, we have proposed the multiregional segmentation method carrying out the skin layers modelling in the form of the binary image. The model carries out the autonomous skin layers extraction while other tissues are suppressed (Fig. 11). The second important function of the SW is an evaluation of the skin layers distance. For this task, the skin model is approximated by the single lines (image skeletonization) allowing for a precise measurement (Fig. 12). Furthermore, SW allows for sequence of measurements for a comparison of the individual skin layers.

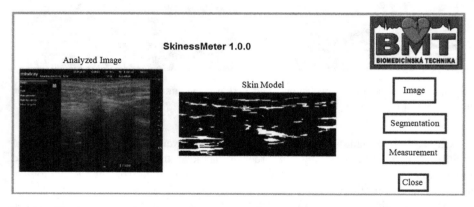

Fig. 11. SW *SkinessMeter 1.0.0:* analysis of native ultrasound images and the skin layers modelling

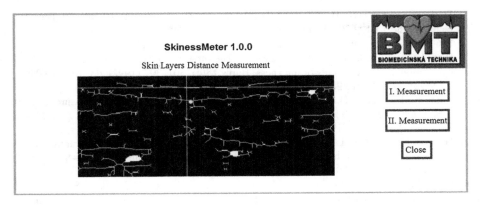

Fig. 12. SW *SkinessMeter 1.0.0:* Approximation of the skin layers by its skeleton and the skin layers measurement

5   Conclusion

In the clinical practice, the objective skin features measurement is relatively time and technically demanding. Our work brings a development of the instrument for objective assessment of selected skin features, respectively of the large areas affected by the hypertrophic scars. The clinical evaluation is frequently different because the ultrasound records are hardly recognizable. This fact is often caused by different manifestation of the skin layers in various image records. It also depends on the particular expertise. Radiologic opinion is commonly different from clinical experts dealing with the hypertrophic scars.

Multiregional segmentation model utilizes a fact that the skin layers have closely concentrated brightness spectrum without significant oscillations. Therefore, the multiregional segmentation method determines the thresholds on the base of the variance minimization. Individual regions represent recognizable tissues into index matrix. In the last step of the segmentation, the skin layers model is extracted from the index matrix while other tissues are suppressed. We have developed the environment SW *SkinessMeter 1.0.0* allowing for the autonomous extraction of the individual skin layers in the form of the mathematical model.

Acknowledgment. The work and the contributions were supported by the project SV4506631/2101 'Biomedicínské inženýrské systémy XII'. This study was supported by the research project The Czech Science Foundation (GACR) No. 17-03037S, Investment evaluation of medical device development.

References

1. Klosová, H., Štětinský, J., Bryjová, I., Hledík, S., Klein, L.: Objective evaluation of the effect of autologous platelet concentrate on post-operative scarring in deep burns. Burns **39**(6), 1263–1276 (2013). doi:10.1016/j.burns.2013.01.020. [cit. 2017-04-29]. ISSN 03054179

2. Štětinský, J., Klosová, H., Kolářová, H., Šalounová, D., Bryjová, I., Hledík, S.: The time factor in the LDI (laser doppler imaging) diagnosis of burns. Lasers Surg. Med. **47**(2), 196–202 (2015). doi:10.1002/lsm.22291. [cit. 2017-04-29]. ISSN 01968092

3. Hambleton, J., Shakespeare, P.G., Pratt, B.J.: The progress of hypertrophic scars monitored by ultrasound measurements of thickness. Burns **18**(4), 301–307 (1992). doi:10.1016/0305-4179(92)90151-J. [cit. 2017-04-29]. ISSN 03054179

4. Katz, S.M., Frank, D.H., Leopold, G.R., Wachtel, T.L.: Objective measurement of hypertrophic burn scar: a preliminary study of tonometry and ultrasonography. Ann. Plastic Surg. **14**(2), 121–127 (1985)

5. Fong, S.S.L., Hung, L.K., Cheng, J.C.Y.: The cutometer and ultrasonography in the assessment of postburn hypertrophic scar—a preliminary study. Burns **23**, S12–S18 (1997). doi:10.1016/S0305-4179(97)90095-4. [cit. 2017-04-29]. ISSN 03054179

6. Kubicek, J., Penhaker, M., Pavelova, K., Selamat, A., Hudak, R., Majernik, J.: Segmentation of MRI data to extract the blood vessels based on fuzzy thresholding. Stud. Comput. Intell. **598**, 43–52 (2015)

7. Kubicek, J., Valosek, J., Penhaker, M., Bryjova, I., Grepl, J.: Extraction of blood vessels using multilevel thresholding with color coding. In: Sulaiman, H.A., Othman, M.A., Othman, M.F.I., Rahim, Y.A., Pee, N.C. (eds.) Advanced Computer and Communication Engineering Technology. LNEE, vol. 362, pp. 397–406. Springer, Cham (2016). doi:10.1007/978-3-319-24584-3_33

8. Kubicek, J., Valosek, J., Penhaker, M., Bryjova, I.: Extraction of chondromalacia knee cartilage using multi slice thresholding method. In: Vinh, P.C., Alagar, V. (eds.) ICCASA 2015. LNICSSITE, vol. 165, pp. 395–403. Springer, Cham (2016). doi:10.1007/978-3-319-29236-6_37

9. Sciolla, B., Cowell, L., Dambry, T., Guibert, B., Delachartre, P.: Segmentation of skin tumors in high-frequency 3-D ultrasound images. Ultrasound Med. Biol. **43**(1), 227–238 (2017)

10. Csabai, D., Szalai, K., Gyöngy, M.: Automated classification of common skin lesions using bioinspired features. In: IEEE International Ultrasonics Symposium, IUS, art. no. 7728752, November 2016

11. Ewertsen, C., Carlsen, J.F., Christiansen, I.R., Jensen, J.A., Nielsen, M.B.: Evaluation of healthy muscle tissue by strain and shear wave elastography - dependency on depth and ROI position in relation to underlying bone. Ultrasonics **71**, 127–133 (2016)

12. Sciolla, B., Delachartre, P., Cowell, L., Dambry, T., Guibert, B.: Multigrid level-set segmentation of high-frequency 3D ultrasound images using the Hellinger distance. In: 9th International Symposium on Image and Signal Processing and Analysis, ISPA 2015, art. no. 7306052, pp. 165–169 (2015)

13. Sciolla, B., Ceccato, P., Cowell, L., Dambry, T., Guibert, B., Delachartre, P.: Segmentation of inhomogeneous skin tissues in high-frequency 3D ultrasound images, the advantage of non-parametric log-likelihood methods. Phys. Proc. **70**, 1177–1180 (2015)

14. Wawrzyniak, Z.M., Szyszka, M.: Layer measurement in high frequency ultra-sonography images for skin. In: Proceedings of SPIE - The International Society for Optical Engineering, 9662, art. no. 96621 (2015)

15. Gao, Y., Tannenbaum, A., Chen, H., Torres, M., Yoshida, E., Yang, X., Wang, Y., Curran, W., Liu, T.: Automated skin segmentation in ultrasonic evaluation of skin toxicity in breast cancer radiotherapy. Ultrasound Med. Biol. **39**(11), 2166–2175 (2013)

Cyber Physical Systems in Automotive Area

Model of a Production Stand Used for Digital Factory Purposes

Markus Bregulla[1], Sebastian Schrittenloher[1], Jakub Piekarz[1],
and Marek Drewniak[2(✉)]

[1] Technische Hochschule Ingolstadt, Zentrum für Angewandte Forschung,
Ingolstadt, Germany
{markus.bregulla, ses1515, jakub.piekarz}@thi.de
[2] Aiut Sp. z o.o., Gliwice, Poland
marek.drewniak@aiut.com

Abstract. In the first part of this paper a typical realisation of a production stand is presented along with a description of stages of its implementation, layers (fields) of realisation and their interconnections and dependencies. In the second part a digital factory concept is presented. The concept is used to support the transfer of knowledge during construction and maintenance of production stands by preparation of a model which includes components, roles and interfaces of a stand. The approach allows to easily and comfortably get access to resources related to specific layers, depending on demands and needs of the user. Remote access which is based on data in a cloud forms a system of services which are available from web browser level.

Keywords: Production · Production stand · Manufacturing · Transfer of knowledge and technology · Digital factory · Cloud-based services · Digitalisation

1 Introduction

Construction of production stands in the industry is a long, arduous and complicated process. The goals are to create a concept of a technological solution, the design of a station and how to build it. The difficulty lays within the use of expert knowledge and strict cooperation of specialists from various fields like mechanics, pneumatics, electrics, electronics or informatics. For this reason there only is a limited number of companies that focus on this type of works.

The solutions that are implemented in industry are mostly based on the usage of tools which are intended for designing, launching and reporting of processes. They are typically used by specialists of the specific field. An example is drawing of mechanical elements and the simulation of their dynamics in 3D tools which are done by a mechanical team or a preparation of a PLC program and the simulations of control algorithms which are performed by a programming team.

Experiences from start-ups, maintenance and modifications of production stands show that conclusions, parts of technology or even selected devices with ready fragments of technological realisation are repeatedly used in consecutive constructions

© Springer International Publishing AG 2017
N.T. Nguyen et al. (Eds.): ICCCI 2017, Part II, LNAI 10449, pp. 195–204, 2017.
DOI: 10.1007/978-3-319-67077-5_19

(e.g. during duplication of stands on production lines or preparation of a stand that is somehow similar to the previous one). For purposes of gathering and transfer of such knowledge it is a good solution to create a system capable of granting easy access to the resources, including documentations, files and written conclusions. The system however, would have to know about interconnections between technological elements. Not only it is necessary for providing resources that are needed during any specific stage of realisation, but also to detect potential changes, that were made within one realisation layer and could impact on others. It is then important to define a model that would contain a description of selected parts of a technology, all its components, and their roles in an installation, interfaces between them and resources that are related to them. In order to provide the capability of universal and scalable transfer of knowledge it is necessary to keep some standard. The Automation Markup Language (AML) standard which recently shows up in the automation branch gains more and more popularity. It allows to describe the machine in neutral XML data format and to create a corresponding model using dedicated editors. The model can then be processed in any way by using it in client-server systems that would work according to user's needs. Additionally, insertion of a system in a cloud could provide a fast realisation of requests and grant easy access to all resources that are currently required.

2 Production Stands in Industry

Industrial production can be performed in variety of structures, from self-standing stands located within an industrial facility to production lines in which the product is prepared during the progression through continuous stations. Based on the type of continuity of a process there are three main types of production defined:

- Continuous ones, in which the flow of a material is continuous. An example of such is the production of a chemical substance or production of a heat [1, 3].
- Batch ones, in which at the input there is a set of parameters and ingredients provided (a batch) which allows to produce a strictly defined amount or number of a product. The set is needed for the performance of a process. This type of production is used in food industry, e.g. in production of a beer or any types of drinks [1, 3, 4].
- Discrete ones, in which the product is produced in an individual way and is countable. This is the most common type of a production, used e.g. in automotive industry [1, 3].

There are many possibilities to control the industrial process. The most common solution is the one based on a control unit. It reads information from inputs, e.g. sensors and field devices, performs a control algorithm and controls (updates) end-point devices on its outputs. The control may be realised by many types of units, e.g.:

- PC computers. Due to the necessity of keeping a specific time determinism which would require the process to be performed in real-time, these solutions are used rarely. The limitation emerges of operation systems.
- Microcontrollers. This solution is used mainly in small systems, for which control via a dedicated controller is possible [2].

- Industrial computers. These devices are constructed in a way that the system can work in real-time so that they are much safer than PC computers. Examples of such units are programmable logic controllers (PLC) realised in systems of local or distributed control systems (DCS). They are the most common solutions that are met in industry [3].

Due to the type of control, end-point devices can be divided in two groups:

- Devices that are controlled in a binary way, e.g. drives or motors controlled logically, devices activated via switches and relays, lamps, indicators, pneumatic actuators.
- Devices that are controlled using communication protocols, e.g. robots, numerical devices (machine tools, nut runners), scanning and identification systems (barcode and data matrix scanners, magnetic card readers).

2.1 Construction

Production stands consist mainly of two big groups of components: groups of field devices and control systems. Their elements may differ due to different realisation techniques and different process-control mechanisms. Typically in the machine part there is mechanics, electrics and optionally systems for control of an additional medium, e.g. pneumatics. The mechanics contains elements like construction frames, mountings, supports or bases. The electrics is composed of wires, contacts, safety switches and other widely understood electrical devices. The sample pneumatics may contain valves, pipes with compressed air and pneumatic actuators. The crucial factor is that mostly all these layers interact with each other which means that they react and influence one each other.

In the system control group there are typically three layers present: software for control algorithms, hardware with periphery and connections that grant communication via networks and protocols. Similarly to the machine group the layers here also interact with each other and there are exact connections and dependencies between them. The scheme of a typical realisation of a production stand is presented in Fig. 1.

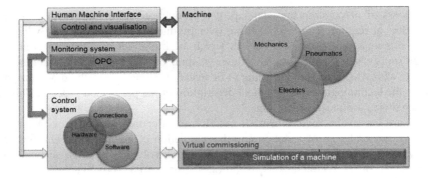

Fig. 1. Typical realisation layers of an industrial stand

In addition to the control system and the machine there are also other groups that may be encountered in realisations of production stands. One of the most common ones are human-machine interfaces (HMI) for real-time control and diagnosis of a process.

The other group contains solutions for monitoring and diagnosis of signals and states of a machine, e.g. technologies based on Object Linking and Embedding for Process Control (OPC), realised e.g. as client-server systems mounted on PCs, remote devices or as embedded on FPGA boards [5]. It gathers signals in external databases for possible post-processing purposes. Additionally there are subsystems for simulation of physical environment, so-called Virtual Commissioning, which are used during processes of start-ups of production stands. They can support the analysis of behavior of mechanical devices and the testing of control algorithms while the physical layer is not ready yet.

2.2   Interconnection of Realisation Layers

The "intersection" of layers mentioned before is a very important factor during the process of designing, realising and maintaining production stands. Each time a new device is added to a station it requires that a mounting, support or base is designed and needs a safe area in which it is going to work. The device usually requires a supply - using e.g. a pneumatic or electrical system - therefore its plugging and connection must be considered. Furthermore, devices are often equipped with auxiliary electrical sensors, e.g. induction or optical for detection of position. These also require their own mechanical and electrical projects, like supports on which they are mounted or plugging and wiring. The device must be included in a control system by attaching it to a dedicated expansion module via an interface or electrically and needs to be included in control algorithms. Finally, for visualisation and diagnostic purposes, input and output signals need to be mapped properly and included in HMI systems, virtual commissioning and technologies for monitoring.

Each time a modification in any of the layers of realisation is performed it may become necessary to make updates in other layers. It is connected not only with analysis whether the changes need to be included in other layers, but also requires all documents to be updated (documentations and instructions).

2.3   Stages of Implementation

The process of preparation and start-up of a production stand starts with a stage of verification of requirements and working over production technology with which the product is going to be produced. Factors which are important here, among others, are type of supply medium for specific devices, time of task realisation, standard and materials with which the stand is going to be made of. Nonetheless, the most important factor is the technological and technical description of the performed process. This and all other stages are presented in Fig. 2.

The answer to the requirements is usually an offer that presents a concept of a solution, a general plan of a station and the simulation of its operation. On these first two stages a flow of documents begins, which involves protocols, technological requirements, plans or conclusions. Additionally, one of the "products" of an offering may be a simulation of a stand, used both for visualisation of a process and emulation

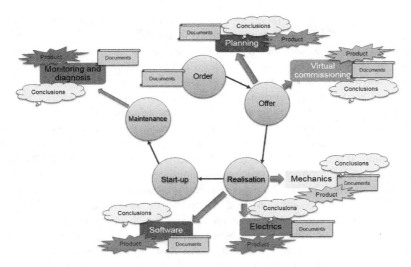

Fig. 2. Stages of construction and maintenance of a production stand

of behaviour of end-point devices of a station. After the project is accepted, a stage of construction of a stand begins. Individual groups of specialists from various fields cooperate with each other in order to create an industrial stand that is working and which meets technological requirements. During this stage some mistakes are inevitable. Such mistakes are usually costly and their corrections require additional time and materials.

In order to avoid them a good professional practice, braced with experience is needed. A reliable help and support is to use simulations and emulations of behaviour of devices which can be started at the stage of an offer. As in previous stages, also in this one a lot of documents and files are generated, including technical documentations and manuals. It is very important to write down all conclusions after mistakes in order to avoid repeating them in the future.

Then the moment of start-up of a production stand comes. The start-up is a stage in which finding and recognition of all abnormalities and mistakes in the station's work need to be done. It is crucial so after the stand is finally in the hands of the local maintenance service it can be kept in a good shape without the necessity to call for the manufacturing company each time an emergency or malfunction occurs.

The process of a station's start-up is a valuable source of knowledge, which is gathered in conclusions and reports and may be archived and processed during the start-ups of new production stands in the future.

3 Digital Factory

Nowadays, due to high competition, companies have to produce efficiently and flexibly and need to be innovative in order to adapt to quickly changing market conditions. Virtual technologies that enter the sector of automation control can fasten and simplify processes of production and therefore can help companies to meet their adaptation requirements. One of the methods that is used in this field is the digital factory.

The digital factory was defined by Kühn as *the upper level for an extensive network of digital models, methods, and tools – inter alia the simulation and 3D visualisation, which is integrated by an overall data management. The target is to completely plan, evaluate and continuously improve all relevant structures, processes and resources related to the product, that are performed in a real industrial plant* [8].

The approach is composed of several mechanisms that support processes of design, implementation and maintenance of industrial installations. One of them is virtual commissioning. It is used on early stages of implementation to show overall function of a station, analyze the production process and simulate it in order to point out potential errors, e.g. mechanical ones. The other one is digital modelling of an installation that can describe the station completely and which can be used as a support for maintenance. The next mechanism is online monitoring which can be used for real-time diagnostics of an installation. In addition, various mechanisms are used that support the transfer of knowledge. They help handling the flow of documents, allow to gather reports and written conclusions and support versioning in order to keep all data up-to-date and to use the knowledge in further implementations.

3.1 Automation Markup Language

The description of a production stand existed ever since the stations are present. The descriptions however were not unified or were focusing only on particular aspects of stations, e.g. their energy efficiency and data related to components with which the cost reduction could have been made. The newest trend is to describe a station in a way that it would support the knowledge during its construction and maintenance.

One of the tools that gains more and more popularity is Automation Markup Language (AML). It is a data format based on XML designed for the vendor independent exchange of plant engineering information. The goal of AML is to interconnect engineering tools from the heterogeneous tool landscape in their different disciplines, e.g. mechanical plant engineering, electrical design, process engineering, process control engineering, HMI development, PLC programming, robot programming etc. It is used to connect information out of various machines and documents, provided in very different formats, in a single data format. Gathering of all these types of information together is the first step in preparation of the description of a machine.

The core of AML is the top-level universal data format called Computer Aided Engineering Exchange (CAEX), which is used to connect the different data formats.

AML stores engineering information following the object-oriented paradigm and allows modelling of real plant components as data objects that encapsulate different aspects.

An object may consist of other sub objects and may be a part of a larger composition or aggregation. It may describe e.g. a signal, a PLC, a control valve, a robot, a manufacturing cell in different levels of detail or a complete site, line or a plant. Typically the objects used in plant automation comprise information on topology, geometry, kinematics and logic, whereas logic comprises sequencing, behavior and control. With all these information a model can be created in the AML Editor. The model can be used in variety of ways, e.g. to provide resources that are needed when changing selected parameters of a specific device or during maintenance actions by looking for dependencies between components. The solution may save a lot of working time and money [9, 10].

The Editor is built of four windows, the InstanceHierarchy, the SystemUnitClassLib, the RoleClassLib and the InterfaceClassLib. The InterfaceClassLib contains relations between objects and their resources that are out of the AML model. The RoleClassLib is used as a class containing possible roles of the elements, e.g. communication. In the SystemUnitClassLib all components are gathered, e.g. pneumatic actuators or electrical switches. It is used as a library of elements with basic properties for modelling in the InstanceHierarchy.

The hierarchy shows the general structure of the model. It can be built up by dragging and dropping of the listed elements from the other three windows and shows up the connections. It however, remains not connected to the libraries.

3.2 Cloud-Based Approach

The idea of preparing knowledge in a way that it can be used immediately, considers another young technology: cloud computing. The approach allows to easily and comfortably get access to resources related to specific layers depending on demands and needs of user. As a part of Internet-based computing, the cloud computing is more and more visible in every branch of life. The same situation takes place in an industrial market. The industry needs fast and reliable solutions for storing, versioning, access control and administration of the data. Cloud computing provides processing resources, networks, servers, storage and other devices on demand [12, 13]. These devices can be provisioned and released with minimal management effort. Companies can scale up, e.g. the space of the storage or the computing power, as computing needs increase and then scale down again as demand decreases.

Cloud computing and storage solutions allow to store and process with the data over various capabilities of the third-party centers. But it is also possible to use own resources, e.g. for the storing of data. This is important for big companies, which are not willing to store their data in foreign countries. The cloud-based service interconnects all project and production data with a web interface, allowing the end-user to get direct access only to necessary and actual files. A versioning system will guarantee that the information is up to date. The web idea is based on the approach that the end-users should not be worried whether the information they work on is the newest. In case of any changes made in a system, the information about it has to be provided to end-users automatically whenever they continue working.

The use of an AML model may be straightly connected to cloud computing and cloud-based storing of the resources. Thanks to this solution, the working conditions may be improved and made more comfortable by granting easy and fast access to all resources. Additionally, the situation when requirements change in future and the involved engineers would have to review the whole project state, can be avoided [14].

3.3 Use Cases

The main idea of using an AML model supported by a cloud storage and computing power is to create a system for supporting engineers at their work. To ensure full platform independence, universality and scalability, a web interface should be used. A typical use case of the proposed system is presented in Fig. 3.

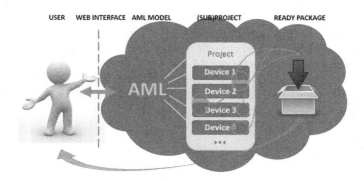

Fig. 3. Example use case of an AML-based system that works in a cloud

Every user has assigned proper privileges and field of interest connected to his or her profession. This allows a program to grant access only for a region of a project connected to the user profession. For example a mechanical engineer will have full access to all CATIA files, results from mechanical calculations, technical requirements and specifications, etc. He or she will however, never see the PLC code, strictly electrical plans and descriptions, production data or pre-project assumptions.

The proposed AML model is strictly tree-structured. The model consist of one or multiple projects. The project can be any task that a company works on, or worked with. Every resource in the project (documents, diagrams, source codes, etc.) is assigned to one or more realisation layers (electrics, pneumatics, mechanics, virtual commissioning, PLC, software, production data). This property allows later on to supply the user with proper data (depending on the user privileges, profession and his or her field of interest).

At first, a user (in this use case a PLC programmer) needs access to the Internet, possibly intranet, with private server. The cloud server provides proper web interface. During logging in the system he or she gets a set of information regarding the user (privileges, profession). Afterwards the user chooses the project and task he or she wants to work on. These information are crucial for further functioning of the program.

In the next step information about user and chosen project are sent further to the module responsible for support of the AML model. To find suitable resources for the user, the program first checks the AML structure for the proper (sub) project.

Every resource that has the PLC property set is added to a list (if the user has an actual right to the resource). For every file a proper information to a repository is sent. Then copies of all files from the request list are collected together, put into a package and sent as a downloadable attachment. The user can start to work immediately, with the full set of the newest available project files. This ensures that the "legacy problem" will not take place. After committing changes, the program informs the repository and copies are transferred to a proper place within a cloud.

The entire solution can support not only works that are related to design and maintenance of the installation.

It can also be integrated with Manufacturing Execution Systems and provide technological and real production data in order to parametrize and optimize the production. This can be especially useful for short-series production [15] required for e.g. prototypes manufacturing.

4   Conclusions

Digitalisation of tasks from the field of automation control gains more and more popularity. It is mainly caused by a development of mechanisms that support the process of creation and maintenance of industrial installations.

The process of installation of industrial equipment is arduous, difficult and requires cooperation of teams from many fields of expertise like mechanics, electrics, pneumatics or programming of control units. What's important, the experiences and technologies that are used during the start-up of a stand are very often used in processes of start-up of another stations.

One of the concepts supporting teams during work in automation control is Digital Factory. The concept assumes several mechanisms. The first is the summarisation of experiences from the designing and construction of a production stand or a line. Secondly, it gathers and versions documents related to technologies in specific realisation layers. Finally, it describes a station using a model that combines physical objects with their roles, common dependencies and interfaces.

The first element needed for the construction of the Digital Factory method is the definition of range of a problem. Firstly, it is necessary to define realisation layers and specify all data formats and project documentations. Secondly, when layers are determined, an analysis of interconnections of specific processing signals, devices and realisation layers can be made.

The Automation Markup Language (AML) standard can be used for preparation of model of a production stand. It provides basic mechanisms for grouping of objects (e.g. end-point devices) into hierarchy instances along with their roles and dependencies.

The processing of a model is performed in cloud storage and computing. This approach has many advantages, from the easiness of remote access to the resources, through the flexibility in storing and management of files to the possibility of versioning of files. Access to the system is realised by the web application which grants access only to files that are related to the user.

Acknowledgements. This work was supported by the European Union from the FP7-PEOPLE-2013-IAPP AutoUniMo project "Automotive Production Engineering Unified Perspective based on Data Mining Methods and Virtual Factory Model" (Grant Agreement No: 612207) and research work financed from funds for science in years 2016-2017 allocated to an international co-financed project (Grant Agreement No: 3491/7.PR/15/2016/2).

References

1. Machine Design (2016). http://machinedesign.com/
2. Urbaniak, A.: Systemy Wbudowane. Politechnika Poznańska, Poznan (2015)
3. White, C.H., Thai, B.W.: Integrated manufacturing logistics: byproducts can be critical. In: Simulation Conference Proceedings, vol. 2, pp. 1310–1315. IEEE, December 1999
4. Selecting a development approach, Office of Information Services, March 2008
5. Cupek, R., Ziebinski, A., Franek, M.: FPGA based OPC UA embedded industrial data server implementation. J. Circ. Syst. Comput. **22**(8), 1350070 (2013)
6. Wedeniwski, S.: Strategy, business model and architecture in today's automotive industry. In: Wedeniwski, S. (ed.) The Mobility Revolution in the Automotive Industry, pp. 75–238. Springer, Heidelberg (2015). doi:10.1007/978-3-662-47788-5_3
7. Monfared, M.A.S., Yang, J.B.: Design of integrated manufacturing planning, scheduling and control systems: a new framework for automation. Int. J. Adv. Manuf. Technol. **33**(5), 545–559 (2007)
8. Kuhn, W.: Digitale Fabrik - Fabriksimulation für Produktionsplaner. Hanser, Wein (2006)
9. Bracht, U., Geckler, D., Wenzel, S.: Digitale Fabrik - Methoden und Praxisbeispiele. Springer, Heidelberg (2011). doi:10.1007/978-3-540-88973-1
10. Drath, R.: Datenaustausch in der Anlagenplanung mit AutomationML - Integration von CAEX, PLCopen XML und COLLADA. Springer, Heidelberg (2010). doi:10.1007/978-3-642-04674-2
11. <AutomationML/>: The Glue for Seamless Automation Engineering, Whitepaper AutomationML, Part 1 - Architecture and general requirements, AutomationML consortium, October 2014
12. Novadex (2016). https://www.novadex.com/de/glossar-artikel/definition-cloud-computing-was-ist-cloud-computing/
13. Armbrust, M., Fox, A., Griffith, R., Joseph, A.D., Katz, R.H., Konwinski, A., Lee, G., Patterson, D.A., Rabkin, A., Stoica, I., Zaharia, M.: Above the Clouds: A Berkeley View of Cloud Computing. Electrical Engineering and Computer Sciences, University of California at Berkeley (2009)
14. Buyya, R., Yeo, C.S., Venugopal, S., Broberg, J., Brandic, I.: Cloud computing and emerging IT platforms: vision, hype, and reality for delivering computing as the 5th utility. Future Gen. Comput. Syst. **25**(6), 599–616 (2009)
15. Cupek, R., Ziebinski, A., Huczala, L., Erdogan, H.: Agent-based manufacturing execution systems for short-series production scheduling. Comput. Ind. **82**, 245–258 (2016)

Improving the Engineering Process in the Automotive Field Through AutomationML

Markus Bregulla and Flavian Meltzer[✉]

Technische Hochschule Ingolstadt, Zentrum für Angewandte Forschung,
Ingolstadt, Germany
markus.bregulla@thi.de, flavian-meltzer@t-online.de

Abstract. For a decade, it has been officially known that the most cost-intensive part of a body-building project is software engineering. The reason for this is the fact that in the engineering process many different types of information from their respective tool chains must come together and be combined. This situation is intensified by the heterogeneous engineering tool landscape that makes it difficult to reuse existing data and information from finished engineering steps without resorting to a paper interface. For this reason, many representatives of the automotive industry came together to solve these problems which resulted in the AutomationML format. AutomationML is an independent data format that allows bridging the gap between the various engineering fields and tool chains, thereby improving the overall process. The goal of this article is to provide an insight into the currently defined AutomationML standard and its possibilities.

Keywords: Production · Transfer of knowledge and technology · Digitalization · Virtual commissioning · AutomationML · Workflow management

1 Introduction

1.1 Development of the Engineering Process in the Automotive Field

Nowadays the theme of Industry 4.0 is very common and describes a core development in the industry which implies an interconnection and data exchange of all actors in a defined working environment [1]. By taking up on this idea existing processes are getting improved and new possibilities are presenting themselves.

The same development is arising in the engineering process because of gradually growing functionality and complexity levels without the availability of more time in implementation. To counter this problem and to reduce costs, existing processes and tool chains are getting analyzed for possible new interfaces and data exchange that could help to speed up the implementation phase.

These solutions are often cut out for a very specific task in the engineering process and therefore, lose the advantage of further possible uses in the whole process. The plethora of tools also further complicates data exchange. Therefore, data should be

© Springer International Publishing AG 2017
N.T. Nguyen et al. (Eds.): ICCCI 2017, Part II, LNAI 10449, pp. 205–214, 2017.
DOI: 10.1007/978-3-319-67077-5_20

saved in a format which enables every tool to read and process it independently of its source format, for its defined working range in the engineering process.

What Industry 4.0 implies for the future work environment AutomationML (AML) clarifies for the future engineering process [2]. With AML a powerful independent data format has been defined which enables every participant in the engineering chain to improve his situation and with this the whole process.

1.2   Finding a Solution

The present engineering process contains many different stages like construction, hardware design, software development and the newest addition of virtual commissioning. Each trade has its different tasks, tools and formats which implies an independent and flexible format is needed [3].

One of the main criteria for this new data format is to avoid striving for new formats and use already existing ones which are complementing the evolving data format AML [3]. AML has been using the CAEX, PLCopen XML and COLLADA data formats for years. Recently, eCL@ass [4] and the OPC architecture [5] have also been introduced.

Further criteria [3] have been defined for the addition of future formats such as:

- The format has to be open source and free of charge
- The format should be standardized to ISO or IEC, respectively to strive for a standardization
- The format should base on XML
- The format must be still well-maintained and the owner must be open for further development suggestions

With this set of criteria AML has a strong foundation through which everyone is enabled to begin his own implementation of this data format [6]. The membership of a great number of industry leaders in the AML consortium and its growth, as well as continuing success with international standardization show that AML will become more important in the future.

2   The Independent Data Format AutomationML

2.1   Architecture of AutomationML

The name Automation Markup Language is only a generic term for the general concept of the format. AML is an independent data exchange format that uses the CAEX format as a central base, which enables AML to achieve its goal, namely to improve the data exchange in the heterogeneous tool landscape [7].

The CAEX format enables the modeling of static object structures through a powerful library concept that enables storing established components, interfaces, requirements and roles.

Furthermore, the CAEX format makes possible the linking of various established data formats, such as PLCopen XML and COLLADA, which can represent specific information content of the engineering process.

AML thus has access to all the advantages that the respective data formats have to offer, resulting in a high degree of flexibility in their application.

Types of information in AutomationML

Through the incorporation of different formats, AML can be applied to most of the existing information flow in the engineering process, thereby resulting in new solutions and workflows.

The information to be stored by AML [7] is as follows:

- **Plant topology information**: The plant topology is determined by the components used by the customer's standard, which comprise the plant. The result is a hierarchical structure of objects which can be modeled and stored using CAEX.
- **Geometry and kinematic information**: Simulation tools have taken on an important role in plant engineering, which have visibly improved the workflow. The geometric information of these simulations as well as the kinematics of the components can be saved and transferred by the COLLADA format.
- **Logic information**: Processing logic plays a major role in plant engineering, which makes handling this information critical. The description of sequences and the behavior of components with the associated logical variables are stored in PLCopen XML.
- **Reference and relationship information**: The CAEX format allows AML to distinguish between references and relationships. Referencing is used to show links between externally stored information and a CAEX object. Relationships are used to show dependencies between different CAEX objects.

These are just some of the many possible data formats and types of information that can be stored in AML.

Libraries in AutomationML

Through the use of basic CAEX concepts complex components or structures can be recreated by modeling their individual elements [8]. However, in order to achieve the efficiency required in the current engineering process, all existing solutions must be readily available for future projects. This is solved in AML through the use of libraries, which cover different aspects of modeling [7].

CAEX Role Concept

The CAEX role concept allows the definition of abstract technical functions which are adapted to the needs during the modeling process [9]. The role concept is the basis for the later use of algorithms as it provides a semantic definition of each object.

The procedure is described by way of example. It can be broken down into the following steps [9]:

1. Modeling of a plant object: The object to be modeled is determined by its intended purpose of information transition and the extent to which the latter is to be incorporated in AML.
2. Define a role: A role that fits the intended use is created.
3. Defining a SystemUnitClass: The new object and its role are grouped into a template that is available for future configurations.

The AutomationML community defines a set of basic role class libraries which are essential for the modeling of core concepts in the AutomationML format. These basic role class libraries must be used as starting point for user-defined role classes.

3   Use Cases

The previous chapter has shown what encompasses AML and thus offered a glimpse of the possibilities provided by this format. In the following chapter two examples will show that AML is not just an idea to play with but a real solution approach.

These are just a few of the long-standing engineering solutions that have come together in recent years. In order to get a better idea of AML based solutions it is recommended to visit the AutomationML community page.

At present many projects are carried out by the AutomationML community and many are to follow due to the vast opportunities provided by the AML implementation [10].

3.1   Transfer of Mechanical Construction Data

In recent years another application has complemented the existing engineering process and has now established itself as fundamental component. This application is the virtual commissioning, which can be executed quickly, economically and purposefully only using data from the already existing engineering process.

To achieve this AML is used for the lossless transmission of information relating to the plant and material flow simulation of virtual commissioning, which is obtained from the existing mechanical design data. The company W.E.S.T GmbH has already created an AML converter for Process Simulate and Robcad, enabling the seamless transition of information to other tools such as RF:Suite [11].

During the conversion process the complete object tree structure from Process Simulate is recreated in AML. These new CAEX objects reference the information transferred from the *.jt format that is used in Process Simulate, which is now stored in COLLADA [12]. This implementation clearly shows that AML is no longer in the test phase but is ready for practical use.

3.2   Hardware Configuration Generation in PLC

For several years, various automobile manufacturers have been striving for automatic sequence and software generation of PLC projects. This is also a topic for AutomationML with PLCopen XML [13].

This use case, however, is not about software generation but the generation of hardware configuration, which still has to be carried out manually, regardless if the necessary data has already been available in paper format for a long time. Siemens AG, together with EPLAN Software & Service GmbH & Co. KG, are already pursuing this approach [14].

After an analysis of the engineering process the following information is obtained. This is required by the PLC programmer in order to be able to create the hardware configuration in the PLC project.

- **Plant structure**: The designation and structure of the entire plant is aligned according to the respective customer standard and additionally displayed in the hardware tree view of the hardware diagram.
- **Device interconnection**: Information about the complete interconnection of the nodes within the configured bus system, in this case PROFINET.
- **Devices**: All devices controlled by the PLC in the plant are accounted for. These also contain information about manufacturer, article number, type, configuration, IP address, start address and F address for safety devices.

Thanks to the wide range of possibilities that CAEX provides the data model can be accurately aligned with the desired data transfer proceeding. The degree of detail of the model can be adjusted as desired, as shown in [15].

User defined SystemUnitClasses are created for this purpose, which store the required information by means of attributes and the AutomationML libraries. Figure 1 shows the recreated setup of a hardware component.

In order to create these SystemUnitclasses consideration had to be given to the interfaces and roles to be used. Particularly the used roles had to be chosen properly, since these will be of great importance for the transfer algorithm used later.

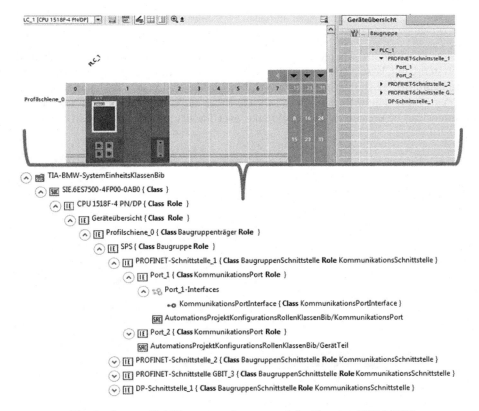

Fig. 1. System UnitClass example generated for Siemens CPU 1518F

The device overview of the TIA portal is used as the basis for the CAEX structure of the SystemUnitClasses. Each interface of a component is represented by an interface in the AML interface library. These interfaces are necessary in order to store information about the connection between devices.

Detailed information, such as the IP address, type, name, etc., are stored by attributes of the corresponding module hierarchy as seen in Fig. 2. The required attributes for the information storage are created according to the requirements.

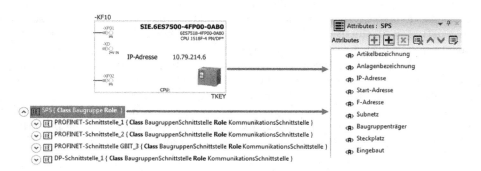

Fig. 2. Created attributes for Siemens CPU 1518F

According to this scheme all hardware components used must be implemented. If all SystemUnitClasses are present the complete hardware plan can be rendered in AutomationML as shown in Fig. 3. By following the customer's designation classification an AutomationML model valid for the entire engineering process of this specific customer can be created, which could take over the data transfer across all trades.

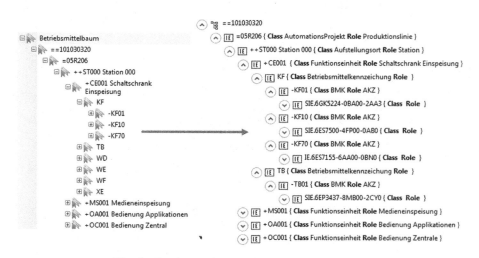

Fig. 3. Hardware plan recreated in AutomationML

The connections between devices are represented by internal links in CAEX, as shown in Fig. 4. By importing this information from the hardware plan the subnet and the connection of the devices can be reproduced whereby this manual operation can be dispensed with.

With the SystemUnitClasses created so far a basic generation of all main modules contained in the hardware plan can be carried out. However, since the components can have main and submodules, this must also be considered. Therefore, the sub modules must also be present in the SystemUnitClass library, as shown in Fig. 5.

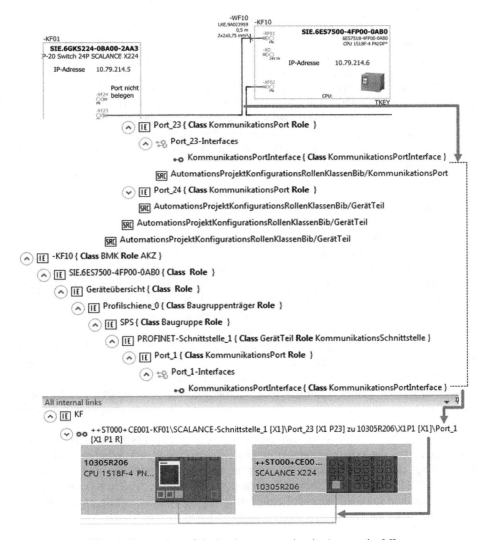

Fig. 4. Recreation of device interconnection in AutomationML

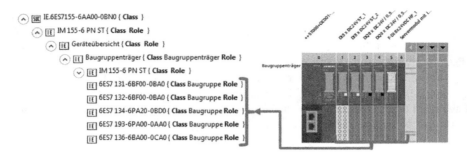

Fig. 5. Submodules in the SystemUnitClass library

The displayed AutomationML model is only one third of the work. The possible interfaces of the engineering tools must now be defined and, on the basis of these interfaces, algorithms must be written which will transmit the necessary information. This will be the next step in the implementation of this use case.

With the implementation of this data exchange the groundwork is done to benefit from the advantages of the AutomationML format. With the implementation of an automatic hardware configuration generation the focus can be shifted towards software generation in which the PLCopen XML sub-format will assume a leading role. This will be a future issue.

4 Conclusion

In this paper we briefly discussed the AutomationML data format and the possibilities of the CAEX format contained in it. On the basis of these possibilities applications which already exist and also applications which will be implemented have been demonstrated. The concept for the generation of a hardware configuration in the PLC software design and the AutomationML model required for this data transmission were both presented in detail. An example was examined to demonstrate the readiness for using AutomationML in the engineering process.

Due to the wide range of supported data formats, which can be found in AutomationML, any information can be stored, no matter in what form it appears. AutomationML will therefore play an increasingly important role in the area of the digital factory and virtual commissioning, as the hurdle of manual data processing is gradually removed, thus advancing the automation of the entire engineering process [16].

AutomationML is a powerful data format with which it is possible to increase speed and efficiency of the existing engineering process. This requires a complete analysis of the existing engineering process in order to find further possible applications for AutomationML. These must then be checked for feasibility, which will be the next step. The gradual implementation of AutomationML in the engineering process results in advantages for all involved parties.

Acknowledgements. This work was supported by the European Union through the FP7-PEOPLE-2013-IAPP AutoUniMo project "Automotive Production Engineering Unified Perspective based on Data Mining Methods and Virtual Factory Model" (Grant Agreement No: 612207) and research work financed from funds for science for years: 2016–2017 allocated to an international co-financed project.

References

1. Ovtcharova, J., Häfner, P., Häfner, V., Katicic, J., Vinke, C.: Innovation braucht resourceful humans Aufbruch in eine neue Arbeitskultur durch virtual engineering. In: Botthof, A., Hartmann, E.A. (eds.) Zukunft der Arbeit in Industrie 4.0, pp. 111–124. Springer, Heidelberg (2015). doi:10.1007/978-3-662-45915-7_12
2. Vogel-Heuser, B.: Herausforderungen und Anforderungen aus Sicht der IT und der Automatisierungstechnik. In: Bauernhansl, T., ten Hompel, M., Vogel-Heuser, B. (eds.) Industrie 4.0 in Produktion, Automatisierung und Logistik, pp. 37–48. Springer, Wiesbaden (2014). doi:10.1007/978-3-658-04682-8_2
3. Hirzle, A.: AutomationML—ein Überblick. In: AutomationML—Fachexperten erklären das Format. SPS Magazin: Zeitschrift für Automatisierungstechnik, Marburg (2014). https://www.automationml.org/o.red/uploads/dateien/1391503893-SPS-Magazin_Whitepaper_AutomationML.pdf
4. AutomationML e.V.: AutomationML whitepaper and eCl@ss integration (2015). https://www.automationml.org. Accessed 10 June 2017
5. AutomationML e.V.: AutomationML whitepaper: OPC Unified architecture information model for AutomationML (2016). https://www.automationml.org. Accessed 10 June 2017
6. Drath, R., Barth, M.: Wie der Umgang mit unterschiedlichen Datenmodellen beim Datenaustausch im heterogenen Werkzeugumfeld gelingt. In: VDI-Berichte 2209, Tagungsband zur Automation 2013, Langfassung auf Tagungs-CD (12 Seiten), pp 339–344. VDI-Verlag, Baden Baden (2013)
7. IEC 62714-1: Engineering data exchange format for use in industrial automation systems engineering—automation markup language—part 1: architecture and general requirements (2014). IEC: www.iec.ch
8. IEC 62714-2: Engineering data exchange format for use in industrial automation systems engineering—automation markup language—part 2: role class libraries (2015). IEC: www.iec.ch
9. Drath, R., Schleipen, M.: Das CAEX-Rollenkonzept. In: Drath, R. (ed.) Datenaustausch in der Anlagenplanung mit AutomationML: Integration von CAEX, PLCopen XML und COLLADA, pp. 51–53. Springer, Heidelberg (2010)
10. AutomationML e.V.: https://www.automationml.org/o.red.c/projects.html. Accessed 10 June 2017
11. Hämmerle, H., Strahilov, A., Drath, R.: AutomationML im Praxiseinsatz: Erfahrungen bei der virtuellen Inbetriebnahme. In: atpedition, pp. 52–64. DIV Deutscher Industrieverlag GmbH/Vulkan-Verlag GmbH, München/Essen (2016)
12. IEC 62714-3: Engineering data exchange format for use in industrial automation systems engineering—automation markup language—part 3: geometry and kinematics (2017). IEC: www.iec.ch
13. AutomationML e.V.: Whitepaper AutomationML part 4: AutomationML Logic (2017). https://www.automationml.org/o.red.c/dateien.html?cat=3. Accessed 10 June 2017

14. AutomationML e.V.: Application recommendations: automation project configuration (2016). https://www.automationml.org/o.red.c/dateien.html?cat=3. Accessed 10 June 2017
15. AutomationML e.V.: AutomationML whitepaper communication (2014). https://www.automationml.org/o.red.c/dateien.html?cat=3. Accessed 10 June 2017
16. Lüder, A., Schmidt, N.: AutomationML—Erreichtes und Zukünftiges. In: AutomationML – Fachexperten erklären das Format. SPS Magazin: Zeitschrift für Automatisierungstechnik, Marburg   (2014).   https://www.automationml.org/o.red/uploads/dateien/1392190786-AutomationML_Artikel13_2014_SPS-Magazin.pdf

Enhanced Reliability of ADAS Sensors Based on the Observation of the Power Supply Current and Neural Network Application

Damian Grzechca$^{(\boxtimes)}$, Adam Ziębiński, and Paweł Rybka

Electronics and Computer Science, Faculty of Automation Control,
Silesian University of Technology, Gliwice, Poland
{damian.grzechca,adam.ziebinski}@polsl.pl,
pawel.rybka@interia.eu

Abstract. Advanced Driver Assistance Systems (ADAS) are essential parts for developing the autonomous vehicle concept. They cooperate with different on-board car equipment to make driving safe and comfortable. There are many ways to monitor their behaviour and assess their reliability. The presented solution combines the versatility of applications (it can be used with almost any kind of sensors), low cost (data acquisition using this method requires only a simple electronic circuit) and requires no adjustments of the sensor's software or hardware. Using this type of analysis, one can determine the device's family, find any over- and under-voltages that can damage the sensor or even detect two-way CAN communication malfunctions. Since the data acquired is complex (and can be troublesome during processing) – one of the best solutions is to cope with the problem by using a variety of neural networks.

Keywords: ADAS · Predictive maintenance · Neural network · CAN · Data acquisition

1 Introduction

Cyber-Physical Systems (CPS) [1] combine the real world with the digital world with the help of computational technologies using the internet infrastructure [2]. CPS can often be used in many areas such as manufacturing [3], transportation, remote monitoring [4] and automotive solutions [5]. Moreover, CPS support the concepts of distributed control by using computing for control systems, efficiency [6] and real time communication [7].

The head-to-head competition among automotive companies continues to grow rapidly in the development of self-driving cars. In order for them to be operated, autonomous vehicles must use numerous data acquisition devices and methods. Advanced Driver Assistance Systems (ADAS) sensors assist in overcoming these difficulties [8]. ADAS [9] collect and process the data about driving parameters such as velocity, acceleration, current gear position etc. from the external environment, thus providing useful information to either alert the driver about a certain road condition or even take control of the car in a critical situation [10]. The most common ADAS

© Springer International Publishing AG 2017
N.T. Nguyen et al. (Eds.): ICCCI 2017, Part II, LNAI 10449, pp. 215–226, 2017.
DOI: 10.1007/978-3-319-67077-5_21

sensors in the embedded systems [11] that are used in such applications are LIDAR – Light Detection and Ranging – mainly used for Emergency Braking Assist (EBA) [12]; stereo cameras that are used for Forward Collision Warning (FCW), Lane Departure Warning (LDW); Intelligent Headlamp Control (IHC); Traffic Sign Assist (TSA); Adaptive Cruise Control (ACC) + Stop&Go; Road Profile Detection (RPD); RADAR used for Blind Spot Detection (BSD) and Rear Cross Traffic Alert (RCTA).

Like any other electronic devices, ADAS sensors are susceptible to problems with communication or other malfunctions. When considering human safety, a system that can assess the reliability of the measurements that are performed by ADAS sensors has to be introduced. The devices have inbuilt software and hardware failure prevention functionalities, but these are dedicated to a particular sensor. This solution is very costly and requires a great deal of time and effort to design another safety circuit every time a new sensor is released. Developing a versatile module that can be implemented with any kind of ADAS sensor that only reprogramming in order to handle certain safety issues would lower the costs of production significantly.

Many methods have been developed in the field of science and technology to meet such requirements. A method called "Motor Current Signature Analysis" (MCSA) is used for motor fault diagnosis [13], which permits common machine faults to be detected, has been developed. Another power supply current analysis method that is used in engineering is analogue fault diagnosis, which is based on ramping the power supply current signature clusters [14]. Although those two methods differ from each other significantly, their main and common feature is to analyse current. Observing the current is widely used in the world of electronics and physics (for example in testing CMOS IC's) because of the potential to detect a large class of manufacturing defects. Small changes in the consumption of energy can be observed over time. Fault detection and diagnosis often uses signal processing methods [15]. Condition-based preventive maintenance solutions have been used to identify a failure before it occurs [16]. The support of sensor measurements can be found in these preventive maintenance approaches [17].

There are two main approaches for the methods to acquire current values – using current sensors that are based on the Hall effect [18] or measuring a voltage drop on a shunt resistor. The advantage of the first solution that no resistor is needed to take the measurements, and therefore, there is no need to select its value or to select a particular value of the instrumentation amplifier's gain according to the sensor's dynamics. On the other hand, Hall effect sensors are less accurate, which can be a sufficient reason to disqualify them from being considered for such a precise application.

When it comes to processing the acquired data, various approaches that differ from each other in their calculation complexity, time consumption and efficiency exist. One of the simplest and easiest classification methods is the k-nearest neighbours algorithm (k-NN) [19]. It is based on the similarities among pre-learnt cases and a new sample – the closer the sample is to a particular cluster, the more likely it is for the new data to be classified as a member of that cluster. The data can be well classified based on a much more complex learning algorithms such as neural networks [20]. The most popular grouping algorithm in this area is a self- organising map (SOM). Although it is designed particularly for the sake of classification, its nature – unsupervised learning – makes it difficult to distinguish any differences among systems that have poor dynamics

– sensors with a low current intake. The other approach to be considered is supervised learning. Some example methods from this subgroup that are commonly used in are the Levenberg-Marquardt algorithm [21], also known as the Damped Least-Squares (DLS), which typically requires more memory and less time, and Bayesian Regularisation [22], which, on the other hand, requires more time, but can result in a good generalisation for small or noisy datasets.

1.1 Measuring Platform

The platform that is used for data acquisition and to control the power supply of ADAS sensors is designed in such a way that a microcontroller (with the help of 74HC595 shift registers) turns on and off the MOSFET transistor (IRLML2502GPBF) that is responsible for the current conductance over a certain sensor. Setting a transistor's gate low, cuts off the current (prevents a sensor from operating) and setting the gate high (3.3 V) – enables the current to flow through the key. Later, the data is measured using a 10 mΩ shunt resistance (R3) that is connected via a filtering circuit (which besides being a beneficial influence, generates a gain error – Eq. (1)) with an instrumentation amplifier (current shunt monitor – INA199) of the gain that is equal to 200 (Fig. 1). The signal is then drawn to the input of the 12-bit analog-to-digital converter of the STM32F3 microcontroller. After the sampling is done, the STM32 device sends the data to the PC for further analysis over the COM port. The current that flows through ADAS device can be calculated from the Eq. (2).

$$Gain\,error\,factor = \frac{1250 * R3}{(1250 * R1) + (1250 * R3) + (R1 * R3)} \quad \text{where } R1 = R2 \quad (1)$$

$$I = \frac{ADC\ Readings * 3.3V}{R3 * INA's\ Gain * ADC\ Resolution} \tag{2}$$

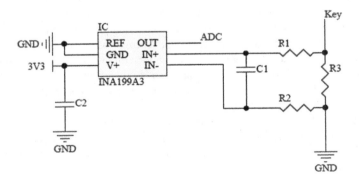

Fig. 1. Current measurement circuit.

1.2   Sampling Method

The most valuable data to analyse is the current response after the power key is turned on. The characteristics for various ADAS sensors differ significantly (for examination purposes, four were taken into account – LIDAR, RADAR and two cameras). As the application has to be versatile, the sampling time to reach a steady state has to be equal for all of the sensors listed above. The statement that the sampling time has to be tuned for the device that takes the longest time to reach the steady state arises from this assumption. The setup times for all of the devices are listed in the Table 1 below:

Table 1. Setup times for ADAS sensors.

Sensor	Time
LIDAR	15 ms
RADAR	22 ms
Camera 1	7 ms
Camera 2	8 ms

Another point of interest occurs when the devices establish the connection over the CAN protocol [23]. For the safety and reliability of the connection in the automotive industry, the protocol uses a differential signal, whose operation requires that a substantial current has to be drawn. Traffic on the CAN bus can, in this case, be monitored by analysing current. An observation of how the devices establish connections has to be performed once again. For example, for the LIDAR, the protocol is as follows:

(a) setup + 0.000 ms
(b) sending CAN frames to the master device + 0.400 ms
(c) stop sending CAN frames + 2.980 ms

Bearing those values in mind, sampling at the third second of the communication can provide information about whether the master device responded to the frames that were sent by the ADAS module or whether the communication failed.

When looking at the setup time, one can easily see that the voltage level of the power source has a huge impact on the shape of the current waveform. This information provides a way to examine whether an over- or under-voltage has occurred. This can not only protect the device from damage, but also provides useful information – the threshold of the reliable working voltage range has been crossed and the data that is being sent by the ADAS module may be distorted.

The last thing to be considered is that the microcontroller's RAM is limited to around 19,000 samples. For this reason, samples have to be distributed between the monitoring setup waveforms and the CAN transmission must be monitored in the correct manner.

1.3   Analysis

Waveform analysis is carried out in two parts – the part that is responsible for signal setting and the part that is responsible for CAN communication. The microcontroller sends 19,000 samples over the COM port to the PC. They are divided into two subsections – 7000 correspond to the first task and 12,000 correspond to the second task. Signal setting samples are processed in order to classify the device family (the current waveform being analysed by the neural network indicates whether LIDAR, RADAR or one of the cameras are connected to the key). After the correct classification of the sensor, the same samples are used to examine whether an over- or under-voltage has occurred. Next, with the information about the device, the algorithm uses the remaining 12,000 samples to assess the communication over the CAN network.

2   Module Family Classification

To perform the successful classification of devices (for example, in order to prevent unauthorised sensors from being connected to our power key, or to simplify the process of connecting), the neural network has to be trained in the differences between the current waveforms of the ADAS sensors. Example waveforms of the four devices are presented in Fig. 2:

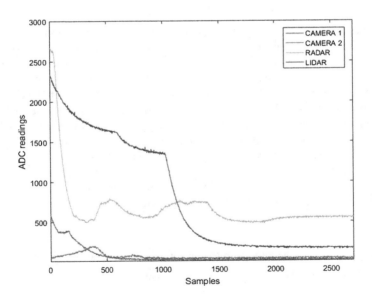

Fig. 2. Current waveforms of ADAS devices

The well-known clusterisation method (the so-called self-organizing map, SOM) was used by the authors in [24]. This experience led to the presented application in the following form. In order to create an unsupervised learning procedure, a set of 520 current waveforms (130 for a single sensor) were generated by actual ADAS devices. The training process, which was performed in the Matlab environment, showed the hits map that is presented in Fig. 3. The colour of the group corresponds to the colour of the waveform in Fig. 2.

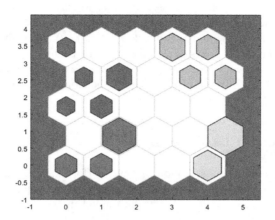

Fig. 3. SOM sample hits (blue clusters represent LIDAR, red – RADAR, green and yellow – camera 1 and camera 2) (Color figure online)

Every waveform was classified correctly and there was not one mismatch. Except for the red and the blue clusters, all of the groups were separated by at least one cell in the topology, which provided reliable and repeatable results of the classifications.

Another approach considers using a neural network with (unlike in the previous method) supervised learning. This means that the example output has to be provided for every sample of the data. The network is trained based on these pairs. One such method is called the Levenberg-Marquardt method. It was designed to output two Boolean values that represent the natural binary code (each value for each sensor). The network output thresholds were set as follows: for a neuron value of 0.25 and less – correction to 0, for 0.75 and more – correction to 1. Using the same input data as in the previous experiment, 364 training, 78 validation and 78 testing samples, the Levenberg-Marquardt method gave the following testing results for the hidden neuron numbers that are listed below (Table 2):

Table 2. The results of the Levenberg-Marquardt method for module family classification.

Hidden neurons	Mean squared error	Misclassified waveforms	Performance
5	6.436e−4	0	100%
10	3.111e−4	0	100%
20	2.499e−4	0	100%

A method that is of the same type as the Levenberg-Marquardt method is Bayesian Regularisation – an algorithm that typically requires more time, but is generally a better choice when considering small or noisy datasets. Using the identical initial conditions and dataset as in the previous example, the results were as follows (Table 3):

Table 3. The results of Bayesian regularisation for module family classification.

Hidden neurons	Mean squared error	Misclassified waveforms	Performance
5	5.578e−5	0	100%
10	1.085e−4	0	100%
20	1.142e−4	0	100%

All three methods for classifying current waveforms appeared to be successful. An SOM algorithm map was generated 100% correctly and when the Levenberg-Marquardt method and Bayesian Regularisation were compared, both algorithms provided excellent results (not a single mismatch was detected). However, the latter was slightly more accurate when minimising the MSE (mean squared error) was taken into account. Its best performance was reached for five hidden neurons with a MSE = 5.578e−5.

3 Over- and Under-Voltage Detection

In order to prevent a device from providing distorted data or being damaged due to an inappropriate voltage value at the power source, it is possible to cut off the current supply when this is necessary. The current waveforms of ADAS modules differ significantly from one another when the supply voltage is changed. This phenomenon can be observed in Fig. 4, where each LIDAR current waveform represents a different supply voltage that varied from 9 V to 15 V (with a step of 0.5 V).

Once again – using neural network algorithms is preferred when analysing a given problem. Unluckily, unlike with the classification problems described in the previous paragraph, no self-organising map can be generated. It has to be assumed that a certain threshold cannot be crossed and that Kohonen's net nature of the unsupervised learning algorithm disables it from being trained with a desirable output.

Thankfully, both the Levenberg-Marquardt method and Bayesian Regularisation could be implemented and the analogue value was used as the output pattern (the network was trained to output the actual value of the voltage). The input data for the neural network consisted of 130 LIDAR current waveforms that varied from 9–15 V. The results of the learning using each of the algorithms are presented in Tables 4 and 5 (90 training, 20 validation and 20 testing samples).

All parameters listed above are crucial to assess performance of a neural network, but in safety applications (and the predictive maintenance module is one), the most important parameter to minimise is the max error. In this case, the best learning algorithm appeared to be Bayesian Regularisation with ten hidden neurons (slightly better than the Levenberg-Marquardt method with five hidden neurons).

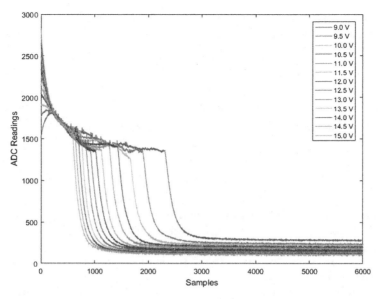

Fig. 4. Current response waveforms for different values of the LIDAR supply voltage – from 15 V (leftmost) to 9 V (rightmost) with a step of 0.5 V.

Table 4. The results of the Levenberg-Marquardt method for over- and under-voltage detection.

Hidden neurons	Mean squared error	Max error	Average error
5	5.871e−4	0.41 V	66 mV
10	2.820e−3	0.75 V	157 mV
20	9.469e−3	0.91 V	216 mV

Table 5. The results of Bayesian regularisation for over- and under-voltage detection.

Hidden neurons	Mean squared error	Max error	Average error
5	7.743e−5	0.43 V	95 mV
10	4.058e−3	0.40 V	68 mV
20	1.274e−2	0.62 V	106 mV

4   CAN Transmission Monitoring

The CAN protocol is an essential part of ADAS devices. CAN transmissions allow them to exchange data between each other. Due to the substantial current usage and the nature of the differential signals that are used in CAN, it is possible to detect ongoing and outgoing data only by monitoring the current usage of a given sensor. Detecting should begin at circa 3,000 ms after turning the device on with the transistor key, as around this time the sensors stop sending outgoing frames and they are waiting for a response from the master. When none is given – the transmission breaks. Both cases (when transmission is up or down) are shown in Fig. 5 below.

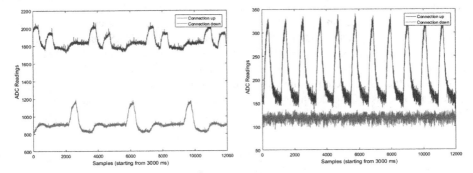

Fig. 5. RADAR's (leftmost) and LIDAR's (rightmost) current waveforms with the CAN transmission up and down

In order to train the neural networks (Levenberg-Marquardt method and Bayesian Regularisation), 200 samples (100 with transmission up and 100 with transmission down) were used. The distribution of the samples was: 140 training, 30 validation and 30 testing samples. The results of training each of the algorithms are presented in Tables 6 and 7.

Table 6. The results of the Levenberg-Marquardt method for CAN transmission monitoring.

Hidden neurons	Mean squared error	Misclassified waveforms	Performance
5	2.421e−3	0	100%
10	1.593e−3	0	100%
20	1.072e−2	0	100%

Table 7. The results of Bayesian regularisation for CAN transmission monitoring.

Hidden neurons	Mean squared error	Misclassified waveforms	Performance
5	1.270e−2	0	100%
10	1.990e−3	0	100%
20	1.290e−2	0	100%

In this case, a 100% performance was also reached. The best (smallest) MSE results, however, were obtained using the Levenberg-Marquardt learning method with ten hidden neurons.

5 Support Predictive Maintenance Module for ADAS

When attempting to enhance the reliability of ADAS sensors, the concept of a support predictive maintenance module must be considered. Such a module would provide data about a sensor's status and the reliability of its readings. For example – over- or under-voltages can cause fluctuations in the output of an ADAS sensor, which is susceptible to a voltage imbalance. Monitoring the current that is used in a transmission via the CAN protocol can also provide significant data about quality of the connection.

The data that is collected in the current diagnostics is then sent to a parent device, where it is processed in order to estimate the credibility of a specific sensor's output (Fig. 6).

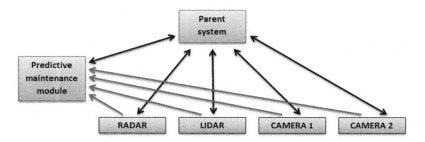

Fig. 6. Block diagram of a support predictive maintenance module

Using the data acquisition methods and monitoring the current waveforms presented in the paragraphs above, we can propose a predictive maintenance module. The module would inform the parent system (via an output register) about its operational status. As the system is designed to handle up to 16 ADAS modules, the 8-bit register looks like this: [s1 s2 s3 s4 c1 v1 v2 t1], where s1-s4 state a number of device (0–15), c1 – classification status, v1 and v2 – voltage status (acceptable level (11–13 V), overvoltage (above 13 V), under-voltage (below 11 V)); t1 – CAN transmission status. For example, the output register [0 1 1 1 0 0 0 1] indicates that for ADAS device no. 7, the module is correctly classified, the source voltage is in the range of 11–13 V but there is a problem with the CAN communication. The block diagram of the predictive maintenance module looks like this (Fig. 7):

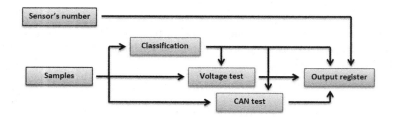

Fig. 7. ADAS predictive maintenance module

After sampling, the waveform is classified as to whether it belongs to any supported type of ADAS device. When the device is recognised – voltage and CAN transmission tests can be run. The binary output of all of the neural networks are sent (alongside the sensor's number) to the register, where they can be read by the parent system later.

6 Conclusion

The research presented in this article has confirmed that diagnostics using current waveform analysis can be successful. Moreover, a support predictive maintenance module to enhance the reliability and safety of ADAS sensors can be implemented at a low cost without having to interfere with the ADAS device's software or hardware. From the variety of methods that can be used to solve the problems discussed in this article, neural networks were chosen. All of the algorithms (including those with supervised learning and those with unsupervised learning) performed their job surprisingly well. The Levenberg-Marquardt learning algorithm appeared to provide the best results in detecting CAN communication. Bayesian Regularisation, however, was the best choice when it came to voltage recognition and (alongside the self-organising map) – device family classification. Although this survey appears to give an excellent outcome, there is still much to be done in this field of engineering as only a fraction of the methods that are available were used and only a fraction of the fields were examined.

Acknowledgements. This work was supported by the European Union from the FP7-PEOPLE-2013-IAPP AutoUniMo project "Automotive Production Engineering Unified Perspective based on Data Mining Methods and Virtual Factory Model" (Grant Agreement No. 612207) and research work financed from the funds for science in the years 2016–2017, which are allocated to an international co-financed project (Grant Agreement No. 3491/7. PR/15/2016/2).

References

1. Khaitan, S.K., McCalley, J.D.: Design techniques and applications of cyberphysical systems: a survey. IEEE Syst. J. **9**(2), 350–365 (2015)
2. Wu, F.-J., Kao, Y.-F., Tseng, Y.-C.: From wireless sensor networks towards cyber physical systems. Pervasive Mob. Comput. **7**(4), 397–413 (2011)
3. Wang, Y., Vuran, M.C., Goddard, S.: Cyber-physical systems in industrial process control. ACM SIGBED Rev. **5**(1), 1–2 (2008)
4. Flak, J., Gaj, P., Tokarz, K., Wideł, S., Ziębiński, A.: Remote monitoring of geological activity of inclined regions – the concept. In: Kwiecień, A., Gaj, P., Stera, P. (eds.) CN 2009. CCIS, vol. 39, pp. 292–301. Springer, Heidelberg (2009). doi:10.1007/978-3-642-02671-3_34
5. Li, R., Liu, C., Luo, Feng.: A design for automotive CAN bus monitoring system, pp. 1–5 (2008)
6. Maka, A., Cupek, R., Rosner, J.: OPC UA object oriented model for public transportation system. Presented at the 2011 Fifth UKSim European Symposium on Computer Modeling and Simulation (EMS), pp. 311–316 (2011)

7. Cupek, R., Huczala, L.: Passive PROFIET I/O OPC DA server. Presented at the IEEE Conference on Emerging Technologies & Factory Automation, ETFA 2009, pp. 1–5 (2009)

8. Jia, X., Hu, Z., Guan, H.: A new multi-sensor platform for adaptive driving assistance system (ADAS). In: 2011 9th World Congress on Intelligent Control and Automation, pp. 1224–1230 (2011)

9. Bengler, K., Dietmayer, K., Farber, B., Maurer, M., Stiller, C., Winner, H.: Three decades of driver assistance systems: review and future perspectives. IEEE Intell. Transp. Syst. Mag. **6** (4), 6–22 (2014)

10. Fildes, B., Keall, M., Thomas, P., Parkkari, K., Pennisi, L., Tingvall, C.: Evaluation of the benefits of vehicle safety technology: the MUNDS study. Accid. Anal. Prev. **55**, 274–281 (2013)

11. Pamuła, D., Ziębiński, A.: Securing video stream captured in real time. Prz. Elektrotech. **86** (9), 167–169 (2010)

12. Ziębiński, A., Cupek, R., Grzechca, D., Chruszczyk, L.: Review of advanced driver assistance systems (ADAS). In: 13th International Conference of Computational Methods in Sciences and Engineering, April 2017

13. Mehala, N., Dahiya, R.: Motor current signature analysis and its applications in induction motor fault diagnosis. Int. J. Syst. Appl. Eng. Dev. **2**(1), 29–31 (2007)

14. Somayajula, S.A., Sanchez-Sinencio, E., Pineda de Gyvez, J.: Analog fault diagnosis based on ramping power supply current signature clusters. IEEE Trans. Circ. Syst. II Analog Digit. Sig. Process. **43**(10), 703–711 (2002)

15. Zhang, X., Kang, J., Xiao, L., Zhao, J., Teng, H.: A new improved kurtogram and its application to bearing fault diagnosis. Shock Vib. **2015**, Article ID 385412, 22 p. (2015). doi:10.1155/2015/385412

16. Mori, M., Fujishima, M.: Remote monitoring and maintenance system for CNC machine tools. Proced. CIRP **12**, 7–12 (2013)

17. Colledani, M., et al.: Design and management of manufacturing systems for production quality. CIRP Ann. Manuf. Technol. **63**(2), 773–796 (2014)

18. el Popovic, R., Randjelovic, Z., Manic, D.: Integrated hall-effect magnetic sensors. Sens. Actuators Phys. **91**(1), 46–50 (2001)

19. Larose, D.T.: k-nearest neighbor algorithm. In: Discovering Knowledge in Data, pp. 90–106. Wiley, Hoboken (2005). doi:10.1002/0471687545.ch5

20. Talaska, T., Dlugosz, R., Wojtyna, R.: Current mode analog Kohonen neural network. In: Mixed Design of Integrated Circuits and Systems, pp. 251–252 (2007)

21. Moré, J.J.: The Levenberg-Marquardt algorithm: implementation and theory. In: Watson, G. A. (ed.) Numerical Analysis. LNM, vol. 630, pp. 105–116. Springer, Heidelberg (1978). doi:10.1007/BFb0067700

22. Burden, F., Winkler, D.: Bayesian regularization of neural networks. In: Livingstone, D. J. (ed.) Artificial Neural Networks. Methods and Applications, pp. 23–42. Humana Press, New York (2009). doi:10.1007/978-1-60327-101-1_3

23. Cupek, R., Ziebinski, A., Drewniak, M.: Ethernet-based test stand for a CAN network. In: 18th IEEE International Conference on Industrial Technology (2017)

24. Grzechca, D.: Soft fault clustering in analog electronic circuits with the use of self organizing neural network. Metrol. Meas. Syst. **18**(4), 555–568 (2011)

ADAS Device Operated on CAN Bus Using PiCAN Module for Raspberry Pi

Marek Drewniak[1(✉)], Krzysztof Tokarz[2], and Michał Rędziński[2]

[1] Aiut Sp. z o.o., Gliwice, Poland
marek.drewniak@aiut.com
[2] Faculty of Automation Control, Electronics and Computer Science,
Institute of Informatics, Silesian University of Technology, Gliwice, Poland
krzysztof.tokarz@polsl.pl,
michal.redzinski@outlook.com

Abstract. In the era of the development of Cyber-Physical Systems, solutions that are based on the transparent and universal possibility to monitor and analyse signals from the measurement devices have become more popular. This is especially crucial in the case of safe and real-time systems, e.g. in the CAN networks that are used in the automotive and manufacturing. During the research, the authors focused on preparing a solution that is based on the Raspberry Pi computer. When used as an independent and non-interfering CAN node, it is capable of monitoring, analysing and controlling Advanced Driving Assistance Systems. The solution can be used as both a gateway between a hermetic, safe system and the external world and also as an onboard element that performs auxiliary data analyses that do not load the main processing unit.

Keywords: ADAS · CAN bus · CANoe · Ethernet · LIDAR · Raspberry Pi

1 Introduction

One of the trends that has become more popular in all systems that are based on communication and computer networks is the development of Cyber-Physical Systems (CPS) [1]. The approach combines the real world (end-point devices and sensors) with the digital world with the help of the internet and computational technologies [2]. The areas in which precursors of CPS can be found are aerospace, the automotive [3], manufacturing [4], transportation, entertainment and mobile communications. The concepts of distributed control, remote reading of data and monitoring [5] statuses for safety and diagnostic purposes were developed especially in those areas [6]. Above all, the last one, mobile communications, has become a source of interest in CPS for a broader group of users due to the plethora of input and output devices, including sensors; sufficient computational power; storage capabilities; the use of multiple communication interfaces and the possibility to develop applications in popular, open, high-level programming languages [7]. Recently, CPS have become more popular in the field of automation control, the automotive and intelligent house systems. Manufacturers systematically strive to find ways to control or read data from end-point devices from any place in the world using multiple technologies, e.g. computational or

© Springer International Publishing AG 2017
N.T. Nguyen et al. (Eds.): ICCCI 2017, Part II, LNAI 10449, pp. 227–237, 2017.
DOI: 10.1007/978-3-319-67077-5_22

storage clouds. Access to specific data sources can be provided from both stationary access centres and any mobile devices [8]. Those latter, which offer comfort through ready-to-use solutions for remote access to a car, the possibility to activate any onboard device or to start selected systems from a smartphone or tablet, are used especially willingly by the automotive industry.

One of the typical communication solutions that is used in automotive systems [9] is the Controller Area Network (CAN) bus [10]. It is used due to its safety, its ability to work in a real-time mode and the automatic control of access to the data bus. However, it is a network that is based on electrical access to the bus and thus, it does not support access to external (remote) devices. Therefore, it is necessary to use a communication coprocessor that would play the role of a mediating unit between a specific device or a car system and the user who would use it remotely. One of the solutions that gives this type of access is the virtualization of a CAN bus into an Ethernet network, which requires a dedicated converter from the CAN standard to Ethernet. In order to achieve higher transparency and universality for access to data on a CAN bus, the authors propose a solution that is based on an external Raspberry Pi computer, which is plugged into a CAN bus with other Advanced Driving Assistance Systems (ADAS) devices [11]. The computer can be used as a gateway to transfer data between the two systems. The proposed solution is based on a ready-to-use PiCAN module that supports the work of a computer on a CAN bus and that is compatible with the Raspberry Pi. The data can then be processed by an external unit [12] in order to avoid unnecessary overloading of the main processing unit.

This paper is organised as follows: Sect. 2 briefly describes the Controller Area Network standard. Sect. 3 introduces the Advanced Driving Assistance System device that was used during the experiment and that uses the light detection and ranging. Section 4 presents a Raspberry Pi computer with the PiCAN module. The laboratory test stand is introduced in Sect. 5 and the results from the experiments are presented in Sect. 6. The last chapter summarises the approach and presents the conclusions and observations after the start-up and the experiment.

2   CAN

One of the most commonly met communication solutions in vehicles is the CAN [13]. The protocol was developed in 1980s by Bosch GmbH [14] and it quickly became widely used in the automotive and manufacturing industries. The main advantages of this solution are the ability to work in a real-time mode, physical control of errors and automatic access of the devices (nodes) to the bus.

CAN uses two wires that are typically marked as CAN_H and CAN_L, which create the bus. Devices that are denominated as nodes or Electronic Control Units (s) are plugged into it serially in a multi-master mode. This means that no supervisory unit is needed to control access to the transmission medium because all of the devices have the same work priority and transmit frames in the form of a broadcast. Physical access to the bus is realised by the bit domination that is set by the individual address of a transmitted message – the lower the ID of a frame the higher its priority.

Currently, there are two types of CAN messages being used – ones that are based on an 11-bit identifier (CAN 2.0A) and ones that are based on a 29-bit identifier (CAN 2.0B). Each frame consists of several fields. The most important ones are the arbitration field, which, among others, contains the identifier; the control field with the definition of data length (DLC); the data field and the data of the control sum.

There are several configurations for the bandwidth of a CAN bus [15] that primarily depend on the length of the wires and the number of nodes that are plugged into the network. The most common implementations are 500 kbps and 1 Mbps. In order to avoid reflections from the ends of the bus, it is necessary to terminate the network with 120-ohm resistors on each of its sides [14].

Due to the popularity and ease of implementation, CAN buses have a variety of versions and standards and their startup, testing, diagnosis and maintenance are supported by many available tools and platforms. One such platform is CANoe by Vector Informatik GmbH [16].

3   LIDAR

The solution that appears to be used most often to detect objects and analyse the surrounding environment is Light Detection and Ranging (LIDAR) [11]. The system is using the reflection of the light from the examined object to calculate its distance based on the time from the emission of a light beam from the emitter through the reflection from the obstacle to its return to the receiver. Examples of such devices and their use cases have been described in various automotive applications [17, 18]. Due to their popularity onboard cars, they are used in ADAS, which support a driver in controlling the vehicle [11]. Short Range Lidar was selected because the authors concentrated their attention on detecting [19] and avoiding obstacles [20] by mobile units, including cars and Autonomous Mobile Platforms (AMPs).

The device allows the distance to be measured from 0.7 to 15 m on three independent channels. The side channels are inclined from the central one by 15°. Additionally, each channel is capable of providing the speed of the approaching object and it measures the level of the reflection of light and is used in e.g. the detection of fog and other aerosols that are suspended in the air. The device is mounted behind the wind screen around rear-view window by car manufacturers. It is used in systems that e.g. support automatic braking while driving in traffic jams.

4   Raspberry Pi and the PiCAN Module

The Raspberry Pi is a small computer created by the Raspberry Pi Foundation, that has one PCB and can run on Linux, RISC OS or Windows 10 operating systems. It is equipped with numerous interfaces and devices, which make it capable of performing a variety of tasks, e.g. controlling via GPIO or interfacing using popular communication standards. The computer can be extended with many ready-to-use solutions, which allow it to be adapted to the desired working conditions.

PiCAN (Fig. 1) is a module that was designed for the Raspberry Pi and is used for communication on a CAN bus. It was designed and produced by SK Pang Electronics Ltd. The main components of the module are an MCP2515 CAN microchip and a TJA1050 CAN transceiver that is manufactured by Phillips, which is capable of operating on both CAN standards with speeds of up to 1 Mbps.

Fig. 1. PiCAN module with 9-pin D-SUB and SJ connectors, the connection circuit and pin configuration

The connection to a CAN bus is possible using a screw terminal, JP2 pins or 9-pin D-SUB connector. The method selected for the start-up was the one via D-SUB, which requires soldering specific SJ connectors onto the PCB, depending on the selected configuration.

After the module is connected to a Raspberry Pi, the configuration of the network interfaces file and parameterisation of the Serial Peripheral Interface using the Raspbian system is required. The process is required in order to initiate the communication between the module and the controller.

The fully-configured setup allows applications prepared in Python and C language to be run. The experiments were based on can-utils – a set of tools that allows data to be received from and transmitted to the bus. The program used for the communication with the LIDAR was prepared using two threads – one for transmitting to and the second for receiving from the bus. The solution was built using open source SocketCAN drivers designed for the Linux kernel.

5   Test Stand

In order to verify the quality of transmission between the LIDAR and the PiCAN module, a test stand with additional possibilities for plugging in nodes was built. Two solutions were selected to compare the transmission. The first one was based on the CAN/USB VN1630A interface produced by Vector Informatik GmbHand the second one was based on the CAN/Ethernet i-7540D gateway by ICP DAS. Signals from the interface and the gateway were forwarded to a CANoe environment where they were decoded and processed, In the case of the Raspberry Pi, the CAN messages were forwarded directly to the PiCAN module, which sent them for further processing using an SPI bus. The setup of devices and networks on the test stand is presented in Fig. 2.

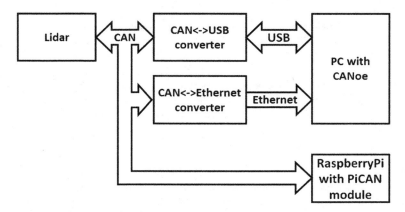

Fig. 2. The test stand to compare LIDAR's work on three different transmission channels

The LIDAR is not a stand-alone device that works onboard a car and therefore continuous communication between it and any other devices working in the system is necessary. This means that messages that come from other nodes must be emulated and transmitted to the LIDAR so it can perform continuous measurements. Generating those emulated frames was prepared based on the technical data provided by the manufacturer of the devices. Although it was necessary to transmit them from only one source, it was prepared for all three variants of the LIDAR's connection. Therefore, both the Raspberry Pi and the PC become sources of the signals that are necessary for the LIDAR to work. The whole system can be used to prepare the functional architecture for autonomous driving [21] using the signals from all of the ADAS devices that are connected by the CAN bus and controlled by an embedded system. Additionally, the OPC UA standard [22] can be used to provide universal access to the measurement data [23].

6 Experiments and Results

The first experiment was run to determine whether there are differences in the transmission between the LIDAR, the CANoe platform and the Raspberry Pi equipped with PiCAN module. For the comparison, besides the standard CAN bus a virtualized CAN into the Ethernet was used. Sample results of the distance measurements from the central beam are presented in Fig. 3.

During the experiment, no significant differences were observed between the measured channels. Insignificant shifts in a series on the X-axes were caused by the different timestamps that were set by the receiving devices. The Raspberry Pi began the registration of timestamps the moment that the LIDAR started to transmit the measurements while CANoe measured from the beginning of the experiment, which included ca. the first 4 s when the LIDAR was still inactive.

Additionally, there were multiple variants that were tested in which the generation of the signals that are needed for the LIDAR to work was realised. For the Raspberry Pi

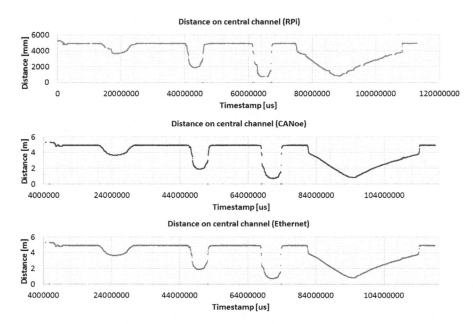

Fig. 3. The comparison of the distance measured on the central channel of the LIDAR for three transmission channels: CAN (CANoe), Ethernet (CANoe) and CAN (PiCAN)

equipped with the PiCAN module, it was observed that messages with different IDs could not be transmitted in the same time intervals. If multiple frames were generated by Raspberry Pi at the same moment of time, only the first one was sent, which was the one that was granted access to the transmission buffer before it entered the bus. Therefore, it was necessary to use fixed delays of 1 ms between forwarding the consecutive frames to the bus.

In order to check the frequency of the frames received by the Raspberry Pi, a series that compared any differences in time between the received frames with the current measurement of distance on all channels of LIDAR was drawn. The results are presented in Fig. 4.

In all of the cases, a period of ca. 20 ms was obtained as the interval between receiving consecutive frames. The result is consistent with the time values that were previously acquired for CAN- and Ethernet-based approaches in CANoe. The noticeable increase of frequency is represented by the yellow points gathered in the lower section of Fig. 4, which is the result of the LIDAR's characteristics of work (e.g. because of the drop of a measurement below the minimum threshold of 70 cm).

A single LIDAR loaded a CAN bus in less than 20%, and therefore it was necessary to verify the correctness and the quality of transmission for a bus that was more loaded. For the purpose of the experiment, a procedure, which cyclically (every 100 ms) transmitted a low priority message, was prepared. The message contains the value of the counter, which is incremented after each attempt to transmit the frame to the bus. The message is then received and decoded by CANoe that is running on the PC and the

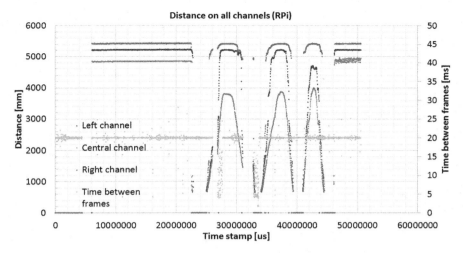

Fig. 4. Time between receiving consecutive frames in relation to current measurements

value of the counter is compared to the previously received one. In the case of data loss due to the overload of the bus, the result of a comparison is more than 1 and it is possible to calculate the number of lost frames. For the bus loading, a procedure was prepared, which transmits a user-defined number of so-called interfering messages with a high priority. Each interfering frame contains 8 bytes of data and has a total length of 108 bytes, which comes out of the CAN 2.0A standard that was used. The results of the first test are presented in Fig. 5.

The minimal time interval in which the forwarding of interfering frames was possible was 10 ms. For the 41 interfering frames that were transmitted cyclically every 10 ms (green series), the traffic on the 500 kbps bus reached ca. 442 800 bits per second. This, combined with the imperfection and the length of wires, resulted in a nearly 100% loading of the bus (blue series). When 42 interfering frames were transmitted, the data loss began (red series). This phenomenon results from the lack of physical access to the bus by data frames with a lower priority than the interfering frames.

The experiment was repeated with a reverted configuration. This time the interfering frames had a lower priority than the frames with the counter. No matter how the bus was loaded, 100% of the transmission frames were transmitted by the Raspberry Pi and received by CANoe. The results of the test are presented in Fig. 6. These phenomena were expected and confirmed that the Raspberry Pi works correctly and efficiently as a CAN node.

The next step was to analyse the influence of interfering frames with a high priority on the LIDAR's work. The results of the experiment are presented in Fig. 7.

The experiment showed that the LIDAR requires less than 20% of the total 500 kbps CAN bandwidth. As was expected, the LIDAR stopped providing measurement data after exceeding 38 interfering frames that were transmitted every 10 ms. This resulted from relatively high IDs of the frames transmitted by the Raspberry Pi

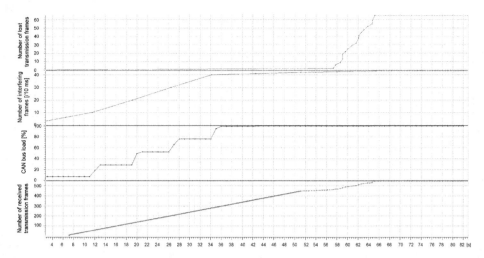

Fig. 5. The influence of the interfering frames with a high priority on the quality of the data received (Color figure online)

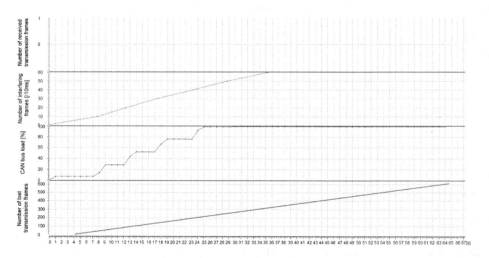

Fig. 6. The influence of the interfering frames with a low priority on the quality of the data received

that are needed for the LIDAR's work. In practice, the data frames could not gain access to the bus because the interfering frames had lower IDs (and a higher priority). The moment the LIDAR stopped receiving the frames that were needed for its work, it also stopped transmitting the measurement frames as a response. The measurement frames, however, would also not be able to gain access to the bus because of the IDs that are higher than those of the interfering frames.

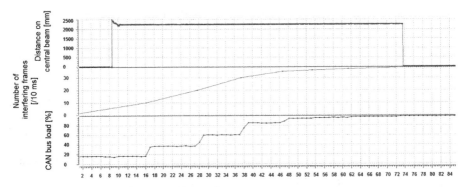

Fig. 7. The influence of the interfering frames with a low priority on the continuity of the LIDAR's work

7 Conclusions

The solution proposed in the paper is based on the use of a Raspberry Pi computer that is equipped with the PiCAN module. The computer communicates with ADAS that are connected to a CAN bus. During the experiments, it was proven that the solution is sufficient to support and control the ADAS devices and that the Raspberry Pi is an effective and efficient node of a CAN bus.

The solution has two significant advantages. The first one is possibility to universally and transparently monitor the traffic on a bus by adding an independent unit that does not interfere with the work of other devices. Besides the possibility to gather data, the unit is also capable of processing the data on its own. The second advantage is the possibility to use the newly added node as a gateway between a closed and safe CAN system and the external world. The solution can be used as a component of a Cyber-Physical System, which provides safe access to data from a hermetic, automotive or industrial system.

The possible areas where the solution can be used is data gathering from various ADAS devices using the Raspberry Pi and processing it for real-time analysis, e.g. to monitor the statuses of devices or to raise an alarm in the event of an abnormal event that is indicated when measuring the surrounding environment. Preparing dedicated applications that are run on a Raspberry Pi allows the transmission of data to other, stand-alone and independent systems using dedicated standards, e.g. OPC UA, which would form a communication component in distributed computing systems. This can be a source for remotely monitoring devices and analysing measurement data. Finally, a Raspberry Pi can be used as a source of the signals that are required for specific devices to work, e.g. in a way in which it was presented by the authors in the solution.

Acknowledgements. This work was supported by the European Union from the FP7-PEOPLE-2013-IAPP AutoUniMo project "Automotive Production Engineering Unified Perspective based on Data Mining Methods and Virtual Factory Model" (grant agreement no: 612207) and research work financed from funds for science in years 2016–2017 allocated to an international co-financed project (grant agreement no: 3491/7.PR/15/2016/2) and supported by Polish Ministry of Science and Higher Education with subsidy for maintaining research potential.

References

1. Cyber-Physical Systems (CPS) (NSF17529) | NSF - National Science Foundation. https://www.nsf.gov/pubs/2017/nsf17529/nsf17529.htm
2. Wu, F.-J., Kao, Y.-F., Tseng, Y.-C.: From wireless sensor networks towards cyber physical systems. Pervasive Mobile Comput. **7**, 397–413 (2011)
3. Khaitan, S.K., McCalley, J.D.: Design techniques and applications of cyberphysical systems: a survey. IEEE Syst. J. **9**, 350–365 (2015)
4. Wang, Y., Vuran, M.C., Goddard, S.: Cyber-physical systems in industrial process control. ACM SIGBED Rev. **5**, 1–2 (2008)
5. Li, R., Liu, C., Luo, F.: A design for automotive CAN bus monitoring system (2008)
6. Cupek, R., Ziebinski, A., Franek, M.: FPGA based OPC UA embedded industrial data server implementation. J. Circ. Syst. Comput. **22**, 1350070 (2013)
7. Marwedel, P.: Embedded System Design: Embedded Systems Foundations of Cyber-Physical Systems. Springer, New York (2010). doi:10.1007/978-94-007-0257-8
8. Jazdi, N.: Cyber physical systems in the context of industry 4.0. Presented at the 2014 IEEE International Conference on Automation, Quality and Testing, Robotics, May 2014
9. Zhou, F., Li, S., Hou, X.: Development method of simulation and test system for vehicle body CAN bus based on CANoe. Presented at the 7th World Congress on Intelligent Control and Automation, 2008, WCICA 2008 (2008)
10. Qianfeng, L., Bo, L., Mingzhao, C.: SJA1000-based CAN-bus intelligent control system design. Tech. Autom. Appl. **1**, 61–64 (2003)
11. Ziebinski, A., Cupek, R., Erdogan, H., Waechter, S.: A survey of ADAS technologies for the future perspective of sensor fusion. In: Nguyen, N.T., Iliadis, L., Manolopoulos, Y., Trawiński, B. (eds.) Computational Collective Intelligence, vol. 9876, pp. 135–146. Springer, Cham (2016). doi:10.1007/978-3-319-45246-3_13
12. Ziębiński, A., Świerc, S.: The VHDL implementation of reconfigurable MIPS processor. In: Cyran, K.A., Kozielski, S., Peters, J.F., Stańczyk, U., Wakulicz-Deja, A. (eds.) Man-Machine Interactions, vol. 59, pp. 663–669. Springer, Berlin (2009). doi:10.1007/978-3-642-00563-3_69
13. Leen, G., Heffernan, D.: Expanding automotive electronic systems. Computer **35**, 88–93 (2002)
14. Bosch, R.: CAN Specification Version 2.0. Rober Bousch GmbH, Postfach (1991)
15. Herpel, T., Hielscher, K.-S., Klehmet, U., German, R.: Stochastic and deterministic performance evaluation of automotive CAN communication. Comput. Netw. **53**, 1171–1185 (2009)
16. CANoe – the Multibus Development and Test Tool for ECUs and Networks. Vector Informatic GmbH (2011)
17. Rasshofer, R., Gresser, K.: Automotive radar and lidar systems for next generation driver assistance functions. Adv. Radio Sci. **3**, 205–209 (2005)
18. Ogawa, T., Sakai, H., Suzuki, Y., Takagi, K., Morikawa, K.: Pedestrian detection and tracking using in-vehicle lidar for automotive application. Presented at the 2011 IEEE Intelligent Vehicles Symposium (IV) (2011)
19. Grzechca, D., Wrobel, T., Bielecki, P.: Indoor location and identification of objects with video surveillance system and WiFi module (2014)

20. Budzan, S., Kasprzyk, J.: Fusion of 3D laser scanner and depth images for obstacle recognition in mobile applications. Opt. Lasers Eng. **77**, 230–240 (2016)
21. Behere, S., Törngren, M.: A functional architecture for autonomous driving. Presented at the Proceedings of the First International Workshop on Automotive Software Architecture (2015)
22. Maka, A., Cupek, R., Rosner, J.: OPC UA object oriented model for public transportation system. Presented at the 2011 Fifth UKSim European Symposium on Computer Modeling and Simulation (EMS) (2011)
23. Cupek, R., Ziebinski, A., Fojcik, M.: An ontology model for communicating with an autonomous mobile platform. In: Kozielski, S., Mrozek, D., Kasprowski, P., Małysiak-Mrozek, B., Kostrzewa, D. (eds.) BDAS 2017. CCIS, vol. 716, pp. 480–493. Springer, Cham (2017). doi:10.1007/978-3-319-58274-0_38

Obstacle Avoidance by a Mobile Platform Using an Ultrasound Sensor

Adam Ziebinski[✉], Rafal Cupek, and Marek Nalepa

Institute of Informatics, Silesian University of Technology, Gliwice, Poland
{Adam.Ziebinski,Rafal.Cupek}@polsl.pl,
Marek.Nalepa@gmail.com

Abstract. The problems of obstacle avoidance occur in many areas for autonomous vehicles. In automotive field, Advanced Driver Assistance Systems modules equipped with sensor fusion are used to resolve these problems. In the case of small mobile platforms, electronic sensors such as ultrasound, gyroscopes, magnetometers and encoders are commonly used. The data obtained from these sensors is measured and processed, which permits the development of automatic obstacle avoidance functions for mobile platforms. The information from these sensors is sufficient to detect obstacles, determine the distance to obstacles and prepare actions to avoid the obstacles. This paper presents the results of research on two obstacle avoidance algorithms that were prepared for small mobile platforms that take advantage of an ultrasonic sensor. The presented solutions are based on calculating the weights of the possible directions for obstacle avoidance and the geometric analysis of an obstacle.

Keywords: ADAS · Detection the obstacle · Obstacle avoidance · Sensors

1 Introduction

Advanced Driver Assistance Systems (ADAS) [1] measure and process data about the external environment using a combination of driving parameters. ADAS provides information to other car modules and informs the driver about certain road conditions. In a critical situation, ADAS takes control of the car [2]. The most common ADAS [3] that are used in such applications are Emergency Breaking Assist (EBA); Forward Collision Warning (FCW), Rear Cross Traffic Alert (RCTA), Lane Departure Warning (LDW) and Blind Spot Detection (BSD).

ADAS modules are often developed as Cyber-Physical Systems (CPS) [4]. CPS are commonly developed as embedded systems to control physical processes with feedback loops, which are able to communicate via the Internet [5] and are used to detect external situation that can affect the computations of the entire system. CPS can be used in many areas such as industrial process control [6], transportation, monitoring and automotive solutions [7].

ADAS solutions are often used to resolve problems by implementing automatic obstacle avoidance in vehicles [8] using sensor fusion [9]. Such a control system has to take many factors into account, distance from an obstacle, its shape and size and the current velocity of vehicle, to name a few. Generally, every obstacle avoidance

© Springer International Publishing AG 2017
N.T. Nguyen et al. (Eds.): ICCCI 2017, Part II, LNAI 10449, pp. 238–248, 2017.
DOI: 10.1007/978-3-319-67077-5_23

algorithm [10] can be considered to be an iterative algorithm in which every iteration consists of three phases – sensory data acquisition, data analysis and computing the control values that are passed to the driving gear. After detecting an obstacle in the vehicle's path, the control system has to perform the appropriate action to avoid the obstacle. First, the control system has to correctly detect an obstacle in front of the vehicle. This task can be accomplished in different ways. One possible solution is to use computer vision. An obstacle can be detected using two cameras that work as a stereovision system [11]. By comparing key point offsets on both images, it is possible to add a third dimension, that is to measure the depth of selected points on image. Using only one camera, there is still the possibility to detect an object in front of the vehicle [12]. When the camera is moving, SFM (structure from motion), can be used. The drawback of all of the methods of computer vision [13] is that it requires much more computing power than other methods. Using the sensor network concept allows the surrounding environment to be observed and analysed and finally for any obstacles and objects to be localised [14]. Sensors that can measure distance are one of main applications in the automotive area. These sensors include ultrasonic distance sensors [15], laser scanners and LIDAR [3] (light detection and ranging). LIDAR sensors and laser scanners are much more expensive than ultrasonic distance sensors, but they are also much more accurate.

In the article, the authors present the results of research on two obstacle avoidance algorithms that were prepared for small mobile platforms that take advantage of ultrasonic sensors.

2   Developing Obstacle Avoidance Algorithms

Two different algorithms prepared in C language are proposed for avoiding obstacles. The first uses a system of weights to calculate the direction in which the vehicle should proceed in order to avoid the obstacle. The second algorithm uses a geometry-based approach to track the edges of the obstacle to estimate the way to avoid the obstacle.

To test the algorithms, a small indoor vehicle was prepared. It is a four-wheeled mobile platform (Fig. 1). The vehicle moves like a caterpillar vehicle, but instead of caterpillar tracks, it has two wheels on each side. Because of its small size, it has only one ultrasonic distance sensor mounted on its front. To measure the distance to obstacles in other directions, the vehicle has to rotate around on its vertical axis and take the measurements during this rotation. The caterpillar-style movement system allows it to rotate easily. The vehicle was also equipped with other sensors such as accelerometer, magnetometer, encoders, gyroscope and camera. All of these sensors were connected to a Raspberry Pi using dedicated standards – GPIO, I2C and SPI [16]. Additionally, the system can be equipped with an FPGA circuit [17], which would allow real-time functions to be developed [18]. The application, which is running on Raspberry Pi, interprets the signals that are collected from the sensors and is responsible for controlling the movement of the mobile platform. The vehicle is supervised, controlled and monitored through a software application on a PC via the internet. The measurements are transmitted to the PC in the XML format [19] in the data field of a TCP frame via WiFi. Next, they are converted to the OPC UA standard [20] by an application on the PC to be analysed and processed in a laboratory test stand for ADAS solutions [21]. The images obtaining from camera should be coding to reduce the amount of stored data [22] in the future. This approach allows the quality of the obstacle avoidance algorithms that are developed to be improved.

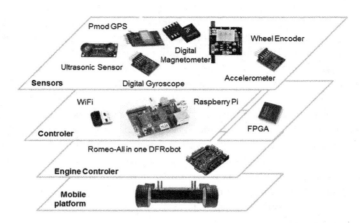

Fig. 1. The hardware architecture of a mobile platform

2.1   Obstacle Detection and Braking

While vehicle is moving, the control system constantly monitors the data from the sensors. This mode can be called the 'supervisor mode'. The most important sensor at this point is the distance sensor. It measures the distance to any obstacles in front of the vehicle. An obstacle is detected when the free space in front of the vehicle is less than a threshold value. For the test vehicle, this value was set at 80 cm.

After detecting an obstacle, the vehicle enters 'brake mode'. In this mode the main task is to brake smoothly in order to avoid colliding with the detected obstacle. On the one hand, the earlier the vehicle stops, the better, but on the other hand, instant stopping could cause a strong jerk. Taking into account the above assumptions, a braking solution was proposed. The control system smoothly decreases the driving speed from the initial detecting distance to the absolute stop distance from obstacle, which had been set to 60 cm on the test vehicle. To sum up, vehicle braking is performed on a 20 cm section. After the vehicle brakes and stops, the correct obstacle avoidance and search for a new route occurs. Two algorithms were proposed for that task.

When an obstacle is being avoided, the vehicle has to track its position in respect to the detected obstacle. This is necessary to correctly recognise when the obstacle has been avoided and it is possible to return to the original route.

The sensors that are mounted on vehicle only provide raw data, which has to be further processed in order to obtain the vehicle's position, heading and other useful information that is necessary to successfully navigate between obstacles. Once an obstacle is detected, a new local coordinates system is determined. According to this coordinates system, the vehicle's heading at the moment of the detection equals 0°.

The on-board 3-axis gyroscope makes it possible to measure heading changes. The gyroscope provides information about the angular velocity for all three axes of the Cartesian system. Angular velocity can be considered to be a derivative of angle (the rate of an angle change in other words). The formula (1) depicts the relationship between the gyroscope indication and the angle of vehicle rotation in respect to one axis only:

$$G_Z = \frac{d\theta}{dt} \tag{1}$$

G_Z is the current gyroscope indication for axis Z and θ is the angle of rotation. To calculate the actual angle using the gyroscope, the indications of that sensor have to be integrated over time. To calculate the vehicle's heading, it is sufficient to take into account only the yaw axis of the gyroscope because the yaw axis of a vehicle only rotates on the horizontal plane. The final formula (2) for calculating the heading is:

$$\theta(t) = \int_0^t G_{Zdt} \approx \sum_0^t G_Z(t)T_s \tag{2}$$

The approximation used in the formula results from the fact that the gyroscope provides discrete data and the integration has to be calculated using numerical integration. In the above formula, a simple rectangle rule for integration is used. T_S means the integration step (the time between subsequent gyroscope readings).

It is worth noting that it is not possible to calculate the absolute angle in relation to the global coordinates system (e.g. associated with the Earth) using the methods presented above due to the lack of an absolute heading reference. This could be solved by using a 3-axis magnetometer as the reference source, but after few tests, was decided not to use a magnetometer because of the strong electromagnetic interference in closed buildings. However, this is not a problem as avoiding an obstacle only requires knowledge about the local position in respect to an obstacle.

Another problem is to correctly measure the distance that was travelled by the vehicle while the obstacle was bypassed. This task was accomplished using the rotary encoders that were mounted on the vehicle. There is one encoder for left side and one encoder for the right side of vehicle. Unfortunately, a rotary encoder can only count the pulses while a wheel is rotating without differentiating between forward and backward movements. In order to calculate the actual distance, one has to subtract every discrete moment of time from the previous pulse counter state from the current state. Knowing the difference and the motor's spin direction (forward/backward), the difference should be either added to or subtracted from the odometer reading of one side of the vehicle. The average of the left and right odometer readings can be considered to be the correct odometer reading for the vehicle.

Another problem is to correctly determine the XY position. If the current heading and the distance travelled are known, it is possible to calculate the position on the XY plane that is associated with the surface on which the vehicle is moving. Formulas (3) and (4) are used to calculate the position.

$$X_t = X_{t-1} + \sin(\theta_t) * (S_t - S_{t-1}) \tag{3}$$

$$Y_t = Y_{t-1} + \cos(\theta_t) * (S_t - S_{t-1}) \tag{4}$$

In the above formulas, X_t and Y_t are the vehicle's coordinates at the current moment of time, whereas X_{t-1} and Y_{t-1} are the coordinates in a previous discrete moment of time. S_t and S_{t-1} is the vehicle's odometer reading in the current and previous moments of time, respectively. The orientation of the coordinates system depends on the selection of the $0°$ heading. Since the heading is reset when the vehicle stops in front of an obstacle, the coordinates system is also reset and vehicle's current position is designated as (0, 0) point, the X-axis runs perpendicular to the current heading ('from left to right') and the Y-axis follows the heading of the vehicle.

2.2   Algorithm 1

The first proposed algorithm is based on calculating the weights. In both algorithms, the first step is to scan the environment for free space. The vehicle first rotates left to a $-60°$ angle and simultaneously measures the distance to other possible obstacles. This process is repeated for the right side (to a $+60°$ angle). The result of these actions is a map that contains the directions (angles) and distances associated with them that were measured. Measurements are performed every few degrees. The next step is calculating the weights that are associated with each direction. The purpose is to sufficiently filter out those directions that could be problematic, e.g. narrow slots between two obstacles where the vehicle would not fit. Filtering is done by calculating the average value of the measured distance and four distances in adjacent directions (both left and right side). This approach prevents the vehicle from avoiding obstacles that are in areas that are too close (because of the width of vehicle). Filtering is only first stage of calculating the weights. The second stage will be discussed because it is not relevant in a case in which the vehicle stops for the first time. The calculations to this point are presented on the diagram in Fig. 2. The obstacle was detected in the middle (short distances) and the filtered weights are on the edges of the obstacle.

After the weights for all of the directions are calculated, the direction with the maximum weight is selected. The higher the weight, the more free space exists in the considered direction. The vehicle then rotates to that direction and tries to move forward for a specified distance, e.g. 20 cm. Then, the vehicle stops and scans the environment for obstacles again. When the vehicle stops when avoiding an obstacle for the second time, calculating the weights is a bit more complicated. To force the vehicle to return to its original route after avoiding an obstacle, an extra step is added to the calculations. It consists of multiplying the filtered weights by the values of a specified polynomial, where the direction is an argument of the polynomial. This method prioritizes directions towards the original direction, so the vehicle eventually returns to original route. Diagram of such polynomial is presented in Fig. 3.

The algorithm ends when two conditions are met – the vehicle is back on original route (that means that the X-axis position of the vehicle is in a small arbitrary range from 0) and farther than the initially measured distance to the obstacle (the Y-axis position is greater than the initially measured distance). The last step includes rotating the vehicle to the original direction and resuming normal movement.

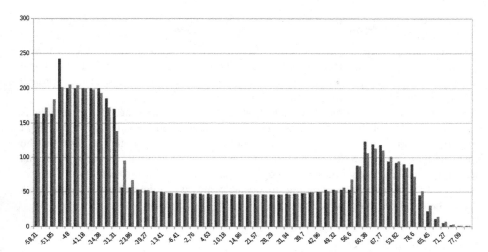

Fig. 2. First stage of calculating the weights. Blue bars represent the distances measured by a sensor depending on the heading (in cm). Red bars represent the weights after filtering the averages (Color figure online)

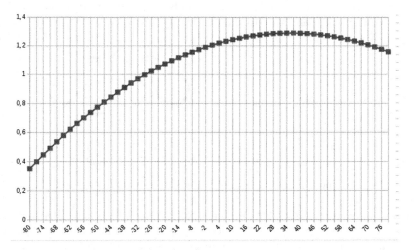

Fig. 3. Diagram of a correction polynomial. The filtered weights are multiplied by the corresponding values of that polynomial to prioritize heading towards the original route

2.3   Algorithm 2

The second algorithm is based on a geometric analysis. After stopping in front of an obstacle, the vehicle rotates to the left and tracks the edge of the obstacle. When more than one point is measured, the line equation of the edge is calculated using trigonometry principles and the last two points. When the next distance is measured, it is compared with the distance that was predicted with the line equation. If the difference between the two is greater than the selected threshold, it is considered to be the end of edge.

Depending on the distance to the obstacle, the necessary clearance angle is calculated. It is the angle that the vehicle must further rotate to safely avoid the edge of the obstacle when driving. It is a function of the width of the vehicle, the desired clearance and the distance to the obstacle. After rotating, the vehicle moves forward. The distance of the movement is equal to the last measured distance to the obstacle. This approach permits the vehicle to reach the edge of the obstacle.

The next action is rotating vehicle towards the original route. If the vehicle is located to the left of the obstacle and the original route, then it rotates 80° to the right with respect to the original direction. If the vehicle is located to the right of the obstacle, then it rotates −80° to the left with respect to the original direction. After this action, the vehicle starts to track the edge of the obstacle as it did initially.

Like the first algorithm, avoiding the obstacle ends when the vehicle is located past the obstacle on the original route.

3   Survey and Testing

Both algorithms were implemented on mobile platform and tested in terms of their effectiveness and time performance. Tests were performed with four types of obstacles in front of the vehicle presented in Fig. 4: a flat obstacle, a flat obstacle with additional depth, a round obstacle and an obstacle course.

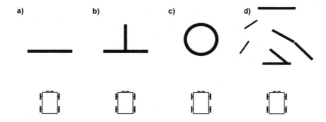

Fig. 4. The different types of obstacles that were used for the survey and tests

Twenty drives were performed for each type of obstacle and each of the algorithms. During each test drive, four parameters were measured: drive time, the distance covered when avoiding an obstacle, the number of changes of direction and the number of forward movements. After completing all 20 drives, the ratio of the correctly finished drives to all of the drives was calculated. This was defined as the percent of drives in which the vehicle correctly avoided an obstacle to all of the drives. Effectiveness was determined using the formula (5).

$$S = \frac{n - p}{n} \times 100\% \tag{5}$$

where S means the effectiveness of a particular algorithm on a given obstacle type, n is the number of attempts and p is the number of aborted attempts (when the algorithm made a wrong decision that caused the operator to stop the vehicle manually).

In order to obtain accurate results, a test bench was specially prepared. The vehicle's starting point was marked on the floor from three sides with sticky tape in order to prevent any displacement between the trials. The environment (shape, size and position of obstacles) was also precisely measured and preserved between trials. Tables 1 and 2 show the results of survey that was performed for the obstacle types that were used. The abbreviation 'NC' in the tables means that the test was 'not completed'. These are the aborted trials. 'T [s]' denotes performed time in seconds, 'D [cm]' is performed distance in centimetres. At the bottom of each table, the mean values, their standard deviations and the overall effectiveness of the algorithms for each obstacle type are calculated.

Table 1. Results of the survey for obstacle types a and b

Trial number	Obstacle type a				Obstacle type b			
	Algorithm A1		Algorithm A2		Algorithm A1		Algorithm A2	
	T [s]	D [cm]	T [s]	D [cm]	T [s]	D [cm]	T [s]	D [cm]
1	31.10	175.81	10.50	166.26	31.00	141.72	12.70	182.89
2	33.30	151.60	11.00	160.44	31.90	129.12	12.10	190.32
3	NC		16.20	216.44	45.80	228.39	NC	
4	34.20	202.16	12.10	184.60	35.80	224.24	NC	
5	31.00	144.70	11.30	164.13	32.30	161.66	12.80	186.48
6	32.50	146.08	11.30	165.53	NC		13.10	181.79
7	NC		11.10	166.78	24.00	120.74	15.90	231.05
8	29.90	168.43	11.40	175.03	32.10	160.17	12.90	209.74
9	NC		11.50	160.80	32.00	150.52	NC	
10	30.90	139.66	11.70	161.52	35.90	231.88	12.60	180.83
11	32.60	150.99	NC		33.20	158.76	13.10	205.85
12	31.80	175.22	10.50	156.72	NC		12.00	185.60
13	34.40	187.40	12.40	178.34	29.00	141.35	12.40	202.45
14	31.60	141.03	12.20	166.65	32.50	154.27	12.70	198.32
15	30.90	152.48	11.90	175.62	NC		NC	
16	NC		11.70	169.82	32.30	172.16	15.30	209.64
17	33.40	205.92	11.40	160.33	NC		12.90	212.08
18	21.50	129.66	12.00	169.71	22.50	124.60	12.20	221.02
19	32.20	166.92	12.20	163.77	NC		12.40	194.09
20	32.90	166.88	12.60	164.28	32.40	195.03	17.20	246.48
Mean	31.51	162.81	11.84	169.83	32.18	166.31	13.27	202.41
Std dev.	2.96	22.34	1.21	13.28	5.25	37.13	1.50	18.90
Effect.	80%		95%		75%		80%	

After an analysis of obtained results, a few conclusions can be drawn. Firstly, algorithm A2 bypassed the obstacle in a shorter time than algorithm A1 for every obstacle type. The difference was significant, and in some cases the times were two or three times shorter. What is interesting is that the total distance covered by the vehicle using algorithm A2 (faster) was a bit longer than for algorithm A1. Despite the fact that algorithm A2 needed much less time to find a successful way to avoid the obstacle, it needed to drive farther.

Another conclusion is that algorithm A2 had a much better effectiveness compared to algorithm A1. This can be seen in Table 3, which is a summary of the averages for all of the trials.

Table 2. Results of the survey for obstacle types c and d

Trial number	Obstacle type c				Obstacle type d			
	Algorithm A1		Algorithm A2		Algorithm A1		Algorithm A2	
	T [s]	D [cm]	T [s]	D [cm]	T [s]	D [cm]	T [s]	D [cm]
1	21.00	112.66	11.50	161.53	30.60	187.32	18.90	241.62
2	30.80	144.72	11.00	157.27	34.50	204.34	16.60	215.29
3	31.60	129.04	12.50	172.31	42.00	210.45	16.80	225.29
4	32.70	144.47	11.30	157.64	NC		NC	
5	29.90	151.14	12.50	172.53	35.60	234.89	16.60	229.35
6	30.60	131.97	12.60	193.69	31.70	188.00	NC	
7	29.50	159.05	13.20	184.87	NC		16.00	205.27
8	31.50	205.87	13.70	192.37	34.20	176.32	17.90	239.52
9	29.30	147.26	12.60	177.95	34.70	190.97	17.20	220.52
10	33.00	203.11	12.90	178.57	NC		17.30	219.35
11	20.50	131.84	14.00	193.28	33.40	193.11	17.40	231.62
12	31.80	197.40	12.20	188.87	42.90	232.24	17.20	226.20
13	29.00	157.09	12.90	179.81	NC		NC	
14	NC		13.30	180.87	NC		18.50	250.84
15	NC		12.00	168.60	32.70	206.40	17.10	204.33
16	32.60	148.68	13.30	182.09	NC		16.80	224.49
17	32.50	187.37	11.50	160.05	43.80	189.68	17.00	225.89
18	22.70	128.80	13.20	180.45	33.20	175.17	17.20	236.52
19	31.20	143.95	13.20	184.43	43.80	220.14	16.90	220.26
20	31.90	139.88	11.40	154.51	32.70	159.80	16.90	216.26
Mean	29.56	153.57	12.54	176.08	36.13	197.77	17.19	225.45
Std dev.	3.96	27.26	0.86	12.55	4.78	21.65	0.70	12.28
Effect.	90%		100%		70%		85%	

Table 3. The summary of all trials

Criterion	Algorithm	
	A1	A2
Mean effectiveness	78.75%	90.00%
Mean time [s]	32.35	13.71
Mean distance [cm]	170.12	193.44

One thing worth noting is the standard deviations of the times that are presented in Tables 1 and 2. Algorithm A2 had much smaller standard deviation in the times than algorithm A1. This can be interpreted as algorithm A2 being able to generate much more reproducible results than algorithm A1 and can therefore be considered to be more 'stable'. This hypothesis is reflected in the effectiveness of both algorithms.

To ultimately point out the better algorithm, the criteria of the survey have to be arranged in the order of their importance. If taken into account that avoiding an obstacle is a crucial safety system for the mobile platform, then effectiveness is the most important factor. It is, of course, more important for a vehicle to navigate safely between obstacles than to do so quickly or using the shortest possible distance. When seen from that perspective, algorithm A2 (based on geometry approach rather than weights calculation) performed better.

4 Conclusions

Work on presented subject led to the successful design and implementation of two different obstacle avoidance algorithms that are suitable for small mobile platforms. Both algorithms were tested on four types of obstacles. Analysis of obtained results led to the conclusion that the geometry-based approach to the problem of obstacle avoidance is more effective than calculating the weights of the headings. Although the geometry-based algorithm permitted the obstacles to be avoided 2.3 mean times faster than the algorithm for calculating the weights, it took a 1.13 longer mean distance to avoid the obstacles. The data from the sensors were used to analyse the current status of the mobile platform. When used with an application on a PC, they can also be used to perform a historical analysis that assesses the quality of obstacle avoidance. The data from various sensors can be forwarded to the learning system, which will permit the route to be optimised. The test stand that was developed will allow other algorithms to be developed using LIDAR and laser scanners. The presented solution will permit the future implementation and development of new computational intelligence applications for a mobile platform to avoid obstacles.

Acknowledgements. This work was supported by the European Union through the FP7-PEOPLE-2013-IAPP AutoUniMo project "Automotive Production Engineering Unified Perspective based on Data Mining Methods and Virtual Factory Model" (Grant Agreement No: 612207) and research work financed from funds for science for years: 2016–2017 allocated to an international co-financed project.

References

1. Bengler, K., Dietmayer, K., Farber, B., Maurer, M., Stiller, C., Winner, H.: Three decades of driver assistance systems: review and future perspectives. IEEE Intell. Transp. Syst. Mag. **6**, 6–22 (2014)
2. Fildes, B., Keall, M., Thomas, P., Parkkari, K., Pennisi, L., Tingvall, C.: Evaluation of the benefits of vehicle safety technology: The MUNDS study. Accid. Anal. Prev. **55**, 274–281 (2013)

3. Ziebinski, A., Cupek, R., Grzechca, D., Chruszczyk, L.: Review of advanced driver assistance systems (ADAS). In: 18th IEEE International Conference on Industrial Technology (2017)
4. Khaitan, S.K., McCalley, J.D.: Design techniques and applications of cyberphysical systems: a survey. IEEE Syst. J. **9**, 350–365 (2015)
5. Wu, F.-J., Kao, Y.-F., Tseng, Y.-C.: From wireless sensor networks towards cyber physical systems. Pervasive Mob. Comput. **7**, 397–413 (2011)
6. Wang, Y., Vuran, M.C., Goddard, S.: Cyber-physical systems in industrial process control. ACM SIGBED Rev. **5**, 1–2 (2008)
7. Li, R., Liu, C., Luo, F.: A design for automotive CAN bus monitoring system (2008)
8. Jia, X., Hu, Z., Guan, H.: A new multi-sensor platform for adaptive driving assistance system (ADAS). In: 2011 9th World Congress on Intelligent Control and Automation, pp. 1224–1230 (2011)
9. Garcia, F., Martin, D., de la Escalera, A., Armingol, J.M.: Sensor fusion methodology for vehicle detection. IEEE Intell. Transp. Syst. Mag. **9**, 123–133 (2017)
10. Sezer, V., Gokasan, M.: A novel obstacle avoidance algorithm: "follow the gap method". Robot. Auton. Syst. **60**, 1123–1134 (2012)
11. Bertozzi, M., Broggi, A.: GOLD: a parallel real-time stereo vision system for generic obstacle and lane detection. IEEE Trans. Image Process. **7**, 62–81 (1998)
12. Yang, C., Hongo, H., Tanimoto, S.: A new approach for in-vehicle camera obstacle detection by ground movement compensation (2008)
13. Budzan, S.: Fusion of visual and range images for object extraction. In: Chmielewski, L.J., Kozera, R., Shin, B.-S., Wojciechowski, K. (eds.) Computer Vision and Graphics. LNCS, pp. 108–115. Springer, Heidelberg (2014). doi:10.1007/978-3-319-11331-9_14
14. Grzechca, D.E., Pelczar, P., Chruszczyk, L.: Analysis of object location accuracy for iBeacon technology based on the RSSI path loss model and fingerprint map. Int. J. Electron. Telecommun. **62**, 371–378 (2016)
15. Strakowski, M.R., Kosmowski, B.B., Kowalik, R., Wierzba, P.: An ultrasonic obstacle detector based on phase beamforming principles. IEEE Sens. J. **6**, 179–186 (2006)
16. Jaskuła, M., Łazoryszczak, M., Peryt, S.: Fast MEMS application prototyping using Arduino/LabView pair. Meas. Autom. Monit. **61**, 548–550 (2015)
17. Mocha, J., Kania, D.: Hardware implementation of a control program in FPGA structures. Prz. Elektrotech. **88**, 95–100 (2012)
18. Ziębiński, A., Świerc, S.: The VHDL implementation of reconfigurable MIPS processor. In: Cyran, K.A., Kozielski, S., Peters, J.F., Stańczyk, U., Wakulicz-Deja, A. (eds.) Man-Machine Interactions, vol. 59, pp. 663–669. Springer, Berlin (2009). doi:10.1007/978-3-642-00563-3_69
19. Cupek, R., Ziebinski, A., Fojcik, M.: An ontology model for communicating with an autonomous mobile platform. In: Kozielski, S., Mrozek, D., Kasprowski, P., Małysiak-Mrozek, B., Kostrzewa, D. (eds.) Beyond Databases, Architectures and Structures. Towards Efficient Solutions for Data Analysis and Knowledge Representation, vol. 716, pp. 480–493. Springer, Berlin (2017). doi:10.1007/978-3-319-58274-0_38
20. Maka, A., Cupek, R., Rosner, J.: OPC UA object oriented model for public transportation system. Presented at the 2011 Fifth UKSim European Symposium on Computer Modeling and Simulation (EMS) (2011)
21. Czyba, R., Niezabitowski, M., Sikora, S.: Construction of laboratory stand and regulation in ABS car system. Presented at the 2013 International Conference on Unmanned Aircraft Systems (ICUAS). IEEE, May 2013
22. Ulacha, G., Stasinski, R.: Improving neural network approach to lossless image coding. Presented at the Picture Coding Symposium (PCS) (2012)

Monitoring and Controlling Speed for an Autonomous Mobile Platform Based on the Hall Sensor

Adam Ziebinski[1], Markus Bregulla[2], Marcin Fojcik[3(✉)],
and Sebastian Kłak[1]

[1] Institute of Informatics, Silesian University of Technology, Gliwice, Poland
Adam.Ziebinski@polsl.pl, Sebastian.Klak@outlook.com
[2] Technische Hochschule Ingolstadt, Ingolstadt, Germany
Markus.Bregulla@thi.de
[3] Western Norway University of Applied Sciences, Førde, Norway
Marcin.Fojcik@hvl.no

Abstract. Cyber Physical Systems are often used in the automotive industry as embedded systems for constructing Advanced Driver Assistance Systems. Further development of current applications and the creation of new applications for vehicle and mobile platforms that are based on sensor fusion are essential for the future. While ADAS are used to actively participate in the controlling a vehicle, they can also be used to control mobile platforms in industry. In the article, the results of tests of different rates of data acquisition from Hall sensors to measure speed for mobile platform are presented. The purpose of the research was to determine the optimal platform parameter to indicate the refresh frequency in such a way that the measurements obtained from a Hall sensor will be reliable and will require less of the available computing power. Additionally, the results from investigations of the precise movement for a specified distance using a Hall sensor for a mobile platform are presented.

Keywords: ADAS · CPS · Data acquisition · Sensor fusion · Hall sensor

1 Introduction

Cyber Physical Systems (CPS) [1] allow embedded functionality to be combined together through their interaction with the physical world using the infrastructure of the Internet [2]. Additionally, they influence the development of the Internet of Things. Moreover, they allow the capabilities of the physical world to be improved through the efficient [3] and distributed communication [4]. CPS are important for the development of future technologies [5] because they increase the functionality, adaptability, efficiency, reliability and autonomy of systems. CPS are often used in the monitoring systems [6], industry solutions [7] as well as in automotive area [8]. Advanced Driver Assistance Systems are commonly used in today's vehicles. Due to the increasing need for mobility, driving has become more and more complex. The industry is using various types of mobile platforms, which are used, for example, to transport various components. The mobile platforms are equipped with different automotive sensors [9]

© Springer International Publishing AG 2017
N.T. Nguyen et al. (Eds.): ICCCI 2017, Part II, LNAI 10449, pp. 249–259, 2017.
DOI: 10.1007/978-3-319-67077-5_24

to ensure the security and control of their movements. In these applications, electronic sensors are the most commonly used. Already proven ADAS modules can also be used for these applications [10]. Due to their low cost and reduced requirements, mobile platforms can be equipped with fewer sensors and weaker processors. For example, platform that was used in this study is equipped in with two three-phase motors, different types of processing units and a set of sensors. The processors are used to monitor the platform as well as to control its movements. Although the mobile platform is not equipped with encoders, its motors are equipped with Hall sensors. These sensors can be used to determine the direction of the rotation of the engine and the rotation speed of engine. In effect, the information from the Hall sensors [11] can be used by the mobile platform to calculate the speed of its movement, the distance travelled and the direction of movement. The actual state of the sensors and the refresh rate of measurements are very important for safely controlling mobile platforms and vehicles [12]. In the article, the authors present the results of research on the influence of the data refresh rate of a mobile platform on the quality of the current data on the example of speed measurements using Hall sensors. This would permit precise movements of a mobile platform.

2   Architecture of the Test Stand

The research station (Fig. 1) consisted of a Forbot 1.4A mobile platform from Roboterwerk, an STM32F3 Discovery board, a Raspberry PI and a PC. The mobile platform (Fig. 2) consisted of two brushless 3-phase BLDC APM-SB03ADK-9 300 W motors (with a maximum frequency of 334 Hz) that were controlled by two SMCI36 motor controllers and these were managed by an STM32F3 Discovery development board (Fig. 3). Communication between the STM32 and SMCI36 was realised using a clock signal – with pulse signals. Control of the engines was realised using the STM32. Each rising edge of the clock signal caused one 'step' of the motor, which means one phase change. Additionally, the STM32 collected measurements from engines. The whole system was controlled by a PC using the Raspberry Pi to manage the mobile platform. During the research, the STM32 controlled the movements of the platform and sent the data that was collected to the PC. Finally, the platform was equipped with ADAS modules [10] and other electronic sensors. The fusion of several sensors that have different working principles such as an accelerometer, gyroscope, magnetometer, Hall sensors permits information about the speed, direction of movement and travelled distance to be calculated. Information from the infrared sensors, radar, lidar, laser scanner and camera permitted any obstacles to be avoided [13], by the indoor localisation [14] and Emergency Brake Assist function. The complete system controlled by Raspberry PI could be equipped for real-time processing [15] with usage of additional embedded system [16]. The signals from all ADAS allowed a functional architecture for autonomous driving to be prepared [17].

The Forbot 1.4A mobile platform was not equipped with encoders. Hall sensors were used instead of encoders. To correctly analyse data that was collected, knowledge about the wheel diameter and the number of Hall sensors state changes per one wheel revolution was required. The wheel diameter of the mobile platform was measured with

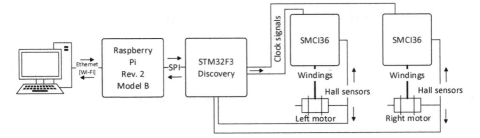

Fig. 1. Architecture of the control system for the Forbot 1.4A mobile platform

a calliper and was equal 15.7 cm. The number of state changes of the Hall sensor per revolution was equal 240 changes. The proposed approach was used for the laboratory test stand for an ADAS solution [18].

Fig. 2. Forbot platform interior

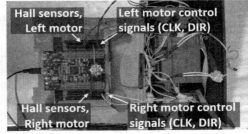

Fig. 3. Connections inside the platform

3 Controlling and Monitoring the 3-Phase Electric Motors on the Test Platform

Controlling electric Brushless DC Motors depends on how they are constructed [19]. BLDC motors have a permanent magnetised rotor and the stator has a variable magnetic field. A BLDC motor's stator is commonly constructed of three windings that are wound so that the magnetic field that is generated by them has the opposite polarity on the opposite sides of the stator. The windings in the motor can be arranged in two ways – into a star or triangle. In order to provide a more stable rotation speed, multiple windings, which are still connected in three phases, are used. Such an approach requires the alternating magnetisation of the rotor. In order to start the rotor, the windings have to be energised in a specific order. The simplest approach is to energise one coil at a time to attract the rotor, but this method can be improved by energising the second coil in parallel with the first one. In this way, the second coil repels the rotor. An example of a time waveform is presented in Fig. 4.

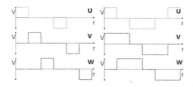

Fig. 4. An example time waveform of windings using the simple approach (left) and the advanced approach with repelling (right).

This motor's construction determines the way in which it is steered, because the position of the rotor has to be known with every commutation. The two most popular methods to synchronise the steering and position of the rotor are the Back EMF [20] or Hall sensors [21]. One of the methods was to use the Hall sensors that were embedded on the mobile platform [22]. An example waveform for a motor with three windings on the sensors during one revolution of the motor's shaft is shown in Fig. 5. There is a specific number (numbers from 1 to 6) of signal state changes for one shaft revelation.

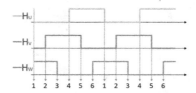

Fig. 5. An example time waveform using the Hall sensors

Specialised SMCI36 controllers, which also use the Hall signals from the motors, were used to provide good control. By connecting these signals between the motor and its controller, the Hall states on the Discovery board could be read. Knowledge about the sequence of the changes on the Hall sensors provided information about the rotation direction of the shaft and knowledge about the number of state changes in a given period of time and the number of pulses per full rotation provided information about the angular velocity. This could then be simply converted into linear velocity using the information about the wheel diameter. The formula for this speed is presented below:

$$v = \frac{P * A}{per * 0.001} \tag{1}$$

where:

v – instantaneous velocity [m/s]
P – number of Hall's pulses since the last velocity calculation
A – length of the circular arc for one Hall's sensor change [m]
per – period between the actualisations of the states [ms]

Hall sensors can be used to collect measurements [9] of the speed and the direction of movement. Taking this into account, the authors began their research on the influence of the data refresh rate by the mobile platform on the acquisition of the current data. While on the one hand, a refreshing rate that is too fast will use much of the processing power of whole platform, on the other hand a rate that is too slow will not give accurate measurements.

4   Experiment

Static algorithms for a slow acceleration to maximal speed while simultaneously measuring the actual speed of the platform were implemented on the STM microcontroller. The motor's controllers were controlled using the signals that were generated on the STM32 board clock. According to scheme shown in Fig. 1, the SMCI36 controllers were controlled by the clock signals from the STM device, which also collected data from the Hall sensors and sent them to the PC, where they were analysed. Data about any changes in the logic level on the Hall sensors with time since the start of research and also a periodic log, which contained the number of Hall sensor state changes for each motor since the last log, were sent to the PC.

The measurement method that was used was based on placing the mobile platform on smooth, even and flat surface. The platform would gradually accelerate from 0 to its maximum velocity while maintaining a constant speed for a specified time. A chart of the ideal velocity in time that should be obtained by the measurements is presented in Fig. 6.

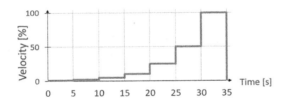

Fig. 6. Chart of the ideal velocity in time

The acceleration algorithm on the STM microcontroller used the following pseudocode:

1. Start
2. Set power on engines
3. Hold constant power for 5 s
4. If velocity is less than 100% go to point 2. Increase the power to the next value
5. Stop

The successive power values were: 1%, 2%, 5%, 10%, 25%, 50%, 75% and 100%. The measurements were taken for different refresh frequencies in milliseconds: 10, 20, 35, 50, 75, 100, 150, 200 and 250. The frequency rate was limited due to a command delay and poor control accuracy. The practical experiments showed that for the correct control of the motors, the maximum frequency of the control commands should not be lower than 10 ms. Information from the Hall sensors were sent to the PC where were saved for analysis.

The results of the experiments are partially presented in the following pictures as the waveforms of speed in time that was read.

A very large variance in the received data is presented in Fig. 7. With an increase of the speed, the dispersion also increased and reached a limit value of about 1.5 m/s.

Fig. 7. Sampling – 10 ms

Figure 8 shows that increasing the sampling time led to an improvement, because for high speeds (near maximum power), the dispersion of the data dropped to about 0.6 m/s.

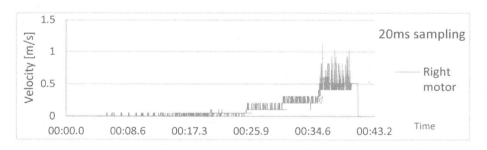

Fig. 8. Sampling – 20 ms

A further increase of the sampling time led to a continuous improvement of the reliability of the data that was read. With the 35 ms sampling (Fig. 9), the dispersion of the data dropped to about 0.35 m/s.

Figure 10 presents a comparison of the readings from both Hall sensors.

Figure 11 shows the readings for the 75 ms sampling.

After increasing sampling time to more than 100 ms (results shown in Fig. 12) the incoming data began to be acceptable. The dispersion of the data for the maximum speed dropped to about 0.1 m/s.

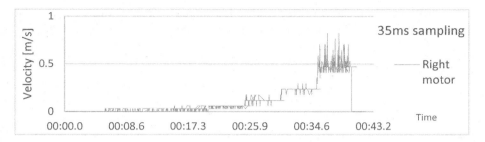

Fig. 9. Sampling – 35 ms

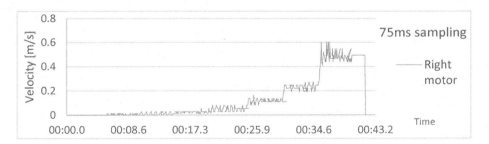

Fig. 10. Sampling – 50 ms

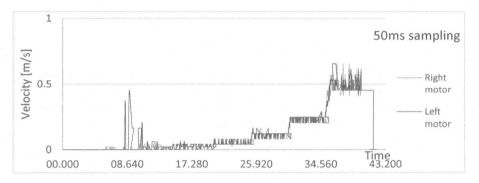

Fig. 11. Sampling – 75 ms

Fig. 12. Sampling – 100 ms

Further increasing the sampling period led to only minor corrections. For the 150 ms sampling time, the dispersion of data was approximately 0.06 m/s (Fig. 13).

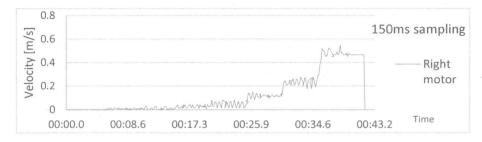

Fig. 13. Sampling – 150 ms

For a sampling time of 200 ms or more, the data variety was very small (Figs. 14, 15), but the time between the velocity calculations was so long, so if something happens to the motors, e.g. motor block, the system would not recognise it for the next 200 ms. With a speed of 0.5 m/s, it was a relatively long time.

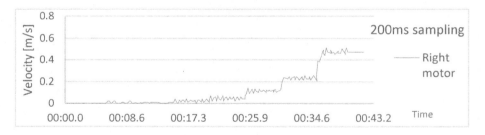

Fig. 14. Sampling – 200 ms

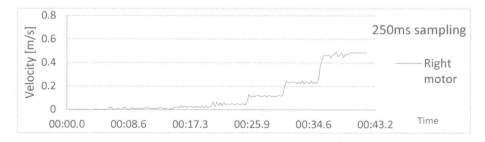

Fig. 15. Sampling – 250 ms

The variance of the measurement data was different for the refresh frequencies at a specified speed. As was shown in Table 1, for a velocity between 1% and 5% with a longer refresh time, the average deviation of the calculated velocity decreased from 0.01 m/s for the periods of 10 ms to 35 ms and to 0.005 m/s for the periods that were higher than 100 ms. For a velocity between 10% and 25%, the data was not so clear, but for a frequency of 10 ms, the results were the worst with a deviation of about 0.04 m/s. When a period equal to or greater than 35 ms was used, the results were similar and the deviations were about 0.015 m/s. For velocities higher than 50%, a sampling time longer than 20 ms seemed to be good; the deviation was about 0.05 m/s, which was about 10% of the average measured velocity.

These experiments show that a frequency of 100 ms was the most reliable one for every speed of the platform.

Table 1. Average velocities and average deviations for a given velocity and calculation period

Velocity		10ms	20ms	35ms	50ms	75ms	100ms	150ms	200ms	250ms	Average
	Average V [m/s]	0,00575	0,00493	0,00575	0,00370	0,00477	0,00329	0,00622	0,00657	0,00657	0,00528
1%	Avg Deviation	0,01086	0,00891	0,00924	0,00606	0,00622	0,00473	0,00490	0,00730	0,00575	0,00711
	Average V [m/s]	0,00904	0,00863	0,00985	0,01191	0,01022	0,00904	0,00934	0,00555	0,00945	0,00922
2%	Avg Deviation	0,01649	0,01436	0,01309	0,01311	0,01098	0,00615	0,00868	0,00454	0,00279	0,01002
	Average V [m/s]	0,02301	0,02301	0,02459	0,02485	0,02469	0,02403	0,02573	0,02177	0,02259	0,02381
5%	Avg Deviation	0,03570	0,02540	0,01282	0,01298	0,00566	0,00901	0,01167	0,01124	0,00555	0,01445
	Average V [m/s]	0,04610	0,04702	0,05151	0,04581	0,04558	0,04581	0,04471	0,04478	0,04519	0,04628
10%	Avg Deviation	0,05082	0,01095	0,01105	0,00945	0,01670	0,00822	0,02075	0,01617	0,00945	0,01706
	Average V [m/s]	0,12633	0,11297	0,11307	0,11339	0,11282	0,11256	0,11287	0,11092	0,11051	0,11394
25%	Avg Deviation	0,04112	0,02366	0,01334	0,01745	0,01594	0,01689	0,01504	0,01578	0,01216	0,01904
	Average V [m/s]	0,22574	0,22348	0,22490	0,22513	0,22367	0,22390	0,22367	0,22307	0,21979	0,22371
50%	Avg Deviation	0,04368	0,03994	0,01954	0,02603	0,02462	0,01668	0,02960	0,02117	0,02239	0,02707
	Average V [m/s]	0,53349	0,48600	0,47484	0,46628	0,46069	0,44984	0,44626	0,44491	0,43793	0,46669
100%	Avg Deviation	0,19549	0,10771	0,06753	0,05480	0,05290	0,03944	0,04956	0,04677	0,04700	0,07347
Avg deviation		0,05631	0,03299	0,02094	0,01998	0,01900	0,01445	0,02003	0,01756	0,01501	

The presented method was used to control the movement of a mobile platform for a specified distance. During the experiments in which the mobile platform drove predetermined distances of 50 to 250 cm along a tape measure in order to check the activity of the platform.

The next behaviour/maintenance that was observed was a different deviation relative to the sampling speed. The worst differences were measured at the fastest sampling rates. According to the motor parameters and tests results, for one full revolution there should be 80 edges from each Hall sensor (U, V and W), which means 240 changes in total. In the maximal power mode, the motors should have had 334 coil voltage changes per second, which would have been 1.4 rotations of the shaft per second. Together, this means that 9–10 Hall state changes should have occurred within 10 ms, which should have been enough to correctly measure the speed.

5   Conclusions

The conducted experiments showed that data refresh is very important and has a big impact on the quality of the data that is obtained especially for measuring velocity. In general, a longer period gave more a stable reading, but for high speeds, calculating the speed once every 0.25 s, when the platform covered a distance of about 14 cm was too risky. Using such a long sampling rate may cause the lack of a proper reaction in case of emergency. The comparison of the Hall sensors and the periodic calculations indicate that operating directly on the Hall readings for test platform was very unstable and ineffective, especially for higher velocities.

One of the reasons for this may be the inaccuracy of the Hall sensors and the speed limitations of hardware elements that were used. Nevertheless, the tests that were carried out shows that in the case of mobile platforms that have many different sensors, but not the most powerful processing unit, an ordinary simple Hall sensor can be used to increase control of platform. This should not be done using only one sensor but when used in combination offers the possibility to better control the platform.

According to the results, the most accurate velocity calculations were obtained using a 100 ms period of data calculations. To improve the stability and reliability of the data that is read, adaptive calculations or a hybrid should be used. It means, that for low speeds, when it is safe, relatively long sampling rate such as 200 ms should be used. For higher velocity the sampling rate should proportionally increase. Additionally, for low speeds when there may be no Hall state switch between each calculation, a function that calculates speed should have additional functionality. Should implemented, so when there is no signal from the sensors, the last good value of speed is used and a flag is set to provide information about the uncertainty of the calculation. In this way, the platform would know not to depend on the Hall sensors and should use the fusion of the sensors, e.g. the accelerometer to measure speed and ultrasound when a stop must be realised before an obstacle. The studied method was tested on a Forbot 1.4A mobile platform. It appears that the best way to stop the platform after a predefined distance is to depend directly on the Hall sensors when the accuracy is about 10 cm, what is sufficiently accurate.

Acknowledgements. This work was supported by the European Union from the FP7-PEOPLE-2013-IAPP AutoUniMo project "Automotive Production Engineering Unified Perspective based on Data Mining Methods and Virtual Factory Model" (grant agreement no: 612207) and research work financed from funds for science in years 2016-2017 allocated to an international co-financed project (grant agreement no: 3491/7.PR/15/2016/2).

References

1. Cyber-Physical Systems (CPS) (NSF17529) | NSF - National Science Foundation. https://www.nsf.gov/pubs/2017/nsf17529/nsf17529.htm
2. Poovendran, R.: Cyber-physical systems: close encounters between two parallel worlds [point of view]. Proc. IEEE **98**, 1363–1366 (2010)

3. Cupek, R., Huczala, L.: Passive PROFIET I/O OPC DA server. Presented at the IEEE Conference on Emerging Technologies and Factory Automation, ETFA 2009 (2009)
4. Maka, A., Cupek, R., Rosner, J.: OPC UA object oriented model for public transportation system. Presented at the 2011 Fifth UKSim European Symposium on Computer Modeling and Simulation (EMS) (2011)
5. Baheti, R., Gill, H.: Cyber-physical systems. Impact Control Technol. **12**, 161–166 (2011)
6. Flak, J., Gaj, P., Tokarz, K., Wideł, S., Ziębiński, A.: Remote monitoring of geological activity of inclined regions – the concept. In: Kwiecień, A., Gaj, P., Stera, P. (eds.) CN 2009. CCIS, pp. 292–301. Springer, Berlin (2009). doi:10.1007/978-3-642-02671-3_34
7. Wang, Y., Vuran, M.C., Goddard, S.: Cyber-physical systems in industrial process control. ACM SIGBED Rev. **5**, 1–2 (2008)
8. Thompson, C., White, J., Dougherty, B., Schmidt, D.C.: Optimizing mobile application performance with model-driven engineering. In: Lee, S., Narasimhan, P. (eds.) SEUS 2009. LNCS, vol. 5860, pp. 36–46. Springer Berlin Heidelberg, Berlin (2009). doi:10.1007/978-3-642-10265-3_4
9. Fleming, W.J.: Overview of automotive sensors. IEEE Sens. J. **1**, 296–308 (2001)
10. Ziebinski, A., Cupek, R., Grzechca, D., Chruszczyk, L.: Review of advanced driver assistance systems (ADAS). Presented at the 13th International Conference on Computer Methods Science Engineering (2017)
11. el Popovic, R., Randjelovic, Z., Manic, D.: Integrated Hall-effect magnetic sensors. Sens. Actuators A Phys. **91**, 46–50 (2001)
12. Proca, A.B., Keyhani, A.: Identification of variable frequency induction motor models from operating data. IEEE Trans. Energy Convers. **17**, 24–31 (2002)
13. Budzan, S., Kasprzyk, J.: Fusion of 3D laser scanner and depth images for obstacle recognition in mobile applications. Opt. Lasers Eng. **77**, 230–240 (2016)
14. Grzechca, D., Wrobel, T., Bielecki, P.: Indoor location and identification of objects with video surveillance system and WiFi module (2014)
15. Kobylecki, M., Kania, D., Simos, T.E., Kalogiratou, Z., Monovasilis, T.: Double-tick realization of binary control program. Presented at the AIP Conference Proceedings (2016)
16. Ziębiński, A., Świerc, S.: The VHDL implementation of reconfigurable MIPS processor. In: Cyran, K.A., Kozielski, S., Peters, J.F., Stańczyk, U., Wakulicz-Deja, A. (eds.) Man-Machine Interactions. AINSC, pp. 663–669. Springer, Berlin (2009). doi:10.1007/978-3-642-00563-3_69
17. Behere, S., Törngren, M.: A functional architecture for autonomous driving. Presented at the Proceedings of the First International Workshop on Automotive Software Architecture (2015)
18. Czyba, R., Niezabitowski, M., Sikora, S.: Construction of laboratory stand and regulation in ABS car system. Presented at the 2013 International Conference on Unmanned Aircraft Systems (ICUAS). IEEE May 2013
19. Rodriguez, F., Emadi, A.: A novel digital control technique for brushless DC motor drives. IEEE Trans. Ind. Electron. **54**, 2365–2373 (2007)
20. Shao, J., Nolan, D., Hopkins, T.: A novel direct back EMF detection for sensorless brushless DC (BLDC) motor drives (2002)
21. Samoylenko, N., Han, Q., Jatskevich, J.: Dynamic performance of brushless DC motors with unbalanced hall sensors. IEEE Trans. Energy Convers. **23**, 752–763 (2008)
22. Pan, C., Chen, L., Chen, L., Jiang, H., Li, Z., Wang, S.: Research on motor rotational speed measurement in regenerative braking system of electric vehicle. Mech. Syst. Signal Process. **66–67**, 829–839 (2016)

Using MEMS Sensors to Enhance Positioning When the GPS Signal Disappears

Damian Grzechca$^{(\boxtimes)}$, Krzysztof Tokarz, Krzysztof Paszek,
and Dawid Poloczek

Faculty of Automatic Control, Electronics and Computer Science,
Silesian University of Technology, Gliwice, Poland
{Damian.Grzechca,Krzysztof.Tokarz}@polsl.pl,
{krzypas819,dawipol566}@student.polsl.pl

Abstract. This paper presents the concept of using embedded MEMS sensors position objects especially when the GPS signal is weak, e.g. in underground car parks, tunnels. Such an approach is important for controlling indoor objects or autonomous vehicles. The signals are acquired by a Raspberry Pi platform with external sensors such as an accelerometer, gyroscope and magnetometer. A self-propelled vehicle was used and several exemplary paths were designed for acquiring signals. It was proven that appropriate signal filtering allows a position to be determined with a small error at a constant velocity condition. Comparing filters such as the moving average, median, Savitzky-Golay and Hampel filters were investigated. Moreover, the system offers a high degree of accuracy in a short time for indoor hybrid positioning systems that also have video processing capabilities. The cyber-physical system can also be used with the existing infrastructure in a building, such as Wi-Fi access points and video cameras.

Keywords: MEMS · Accelerometer · Gyroscope · Positioning · GPS

1 Introduction

Cyber physical systems (CPS) are important for the development of future technologies [1] because they allow the capabilities of the physical world to be improved through the use of computational, communication and control systems. Moreover, many examples of CPS can be found in the automotive industry [1–3].

Advanced driver assistance systems [4] provide additional information from the environment around a vehicle to support a driver. One of the main problems is obtaining the actual current geographical position of a vehicle even if the GPS signal cannot be acquired. In order to enhance the position of an object in motion, a set of MEMS sensors can be applied by creating a cyber physical system for further signal processing. There are many different techniques for object positioning that are based on the signal strength indicator, wave propagation latency, phase shift, etc. [5, 6]. Generally, positioning can be classified into positioning outside of buildings (outdoor positioning) and positioning inside buildings (indoor positioning systems, IPS) [7]. The quality of a positioning system is determined by five main metrics: accuracy, precision,

© Springer International Publishing AG 2017
N.T. Nguyen et al. (Eds.): ICCCI 2017, Part II, LNAI 10449, pp. 260–271, 2017.
DOI: 10.1007/978-3-319-67077-5_25

coverage, resolution and latency – location updates. IPS can be used in many different sectors, e.g. tracking kids in crowded places, finding stolen objects, locating patients in a hospital or prisoners in a prison etc. It also can be useful in tracking emergency services workers in buildings where they are providing assistance as well as when an unexpected, dangerous situation occurs [8]. Such systems can also be useful in enhancing vehicle positioning, which is important in places where a GPS signal does not reach or is too weak, e.g. underground car parks, tunnels.

The performance of a system is measured by the following metrics: accuracy, availability, coverage area, scalability, cost and privacy. Accuracy is described by the average Euclidean distance between the estimated position and the true position of an object. Availability is the ratio of the time during which the positioning system is available (provides information about the location of the object) to the total time in which the tracking process is performed. Coverage is the area in which the system is reachable, i.e. three levels are distinguished: local – an area that is not extendable, scalable – adding hardware increases the area covered and global – worldwide. Scalability is the degree to which the system guarantees a normal positioning function when it is scaled in one of two dimensions – geography and number of users. Cost can be related to any aspect of a system that is required for the system to work correctly. Cost can be understood in terms of money, time, space or energy. Privacy is important when data are collected, stored or processed and how the data are kept safe against unauthorized access or theft.

The following technologies are used to position objects inside buildings: Radio Frequency Identification (RFID), Ultra Wideband (UWB) [9], Infrared (IR), Ultrasonic, ZigBee, Wireless Local Area Network (WLAN), Cellular Based, Bluetooth [10], Image Based Technologies and Pseudolites [8, 11]. These technologies can be combined with micro electro-mechanical system (MEMS) sensors, which do not require any additional hardware or network infrastructure in the building (additional sensors are only located in the equipment of the tracked vehicle). The most promising technology is UWB, which allows objects to be tracked with centimetres accuracy [5]. This paper focuses on MEMS sensors, Wi-Fi and CCTV under the assumption that the object being tracked moves in buildings with wireless network infrastructure and CCTV or in a location where the GPS signal is poor, i.e. the object can use the approximated location that is determined by the MEMS sensors in relation to the last well-known position that was determined by either the GPS or CCTV.

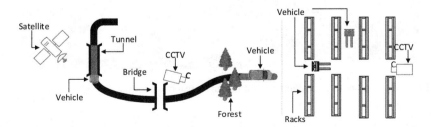

Fig. 1. MEMS sensors can be used in locations where the GPS signal is poor or when the video camera loses the position of the object. (Color figure online)

A green vehicle in the Fig. 1 is moving through a tunnel, under a bridge and through a forest – the reference path can be detected from the CCTV. Production or storage halls, on which the vehicles are moving, are equipped with CCTV system which can be used to localize the objects. In some cases, the object (blue vehicle in the Fig. 1) can be obscured (e.g. by rack) and then the MEMS sensors can be used to **short-term positioning of the object** in order to reduce e.g. risk of collision with another vehicle. Such a case is investigated in this paper.

2   A Reference Path Based on Video Processing

In order to obtain reference data related to the vehicle's movement, an analysis of an image based on the information that was received from a camera located above the movement area of the object was conducted. The image was registered using a GoPro Hero3 camera, which was placed three meters above the area on which the object was moving. The diagram below presents the idea of the image processing for detecting movements and presenting their characteristics. Such an approach allows a reference path to be calculated for further investigation.

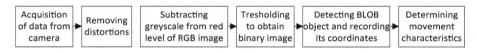

Fig. 2. Diagram of the image analysis algorithm.

In order to remove image distortions, a built-in function of the MATLAB program was used, the functioning of which is based on Zhang's camera calibration algorithm, which is described in article [12]. The juxtaposition in which it is possible to observe the image before and after removing the distortions is presented in Fig. 3. It can be observed that after removing the distortions, the image is flatter and larger.

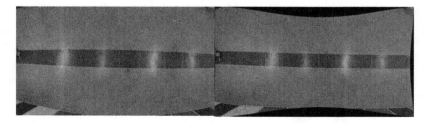

Fig. 3. Before and after reducing the curvature (distortion introduced by camera lenses).

One can see the path in the middle of Fig. 3 (left picture). After applying the function to eliminate the lens distortion, the image on the right was obtained. The black regions on the processed image are an effect of flattening. Hence, a reference system was obtained for further comparison with the signal processing from MEMS sensors. For reference path determination an algorithm shown in Fig. 2 has been applied.

3 Preprocessing the MEMS Sensors Signal

Today, MEMS sensors are embed in mobile phones, microcontrollers, etc. because the production cost is very low. On the other hand, there are many papers related to indoor positioning based on accelerometer, gyroscope and RSSI index [9, 10, 13]. The authors are going to use such approach to enhance vehicle position inside "difficult" environment when GPS signal is weak or it is lost.

3.1 Measuring Platform

The measuring platform was composed of a Raspberry Pi platform (type A) equipped with MEMS sensors (Fig. 4). The following sensors were connected to the Raspberry Pi: an accelerometer (ADXL345), a gyroscope (L3G4200D) and a magnetometer (HMC5883L). The sensors were connected to the Raspberry Pi by a I2C bus. Each of the sensors was 3-axis, which means that the sensors provide information from the X, Y and Z axes. The measuring platform was equipped with a wireless card (TP-LINK TL-WN722 N) in order to use a Wi-Fi signal. The Raspberry Pi registered about 100 readings per second from each MEMS sensor. The wireless network was scanned five times per second on average. The measuring station was in a room located in a brick building (brick and reinforced concrete structure). The station was equipped with three access points and an image capture device.

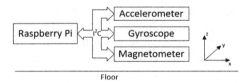

Fig. 4. The measuring platform – orientation of the sensors.

A further experiment was performed using a self-propelled vehicle (Fig. 5). The frame of the vehicle was made of metal. The electronic devices were separated from the metal construction of the vehicle using low conductivity electrical material. The drive of vehicle was made of two stepper motors. The motors were powered by a 12 V battery. The 0.13 m front wheels were made of plastic and rubber (to reduce slippage). One self-steering wheel without a drive was mounted on the back of the platform. The weight of the complete measurement platform was 1980 g. The maximum speed of the vehicle was about 0.7 m/s. The movement of the vehicle was controlled by an Arduino MEGA 2560 platform with an attached motor shield. The anticipated movement of the vehicle was predefined in the source code of the Arduino board.

Fig. 5. The measuring platform – the vehicle.

The data that was measured by the Raspberry Pi was exchanged with the main system using XML files [14]. For fast data processing, the system will be equipped with FPGA solution in the future [15].

3.2   Data Processing for Enhancing the Position

The signals that are acquired from MEMS sensors are noisy and they require processing in order to get valuable information for further investigation. This research was focused on the following filters with a moving window of different widths: a median filter [16], a moving average filter [17], a Hampel filter [16, 18] and a Savitzky-Golay filter [19]. The authors assert that appropriate filtration should enhance the accuracy of the position of an object in motion.

The filters were also interconnected to make the best use of their features. One parameter of the filters was the width of the window k. The value of the ith sample was calculated using the Eq. (1).

$$s'_i = filter\left(s_{i-\left\lfloor\frac{k}{2}\right\rfloor}, \ldots, s_{i-1}, s_i, s_{i+1}, \ldots, s_{i+\left\lfloor\frac{k}{2}\right\rfloor}\right) \tag{1}$$

There is a discussion outside this work as to whether a window width that is too small does not smooth the signal well enough, whereas a window that is too large leads to excessive smoothing of the data and causes information loss. Attention should also be paid to the latency that is generated by the filters, which is very important for a vehicle that is moving at a higher speed. Most k-length symmetric filters have a delay of 0.5 k– 0.5 samples. This effect is closely dependent on the implementation of the filter.

The analysis of the movements of an object can be carried out in two ways. The object can be tracked on a sequence of images that are captured with a video camera (described in Sect. 2) or based on readings from MEMS sensors. The Wi-Fi signal that was mentioned earlier is only used to determine the position of an object. However, it is made on the scale of a building not a room. This is due to the fact that in a small room, the Wi-Fi signal strength is almost identical for each of the reference points. The poor reliability of a Wi-Fi signal is connected with the variable signal gain of wireless network adapters that are used and the access points.

The effect of the use of filters is presented based on the measurements from the X-axis of the accelerometer. Other data from the MEMS sensors were processed in a similar manner. Figure 6 shows the original data from the accelerometer. The data were very noisy and this could cause errors when calculating the velocity of the object and determining the track on which the object was moving. In the first step, the reaction of a filter depending on the window width was determined. Figure 7 shows the data that was obtained after using the moving average filter with the different window widths. The window width was determined based on the frequency of the measurements that were collected. The width of a window depends on factors such as the quality of the sensors, the velocity of the objects and the frequency with which the measurements are collected.

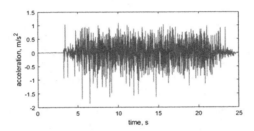

Fig. 6. Measurements from X-axis of the accelerometer.

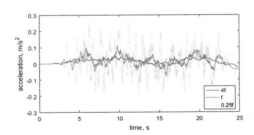

Fig. 7. Moving average filter with different window widths.

Different filters with the same window width are presented in Fig. 8. As can be seen in the waveform of the signals for the moving average filter and the Savitzky-Golay filter, they are similar because these filters belong to the same group of filters. However, significant differences are seen when these filters are compared with the median filter. In the case of the median filter, the changes occur more rapidly than for the average filters; this filter is also less sensitive to local outliers.

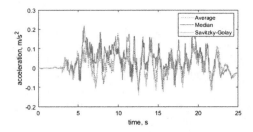

Fig. 8. Results of the following filters: average filter, median filter, Savitzky-Golay filter.

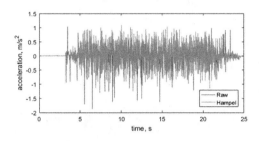

Fig. 9. Results of the Hampel filter.

Figure 9 shows the essence of the action of the Hampel filter. This filter only suppresses values that significantly stand out from the rest of the values within the window. As can be seen, this filter produced smooth single values. The filters mentioned above can be interconnected in order to best use their features. The effects of the fusion part of these filters are presented next along with an analysis of the velocity and track on which the object was moving.

4   Position Determination

Obviously, it is possible to calculate the acceleration of an object based on the data from the MEMS sensor – accelerometer. Velocity is defined as the integral over time of the acceleration. The velocity at point P can be calculated numerically from the equation:

$$v_P = \int_{t_0}^{t_1} a(t)dt + v_0 \tag{2}$$

where a(t) is acceleration in the time domain, t_0 – is the starting point and t_1 – the ending point, v_0 – is the initial velocity.

$$a(t) = \sqrt{a_x^2(t) + a_y^2(t)} \tag{3}$$

where a(t) is the equivalent acceleration, a_x and a_y are the accelerations at the x and y axes, respectively. Due to the complex problem connected with necessity of excluding the acceleration of the earth acting on the object and a very short distance within the vehicle moves, the problem of the vehicle's tilt has been omitted in this paper and it is assumed that the vehicle moves on a flat floor perpendicular to the z-axis, which leads to formula (3). In fact, it is necessary take into account that during the vehicle's movement, the earth's gravity can have an impact not only on Z axis but also, to a certain extent, on axis X and Y (depending on the tilt of the vehicle).

The test scenario asserts that the data is acquired at the maximum speed and at constant intervals. The initial velocity of the object is 0. Distance is defined as the integral over time of the velocity. The distance at point P (within time interval $\Delta t = t_1 - t_0$) with respect to the last well-known position for time t0 can be calculated from (4).

The approximate direction of moving objects was calculated from the gyroscope (turn on axis Z) and the accelerometer (the centrifugal force) data with the appropriate weights α and $1 - \alpha$, respectively. Exemplary graphs of the speed, distance and path are presented in Figs. 10, 11 and 12.

$$S_p = \int_{t_0}^{t_1} v(t)dt \tag{4}$$

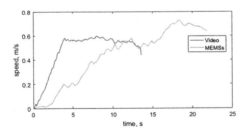

Fig. 10. The speed of the object vs. time for the video and sensors.

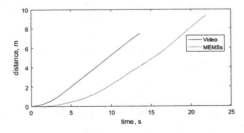

Fig. 11. The distance of the object vs. time for the video and sensors.

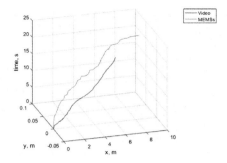

Fig. 12. The track on which the object was moving.

The worst case scenario is presented in Table 1 where a comparison of the average speed and the distance depending with respect to filtering methods are presented. Acquisition time of the movement is 13.5 s (position P).

Table 1. Average speed and distance depending on the different filters (MA – moving average filter, MM – moving median filter, H – Hampel filter, SG – Savitzky-Golay filter)

Type of filter	Window width	Average speed [m/s] at position P_{end}	Distance at position P_{end} [m]	Average speed [m/s]
Reference (CCTV)	–	**0.4729**	**7.468**	–
Moving Avg.	0.5 f	0.1757	2.345	0.2477
Moving median		**0.4332**	**5.934**	0.6757
H+SG		0.2970	4.005	0.4140
H+MA		0.3035	4.089	0.4291
H+MA	0.25 f	0.3210	4.326	**0.4526**
–	–	0.2784	3.812	0.4273

According to Table 1, the total reference distance was 7.468 m and the average speed was 0.4729 m/s. The object starts its move, than accelerate (the movement characteristic is not constant) and reaches constant speed after approx. 9 s. The best results are bold in the Table 1. The most accurate average speed in position P_{end} (0.4332 m/s) was achieved for moving median (error of 8.40%), total distance from P_0 to P_{end} has least error of 1.5340 m (20.54%) also for moving median. The average speed along the path was 0.4526 m/s for the combination of the moving average with the Hampel filters (H+MA). The window width equalled 0.25 f.

Table 2 presents the distance error after 3 s of no reference signal (GPS signal or CCTV camera) starting at point P_i. The error is calculated as a distance between the reference point (from CCTV) and the position obtained from the MEMS sensors. The first column contains the abbreviation for the filter type, the next – the starting point

Table 2. Average position error for several starting points at a constant time interval. The window width was 0.5 f.

Type of filter	Starting points P_i [m]				
	$P_0 = 0$ m ($T_{P(0)} = 0$ s)	$P_1 = 0.8015$ ($T_{P(1)} = 3$ s)	$P_2 = 2.6624$ ($T_{P(2)} = 6$ s)	$P_3 = 4.6278$ ($T_{P(3)} = 9$ s)	$P_4 = 6.5273$ ($T_{P(4)} = 12$ s)
	Calculated distance [m]/position error [%] after 3 s measurements				
MA	0.0650/91.90	2.4694/7.25	4.7700/3.07	6.5735/0.71	**7.4379/0.02**
H+MA	0.0830/89.64	2.5712/3.42	4.8651/5.13	6.7341/3.17	7.4293/0.13
H+SG	0.0832/89.62	2.5323/4.88	4.8814/5.48	6.7295/3.10	7.4207/0.25
Reference [m]	0.8015	2.6624	4.6278	6.5273	7.4390

(where the video camera lost sight of the vehicle – the last well known position). As it can be observed that the maximum error of 0.7365 m (91.90%) occurred when the vehicle accelerates but the error decreases when the vehicle reaches constant speed. The minimum error of 0.016 m (0.02%) is for the moving average.

Table 3 presents the distance error between the reference point and the position obtained from the MEMS sensors with several time intervals. The first column contains the abbreviation of the filter type, the next – the position error after five time intervals (video camera lost sight of the vehicle at P_4). The position error increased with the time interval.

Table 3. Average position error for P_4 with several time intervals. The window width is 0.5*f*.

Type of filter	Calculated distance [m]/position error [%] for starting point P_4 with changing time interval				
	$\Delta T = 0.5$ s	$\Delta T = 1.0$ s	$\Delta T = 1.5$ s	$\Delta T = 2$ s	$\Delta T = 2.5$ s
MA	4.9627/0.30	5.2717/0.07	5.5966/0.23	5.9102/0.15	6.2400/0.36
H+MA	4.9716/0.48	5.3041/0.55	5.6543/1.26	5.9957/1.60	6.3566/2.23
H+SG	4.9720/0.49	5.3040/0.54	5.6547/1.27	5.9941/1.57	6.3542/2.19
Reference distance [m]	4.7878	4.9493	5.1097	5.2680	5.4258

The speed that was calculated by the MEMS sensors was equal to the actual speed (video camera as a reference) after 11.45 s until the end of movement, which can be seen in Fig. 10. The average speed in this range varied by only about 0.0035 m/s, i.e. 0.65% (maximum difference was 0.03 m/s).

The frequency with which the position of the object was updated based on the MEMS sensors was about 100 Hz, which at the maximum speed of the moving object (0.7 m/s) provided the location with a 0.007 m precision. Analogically, the precision of the positioning that was based on the records of the video camera with a record speed 30 fps (30 Hz) was 0.023 m.

5 Conclusion

Positioning objects is possible using the presented CPS, i.e. a vehicle equipped with a microcontroller platform, MEMS sensors and the appropriate signal processing. However, the accuracy of the approach is limited and can only be applied within a short time interval. For a hybrid system, i.e. CCTV and MEMS sensors, the precise position can be obtained using a camera and the support (when the object is under/behind obstacles) of a filtered (by using very simple and fast algorithms that can be implemented in weaker devices) signal from accelerometer. The coverage of a system based on MEMS sensors is 100% due to the fact that the sensors are mounted on the vehicle. MEMS sensors require stable mounting on moving objects. Such positioning techniques can also be used as additional elements for autonomous movement in Advanced Driver Assistance Systems (ADAS). Although the position that was obtained from the video camera was characterized by a greater accuracy, there are also problems with the angles of observation. A hybrid system with a camera wide angle of observation will be investigated in the near future.

Acknowledgements. This work was supported by the European Union from the FP7-PEOPLE-2013-IAPP AutoUniMo project "Automotive Production Engineering Unified Perspective based on Data Mining Methods and Virtual Factory Model" (grant agreement no: 612207) and research work financed from funds for science in years 2016–2017 allocated to an international co-financed project (grant agreement no: 3491/7.PR/15/2016/2) and supported by Polish Ministry of Science and Higher Education with subsidy for maintaining research potential.

References

1. Baheti, R., Gill, H.: Cyber-physical systems. The Impact of Control Technol. **12**, 161–166 (2011)
2. Kao, H.-A., Jin, W., Siegel, D., Lee, J.: A cyber physical interface for automation systems—methodology and examples. Machines **3**(2), 93–106 (2015)
3. Ziębiński, A., Cupek, R., Grzechca, D., Chruszczyk, Ł.: Review of advanced driver assistance systems (ADAS). In: 13th International Conference on Computational Methods in Sciences and Engineering, April 2017
4. Ziebinski, A., Cupek, R., Erdogan, H., Waechter, S.: A survey of ADAS technologies for the future perspective of sensor fusion. In: Nguyen, N.-T., Manolopoulos, Y., Iliadis, L., Trawiński, B. (eds.) ICCCI 2016. LNCS, vol. 9876, pp. 135–146. Springer, Cham (2016). doi:10.1007/978-3-319-45246-3_13
5. Pittet, S., Renaudin, V., Merminod, B., Kasser, M.: UWB and MEMS based indoor navigation. J. Navig. **61**(03), 369–384 (2008)
6. Budniak, K., Tokarz, K., Grzechca, D.: Practical verification of radio communication parameters for object localization module. In: Gruca, A., Brachman, A., Kozielski, S., Czachórski, T. (eds.) Man–Machine Interactions 4. AISC, vol. 391, pp. 487–498. Springer, Cham (2016). doi:10.1007/978-3-319-23437-3_41
7. Grzechca, D., Chruszczyk, Ł.: Location and identification wireless unit for object's monitoring in a protected area. In: 12th International Conference on Data Networks, Communications, Computers (DNCOCO 2013), pp. 186–191 (2013)

8. Alarifi, A., et al.: Ultra wideband indoor positioning technologies: analysis and recent advances. Sensors **16**(5), 707 (2016)

9. Chruszczyk, Ł., Zając, A., Grzechca, D.: Comparison of 2.4 and 5 GHz WLAN network for purpose of indoor and outdoor location. Int. J. Electron. Telecommun. **62**(1), 71–79 (2016)

10. Grzechca, D.E., Pelczar, P., Chruszczyk, L.: Analysis of object location accuracy for iBeacon technology based on the RSSI path loss model and fingerprint map. Int. J. Electron. Telecommun. **62**(4), 371–378 (2016)

11. Mautz, R.: Indoor positioning technologies. Habilitation Thesis submitted to ETH Zurich (2012)

12. Burger, W.: Zhang's camera calibration algorithm: in-depth tutorial and implementation. month (2016). Technical report HGB16-05, 16 May 2016, Department of Digital Media, University of Applied Sciences Upper Austria, School of Informatics, Communications and Media, Softwarepark 11, 4232 Hagenberg, Austria. www.fh-hagenberg.at, https://www. researchgate.net/profile/Wilhelm_Burger/publication/303233579_Zhang's_Camera_ Calibration_Algorithm_In-Depth_Tutorial_and_Implementation/links/ 5739ade408ae9f741b2c816f/Zhangs-Camera-Calibration-Algorithm-In-Depth-Tutorial-and-Implementation.pdf

13. Grzechca, D., Wróbel, T., Bielecki, P.: Indoor localization of objects based on RSSI and MEMS sensors. In: 2014 14th International Symposium on Communications and Information Technologies (ISCIT), pp. 143–146. IEEE (2014). doi:10.1109/ISCIT.2014. 7011888

14. Cupek, R., Ziebinski, A., Fojcik, M.: An ontology model for communicating with an autonomous mobile platform. In: Kozielski, S., Mrozek, D., Kasprowski, P., Małysiak-M-rozek, B., Kostrzewa, D. (eds.) BDAS 2017. CCIS, vol. 716, pp. 480–493. Springer, Cham (2017). doi:10.1007/978-3-319-58274-0_38

15. Cupek, R., Ziebinski, A., Franek, M.: FPGA based OPC UA embedded industrial data server implementation. J. Circuits Syst. Comput. **22**(08), 18 (2013)

16. Pearson, R.K., et al.: The class of generalized Hampel filters. In: 2015 23rd European Signal Processing Conference (EUSIPCO), pp. 2501–2505. IEEE (2015). doi:10.1109/EUSIPCO. 2015.7362835

17. Smith, S.W.: The scientist and engineer's guide to digital signal processing (1997). http:// www.dspguide.com/CH28.PDF

18. Pearson, R.K., Neuvo, Y., Astola, J., Gabbouj, M.: Generalized Hampel filters. EURASIP J. Adv. Signal Process. **2016**(1), 87 (2016)

19. Schafer, R.: What is a Savitzky-Golay Filter? [Lecture Notes]. IEEE Sig. Process. Mag. **28**(4), 111–117 (2011)

Application of OPC UA Protocol
for the Internet of Vehicles

Rafał Cupek[1], Adam Ziębiński[1], Marek Drewniak[2],
and Marcin Fojcik[3(✉)]

[1] Faculty of Automation Control, Electronics and Computer Science,
Institute of Informatics, Silesian University of Technology, Gliwice, Poland
{Rafal.Cupek,Adam.Ziebinski}@polsl.pl
[2] Aiut Sp. z o.o, Gliwice, Poland
Marek.Drewniak@aiut.com
[3] Western Norway University of Applied Sciences, Førde, Norway
Marcin.Fojcik@hvl.no

Abstract. Nowadays, Advanced Driver Assistance Systems (ADAS) support drivers of vehicles in emergency situations that are connected with vehicular traffic. They help to save people's lives and minimise the losses in accidents. ADAS use information that is supported by a variety of sensors, which are responsible for tracking the vehicle's surroundings. Unfortunately, the range of the sensors is limited to several dozen metres and even less in the case of obstacles. This shortens the time for a reaction and, therefore, there may not be enough time to avoid an accident. In order to overcome this drawback, vehicles have to share the information that is available in ADAS. The authors investigated different vehicle-to-vehicle communication possibilities. Based on an analysis of the state of the art, the authors present an original concept that is focused on applying the OPC UA (IEC 62541) communication protocol for services that correspond to the Internet of Vehicles concept.

Keywords: Internet of Vehicles · Vehicle-to-vehicle communication · Advanced Driver Assistance Systems (ADAS) · Smart car · OPC UA (IEC 62541)

1 Introduction

Modern vehicles are equipped with more and more advanced electronic devices that are designed to improve the safety of travelling. Advanced Driver Assistance Systems (ADAS) [1] process the information that is gathered by many different types of sensors such as LIDAR, GPS, cameras, radars and others [2]. The information provided by the sensors is then processed [3] by ADAS processing units in order to support a driver in making the correct decisions in the event of an emergency, or they can even act as autonomous control systems in order to avoid accidents or reduce the effects of collisions.

Despite increasingly sophisticated sensors, all ADAS receive only local information. Although many other drivers have ADAS in their vehicles, they do not share any information among the ADAS. Although most of the ADAS sensors exchange

© Springer International Publishing AG 2017
N.T. Nguyen et al. (Eds.): ICCCI 2017, Part II, LNAI 10449, pp. 272–281, 2017.
DOI: 10.1007/978-3-319-67077-5_26

information using a local sensor network, unfortunately, the information cannot be exchanged between vehicles for the following reasons:

- Individual manufacturers of automotive electronics utilise on-board communication in different ways. The differences are not only between vehicle brands, but even between subsequent versions of the same product.
- The information that is transmitted by the advanced equipment does not contain any meta information that would facilitate its interpretation, or even help in finding specific information within the content of the transmitted message.
- The ADAS were designed to communicate within a local area (a single vehicle). They do not support the exchange of information in mobile systems that have dynamically changing connections.

The proposed concept is focused on applying the OPC UA (IEC 62541) communication protocol for the services that correspond to the Internet of Vehicles concept. The authors investigated different vehicle-to-vehicle communication possibilities as well as their advantages and disadvantages as well. Based on an analysis of the state of the art, the authors present a concept of vehicle-to-vehicle communication that is based on the OPC UA protocol. The main focus is on applying the communication services that are available by OPC UA in order to create interoperable interfaces between different ADAS that are supported by different manufactures. The dynamic server discovery option is used to find neighbouring ADAS. The OPC UA subscriptions are used to supply the required information and to optimise the communication performance.

The remainder of this paper is organised as follows: Sect. 2 is focused on a review of related vehicle-to-vehicle communication solutions. In Sect. 3, the authors present the concept of an external communication system for ADAS that is based on OPC UA servers. The OPC UA servers that are installed in neighbouring vehicles expose the ADAS information along with the meta information that enables it to be understood and used. The conclusions are presented in Sect. 4.

2   Challenges in Vehicle-to-Vehicle Communication

Vehicle-to-vehicle communication integrates wireless networks into vehicles for communication between mobile users on the road [4]. Because mobile users are moving, the network topology changes rapidly [5]. Vehicular ad hoc networks [6] can be made independent from the fixed infrastructure through the use of mobile wireless devices and the creation of mobile nodes for exchanging information. Besides radio interface short-range wireless, an OnBoard Unit (OBU) that allows ad hoc networks to be created is often used. The steady development of ad hoc networks has had a big impact on the evolution of Intelligent Transportation Systems (ITS). By obtaining information from Vehicular Ad Hoc Networks by RoadSide Units (RSU), ITS allow information to be forwarded to a vehicular network or transportation agency where it is further processed in order to organise road traffic [7]. Because of the current infrastructure, vehicles do not always have wireless access to RSU. In this case,

VANET allow vehicle-to-vehicle (V2V), Vehicle-to-Roadside (VRC) and Vehicle-to-Infrastructure (V2I) communications.

One can distinguish several ad hoc network architectures (Fig. 1) such as pure cellular (V2I), pure ad hoc (V2V) and hybrid (V2I & V2V). Pure cellular architecture uses permanent cellular gateways and WLAN access points or base stations at traffic junctions to gather traffic information. In a pure ad hoc architecture, the nodes are engaged with each other via cellular towers and wireless access points. Hybrid architecture uses both pure cellular and pure ad hoc architectures. The topologies of ad hoc networks always vary because they are based on the direction, speed and transmission range of a vehicle.

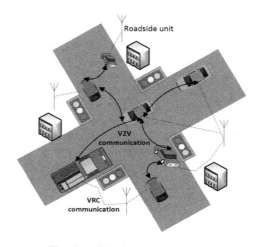

Fig. 1. Vehicular ad hoc network

Numerous wireless access technologies are used to build vehicular ad hoc networks [6]. Cellular gateways use cellular systems (2/2. 5/2. 75/3G). To enable V2V or V2I communication, a Wireless local area network (WLAN) or wireless fidelity (Wi-Fi) on standards IEEE 802.11 bgn are used. Mobile-WiMAX or IEEE802.16e are often used for multimedia, video and voice over internet protocol (VoIP) applications. In area, V2V and VRC for short-to-medium range communication, Dedicated Short Range Communications (DSRC) is used. DSRC are introduced in the ASTM E2213-03 standard. Depending on the region, they have different specifications, e.g. the data transmission rate is 1 or 4 MBits/s in Japan, 250–500 Kbits/s in Europe and 3–27 Mbits/s in the US. For long- and medium-range communication, combined wireless access technologies such as GSM-2G/GPRS-2.5G, UMTS-3G, infrared communication and wireless systems in the 60 GHz band, which is adapted to EEE802.11p, are used. The IEEE 1609 specifications were developed to cover additional layers in the protocol suite. IEEE 802.11p and IEEE 1609.x are referred to as wireless access in vehicular environments (WAVE). In vehicle-to-vehicle communication, many routing protocols [8] can be distinguished in both local V2V and V2I communication. The routing protocols in local V2V are based on topology, position, cluster, geo-cast, broadcast and multicast. The routing protocols in local V2I are based on the static infrastructure and mobile infrastructure.

The solutions for vehicle-to-vehicle communication allow drivers to obtain up-to-the-minute information from the immediate vicinity. The sensors of the infrastructure gather information via V2I communications, then process and analyse the information from moving vehicles. Vehicles can also exchange the information that is gathered by its own sensors with other vehicles via V2V communication. V2I and V2V communication can be utilised for both non-safety and safety applications. Non-safety applications often use value-added services that have information about fuel stations, toll charges, closed roads and detours or location-based services. In safety applications, information that the vehicle is quickly approaching the site of an accident, the end of traffic jams, the approach of thick fog or strong rain, etc. are often important. Often obtaining this information by modern human machine interaction systems [9] in a timely manner permits drivers to avoid a mishap and alerts other vehicles about problems that are arising on the road.

Solutions in local vehicle-to-vehicle communication affect the development of a large number of applications, e.g. collision avoidance at intersections [10], sign extension, vehicle diagnostics and maintenance, public safety [11] and information from other vehicle applications. In the last type of applications, the following can be distinguished: vehicle-based road condition warnings, cooperative forward collision warnings, emergency electronic brake lights, highway merge assistant, visibility enhancer, adaptive cruise control, pre-crash sensing, highway/rail collision warning, vehicle-to-vehicle road feature notification, lane change warning [12], blind spot warning and cooperative vehicle-highway automation system [13]. Many applications can use the information from their own and the sensors of other moving vehicles, which allows the localization to be calculated [14] as well as the minimum safe distance to other vehicles. Safety applications can be disseminated by two types of safety messages [15] – event-driven messages and periodic messages. Event-driven messages contain the type of event, the location and the time. These messages are only sent when hazardous conditions are detected. Periodic messages contain important data to prevent unsafe situations, including speed, location, direction and other data.

Another challenge is security with vehicle-to-vehicle communication systems. On-board communication protocols are attractive to attackers [16]. The on-board communication systems often do not include authentication and data integrity [17]. There are numerous solutions for increasing security [18], e.g. using remote verification [19], a public key infrastructure [20] and certificates or a Message Authentication Code.

3   Vehicle-to-Vehicle Communication Based on the OPC UA Protocol

In order to solve many problems in vehicle-to-vehicle communication mentioned in Chapter two, the authors examined the application of the OPC UA (IEC 62541) for data exchange between vehicles. OPC UA [21] is maintained by the OPC Foundation.

This solution supports communication via Web Services [22] or a modified TCP/IP protocol and permits a service-based, client-server architecture. OPC can be used to exchange information between sensors, remote controls, actuators or other systems. Additionally, it can support information about the system features including a description of the available services.

The OPC UA is an object-oriented architecture that allows the flexible flow of information and easy data exchange between different systems. Primarily, OPC was designed for communication in industrial applications but due to its ability for the real-time management of large amounts of data, it can be applied in other application areas including transportation systems [23].

The OPC UA address space contains information about the organisation of the available information. It can be managed by the servers and shared with connected clients. Additionally, clients have the possibility to discover the object data structures. OPC UA services are organised into ten Service Sets: Discovery, Session, NodeManagement, SecureChannel, Query, View, Method, Attribute, Subscription and MonitoredItem. These service sets contain 37 services that represent all of the possible interactions between the UA client and the server applications. Twenty-one services are used to manage the communication infrastructure.

The discovery service supports connecting a new OPC UA server to the network of servers that is managed by the discovery server and then to help clients find information from the connected OPC UA server. The idea of the discovery services in OPC UA is shown in Fig. 2. In the next step, the connected OPC UA client subscribes to the information that it requires. The mechanism for establishing a secure connection in OPC UA is illustrated in Fig. 3.

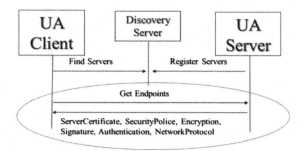

Fig. 2. Discovery services in OPC UA

For the actual information exchange, 16 services are used. A Client can browse the server's address space to search for data. A connection can be realised using references if some of the information is stored on another server. The exchange of information begins after a secure connection between the Client and the Server is established.

The connection can be realised in a specific mode, which is called a subscription, which allows for both flexible and efficient communication in open systems. The OPC UA-based gateway allows secure data communication between ADAS applications and

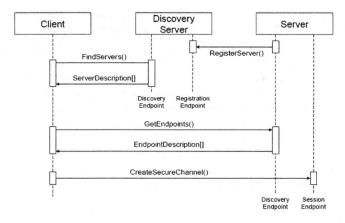

Fig. 3. The mechanism for establishing a secure connection in OPC UA

the ECU to be obtained. It allows access to the ADAS data and the structure of the information from the OPC UA server.

To effectively use the information available in the ADAS that are installed in other vehicles, the vehicles that can be potential sources of valuable information have to be identified first. The first and most obvious criterion is distance. Information from the vehicles that are located nearer is more important for security than the information from distant vehicles. A simple calculation of the distance by comparing the position measured by the GPS in the own and the neighboring vehicles can be done. Then, the selection from among all of the information that is managed by the other ADAS in order to obtain only the information that can increase security has to be made. Finally, establishing a connection with the selected vehicles and beginning to monitor the required information can be performed. The exemplary, considered use case is shown in Fig. 4.

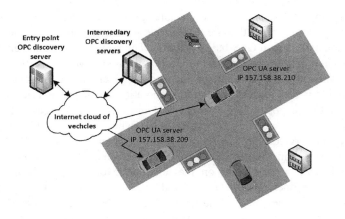

Fig. 4. Use case of the vehicular discovery service in OPC UA

The advantage of an internet connection is the potential ability to obtain information from any vehicle in the world that is equipped with the proposed communication capabilities and has an active connection to the internet; however, information overload is also a disadvantage. Finding only the relevant information sources from the neighboring vehicles is necessary. Of course, it is not possible to check the GPS location of all of the vehicles that are connected to the internet in real time. A mechanism is proposed that will support the pre-selection from the entire cloud of vehicles in which only a very limited set of vehicles are to be tracked and are located in the nearest vicinity of the vehicle. Only information from the selected vehicles will be transmitted and then processed by ADAS. In order to solve this problem, using a dynamic identification mechanism, which is similar to the discovery service that is available in OPC UA, is proposed. This mechanism is shown in Fig. 5.

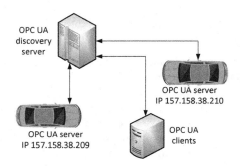

Fig. 5. Vehicular discovery service based on OPC UA

The discovery server has a publicly known IP address and other servers can easily find it and register themselves using the discovery service. The discovery server manages a list of registered servers and exposes it to interested OPC UA clients. In the proposed solution, registration on the discovery server will be performed in two steps. First, the OPC UA server that is located in the vehicle connects to the main discovery server in order to check the list of intermediary discovery servers. Then, it selects the intermediary server or servers that are responsible for entered area based on the criteria of their location and its own GPS-based position. Then, the OPC UA server registers as a potential source of information on the intermediary discovery server(s) and in this way will be available for other vehicles entering the given area as presented in Fig. 6. Since the vehicle is moving, this registering operation has to be repeated cyclically.

In contrast to the positioning systems that are based on wireless networks, the intermediary discovery servers can be located anywhere in the world since the GPS coordinates of the serviced area are the only parameters and do not require the physical location of the server in the serviced area. The discovery servers are virtual because they can be located at any point on the network. Moreover, the structure of the intermediary servers does not have to be fixed and can be adjusted to the actual traffic by adding or removing new intermediary servers operating in a given area in a way that corresponds to the actual traffic requirements.

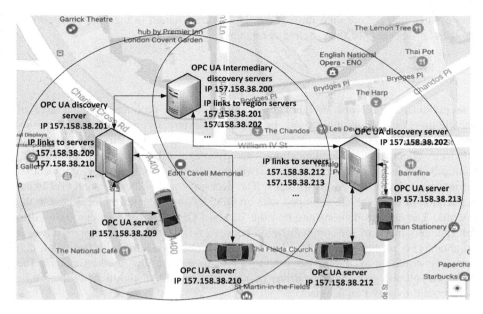

Fig. 6. Registering a source of information on the OPC UA discovery server

OPC UA discovery services are used by OPC UA clients that are searching for information from other systems. The main discovery server provides a list of intermediary discovery servers that expose the GPS coordinates that will be used to select the relevant regions. Based on the GPS position of the nearest intermediary server, each vehicle can find a list of vehicles that are present in its surroundings that can be reached by OPC UA communication. Clients themselves make the decision of which server located in which vehicle should be used. This mechanism is highly scalable if more than one level of intermediary discovery servers needs to be added. In such a case, the selection of the relevant ADAS will be based on tree search algorithms but this aspect is out of the scope of the authors' research. The structure of the intermediary discovery servers can also be changed dynamically. An OPC UA client only needs to know the entry point into the system. Each time that an OPC UA client needs to renew the information of neighbouring vehicles, it connects to the main entry point and finds the relevant OPC UA servers.

The next issue is how an OPC client can properly select and understand the information that is available on the OPC server's side. The assumption was that the OPC UA client knows the IP addresses of the neighbouring vehicles but the structure of the monitoring equipment being used by the other vehicles is unknown. The client still does not know what information is useful for processing. As was pointed in the second Chapter, the on-board vehicle communication does not support any mechanisms that can help to identify the information that is being transmitted. Clients can find the available and useful information using the OPC UA browsing services. The application of OPC UA subscription mechanism allows a customer to select the important data and also ensures that the required information is only transmitted in the case of important changes in the value of the required data source.

4 Conclusions

The publication presents an innovative concept that is proposed for the area of vehicle-to-vehicle communication. In order to externally exchange information between ADAS, the authors proposed a communication infrastructure based on OPC UA client-server communication. The advantage of this approach compared to solutions that are based on wireless networks is that no additional communication infrastructure has to be added to roads. The necessary and sufficient condition is the connection of each vehicle equipped with an OPC UA-based gateway for access to the internet. The authors demonstrate that the standard OPC UA services can be used to fulfil the requirements for communication between ADAS. Of course, a prerequisite is reliable access to the internet. As a benefit, ADAS obtains additional information that normally would be available after driving a few dozen or a few hundred meters. The proposed solution gives both the driver of a vehicle and ADAS significantly more time to react and to avoid an accident.

Acknowledgements. This work was supported by the European Union from the FP7-PEOPLE-2013-IAPP AutoUniMo project "Automotive Production Engineering Unified Perspective based on Data Mining Methods and Virtual Factory Model" (grant agreement no: 612207) and research work financed from funds for science in years 2016–2017 allocated to an international co-financed project (grant agreement no: 3491/7.PR/15/2016/2).

References

1. Ziebinski, A., Cupek, R., Grzechca, D., Chruszczyk, L.: Review of advanced driver assistance systems (ADAS). In: 18th IEEE International Conference on Industrial Technology (2017)
2. Bengler, K., Dietmayer, K., Farber, B., Maurer, M., Stiller, C., Winner, H.: Three decades of driver assistance systems: review and future perspectives. IEEE Intell. Transp. Syst. Mag. **6**, 6–22 (2014)
3. Ziebinski, A., Swierc, S.: Soft core processor generated based on the machine code of the application. J. Circ. Syst. Comput. **25**, 1650029 (2016)
4. Faezipour, M., Nourani, M., Saeed, A., Addepalli, S.: Progress and challenges in intelligent vehicle area networks. Commun. ACM **55**, 90 (2012)
5. Yang, Q., Lim, A., Li, S., Fang, J., Agrawal, P.: ACAR: adaptive connectivity aware routing for vehicular ad hoc networks in city scenarios. Mobile Netw. Appl. **15**, 36–60 (2010)
6. Al-Sultan, S., Al-Doori, M.M., Al-Bayatti, A.H., Zedan, H.: A comprehensive survey on vehicular Ad Hoc network. J. Netw. Comput. Appl. **37**, 380–392 (2014)
7. Andrzejewski, G., Zając, W., Kołopieńczyk, M.: Time dependencies modelling in traffic control algorithms. Presented at the International Conference on Transport Systems Telematics (2013)
8. Sharef, B.T., Alsaqour, R.A., Ismail, M.: Vehicular communication ad hoc routing protocols: a survey. J. Netw. Comput. Appl. **40**, 363–396 (2014)

9. Kasprowski, P., Harezlak, K., Niezabitowski, M.: Eye movement tracking as a new promising modality for human computer interaction, May 2016
10. Tung, L.-C., Mena, J., Gerla, M., Sommer, C.: A cluster based architecture for intersection collision avoidance using heterogeneous networks, June 2013
11. Maslekar, N., Mouzna, J., Boussedjra, M., Labiod, H.: CATS: an adaptive traffic signal system based on car-to-car communication. J. Netw. Comput. Appl. **36**, 1308–1315 (2013)
12. Dang, R., Ding, J., Su, B., Yao, Q., Tian, Y., Li, K.: A lane change warning system based on V2V communication, October 2014
13. Gradinescu, V., Gorgorin, C., Diaconescu, R., Cristea, V., Iftode, L.: Adaptive traffic lights using car-to-car communication, April 2007
14. Obst, M., Mattern, N., Schubert, R., Wanielik, G.: Car-to-Car communication for accurate vehicle localization: the CoVeL approach, March 2012
15. Olariu, S., Weigle, M.C.: Vehicular Networks: From Theory to Practice. CRC Press, Boca Raton (2009)
16. Koscher, K., Czeskis, A., Roesner, F., Patel, S., Kohno, T., Checkoway, S., McCoy, D., Kantor, B., Anderson, D., Shacham, H., Savage, S.: Experimental security analysis of a modern automobile (2010)
17. Koushanfar, F., Sadeghi, A.-R., Seudie, H.: EDA for secure and dependable cybercars: challenges and opportunities (2012)
18. Pamuła, D., Ziębiński, A.: Securing video stream captured in real time. Przegląd Elektrotechniczny. R. **86**(9), 167–169 (2010)
19. Buk, B., Mrozek, D., Małysiak-Mrozek, B.: Remote video verification and video surveillance on android-based mobile devices. In: Gruca, D.A., Czachórski, T., Kozielski, S. (eds.) Man-Machine Interactions 3. AISC, vol. 242, pp. 547–557. Springer, Cham (2014). doi:10.1007/978-3-319-02309-0_60
20. Bißmeyer, N., Stübing, H., Schoch, E., Götz, S., Stotz, J.P., Lonc, B.: A generic public key infrastructure for securing car-to-x communication. Presented at the 18th ITS World Congress, Orlando, USA (2011)
21. Cupek, R., Huczala, L.: Passive PROFIET I/O OPC DA Server. Presented at the IEEE Conference on Emerging Technologies & Factory Automation, 2009. ETFA 2009 (2009)
22. Mrozek, D., Malysiak-Mrozek, B., Siaznik, A.: search GenBank: interactive orchestration and ad-hoc choreography of Web services in the exploration of the biomedical resources of the National Center For Biotechnology Information. BMC Bioinform. **14**, 73 (2013)
23. Maka, A., Cupek, R., Rosner, J.: OPC UA object oriented model for public transportation system. Presented at the 2011 Fifth UKSim European Symposium on Computer Modeling and Simulation (EMS) (2011)

Feasibility Study of the Application of OPC UA Protocol for the Vehicle-to-Vehicle Communication

Rafał Cupek[1], Adam Ziębiński[1], Marek Drewniak[2],
and Marcin Fojcik[3(✉)]

[1] Faculty of Automation Control, Electronics and Computer Science,
Institute of Informatics, Silesian University of Technology, Gliwice, Poland
{Rafal.Cupek,Adam.Ziebinski}@polsl.pl
[2] Aiut Sp. z o.o, Gliwice, Poland
Marek.Drewniak@aiut.com
[3] Western Norway University of Applied Sciences, Førde, Norway
Marcin.Fojcik@hvl.no

Abstract. Advanced Driver Assistance Systems (ADAS) support drivers of vehicles in emergency situations and help to save people's lives and minimise the losses in accidents. The information from car sensors are collected locally by Controller Area Network (CAN) and then processed by ADAS. The authors focus on a CAN/OPC UA (IEC 62541) gateway that can support vehicle-to-vehicle communication in order to enable cooperation between ADAS in neighbouring vehicles. The authors present proof of the concept of the CAN/OPC UA gateway and propose a relevant OPC UA address space that contains the type hierarchy that allows meta information about sensors to be shared between ADAS. The research part focuses on the communication parameters including reliability, communication delays and the stability of the connection.

Keywords: Internet of vehicles · Vehicle-to-vehicle communication · Advanced Driver Assistance Systems (ADAS) · Smart cars · Control Area Network (CAN) · OPC UA (IEC 62541)

1 Introduction

Advanced Driver Assistance Systems (ADAS) [1] process the information that is supplied by advanced electronic devices such as LIDAR, GPS, cameras, radars and others [2]. The Control Area Network (CAN) [3] is commonly used as the communication interface. ADAS use the data from CAN to support a driver in making the correct decisions in the event of an emergency and they can even act as autonomous control systems in order to avoid accidents or reduce the effects of collisions. Unfortunately, CAN, which is the standard in onboard vehicle communication, only supports local communication between car sensors and ADAS processing units.

CAN networks are also widely used in process control and many other industrial areas [4]. The distributed applications that are based on real-time communication

© Springer International Publishing AG 2017
N.T. Nguyen et al. (Eds.): ICCCI 2017, Part II, LNAI 10449, pp. 282–291, 2017.
DOI: 10.1007/978-3-319-67077-5_27

systems are becoming more and more complicated. This has forced research into network monitoring systems that will support the communication and diagnosis of such systems [5]. There are a number of tools that are available that support the testing and development of safety CAN-based distributed real time systems [6]. Safety-critical applications [7] require real-time analysis. They are often realised as fast data processing solutions that have the properties of the application-specific embedded systems [8]. FPGA-based solutions can be found, e.g. in the cyber security protocol for Automated Traffic Monitoring Systems [9].

The sensors inside a vehicle can provide measurements to the Application Unit (AU) using a Control Area Network (CAN) and can then share it by communicating with the On Board Units (OBU). A CAN bus [3] is a serial data communication protocol that is used for data exchange among numerous Electronic Control Units (ECUs) [10], especially in solutions for ADAS.

Unfortunately, the CAN protocol has many obstacles that prevent information exchange between vehicles:

- Manufacturers of automotive electronics do not support CAN communication interoperability. In most cases, the information transmitted by CAN frames remains a trade secret of vehicle manufacturers or at most is available in the specialised technical documentation that is available for service purposes.
- The addressing mechanism that is used by a CAN network does not provide any support to assist in interpreting the transmitted information. CAN messages do not support any meta information that describes the meaning of the transmitted data.
- Despite later extensions to the CAN protocol that were related to its application in building automation or industrial control systems (CANopen), the CAN protocol still does not support the exchange of information in mobile systems.
- The information that is transmitted from sensors via a CAN network must be delivered to ADAS processing units in real time, but the CAN protocol does not support any mechanism that helps to verify the time that the transmitted information was produced.

The above limitations are the result of the standard properties of a CAN protocol. This means that despite the fact that the advanced electronic sensors that are equipped with the CAN communication interface are ubiquitous in today's vehicles, they cannot exchange any information outside of a vehicle. The aim of this feasibility study was to investigate the possibility of sharing information between vehicles using the data transmitted by CAN networks between ADAS and sensors using a CAN/OPC UA gateway [11]. The presented research is focused on two main issues:

- presenting the data that is transmitted by a CAN network together with the meta information that will support interpreting the meaning of transmitted messages including the ability to automatically interpret their content by other ADAS. For this task, the browsing services of OPC UA [12] were used.
- allowing ADAS to exchange information between vehicles by providing communication mechanisms that will provide selective transmission mechanisms (only the information that is needed) and will control the timing of transmissions. The use and

verification was done through an experiment using the subscription-based communication that is available for the OPC UA protocol [13].

The remainder of this paper is organised as follows: Sect. 2 is focused on the address space of OPC UA server, which is dedicated to present the information from ADAS to potential OPC UA clients. Section 3 shows the results of the research and laboratory tests including reliability, communication delays and the stability of the connection. Conclusions are presented in Sect. 4.

2   ADAS Address Space Exposed by the CAN/OPC UA Gateway

Information about the organisation or structure of the transmitted data is not included in CAN messages. In this case, another system that receives the data from a CAN network [14] must be properly configured to be able to read the transmitted data correctly. To solve this drawback, the authors propose a solution based on a CAN/OPC UA gateway that is dedicated to vehicle-to-vehicle communication. The gateway explores the data being exchanged by the sensors and ADAS add the meta information making them understandable for other ADAS and provides it to the OPC UA [15] clients that are connected to ADAS that are located in neighbouring vehicles. The novelty of the proposed solution is the concept of using the OPC UA internet connection [16] for vehicle-to-vehicle communication without the need for any additional roadside infrastructure.

The OPC UA address space type definition mechanisms [17] allow the object-oriented data structures that their clients need with the possibility to discover the object data structures that are managed by the servers to be created. The OPC UA address space includes knowledge about the organization of the available information and presents this knowledge to any application that is connected on the client side. A client can use the object-oriented data types to discover their definitions, which are stored on the server. The references and cross-server references are part of the type definition. Types organize the data into a structure that includes object-oriented mechanisms such as subtyping and type inheritance. Types can be used for objects, variables, data and reference definitions.

Based on the type definitions, the OPC UA server supports access to "live" CAN data and meta information about the data including its structure and meaning and makes them available to other ADAS. The live data are obtained directly from the messages transmitted by the CAN network while the meta information is supported by the address space of the OPC UA server. The proposed address space of the ADAS defines its own types, which are related to functionality of the ADAS and include information that is available by CAN from the sensor network of a car. The OPC UA address space defines all possible information and meta information available to a client. The actual data that is transmitted depends on the subscription made by client.

In the considered experimental test stand, ADAS that works as the supplier of information exposes the measurements that were performed by its LIDAR device. This information is available via the local CAN bus. The position of the vehicle that is obtained by its GPS is also important and available via the OPC UA server. All information is exposed together with the meta information that describes its mining and the technical details that are necessary for its correct interpretation by the OPC UA clients. The view of the OPC UA address space that is visible via the generic OPC UA client (in the experimental case the UAExpert from Unified Automation) is presented in Fig. 1. This information is visible to all of the clients that are connected to the considered OPC UA server. In order to make the structure of information more transparent, the object-oriented hierarchy of OPC UA types was created for vehicle-to-vehicle communication as shown in Fig. 2.

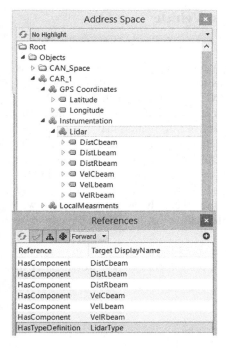

Fig. 1. ADAS devices exposed by OPC UA address space

Fig. 2. An object oriented hierarchy of OPC UA types dedicated for vehicle-to-vehicle communication

A client has the ability to connect to more than one OPC UA server at the same time. Each connection is simply the creation of one OPC UA session that has specified communication parameters. The advantage of a subscription mechanism is not only that a customer can select the important data but also that this information is only transmitted in the case of important changes in the value of the required data source. The location that is monitored by the GPS changes more slowly than the vehicle's speed or its distance to the next vehicle. This is why the communication with the other vehicle was split into two different sessions – one that was dedicated to the vehicle location (data based on the GPS measurements) and the second dedicated to information about the speed and distance (data measured by LIDAR). The information was transmitted from the LIDAR together with its OPC attributes such as its time stamp and quality. The time stamp permits the time that the information was produced to be determined. A quality status of "good" means that the connection is active and that the information is up to date.

In the experimental part described in the next section, the timing of the transmission was verified. The goal was to determine the scope of the applicability of OPC UA communication for the exchange of information between the ADAS in neighbouring vehicles.

3   OPC UA Performance in Vehicle-to-Vehicle Communication

In order to verify the applicability of the OPC UA protocol, a laboratory test stand was prepared as shown in Fig. 3. The main goal was to assess how the introduction of a vehicle-to-vehicle gateway that is based on OPC UA will reduce the possibility of using information from the sensors in one of vehicles by ADAS in another vehicle. The second goal was to assess any delay in the delivery of information that was calculated as the time that was necessary for vehicle-to-vehicle transitions including aspects of its repeatability and the stability of the connections. The information was produced by the LIDAR used by the ADAS in one vehicle, then it was transmitted locally by the CAN network, and next it was exposed to all interested clients on the OPC UA server and finally it was received by the OPC UA client in the other vehicle. An assumption was made that the distant client would provide the information that it received directly to its ADAS system.

It is impossible to compare the time that is required for the complete transmission of information from its source to the destination. Such a comparison would differ in

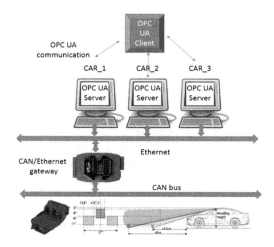

Fig. 3. Laboratory test stand

significant ways and would depend on the sensors being used and on the processing algorithms being used by the ADAS. Instead, any deterioration in the communication parameters that are introduced by the vehicle-to-vehicle transmission is detected. In the experiment, an assumption was made that a communication channel started in the CAN network segment in one of the communicating vehicles and ended at an OPC UA client that was connected to the ADAS in the second vehicle. The LIDAR was selected as the reference source of information. The typical LIDAR generates information about speed and distance that is repeated periodically within a period of about 20 ms. In an actual traffic situation, each vehicle is equipped with its own LIDAR, which is located behind the windscreen. In the experimental scenario, only one LIDAR was used by all three of the OPC UA servers that were situated on the three vehicles as is shown in Fig. 3. This was because it was necessary to receive information from the sensors that could easily be compared among all of the tested vehicles. Since the experiment focused on the communication parameters but not on the distance and speed of the vehicle, this change did not affect the results.

The first step was to convert the information from the CAN network to the Ethernet standard. The standard CAN/Ethernet converter (i-7540D – IP COM) was used. The converter adds its timestamp (counted in micro seconds) to all of the frames that are transmitted between the CAN network and the Ethernet. In this way, each vehicle receives exactly the same information from the LIDAR together with the timestamp that is added by the converter. A separate PC was used to track this information. From all of the frames that were transmitted by the CAN network, only two with ID messages that were equal $-0 \times 1E6$ for speed and $0 \times 1E5$ for distance were selected. In Fig. 4, it can be observed that the frames were transmitted very regularly every 20 ms, which resulted in a value of about 50 frames per second as is shown in the figure. The value of the distance and speed (measured in our laboratory) can also be observed. Such information is typically available for standard ADAS that use LIDAR.

Next, the information that was collected from the LIDAR by the OPC UA servers (using the CAN/Ethernet gateway) was registered. According to the rules of the

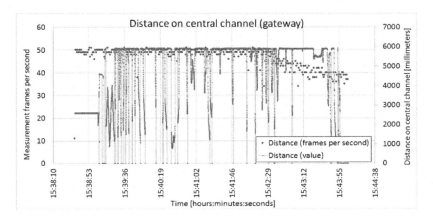

Fig. 4. Distance frames transmitted by LIDAR

OPC UA protocol, the server checks the information source (in the experimental case, the frames transmitted from the CAN/Ethernet gateway) only when the OPC Client is connected and requires this information. For optimisation purposes, the transmissions are grouped into sessions. During the experiment, the OPC UA client created one session for each of the three OPC UA servers in the three vehicles (three PC stations in the laboratory tests). In each session, the client required the speed, distance and timestamp (added by the CAN/Ethernet gateway). The required sampling period of the server was equal to 200 ms. When new information was not available for the server, it did not send anything to the connected clients. The way in which the information was collected by the OPC UA servers in the three vehicles is presented in Fig. 5.

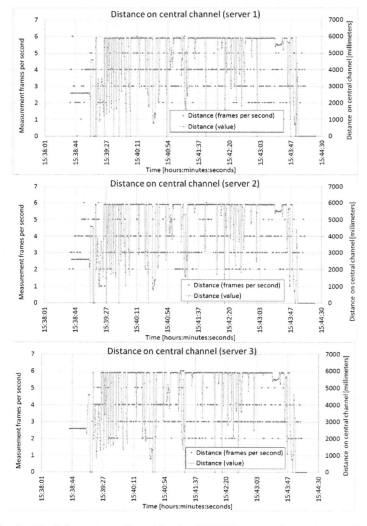

Fig. 5. The information collected by the OPC UA servers in the three vehicles

The figure shows the frequency of access for the information and values of the distance. Because each OPC UA server was connected to the same LIDAR that was available by the CAN/Ethernet gateway, all three trends look similar. The small differences could have been caused by the TCP/IP connection between the server and the converter. The principle of the connection between an OPC UA server and CAN/Ethernet gateway is not based on the broadcast mode as is the case with devices that are directly connected to a CAN network; it requires that three independent client-server communication channels be established under the TCP/IP protocol.

Figure 6 presents the information that was observed by the OPC UA client. During the experiment, an observation was made that the number of records that were registered by all three servers was equal to the number of records that were registered by the client. This led to the conclusion that in the case of the experiment, there was no frame loss in the communication that was established between the OPC UA client and the servers. Therefore, it can be assumed that remote ADAS can see exactly the same information as local ADAS. These results were expected since the OPC UA protocol is equipped with mechanisms that support the integrity of the information, which has been tested by the many real-time communication scenarios that have been prepared for industrial control systems.

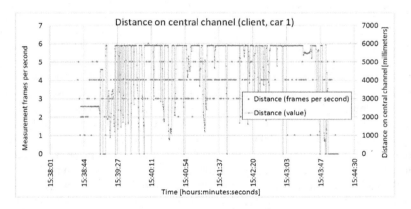

Fig. 6. The information that was observed by the OPC UA client

After the first phase of the experiment, the next research issue had to be verified. In order to verify any delay in the transmission and the stability of the connection, two relationships between the data that was collected in the experiment were checked. Figure 7 shows the time delay of the transmission that was measured based on the time difference between information being sent by the server and received by the client.

Because the PC clock of the client and server were not synchronized, there was a constant difference that was caused by a time shift between the PC clocks during the experiment. The value of the difference was estimated to be 178 ms (between the first and second server) and 787 ms (between the second and third server). The other time differences were caused by various communication delays. The maximum value, average value and standard deviation of the transmission time are shown in Table 1.

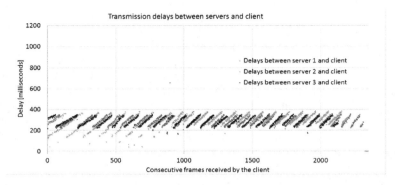

Fig. 7. Time delay of the transmission

Table 1. Transmission time between servers and client

Server name	Maximum [ms]	Average [ms]	Standard deviation [ms]
Server 1 - client	358.59	282.09	34.68
Server 2 - client	385.42	288.08	75.512
Server 3 - client	359.62	284.03	35.62

4 Conclusions

The publication presents proof of the concept of a CAN/OPC UA gateway that is dedicated for vehicle-to-vehicle communication. The authors propose the relevant OPC UA address space that contains the type hierarchy that allows meta information about sensors to be shared between ADAS. In the experimental part, the authors demonstrated that a standard OPC UA solution can be used and that it fulfils the connections that are required for ADAS. The results presented in section three show that any delay in the transmissions is insignificant and that the communication was stable. ADAS obtain additional information that normally would be available after driving a few dozen or a few hundred meters. This gives a driver significantly more time to react and to avoid an accident. Although the presented experimental studies were conducted on a PC-based architecture, the OPC UA is highly scalable and can be transferred to the embedded systems that are used in vehicles.

Acknowledgements. This work was supported by the European Union from the FP7-PEOPLE-2013-IAPP AutoUniMo project "Automotive Production Engineering Unified Perspective based on Data Mining Methods and Virtual Factory Model" (grant agreement no: 612207) and research work financed from funds for science in years 2016–2017 allocated to an international co-financed project (grant agreement no: 3491/7.PR/15/2016/2).

References

1. Bengler, K., Dietmayer, K., Farber, B., Maurer, M., Stiller, C., Winner, H.: Three decades of driver assistance systems: review and future perspectives. IEEE Intell. Transp. Syst. Mag. **6**, 6–22 (2014)
2. Ziebinski, A., Cupek, R., Grzechca, D., Chruszczyk, L.: Review of advanced driver assistance systems (ADAS). In: 13th International Conference of Computation Methods in Science and Engineering (2017)
3. Bosch, R.: CAN specification version 2.0. Rober Bousch GmbH, Postfach. 300240 (1991)
4. Farsi, M., Barbosa, M.B.M.: CANopen Implementation: Applications to Industrial Networks. Research Studies Press, Baldock (2000)
5. Lesser, V.R., Corkill, D.G.: The distributed vehicle monitoring testbed: a tool for investigating distributed problem solving networks. AI Mag. **4**, 15 (1983)
6. Thompson, H.A., Benitez-Perez, H., Lee, D., Ramos-Hernandez, D.N., Fleming, P.J., Legge, C.G.: A CANbus-based safety-critical distributed aeroengine control systems architecture demonstrator. Microprocess. Microsyst. **23**, 345–355 (1999)
7. Amir, M., Pont, M.J.: Improving flexibility and fault-management in CAN-based "Shared-Clock" architectures. Microprocess. Microsyst. **37**, 9–23 (2013)
8. Ziębiński, A., Świerc, S.: The VHDL implementation of reconfigurable MIPS processor. In: Cyran, K.A., Kozielski, S., Peters, J.F., Stańczyk, U., Wakulicz-Deja, A. (eds.) Man–Machine Interactions. AISC, vol. 59, pp. 663–669. Springer, Berlin (2009). doi:10.1007/978-3-642-00563-3_69
9. Chattopadhyay, A., Pudi, V., Baksi, A., Srikanthan, T.: FPGA based cyber security protocol for automated traffic monitoring systems: proposal and implementation, July 2016
10. CANoe – the Multibus Development and Test Tool for ECUs and Networks. Vector Informatic GmbH (2011)
11. Maka, A., Cupek, R., Rosner, J.: OPC UA object oriented model for public transportation system. In: 2011 Fifth UKSim European Symposium on Computer Modeling and Simulation (EMS) (2011)
12. Lange, J., Iwanitz, F., Burke, T.: OPC from Data Access to Unified Architecture. OPC Foundation, Scottsdale (2010)
13. OPC Foundation: Unified Architecture. https://opcfoundation.org/about/opc-technologies/opc-ua/
14. Li, R., Liu, C., Luo, F.: A design for automotive CAN bus monitoring system. In: 2008 IEEE Vehicle Power and Propulsion Conference (2008)
15. Mahnke, W., Leitner, S.-H., Damm, M.: OPC Unified Architecture (2009)
16. Unified Architecture - OPC Foundation. https://opcfoundation.org/about/opc-technologies/opc-ua/
17. Cupek, R., Ziebinski, A., Fojcik, M.: An ontology model for communicating with an autonomous mobile platform. In: Kozielski, S., Mrozek, D., Kasprowski, P., Małysiak-Mrozek, B., Kostrzewa, D. (eds.) BDAS 2017. CCIS, vol. 716, pp. 480–493. Springer, Cham (2017). doi:10.1007/978-3-319-58274-0_38

Data Mining Techniques for Energy Efficiency Analysis of Discrete Production Lines

Rafal Cupek[1], Jakub Duda[1($\boxtimes$)], Dariusz Zonenberg[2],
Łukasz Chłopaś[2], Grzegorz Dziędziel[2], and Marek Drewniak[2]

[1] Institute of Informatics, Silesian University of Technology, Gliwice, Poland
{Rafal.Cupek,Jakub.Duda}@polsl.pl
[2] AIUT Sp. z o.o. (Ltd.), Gliwice, Poland
{dariusz.zonenberg,lukasz.chlopas,grzegorz.dziedziel,
marek.drewniak}@aiut.com.pl

Abstract. Machine-level energy efficiency assessment supports the rapid detection of many technological problems related to a production cycle. The fast growth of data mining techniques has opened new possibilities that permit large amounts of gathered energy consumption data to be processed and analyzed automatically. However, the data that are available from control systems are not usually ready for such an analysis and require complex preparation – cleaning, integration, selection and transformation. This paper proposes a methodology for energy consumption data analysis that is based on a knowledge discovery application. The input information includes observations of the production system behavior and related energy consumption data. The proposed approach is illustrated on the use case of an energy consumption analysis that ws prepared for an automatic production line used in electronic manufacturing.

Keywords: Energy efficiency · Data mining · Manufacturing Execution System (MES) · Industrial communication · OPC UA (IEC 62541)

1 Introduction

As discrete production lines become more and more complicated, predictive maintenance activities have become a vital task for the engineers responsible for production support. Many potential technological and technical problems can be detected based on early symptoms that are noticeable in changes in energy consumption first [1]. Progressive faults can be caused by the aging or deterioration of the operating environment. Often progressive faults are noticed too late. Slow changes can especially be seen in greater energy consumption. So use of the timely monitoring of conditions and modern methods for fault diagnosis are required [2]. These provide the opportunity to detect maintenance or technological problems on the machine level by observing and processing information about energy consumption [3]. The current data can be compared with the information related to the energy consumption profiles in an appropriate production context. This comparison allows maintenance activities to be planned in advance and, as a consequence, reduces the losses related to production breakdowns.

© Springer International Publishing AG 2017
N.T. Nguyen et al. (Eds.): ICCCI 2017, Part II, LNAI 10449, pp. 292–301, 2017.
DOI: 10.1007/978-3-319-67077-5_28

One of the main challenges related to predictive maintenance activities is the limited hardware resources that are available in PLC (Programmable Logic Controller) based control systems. Despite emerging hardware solutions that permit advanced data processing directly on the control system level [4], in most cases, the information related to energy consumption has to be sent to Business Intelligence systems for further processing. In such a case many problems related to the collection of real-time process information in distributed and large-scale data acquisition systems [5] have to be solved. The next issue is the discrimination of different production variants, which can be solved using a data mining approach [6].

This paper presents a new algorithm to automatically classify the energy consumption data that is relevant to machine production cycles. The proposed method is based on observing the behavior of a machine (control signals) and can be used in the case of mass-customized production. A predictive maintenance diagnostic system uses the results of the proposed clustering algorithm to automatically assess energy efficiency on the production station level. The results are further used for the early detection of machine faults that are visible by anomalies and changes in energy consumption. In the experimental part, the authors illustrate the proposed approach through an example of the automatic processing of information relevant to the production system behavior and its energy consumption that was created for a fully automatic electronic production line called "OKO" which is owned by the AIUT company.

The rest of this paper is organized as follows: the second section is focused on the data mining methods that can be applied for energy consumption information analysis with a special focus on multivariate discrete production systems. The third section presents the main components that are used in the proposed approach. Section four gives some details for the considered use case, and briefly describes the system that was prepared for the production line "OKO". The conclusions are presented in section five.

2   Application of Data Mining Methods for Energy Consumption Analysis

Han, Kimber et al. provide a definition of data mining as "as process of discovering interesting patterns and knowledge from large patterns of data" [7], also known as knowledge discovery from data, which they described as an iterative sequence of: data cleaning, data integration, data selection, data transformation, data mining, pattern evaluation, and knowledge presentation This definition directly applies to the research presented in this paper, and to the challenges of energy efficiency analysis in general. The presented research emphasizes data transformation and mining - building tools for efficient knowledge discovery.

The existing literature on data mining and energy efficiency often concentrates on electrical energy consumption in the context of buildings as presented by Zhao et al. in [8]. A detailed overview of the measures, methods and techniques for energy efficiency optimization in discrete manufacturing was provided by Duflou et al. in [8], while Cannata et al. presented and discussed an energy efficiency driven process in [9].

The initial goals for the presented research were to gather and analyze energy related measures from a real production line, by looking for anomalies and interesting patterns in energy consumption and efficiency. For this purpose, the "OKO" production line – a robotic line used in testing, total quality management (TQM) and final assembly of natural gas use GPRS telemetry devices was selected.

However, the authors wanted to abstract the analysis from this particular line - the findings, conclusions and developed tools should be universally applicable to discrete production lines, especially in the automotive industry. Another aim was to look for easily identifiable markers for abnormal energy consumption that can be determined on-the-fly in a production environment, using only PLC processing.

For the purpose of the research, the following algorithm (presented in Fig. 1) was proposed:

1. Gather production process data and energy efficiency data snapshots
2. Store the data snapshots in the source database
3. Cleanup the data and transform it from the source format (snapshots) to energy efficiency data
4. Apply data mining analysis
5. Provide the results to the clients

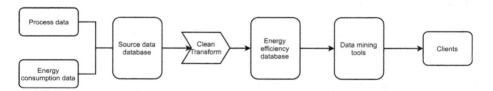

Fig. 1. Proposed algorithm for energy efficiency data analysis

To accomplish this, the communication setup presented in Fig. 2 was developed. The production data that was relevant to energy consumption and the actual measures (electrical energy ad compressed air) were read from PLC using an industrial network by an OPC UA compliant server. These values were made available in an Ethernet network using the OPC UA data access protocol, and read by client software with historian capabilities, which stored all of the relevant data in an SQL database. For efficiency and ease of use the Unified Automation OPC UA SDK and MS SQL database were used for the historian software.

Finally the gathered historical data can be batch processed in data mining environment – for this purpose RapidMiner was selected.

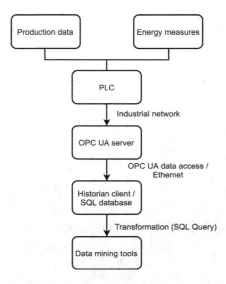

Fig. 2. Communication setup

Although this setup may look excessive at first (the OPC UA server could theoretically be integrated with the SQL historian for this particular line), it allows the data gathering process to be easily scaled by adding more PLCs (it is not unusual to have a hundred PLCs per production line) and distributed data servers while maintaining the central data processing point. As there are multiple telemetry devices processed in parallel in the OKO line at each moment (up to five in the testing stations, one in the laser station and up to 20 in the plotter), it is virtually impossible to determine energy consumption per product. Therefore, the decision was made to analyze the production cycles - especially the robot movements - instead of the product cycles. This approach supports comparison of the energy footprint for repeated actions over long periods of time, which should allow any anomalies or efficiency deterioration to be easily detected.

3   Energy Consumption Analysis for the Fully Automatic Electronic Production Line "OKO"

The data provided by the PLC control routines is just a snapshot of the state of the line devices and meters at a given point in time (as presented in Fig. 3). For electrical energy and compressed air it is the instantaneous consumption and the state of the global meters. Because these are not directly applicable for energy efficiency analysis, the data needs to be transformed into information that describes the production cycles (presented in Table 1).

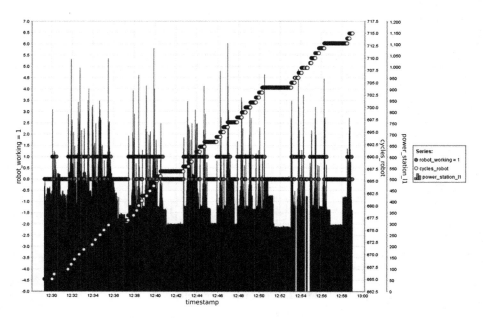

Fig. 3. Source data – instantaneous energy consumption, robot busy status and robot cycle counter shown as a function of time.

Table 1. Comparison of the available source data and the data required for a production cycle energy analysis.

Source data	Transformed data
Energy meter	Energy use per cycle
Air flow meter	Air use per cycle
Robot cycle counter	Cycle time
Device state bits	Robot movement type

4   Data Analysis by RapidMiner Tool

There are several ways to apply the data transformations that are required for analysis. One approach could be based on OPC UA events handling directly in historian software. However besides problems with data verification (source data would not be stored) this method also requires the production events to be coded into the client software as subscription event conditions, thus making it prone to errors and difficult to change, especially when a process is evolving.

Another possibility is to use the data mining environment logic conditions – RapidMiner provides many functional blocks for manipulating data that make it possible to perform all of the transformations this way. The main drawback for this approach is that the transformation method is tied to this particular environment.

Eventually SQL query was selected as the method for data transformation. This is a proven industry standard that makes transformation relatively easy to prepare and modify if necessary. An SQL transformation query can also be used (with some changes reflecting differences in SQL dialects) in different databases and can be executed periodically as a 'stored procedure'.

The following conditions were used in the data transformation query:

- Select only the data for the line "automatic mode" (manual mode where the operators can interfere with a process is available for testing purposes)
- Determine the beginning and end time stamp of the robot work cycle that were to be used to compute the production cycle length
- Measure any changes for all of the energy counters (for both the electrical energy and air consumption that needed to be computed based on the cycle start and end timestamps)
- Transform the robot status bits into single enumeration

To ensure that the device state signals and energy measures were interpreted correctly, the authors also needed to develop verification methods. Besides the manual observation of the line, the behavior signals were also checked using SQL queries and RapidMiner analysis. One of the main challenges was the discovery that various signals were either probed by the PLC or reported by devices with different frequencies. For example, the power consumption peaks that are characteristic for the start of a robot's movement could be separated from the PLC signal "robot activity" by almost one hundred milliseconds.

Figure 4 shows that there was more than a second's difference between the change of the robot cycle counter and the change of the robot activity bit. Further examination showed that, in some cases, the robot activity bit did not change from 1 to 0 between

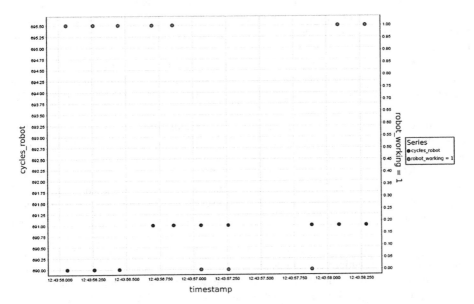

Fig. 4. Plot comparing the robot cycle counter and robot activity status.

cycles (or changed for a period shorter than the PLC scan resolution) and the only way to determine the cycle change was to look at both the cycle counter and activity bit.

Although the SQL transformation query could be applied each time an analysis was performed, in order to speed up the process, the results were stored into a comma-separated value file (CSV). While this format was used for the convenience of the analysis, for larger installations it should be replaced by a regular SQL database (energy efficiency database). Using two databases (source and transformed data) with the SQL query for the data transformation executed periodically as a database "stored procedure" would also allow old data to be removed from the source database, thus preventing it from excessive growth. Storing the transformed data greatly accelerated the execution of subsequent processes with different parameters as the transformation query took several minutes to complete and reduced the working data set for 4,000 production cycles from the gigabytes of the source snapshot data to less than 500 kB.

The initial blocks of the Rapid Miner process apply some additional filtering conditions. In general, "constant" conditions (such as the detection of station automatic mode) were used in the SQL and "dynamic" conditions (such as narrowing the time range or eliminating potentially erroneous cycles) were used in Rapid Miner processing – this permitted fast iterations with the changing processes.

The energy consumption data, which was processed without additional filtering conditions ("dynamic" conditions), exhibited several types of erroneous cycles:

- Excessive use of air
- Extremely long robot work times
- Excessive electrical energy use correlated with long robot work cycles
- Cycles with no energy consumption or no compressed air consumption

After successfully transforming, importing and filtering data, the authors looked for correlations between the production cycles and energy use. The data mining environment permits various aspects of energy efficiency to be quickly analyzed in relation to the production data.

As an example, Fig. 5 shows a plot of compressed air (blue) and electrical energy (red) use as a function of robot cycle time with a linear correlation between electrical energy use and the robot cycle length and no obvious correlation between cycle length and compressed air use.

In the case of the "OKO" line, the robot movement types could be interpreted as production variants, so it would be particularly interesting to find out whether a production variant can be determined only by its energy footprint.

An "OKO" line robot performs four main actions – it moves a new device from the input palette to one of the testing stations, it moves a new device from an additional supply drawer to one of the testing stations, it moves a device from a testing station to a laser stand (test succeeded) and it moves a device from the testing station to a drop out drawer.

Depending on the testing station being used, there are six different variants of a robot's movements, which result in 24 program variants altogether. Theoretically, all of these variants could influence media consumption for both electrical energy and compressed air. Other interesting factors are the cycle length, type of robotic arm movement and laser use.

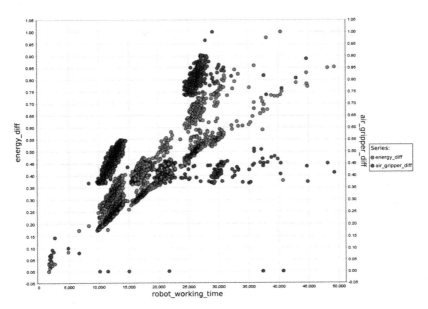

Fig. 5. Air and electrical energy consumption in function of robot working time (Color figure online)

However, the research showed that there is very little difference in energy consumption for the same robot movement type using the different testing stations. Taking this into account and considering the fact that the analysis for 24 variants significantly lowers sample size per variant, the authors decided to concentrate on four robot movements. Although the K-means algorithm was used and tested with various parameters (different distances, use of "good start values" and changes in the number of iterations), the only significant change was that the quality of clustering improved with the number of iterations that were applied. The next step was to compare clustering by energy consumption (Fig. 6) with clustering by the actual robot movement type or the production variant (Fig. 7) to determine whether this method could be successfully used to identify production variants.

Clustering is represented by the color parameter. Both compressed air and electrical energy were taken into consideration.

In both cases one cluster was identified correctly, as showing a significantly higher energy consumption (air and electrical), and this cycle was also longer. This reflects the robot movement type "test succeeded" which involves the laser engraving of elements. For the other types of robot actions, the energy footprint was not directly related to robot movement type.

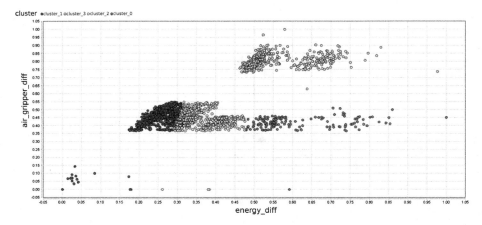

Fig. 6. K-means clustering by energy consumption (Color figure online)

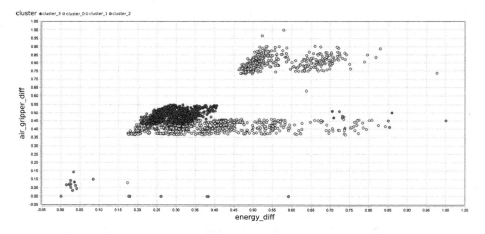

Fig. 7. K-means clustering by robot movement type (Color figure online)

5 Conclusions

In this paper, the authors have described and proposed algorithms that are dedicated to the automatic classification of the energy consumption data that is relevant to machine production cycles. The presented method is based on observing the behavior of a machine (control signals) and can be used in the case of mass-customized production. A predictive maintenance diagnostic system uses the results of the proposed clustering algorithm to automatically assess energy efficiency on the production station level energy consumption model that can be used to both present and process energy efficiency data using data mining tools. The authors applied the proposed algorithm to aggregate a large input data set that was obtained from a control system into information that was relevant to data mining algorithms. Then, data mining analyses were applied. The results were evaluated on the production line called "OKO". It was proven that the proposed approach permits abnormalities related to energy consumption to be detected and classified in the context of the production cycle.

Acknowledgment. This work was supported by the European Union from the FP7-PEOPLE-2013-IAPP AutoUniMo project "Automotive Production Engineering Unified Perspective based on Data Mining Methods and Virtual Factory Model" (grant agreement no. 612207) and research work financed from the funds for science in the years 2016–2017, which are allocated to an international co-financed project (grant agreement no. 3491/7.PR/15/2016/2).

References

1. Peng, Y., Dong, M., Zuo, M.J.: Current status of machine prognostics in condition-based maintenance: a review. Int. J. Adv. Manuf. Technol. **50**(1), 297–313 (2010)
2. Bunse, K., Vodicka, M., Schönsleben, P., Brülhart, M., Ernst, F.O.: Integrating energy efficiency performance in production management–gap analysis between industrial needs and scientific literature. J. Clean. Prod. **19**(6), 667–679 (2011)
3. Cupek, R., Drewniak, M., Zonenberg, D.: Online energy efficiency assessment in serial production-statistical and data mining approaches. In: 2014 IEEE 23rd International Symposium on Industrial Electronics (ISIE), pp. 189–194 (2014)
4. Ziębiński, A., Świerc, S.: The VHDL implementation of reconfigurable MIPS processor. In: Cyran, K.A., Kozielski, S., Peters, J.F., Stańczyk, U., Wakulicz-Deja, A. (eds.) Man-Machine Interactions. Advances in Intelligent and Soft Computing, vol. 59, pp. 663–669. Springer, Heidelberg (2009). doi:10.1007/978-3-642-00563-3_69
5. Flak, J., Gaj, P., Tokarz, K., Wideł, S., Ziębiński, A.: Remote monitoring of geological activity of inclined regions – the concept. In: Kwiecień, A., Gaj, P., Stera, P. (eds.) CN 2009. CCIS, vol. 39, pp. 292–301. Springer, Heidelberg (2009). doi:10.1007/978-3-642-02671-3_34
6. Figueiredo, V., Rodrigues, F., Vale, Z., Gouveia, J.B.: An electric energy consumer characterization framework based on data mining techniques. IEEE Trans. Power Syst. **20**(2), 596–602 (2005)
7. Han, J., Kamber, M., Pei, J.: Data Mining: Concepts and Techniques, 3rd edn. Morgan Kaufmann, Burlington (2011)
8. Duflou, J.R., Sutherland, J.W., Dornfeld, D., Herrmann, C., Jeswiet, J., Kara, S., et al.: Towards energy and resource efficient manufacturing: a processes and systems approach. CIRP Ann. Manuf. Technol. **61**(2), 587–609 (2012)
9. Cannata, A., Karnouskos, S., Taisch, M.: Energy efficiency driven process analysis and optimization in discrete manufacturing. In: 35th Annual Conference of IEEE Industrial Electronics, IECON 2009, pp. 4449–4454. IEEE (2009)

Internet of Things - Its Relations and Consequences

Different Approaches to Indoor Localization Based on Bluetooth Low Energy Beacons and Wi-Fi

Radek Bruha and Pavel Kriz[(✉)]

Department of Informatics and Quantitative Methods,
Faculty of Informatics and Management, University of Hradec Kralove,
Hradec Kralove, Czech Republic
Pavel.Kriz@uhk.cz

Abstract. Thanks to Global Navigation Satellite Systems, the position of a smartphone equipped with the particular receiver can be determined with an accuracy of a few meters outdoors, having a clear view of the sky. These systems, however, are not usable indoors, because there is no signal from satellites. Therefore, it is necessary to use other localization techniques indoors. This paper focuses on the use of Bluetooth Low Energy and Wi-Fi radio technologies. We have created a special mobile application for the Android operating system in order to evaluate these techniques. This application allows localization testing in a real environment within a building of a university campus. We compare multiple approaches based on K-Nearest Neighbors and Particle Filter algorithms that have been further modified. The combination of Low Energy Bluetooth and Wi-Fi appears to be a promising solution reaching the satisfying accuracy and minimal deployment costs.

Keywords: Indoor localization · Indoor positioning · Bluetooth Low Energy · Internet of Things · iBeacon · K-Nearest Neighbors · Particle Filter

1 Introduction

In today's world, it is hard to imagine our lives without a wide range of various modern technologies, such as smartphones. We often use them instead of conventional maps for navigation. A smartphone is also able to localize itself thanks to the Global Navigation Satellite System (GNSS), enabling mobile applications to navigate the user to the destination.

The localization accuracy is typically several meters for the GNSS. The American GPS and the Russian GLONASS are the two globally available systems. It is also possible to use the Assisted GPS (A-GPS) to speed up the localization. The A-GPS downloads the most recent information regarding the satellites' orbital trajectories using a cellular network instead of downloading it via a slow down-link from the satellites.

© Springer International Publishing AG 2017
N.T. Nguyen et al. (Eds.): ICCCI 2017, Part II, LNAI 10449, pp. 305–314, 2017.
DOI: 10.1007/978-3-319-67077-5_29

But the GNSS has its drawback – it is inapplicable indoors. Without the view of the sky, there is no signal from the satellites. Therefore, it is necessary to use other localization methods indoors. These methods can, for example, utilize existing sources of radio signals such as IEEE 802.11 (Wi-Fi) Access Points (AP). But the density of Wi-Fi transmitters in the building may not be sufficient for the high quality localization. Fortunately, they can be supplemented by additional transmitters such as Bluetooth Low Energy (BLE) transmitters, called *Beacons*. A BLE beacon transmits, similarly to a Wi-Fi AP, a unique ID that can be used for unambiguous identification of the transmitter. Modern BLE beacons can communicate with each other via a BLE mesh network and support remote management from the cloud [1].

This paper focuses on the indoor localization using Bluetooth Low Energy and Wi-Fi wireless technologies. Several approaches based on K-Nearest Neighbors and Particle Filter algorithms will be compared. The solutions will be tested in a real environment within the building of the university campus.

The rest of this paper is organized as follows. Section 2 briefly describes the existing positioning techniques and related work. We formulate the problem in Sect. 3. Section 4 describes the radio localization techniques in detail and the proposed solution. Several details regarding the implementation of the mobile application are shown in Sect. 5. We present the results of the testing in Sect. 6. Section 7 concludes the paper.

2   Related Work

There are many approaches to indoor localization. Their feasibility is largely influenced by the sensors and receivers inside smartphones. Furthermore, the quality and availability of the data from the sensors and receivers provided to the applications via an Application Programming Interface (API) determine the achievable precision of the localization.

The three following sensors belong to the basic equipment of smartphones. Firstly, an *accelerometer*, which uses gravity as a reference axis and provides information about the phone's acceleration in each axis. It is used for tilt and movement detection. A linear accelerometer is a special variant that excludes the force of gravity. Secondly, a *gyroscope* provides information on the rotation of the phone in the space in all three axes, namely, the speed and direction of rotation. Thirdly, a *magnetometer* (an electronic compass) measures the Earth's magnetic field and its direction. The combination of these three sensors is referred to as an Inertial Measurement Unit (IMU). It can be used to determine the absolute position of a smartphone in space using so called Sensor Fusion [2]. In the context of the indoor localization, the IMU is used in approaches known as Pedestrian Dead Reckoning (PDR). The PDR uses the IMU sensors for measuring the number of steps (the distance estimation), and detecting changes in the direction of walking [3].

For the radio-based indoor localization, the built-in receivers of IEEE 802.11 (Wi-Fi), cellular networks and Bluetooth (optimally in the Low Energy variant) are the only feasible sources of data. A smartphone is able to detect nearby transmitters thanks to these built-in receivers. The distance from a particular transmitter can be determined by several methods, such as by measuring the signal strength using the *Received Signal Strength Indicator* (RSSI). This method provides a good estimate in open spaces without obstacles. However, indoors, while signal passes through and reflects off various obstacles it is not accurate. The results of the RSSI-based indoor distance estimation should be treated with caution. *Time of Arrival* (ToA, Time of Flight) is a more accurate method of the distance estimation because it is not affected by the attenuation of the signal [4]. But ToA is not supported by smartphones, except for GNSS receivers.

There are three popular approaches to the radio-base localization: *K-Nearest Neighbors* [5], *Multilateration* [6], and a *Particle Filter* [7]. Each method has its advantages and disadvantages. The methods can also give different results in different environments. The K-nearest Neighbors (KNN) approach is based on fingerprints describing (based on measurements) the environment at a given point in space. The KNN has the advantage that it can be implemented even without knowledge of the exact positions of the transmitters. On the other hand, Multilateration and the Particle Filter usually work with known positions of transmitters.

These methods are elaborated by many authors who are trying to improve them. For example, Kumar et al. apply Gaussian regression in fingerprint-based localization algorithms [8]. Song et al. focus on advanced methods of selecting a subset of all available transmitters from measurement [9]. The RSSI may also be replaced by Channel State Information (CSI) according to Yang et al. [10].

In this paper we focus on the comparison of several approaches based on the KNN and the Particle Filter and their modifications (see Sect. 4). The results of evaluation in a real-world environment are discussed.

3   Problem Formulation

The aim of this work is to implement multiple (variously modified) indoor localization approaches and evaluate their results in real environment. The 3rd floor of the building of the Faculty of Informatics and Management, University of Hradec Kralove, was chosen as the test environment.

There are 4 existing dual-band (2.4 GHz and 5 GHz) Wi-Fi access points on this floor and 17 additional Bluetooth Low Energy beacons (made by Estimote) have been deployed there. The beacons have been firmly attached to the ceiling. Their *Advertising Interval* has been set to 100 ms and *Transmitter Power* to −4 dBm.

Series of measurements will be made in this environment. Several different indoor localization algorithms will then be evaluated at 9 testing positions and their accuracy will be compared. We will consider various modifications of the KNN and Particle Filter algorithms based on Wi-Fi signals, Bluetooth Low Energy signals and the combination of both technologies. Fig. 1 shows the positions of Wi-Fi access points ⓦ, Bluetooth Low Energy beacons ⑴ to ⑰ and testing points A• to I•.

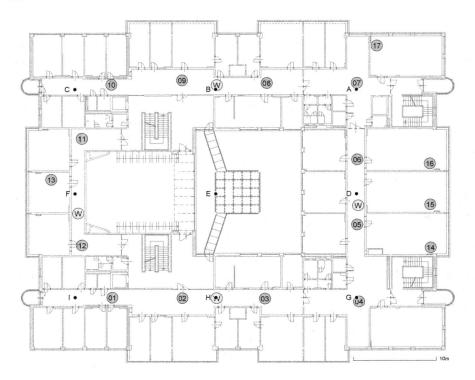

Fig. 1. Floor plan with Wi-Fi access points, BLE beacons and testing points

4   Indoor Localization Techniques and Their Modifications

The following subsections present two basic indoor localization techniques that
we evaluate in this work. Various modifications of these techniques (possibly
improving the accuracy) will also be described.

4.1   K-Nearest Neighbors

The first technique of radio-based indoor position estimation uses the K-Nearest
Neighbors algorithm [5]. This algorithm searches several most similar neighbors –
radio fingerprints. In order to be applicable in a new building it is necessary
to walk through the whole building first and to measure the strengths of radio
signals at reference points with known positions. At each point denoted by B_i an
R_i vector of the RSSIs for every transmitter (called the *fingerprint*) is created.
The record consisting of B_i and R_i is stored in the database. This process is
called a *learning-phase* or an *offline-phase*.

During the *localization-phase* (*online-phase*), the signals from transmitters
are received at an unknown position. These signals are compared to the previously measured data (fingerprints) stored in the database. In case the measured

signals match an existing R_i fingerprint we will say that the estimated position is the B_i position of the existing R_i fingerprint measured during the learning-phase. Of course, the exact match is very unlikely. Therefore, we sort existing fingerprints by their degree of similarity to the currently measured data. Then we pick the k most similar fingerprints. The degree of similarity among fingerprints is usually calculated using the Euclidean or Manhattan (city-block) distance in Signal Space [11]. We obtain an estimated position by calculating the center (the mean of coordinates) of the similar fingerprints. The choice of the k (number of similar fingerprints to be picked) is an interesting topic. One might think that the higher k would lead to better results, i.e. a higher estimation accuracy. However, it turns out this hypothesis is false and the best results are typically provided by k to be 3 [12]; the same value is used in our experiments. Another interesting topic is the possibility of reducing the number of collected fingerprints while keeping the accuracy to be high [13].

The KNN algorithm has first been modified to be able to select the type of signals from the fingerprint; so it is possible to choose among (1) Wi-Fi signals, (2) Bluetooth Low Energy signals or (3) signals combined from both technologies into a single set. The combined fingerprint is defined as

$$R_i = R_i^W ⌢ R_i^B \tag{1}$$

where R_i^W is a vector of Wi-Fi RSSIs and R_i^B is a vector of BLE RSSIs. These two vectors are concatenated to a single R_i vector. The k nearest neighbors are then found in a set of fingerprints denoted as R. In our work we call such a modified algorithm the *KNN1*. The KNN1 compares two fingerprints using the Euclidean distance.

Furthermore, we have made a second modification of the KNN algorithm called the *KNN2*. This algorithm can also combine both technologies together but searches for the nearest neighbors independently in two sets by a matching technology. We consider these R_i^W and R_i^B fingerprints to be independent in every place i. The nearest neighbors are found twice; (1) in an R^W set of Wi-Fi fingerprints and (2) in an R^B set of BLE fingerprints. Then the $2 \cdot k$ neighbors are sorted by the distance and the k nearest neighbors are chosen. We can also prioritize individual neighbors based on their technology. For example, we can prioritize Bluetooth Low Energy by setting a higher weight to these neighbors during the calculation of the neighbors' center. The KNN2, in contrast to the KNN1, compares two fingerprints using the Manhattan distance.

4.2 Particle Filter

The second indoor localization technique is based on a Particle Filter (PF) [7]. It gradually refines the estimated position based on continuous measurement. This approach takes into account random perturbations in the RSSI.

In the first iteration of the algorithm, particles with zero weight are generated evenly throughout the space. Some of the particles are identified as possible locations based on the measurement (observation). We increase the weight of such particles. Subsequently, the particles having zero weight are removed. We generate new random particles near the estimated position, i.e. close to the remaining particles. During the next iterations, clusters of highly-probable estimated positions will be formed. The clusters should be reduced gradually into a single position which will be considered the result of the position estimation algorithm.

The Particle Filter typically uses an estimated distance from transmitters based on the RSSI. As it turns out, the estimated distance is very inaccurate indoors. Thus, the algorithm has been modified to take advantage of the data measured during the KNN's learning-phase – the fingerprint database. For each particular signal received from a transmitter a set of fingerprints having similar signal-strength is found. Considering the positions of fingerprints in the set, the area where the observer is found (regarding the particular transmitter) is determined. This process is repeated for each transmitter. It gradually increases the weight of particles in these areas. The positions of particles having high weight corresponds to the estimated position(s) of the observer. Another measurement is then performed and positions are gradually adjusted and refined. In our work we call such a modified algorithm *PF*.

5   Mobile Application

We have created a special mobile application for the Android operating system in order to evaluate indoor radio-based localization techniques in real environment. A person equipped with a smartphone and the application walks through the building and performs measurements in a predefined grid. The measurements (fingerprints) are then stored in the database.

The application is designed for the Android mobile platform version 4.4 (SDK ver. 19) and higher. Due to the lack of Bluetooth Low Energy support in earlier versions than 4.3 [14], it is not feasible to support much older versions of the Android platform. The application also requires a Graphic Processing Unit (GPU) supporting OpenGL ES version 2.0 or higher, which is used for displaying maps of the building with the aid of the *Rajawali* library[1]. We use the *Estimote SDK*[2] for scanning the nearby Bluetooth Low Energy devices (Beacons).

The application stores all data internally within the integrated SQLite database. The simplified database schema is shown in Fig. 2. We have also implemented the synchronization of the collected data among users (their smartphones) using the Couchbase database. It stores individual records in the JSON format and offers advanced synchronization techniques. Synchronization is performed as follows. Internal records from the SQLite database are converted into the local Couchbase Lite database[3]. Couchbase Lite is then synchronized to the server-side Couchbase database. In another device, records in the local

---

[1] https://github.com/Rajawali/Rajawali.
[2] https://github.com/Estimote/Android-SDK.
[3] https://github.com/couchbase/couchbase-lite-android.

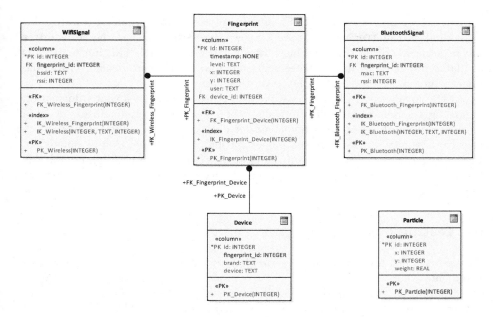

Fig. 2. Simplified SQLite database schema

(already synchronized) CouchBase Lite database are converted back into the local SQLite database. The system architecture is shown in Fig. 3.

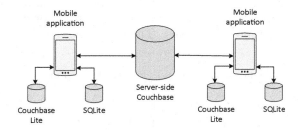

Fig. 3. System architecture overview

During the measurement the application shows how much individual measurements differ from each other. Thus, the person performing the measurement can immediately see whether the measurements are correct or whether something has changed significantly. These changes may indicate errors in measurements or changes in the environment (e.g. removed or new transmitters).

Similarity/difference of measured fingerprints is based on the average values of the RSSI. Differences in RSSIs are shown from two points of view. First, differences among transmitters of a particular technology are considered. For example, we can detect differences in Wi-Fi signals only. Second, signals of a particular individual transmitter are compared. So we can detected that the transmitter has been moved or its transmitting power has been changed. This information is available not only during measurement, but also later when viewing individual fingerprints and it may be displayed on the map.

In the localization phase the application also displays relevant information regarding the selected localization algorithm. For the Particle Filter, an area with our probable position is displayed for each of the transmitter in range. A user can visually check where the highest probability (intersection of signals) is. In the case of the KNN algorithm the closest neighbors are highlighted and the user can manually review the degree of similarity between an existing fingerprint and signals currently measured.

6   Testing and Results

The experiment took place in the corridors of the third floor. Individual measurements were carried out in a grid where each spot was one meter far from another one. It was important to measure the fingerprints precisely and consistently; three measurements were performed in one spot. Measurements were carried out with the aid of the Galaxy S3 Neo smartphone having the Android 4.4.2 (SDK 19) operating system.

The localization accuracy was evaluated in nine locations labeled A to I (see Fig. 1), where the position was estimated three times for each algorithm. The results are shown in Table 1 and Fig. 4. Three algorithms KNN1, KNN2 and PF (see Sect. 4) were tested, each in three variants: using Wi-Fi signals (W), Bluetooth Low Energy signals (B), and the combination of both types of signals (C). The position estimation error (in meters) for each location and each algorithm is presented in Table 1. The error is also shown in the boxplot

Table 1. Localization error (m) for each location and algorithm

	A	B	C	D	E	F	G	H	I	Avg.
KNN1(B)	5.31	18.03	34.53	19.00	17.93	13.03	32.00	5.48	11.76	17.45
KNN1(C)	1.54	1.34	1.44	1.87	3.71	0.01	0.77	2.30	1.24	1.58
KNN1(W)	5.35	1.33	1.88	3.69	3.69	2.01	1.65	1.53	14.71	3.98
KNN2(B)	1.06	1.67	0.20	0.87	5.15	1.99	0.37	1.97	9.02	2.48
KNN2(C)	1.47	3.87	5.21	2.29	6.03	1.94	2.92	2.51	11.67	4.21
KNN2(W)	3.85	7.77	14.89	7.57	21.09	2.00	7.40	3.34	15.45	9.26
PF(B)	8.77	2.07	14.90	7.58	9.36	1.16	14.30	4.97	17.73	8.98
PF(C)	9.97	1.86	13.97	4.99	7.54	0.95	12.67	4.92	15.77	8.07
PF(W)	12.75	1.00	12.80	5.18	9.73	2.50	10.17	0.00	13.55	7.52
Avg.	5.56	4.33	11.09	5.89	9.36	2.84	9.14	3.00	12.32	7.06

(see Fig. 4) where the algorithms are ordered by the average error – the smaller the error is the better the algorithm performs.

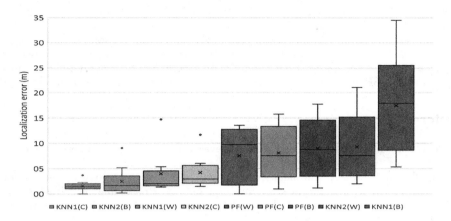

Fig. 4. Localization error chart

The results show that the KNN1(C) – the K-Nearest Neighbors algorithm based on the combination of Wi-Fi and BLE, using one set of signals and Euclidean distance – achieves the highest average accuracy of 1.58 m. The testing point (location) E gives the worst results due to poor signal coverage (both Wi-Fi and BLE) in the center of the building's atrium. The result proves our expectation that an algorithm based on combination of the two technologies will perform very well. Algorithms using the Particle Filter achieved quite a poor accuracy, which may be caused by the bad RSSI-based distance estimation (influenced by reflections and attenuation of the signal) indoors.

7 Conclusion

Aim of the work was to design and implement a mobile application that will be able to estimate the user's position within the building. We used it to further evaluate variously modified indoor localization algorithms and discussed the results. The application is able to effectively synchronize the fingerprint database among multiple mobile devices and provides advanced visualizations of the estimated position with regard to a particular localization algorithm.

The application implements two of the most widely used indoor localization algorithms; the K-Nearest Neighbors and Particle Filter, in 9 variants. The evaluations showed that the Nearest Neighbor algorithm combining Wi-Fi and Bluetooth Low Energy performs the best in the real-world environment (the campus building). The deployment scenario shows a feasible way to deploy additional Bluetooth Low Energy beacons in the building. It should also be noted that the localization error depends on the specific place. The accuracy is worse (sometimes in an order of magnitude) in places with poor coverage than in other places.

The application is robust enough to be further expanded in order to evaluate new indoor localization approaches or current approaches in a new environment.

Acknowledgements. The authors of this paper would like to thank Tereza Krizova for proofreading. This work and the contribution were also supported by a project of Students Grant Agency – FIM, University of Hradec Kralove, Czech Republic (under ID: UHK-FIM-SP-2017). Radek Bruha is a student member of the research team.

References

1. Estimote: Mesh networking - Estimote developer. http://developer.estimote.com/managing-beacons/mesh/
2. Siddharth, S., Ali, A., El-Sheimy, N., Goodall, C., Syed, Z.: A robust sensor fusion algorithm for pedestrian heading improvement. In: Proceedings of 24th International Technical Meeting of The Satellite Division of the Institute of Navigation (ION GNSS 2011). p. 1337-0 (2001)
3. Racko, J., Brida, P., Perttula, A., Parviainen, J., Collin, J.: Pedestrian dead reckoning with particle filter for handheld smartphone. In: 2016 International Conference on Indoor Positioning and Indoor Navigation (IPIN), pp. 1–7. IEEE (2016)
4. Abu-Shaban, Z., Zhou, X., Abhayapala, T.D.: A novel TOA-based mobile localization technique under mixed LOS/NLOS conditions for cellular networks. IEEE Trans. Veh. Technol. **65**(11), 8841–8853 (2016)
5. Yiu, S., Dashti, M., Claussen, H., Perez-Cruz, F.: Wireless RSSI fingerprinting localization. Sig. Process. **131**, 235–244 (2017)
6. Matharu, N.S., Buttar, A.S.: An efficient approach for localization using trilateration algorithm based on received signal strength in wireless sensor network. Int. J. Comput. Sci. Netw. Secur. (IJCSNS) **16**(8), 116 (2016)
7. Sanguino, T.D.J.M., Gómez, F.P.: Toward simple strategy for optimal tracking and localization of robots with adaptive particle filtering. IEEE/ASME Trans. Mechatron. **21**(6), 2793–2804 (2016)
8. Kumar, S., Hegde, R.M., Trigoni, N.: Gaussian process regression for fingerprinting based localization. Ad Hoc Netw. **51**, 1–10 (2016)
9. Song, C., Wang, J., Yuan, G.: Hidden Naive Bayes indoor fingerprinting localization based on best-discriminating AP selection. ISPRS Int. J. Geo-Inf. **5**(10), 189 (2016)
10. Yang, Z., Zhou, Z., Liu, Y.: From RSSI to CSI: indoor localization via channel response. ACM Comput. Surv. (CSUR) **46**(2), 25 (2013)
11. Machaj, J., Brida, P.: Performance comparison of similarity measurements for database correlation localization method. In: Nguyen, N.T., Kim, C.-G., Janiak, A. (eds.) ACIIDS 2011. LNCS, vol. 6592, pp. 452–461. Springer, Heidelberg (2011). doi:10.1007/978-3-642-20042-7_46
12. Honkavirta, V., Perala, T., Ali-Loytty, S., Piché, R.: A comparative survey of WLAN location fingerprinting methods. In: 6th Workshop on Positioning, Navigation and Communication, WPNC 2009, pp. 243–251. IEEE (2009)
13. Gu, Z., Chen, Z., Zhang, Y., Zhu, Y., Lu, M., Chen, A.: Reducing fingerprint collection for indoor localization. Comput. Commun. **83**, 56–63 (2016)
14. Android.com: Bluetooth low energy - Android developers. https://developer.android.com/guide/topics/connectivity/bluetooth-le.html

Towards Device Interoperability in an Heterogeneous Internet of Things Environment

Pavel Pscheidl[1], Richard Cimler[2(✉)], and Hana Tomášková[1]

[1] Faculty of Informatics and Management, University of Hradec Kralove, Hradec Kralove, Czech Republic
[2] Faculty of Informatics and Management, Center for Basic and Applied Research (CBAR), University of Hradec Kralove, Hradec Kralove, Czech Republic
`richard.cimler@uhk.cz`

Abstract. Internet of Things can be viewed as a global network of devices with capabilities of autonomous communication. This is one step ahead of current state, where automatic mutual service communication without human interference is limited. This paper evaluates possibilities of extending nowadays commonly used protocols to achieve higher level of autonomous device communication. It is presumed that fully automatic machine to machine communication and understanding is not yet technologically possible. Several ways of extending current communication protocols by semantic metadata are described and their impact on device communication evaluated.

Keywords: Internet of Things · Virtual device · Meta-data · Protocols · Device Interoperability

1 Introduction

A broad vision for the Internet of Things represents a network of interconnected devices, cooperating towards individual or common goals. The currently proposed [12,26], yet unfulfilled possibilities of the Internet of Things are mostly seen in broader access of devices to data or information provided by other devices in the network. However, the idea of IoT also allows sharing more general resources like know-how or computational capacity. Therefore, a more general understandings introduce the idea of services [1], where each device provides behavior defined by it's public interface.

A network of interconnected devices providing services to each other is not introduced by Internet of Things. In fact, there are decades of computer elements interconnection evolution demonstrated by [22]. The whole concept of the Internet of Things can be purely seen as yet another evolution of mutual computer system interaction driven by newly discovered technology use cases. A promising next step is to build scalable interaction models on top of this basic network connectivity and thus focus on the application layer [13]. The added value and major difference of the Internet of Things lies in autonomous

© Springer International Publishing AG 2017
N.T. Nguyen et al. (Eds.): ICCCI 2017, Part II, LNAI 10449, pp. 315–324, 2017.
DOI: 10.1007/978-3-319-67077-5_30

interoperability of virtual devices [16]. Specialized gateways, service brokers and other design patterns requiring human intervention are the current way of device interconnection. Often, specialized projects incorporating a constant set of various SMART devices cooperating together are presented as Internet of Things projects [10].

The Internet of Things without the improved capabilities of mutual device communication degrades to a network of Internet-enabled devices. Goal of this paper is to evaluate current possibilities and propose specific ways of moving towards network of inter-operable devices.

In Sect. 2, challenges and rationale behind the automation of device communication is explained. Addition of metadata description of existing resources as a next step towards device interoperability is proposed in Sect. 3. Computational complexity, reducing computational overhead and heterogeneous environment querying problems are also involved.

2   Device Interoperability in Heterogeneous Environment

Nowadays, services spread across the network are perceived as black boxes, specialized in solving problems from selected domain [6]. Services found in today's networks, most notably in the Internet, are vastly heterogeneous. Not only they differ in protocols used across ISO/OSI reference model, differences are also found in service interfaces, used data types and data formats. Such services are granted a different level of accessibility [20].

Interconnecting services, despite it's evolution, mostly works for small groups of devices with low or no heterogeneity. Direct way of connection, where client application directly connects to a set of services, was soon recognized to be unusable in many cases [24]. Therefore, additional design pattern in service interconnection were developed [7] and are still being deployed nowadays. Gateways, Service brokers, Enterprise service buses and others are used to decouple communication between network nodes, provide orchestration services and for overall client isolation from complex service provider environment. Yet still, some components of the whole system has to be explicitly told about other service's contract, which implies involvement of human factor.

As stated before, the Internet of Things is aimed to become a large system of interconnected services. Large volume of service are already present in the global Internet network, but can not be understood by other devices. A very similar situation occurred in the domain of World Wide Web, where attempts were made to create a way for machines to understand existing web-based services [3] without the need of re-creating them, widely known as "The Semantic Web". One of the resulting protocols was RDF [15].

Metadata are used to achieve machine-to-machine understanding of a service. They carry representation of domain knowledge describing the information domain to which the underlying data belong [22]. This knowledge may then be used to make inferences about the underlying data.

A "thing" connected to a network may require data, later turned into information. This represents basic use case in the Internet of Things. A "thing" may also provide potentially useful data for other devices. By "thing", a virtual device providing services to other "things" or devices in the same logical network is often understood [25].

3   The Role of Metadata in Device Intercommunication

Attributes of the Internet of Things can be seen as an evolution of the idea of service oriented architecture. In service oriented architecture, the components of an application are autonomous services that execute independently and communicate via messaging [5]. In the Internet of Things, there are many services provided by many devices. Those devices are no longer part of one coherent subsystem [9], each device typically works towards different goals. Still, the idea of obtaining data to achieve certain goal persists. There are few basic steps outlining device's way to obtain data from another service providing device.

– Service discovery. Which services are available?
– Service filtering. Which services meet the requirements of current use case?
– Service invocation. Requesting and obtaining data.

Currently, in the environment of World Wide Web and in the era of interconnected services in general, automatic mutual discovery of service and further filtering is yet another challenge to automate. Nowadays, semi-automatic services following the previously mentioned service broker pattern [24] are being deployed, mainly in smaller networks. Common service broker's ability to function in homogeneous, open environment of the Internet is limited [18]. Improving autonomous service discovery and filtering is the key to improve degree of automation in the domain of service communication.

3.1   Meta-data in Service Discovery and Filtering

Typically service output is represented by a structured data hierarchy. Input and output data, no matter the structure, can be identified by a limited set words describing certain domain [21]. In fact, the service itself can be categorized into certain problem domain with it's own set of set of words identifying the problem domain. Identifying data in request/response is nowadays an implicit part of application-level protocols. Services following contract of REST or SOAP are widely known examples of data identifying strings being naturally present in the messages sent over network [8].

Thus, the problem of including meta-information about transferred data is narrowed simply into the problem of choosing the right protocol. Even if the metadata are not part of the message explicitly, any other attribute able to identify each attribute of the message is sufficient, for example attribute order or linking to a separate metadata-providing service in message headers.

To move towards better mutual device interconnection, creating a way to describe services by means of semantic information is crucial. A great deal of

```
 1  {
 2      "loc": {
 3        "lat": 47.60621,
 4        "long": -122.33207
 5      },
 6      "place": {
 7        "name": "Seattle",
 8        "state": "WA",
 9        "stateFull": "Washington",
10        "country": "US",
11        "countryFull": "United States",
12        "region": "usnw",
13      }
14  }
```

Fig. 1. Service-specific code

data on the web is already available through REST and SOAP-based access points but this data carries no markup to relate it to semantic concepts [2]. Words from a dictionary are used to describe attributes of the message and represents shared meaning to both service provider and the client. A dictionary may be complemented with a thesaurus to handle synonyms potentially present in metadata. A trivial, but insufficient solution is to describe the message attributes directly. In such simplified case, original value descriptors are replaced by corresponding ones from the shared dictionary. The client, once the response has been sent by the service, directly searches the response for desired data. In key-value protocols, the original keys are simply replaced, see differences between Figs. 1 and 2.

```
 1  {
 2      "location": {
 3        "latitude": 47.60621,
 4        "longitude": -122.33207
 5      },
 6      "place": {
 7        "name": "Seattle",
 8        "stateCode": "WA",
 9        "state": "Washington",
10        "countryCode": "US",
11        "country": "United States",
12        "regionCode": "usnw",
13      }
14  }
```

Fig. 2. Response updated with keys replaced by dictionary terms

This approach does not suit the heterogeneous environment of the Internet very well. Firstly, many services are already in existence and commonly, there are $n > 1$ clients already adapted to their contract. Changing the message by modifying the way attributes are identified may cause client incompatibility. This implies the need for the service to be backwards compatible for "direct-access" clients with contracts being statically present ("hard-coded") into their execution logic. Secondly, there is no need to imply existence of one universal dictionary describing all the domains. The existence of several dictionaries is considered an advantage for diversification and security reasons.

Both problems can be solved by describing the structure of service's messages independently, without modifying the original structure. Evading the need for constant re-implementation of the service, the document-describing key-value set exists separately, describing each attribute in the original document. This approach allows multiple project-describing documents to exist, but with each document describing the original message, the volume of data transferred and processed by both server and client grows. Two basic approaches can be identified

1. Appending metadata to original document,
2. Distributing service metadata separately.

The first approach results in appending metadata to original response per each request and is the most simple method. Client, receives both original response and metadata with each request. Appending metadata every time results in a fixed expense (overhead) for each response served, further explained in Sect. 3.2. Appending metadata into original response also requires special syntactic elements to differentiate original message from it's description.

A more advanced approach is to distribute the metadata separately. When service responds, the original message is complemented with

1. A reference to the metadata document,
2. A unique mark, identifying the last modification of the metadata.

Time-stamp can be considered a suitable unique mark, due to it's space requirements. In memory, up to 64 bits is required to hold a time-stamp value (an unsigned integer up to 2^{64}). The client evaluates the unique mark first. If the client already owns metadata document with the same time-stamp, no further steps are required and the document can be processed. This reduces the additional computational demands described in the Sect. 3.2. Usually, the metadata documents have to be processed by the client and served by the service provider in case the client calling the service for the very first time, when client has no knowledge about the service called.

3.2   The Cost of Metadata Presence

Calculating the additional cost of metadata presence means to evaluate the computational cost of parsing, storing and retrieving them. The message structure

```
1  {
2      "loc": {
3         "lat": "latitude",
4         "long": "longitude"
5      },
6      "place": {
7         "name": "name",
8         "state": "stateCode",
9         "stateFull": "state",
10        "country": "countryCode",
11        "countryFull": "country",
12        "region": "regionCode",
13     }
14 }
```

Fig. 3. Proposed metadata tree structure, describing the keys from original document by words from a shared dictionary

of application-level protocols can be represented as an acyclic graph G_R. For each response type, another topologically equivalent graph G_M is constructed, representing meta-information about vertices in the original graph G_R. Each vertex of the newly constructed graph is n-dimensional structure containing meta-information about the node in the original response graph. Not every node must be necessarily fully described by metadata, an empty record may be present instead, symbolizing there is either no need to describe it with metadata or there are no description available (Fig. 3).

Provided the metadata document graph G_M has the same cardinality ($|G_M| = |G_R|$) and the topology of both graph is equal, parsing the document with complementary metadata introduces no bigger overhead than walking through the structure of the original document, which is the response of the service. In other words, let O_R be the computational complexity of service's response and d the number of attached metadata documents. Then, the final computational complexity O_F of receiving and understanding service's response can be expressed as

$$O_F = \sum_{i=1}^{d+1} O_R,$$

multiplying original computational cost with additional d number of passes of metadata processing. The resulting linear growth is only present when service parameter changes and metadata documents must be assembled again.

4   Implementation Design Patterns

The worldwide network has been designed as decentralized and this core architectural principle remains untouched nowadays, despite many counter-attempts

[11]. It is crucial to state that no structure is enforced when a system of vocabularies and thesauri is introduced. Anyhow, it may lead to emergence of many patterns regarding mutual device interconnection. Some of those patterns based on currently working real-world scenarios are described in this section.

4.1 Service Mapping and Metadata Interchange

By using added metadata as resources proposed in this research, no service topology or design patterns are enforced. For any network peer to discover service's possibilities (e.g. what data does the called device provide), simply the metadata tree has to be downloaded and analyzed. The metadata can be generated by the service once the service is started and cached, therefore no actual service call must be done before analyzing it's fitness for the client. Assuming there is a cost of the service execution, serving cached metadata with no execution cost is a better alternative.

4.2 Dedicated Services for Filtering and Discovery

A typical use case for a client is to obtain a list of n services providing the desired kind of data. Network peers can be queried by clients for service discovery. Filtering is based on metadata indexing. The indexing peer retrieves and stores metadata for known services, potentially for each and every one of them. The data returned by the service may be described with words from completely different vocabulary. Therefore, conversion to a common language is made. Then, meta-data are indexed for faster search by the peer which acts as a service broker.

Whenever a client asks the indexing peer for a list of services, the peer searches it's local indexed database with metadata already transformed into a language common to the client and returns addresses of services together with the service metadata of such services. The metadata retrieval describes the remote service to the client. By means of the metadata, the client makes sure its query has the right structure accepted by the service. Also, the client can remember the metadata, avoiding further queries to the indexing peer - at least until exploration of new services is required.

The filtering first occurs when the indexing peer returns only those services which provide specified metadata. In practice, a graph-oriented database provides quick way of filtering in such use cases [17]. Secondly, the client itself can further pick final set of services to query based on various criteria, for example criteria found in metadata, based on the address and reachability.

This functionality can be summed up as a more liberate and decentralized variant service brokers mentioned in Sect. 4.1. The freedom and main advantage is in the query language itself - there is no specific query language enforced. Each network of devices may set it's own language.

4.3 Resource Saving with Precise Query Specifications

It has to be noted client may only need a subset of data a service endpoint hidden behind an address offers. By obtaining metadata document first, the client can then query only for the values required. Expressing query as a graph, selecting only the attributes required is latest trend in API building [19]. This is especially beneficial for services where obtaining a value is resource-consuming.

Fig. 4. Querying a service by means of a translation service

Whole query demonstrated by Fig. 4 is started by the client device, querying a translator service residing in the middle for data desired. The translator service itself can be horizontally scaled in an independent way based on query load. This approach fits into current trend of horizontal scaling with containers, where service instances can be spawned on demand [23]. The translator service however connects to services potentially communicating by means of completely different protocol. Currently, not only the protocol has to be known to the translator when the service is built, but also the service contract must be known at the time the translator service is built. The translator provides both request and response translation, including metadata, if required. A scenario where client may do the translation on it's own expense, or use the translation service only for a certain subset of called resources is also a viable choice.

5 Conclusion

A way of interconnecting large amount of service-providing devices in a heterogeneous environment has been presented. It is proposed to utilize domain-oriented dictionaries complemented by thesauri to describe services provided by interconnected devices.

As a recommended solution, describing a service with a general metadata document is advised. The metadata document describes the service's response by means of words present in a specific dictionary. Thesaurus can then be built and used to translate among vocabularies, making services understand each other. The introduction of metadata comes with acceptable linear computational complexity growth with each metadata document added. By simply Versioning the services, repeated metadata parsing can be avoided easily. Meta-data, once generated, can be also easily cached and served statically by the service. The server reply for metadata requires no additional computational cost the metadata are generated only once per service contract change.

Especially in the upcoming Internet of Things era, there are other issues with security, trust, reliability and possibly more [14]. Such factors may affect the process of service discovery and data retrieval. These factors represent separate challenges. For example, a service with low trust rating may not be used, so can be avoided service with high latency.

With metadata already built inside services, emphasis can be put on improving the ways of service discovery. Most notably a proper mechanism to interlink distributed services into a global service space for service discovery and composition [4]. Creation of new kinds of intermediate services, providing filtering and discovery capabilities for common clients is also possible.

Acknowledgment. The support of the Specific Research Project at FIM UHK is gratefully acknowledged.

References

1. Atzori, L., Iera, A., Morabito, G.: The internet of things: a survey. Comput. Netw. **54**(15), 2787–2805 (2010)
2. Battle, R., Benson, E.: Bridging the semantic web and web 2.0 with representational state transfer (rest). Web Semant.: Sci. Serv. Agents World Wide Web **6**(1), 61–69 (2008)
3. Berners-Lee, T., Hendler, J., Lassila, O., et al.: The semantic web. Sci. Am. **284**(5), 28–37 (2001)
4. Chen, W., Paik, I., Hung, P.C.: Constructing a global social service network for better quality of web service discovery. IEEE Trans. Serv. Comput. **8**(2), 284–298 (2015)
5. Dragoni, N., Giallorenzo, S., Lafuente, A.L., Mazzara, M., Montesi, F., Mustafin, R., Safina, L.: Microservices: yesterday, today, and tomorrow. arXiv preprint arXiv:1606.04036 (2016)
6. Erl, T.: Service-Oriented Architecture (SOA) Concepts, Technology and Design. Prentice Hall, Upper Saddle River (2005)
7. Erl, T.: SOA Design Patterns. Pearson Education, Harlow (2008)
8. Fielding, R.T., Taylor, R.N.: Principled design of the modern web architecture. ACM Trans. Internet Technol. (TOIT) **2**(2), 115–150 (2002)
9. Fowler, M., Lewis, J.: Microservices. Viittattu **28**, 2015 (2014)
10. Gea, T., Paradells, J., Lamarca, M., Roldan, D.: Smart cities as an application of internet of things: experiences and lessons learnt in Barcelona. In: 2013 7th International Conference on Innovative Mobile and Internet Services in Ubiquitous Computing (IMIS), pp. 552–557. IEEE (2013)
11. Gehlbach, S., Sonin, K.: Government control of the media. J. Publ. Econ. **118**, 163–171 (2014)
12. Gubbi, J., Buyya, R., Marusic, S., Palaniswami, M.: Internet of things (IoT): a vision, architectural elements, and future directions. Future Gener. Comput. Syst. **29**(7), 1645–1660 (2013)
13. Guinard, D., Trifa, V., Mattern, F., Wilde, E.: From the internet of things to the web of things: resource-oriented architecture and best practices. In: Uckelmann, D., Harrison, M., Michahelles, F. (eds.) Architecting the Internet of Things, pp. 97–129. Springer, Heidelberg (2011). doi:10.1007/978-3-642-19157-2_5
14. Hernández-Ramos, J.L., Jara, A.J., Marın, L., Skarmeta, A.F.: Distributed capability-based access control for the internet of things. J. Internet Serv. Inf. Secur. (JISIS) **3**(3/4), 1–16 (2013)

15. Kahan, J., Koivunen, M.R., Prud'Hommeaux, E., Swick, R.R.: Annotea: an open RDF infrastructure for shared web annotations. Comput. Netw. **39**(5), 589–608 (2002)
16. Maximilien, E.M., Wilkinson, H., Desai, N., Tai, S.: A domain-specific language for web APIs and services mashups. In: Krämer, B.J., Lin, K.-J., Narasimhan, P. (eds.) ICSOC 2007. LNCS, vol. 4749, pp. 13–26. Springer, Heidelberg (2007). doi:10.1007/978-3-540-74974-5_2
17. Mitra, P., Wiederhold, G., Kersten, M.: A graph-oriented model for articulation of ontology interdependencies. In: Zaniolo, C., Lockemann, P.C., Scholl, M.H., Grust, T. (eds.) EDBT 2000. LNCS, vol. 1777, pp. 86–100. Springer, Heidelberg (2000). doi:10.1007/3-540-46439-5_6
18. Ngan, L.D., Kanagasabai, R.: Owl-s based semantic cloud service broker. In: 2012 IEEE 19th International Conference on Web Services (ICWS), pp. 560–567. IEEE (2012)
19. Nogatz, F., Seipel, D.: Implementing GraphQl as a query language for deductive databases in SWI-Prolog using DCGs, quasi quotations, and dicts. arXiv preprint arXiv:1701.00626 (2017)
20. Olifer, N., Olifer, V.: Computer Networks: Principles, Technologies and Protocols for Network Design. Wiley, Hoboken (2005)
21. Rochwerger, B., Breitgand, D., Levy, E., Galis, A., Nagin, K., Llorente, I.M., Montero, R., Wolfsthal, Y., Elmroth, E., Caceres, J., et al.: The reservoir model and architecture for open federated cloud computing. IBM J. Res. Dev. **53**(4), 1–4 (2009)
22. Sheth, A.P.: Changing focus on interoperability in information systems: from system, syntax, structure to semantics. In: Goodchild, M., Egenhofer, M., Fegeas, R., Kottman, C. (eds.) Interoperating Geographic Information Systems. The Springer International Series in Engineering and Computer Science, vol. 495, pp. 5–29. Springer, Boston (1999). doi:10.1007/978-1-4615-5189-8_2
23. Sotiriadis, S., Bessis, N., Amza, C., Buyya, R.: Vertical and horizontal elasticity for dynamic virtual machine reconfiguration. IEEE Trans. Serv. Comput. **PP** (2016). http://ieeexplore.ieee.org/abstract/document/7762944/
24. Stal, M.: Web services: beyond component-based computing. Commun. ACM **45**(10), 71–76 (2002)
25. Sundmaeker, H., Guillemin, P., Friess, P., Woelfflé, S.: Vision and challenges for realising the Internet of Things. In: EUR-OP, vol. 20 (2010)
26. Uckelmann, D., Harrison, M., Michahelles, F.: An architectural approach towards the future internet of things. In: Uckelmann, D., Harrison, M., Michahelles, F. (eds.) Architecting the Internet of Things. Springer, Berlin (2011). doi:10.1007/978-3-642-19157-2_1

Hardware Layer of Ambient Intelligence Environment Implementation

Ales Komarek, Jakub Pavlik, Lubos Mercl, and Vladimir Sobeslav[✉]

Faculty of Informatics and Management, University of Hradec Kralove,
Rokitanskeho 62, Hradec Kralove, Czech Republic
{ales.komarek,jakub.pavlik.7,lubos.mercl,vladimir.sobeslav}@uhk.cz

Abstract. Ambient Intelligence is growing phenomena caused by advances of speed and size of computational hardware. It is possible to collect data from various sensors, devices or services and react to evaluated event and start the predefined process. These systems are closely connected with Cloud based services as they provide publicly available interface for management and visualization, computational and storage capabilities for forecasts and other advanced data processing to create ambient intelligence environment. The true ambient environments react to the presence of people and the sensors and actuators have become part of the environment.

This article presents RoboPhery project aimed to provide abstraction layer for interfacing any low cost hardware sensors and actuators with MQTT and Time-series Database bindings, that can serve as sensory network for Ambient Intelligence environments. The service architecture is designed to be so simple at hardware level to support single-board microcontrollers like ESP2866, ESP32 modules as well as single-board computers based on ARM or x86 architectures. The communication among individual devices is handled by the standard MQTT messages. The same devices can be configured to support multiple use-cases on configuration level to lower the operational costs of the solution.

Keywords: Automation · Sensor · Actuator · Service-oriented architecture · Ambient Intelligence

1 Introduction

The RoboPhery project makes it possible to utilize inexpensive Linux and MicroPython compatible micro-controllers, hardware actuators and sensors to create fully autonomous Ambient Intelligence platform. This platform can perform wide range automation use cases. These range from simple environmental automations, surveillance, to complete ambient intelligence environments.

First chapter introduces global architecture of autonomous agents with communication protocols for messages to the overlaying event driven controllers, dashboards or any other services. These systems use scalable and modular architecture to minimalize the computational overhead which allows efficient use of available hardware resources.

© Springer International Publishing AG 2017
N.T. Nguyen et al. (Eds.): ICCCI 2017, Part II, LNAI 10449, pp. 325–334, 2017.
DOI: 10.1007/978-3-319-67077-5_31

The next part covers the autonomous agents in more details. Models of hardware interfaces and corresponding modules. It also covers description of the virtual modules that provide the simple thresholds or fuzzy logic reasoning for autonomous control.

The final chapter shows simple interaction with overlay control system that can communicate with multiple autonomous systems, gather vital information and even publish arbitrary actions on demand.

The control system contains time-series database, dashboard and reasoning platform which can detect hardware malfunctions of autonomous agent systems and perform necessary steps to repair it. With the usage of virtual models that provide high-level access to individual physical modules of individual agents.

1.1   Ambient Intelligence Environments

The main and basic ideas of Ambient Intelligence is an interconnecting an environment and technologies (mainly interconnected sensors and devices). This whole system can be designed to take care of this environment based on real-time and historical information (data). Ambient Intelligence inherits aspects of many areas of Computer Science, see Fig. 1, but should not be confused with any of those technologies [1, 14, 15].

An example of ambient intelligence is a Home environment, where is possible to solve some a lot of challenges using ambient intelligence [9]. Ambient intelligence was originally pointed by Eli Zelkha at Palo Alto Ventures in the late 1990s. This was a future vision of consumer electronics and computing and time frame for this was 2010–2020 [1, 20].

In an ambient intelligence, devices are interconnected to people support in life activities everyday, tasks and rituals in a natural way that uses information from physical environment collected by network connected devices and uses cloud power to drive the intelligence that can react and learn to the information coming from the environment [1, 5, 19].

The physical devices are constantly smaller and integrated into environment and the technology is mainly visible for end users just only via user interfaces (dashboards) [10].

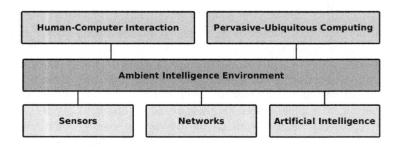

Fig. 1. Ambient Intelligence relationship to other computing science areas

The ambient intelligence is built on a pervasive and ubiquitous computing, profiling, context awareness, and design on human-centric computing. Also it is characterized by interconnected systems and technologies [1,5,15,20] (Table 1).

Table 1. Main Ambient Intelligence environment properties

Property	Description
Embedded	A lot of interconnected devices in the environment
Context aware	React to current context
Personalized	Meet the requirements of users
Adaptive	React to current state and context in the environment
Anticipatory	Anticipate user desires

1.2 Information Flows

There is a lost of was for building an Ambient Intelligence systems or environments, but the most often there are needed sensors and devices to surround users of an environment (interactors) with technology [5].

The technology collects and provides accurate data to the system. The collected data are transferred over the network and processed by middleware, which can collect data from multiple devices [4,18] (Fig. 2).

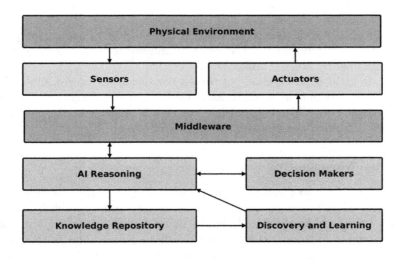

Fig. 2. Flow of information in Ambient Intelligence system

An decision-making can be more beneficial to the people in the environment, when the system will provides higher level of reasoning and will accomplish diagnosis and advise, or system will assist humans with responsibility for intervention [4,16].

High level Decision Making service are composed of a Knowledge Repository, where the events from environment are collected and stored. Thanks of this, AI Reasoner applies for example spatio-temporal reasoning to make decisions based on collected data.

Decision trigger actions to be performed in the environment and this is enabled via Actuators. The Knowledge discovery uses machine learning techniques to learn from the acquired data and information in order to update the AI Reasoner with the gained experience of the system. A typical information flow of Ambient Intelligence system is depicted in Fig. 3.

2   Ambient Intelligence Environment System

RoboPhery project provides software drivers for a wide variety of commonly used sensors and actuators through both wired and wireless interfaces. Platform is designed for multiple interfaces to be stacked (GPIO, IC or SPI over IQRF or Bluetooth) to provide seamless integration with large variety of hardware platforms. The software drivers interact with the underlying hardware platform (or micro-controller), as well as with the attached sensors, through API calls to the interfaces [1,3,8,17].

The actuators are controlled and the data is collected on RoboPhery Managers and this functionality forms a hardware foundation for ambient intelligence environment. The event-driven control system, data storage and data processing systems are provided in form of external cloud services or handled by computer with more processing power [7].

On the Fig. 3, you can see components of our proposed ambient intelligence system and their relationships. The central component is the message bus, in our case provided by MQTT broker [7,11].

The basic architecture consists of time-series database, user dashboard and event-driven control engine, machine learning services and the on premise Robo-Phery Manager. The actual RoboPhery service is written in Python in way compatible with MicroPython used on micro-controllers.

2.1   Global Message Bus

MQTT is a machine-to-machine connectivity protocol in area of "Internet of Things" [12]. MQTT was designed as an extremely simply publish/subscribe messaging transport protocol and thanks of this, it can be used for connections with remote locations where a small code footprint. This is reason, why requirements and/or network bandwidth are low. The MQTT broker can handle thousands of messages per second, supports high-availability setups for both high performance and stability. Individual autonomous agents and cloud-based Control system along with time-series databases are connected to this common message bus [11].

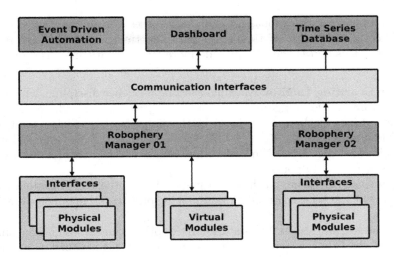

Fig. 3. Ambient Intelligence environment system

2.2 Cloud-Based Controller

As a architecture can be used an Event-driven architecture (EDA), which is message-driven architecture, where all components are reacting to events [13].

Events are significant changes in a state of components in the environment. For example, when a user turns on a computer, the state of computer changes from "off" to "on" and this change is captured as an event [7]. After this system architecture will treat this state change as an event which can affect other components in the environment. Also based on this change can be created message called the event notification. Events just occur, notification can be provide to the environment.

Time-Series Database. A time series databases are optimized for time series data handling and storage, where numbers are indexed by time (a date time or a date time range) [6,8]. A time series of storage prices is called a price curve and energy consumption is called a load profile.

Database queries and transactions are used and useful for analyzing, but these operations should be specialized for time-series data for correct, reliable and effective implementation of an time-series database. Time-series Databases are optimized for time series data.

Complex softwares for time series data may not be practical with regular relational database systems. Traditional flat file databases are not appropriate, whe the data and a count of transactions reaches a maximum threshold determined by the capacity of individual servers. Queries for historical data, with time ranges and time zone conversion are difficult in a traditional relational database.

The problem for traditional relational databases is that they are often not modeled correctly to respect time series data and time-series databases impose a model and this allows them to provide more features for time series data.

Machine Learning Engine. Machine learning is science discipline, which deals with algorithms for learning from data and making predictions. These algorithms make data-driven predictions or decisions through a model, which is built based on sample inputs.

Machine learning is used in case, when is not possible to program explicit algorithm with a solid performance.

As a science discipline, machine learning is close connected to computational statistics, which is focused on prediction-making on computers or computational systems.

Machine learning is used in data analytic for making a design for complex model and algorithm, which help with prediction. Also it is called as predictive analytics.

This analytical model helps researchers, data scientist, engineers and analysts with research of real condition of environment. This model should also produce reliable and repeatable decisions, which should be consistent.

The one of most important activities is uncovering of hidden insights through learning from historical relationships and trends in data.

2.3   Ambient Intelligence Environment Middleware

Autonomous unit is Python service, which communicates with hardware peripherals and sending and receiving data from external communication sources. RoboPhery unit consists of several objects. Communication objects handle sending and receiving messages from the upper layer services or other autonomous units. Interface objects handle abstraction to hardware communication at device level. Modules encapsulate individual hardware sensors and actuators. Finally the RoboPhery manager serves as central service that connects all other models within the autonomous unit.

When data are collected from sensors, it is important to data will be transfered and stored in correct state to the highest part of system, e.g. to time-series database.

Message Bus mainly take care about communication between agent in Autonomous Agent System, because data can't be easily transfered from sensor (agent) to database directly. Message Bus also can aggregate data to bigger units or make some basic transformations.

Autonomous services take care about conditions from sensors and values, which are captured. There are predefined conditions and when captured values are identical with same condition, monitoring agent send a message via message bus to reacting agent, which will performs predefined action (Fig. 4).

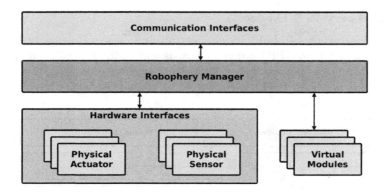

Fig. 4. Ambient Intelligence environment interactor

3   Sample RoboPhery Deployments

This chapter shows several use-cases of the usage of RoboPhery as hardware interfacing layer. Following Figures shows simple device configuration setups for various wired and wireless interfaces.

3.1   ESP2688 ModeMCU Micro-controller

Figure 5 shows the RoboPhery manager service running on the ESP2688 ModeMCU micro-controller with multiple sensors connected to interfaces present directly on the target device. The device has low computational power and low power consumption making it suitable for remote wireless probes with several attached sensors.

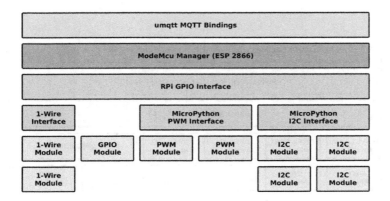

Fig. 5. ESP2688 micro-controller direct interfaces

3.2 Raspberry Pi 2 Device

Following figure shows RoboPhery service running on the Raspberry Pi device with MCP23008 I2C to GPIO expander boards and multiple sensors connected to individual interfaces. You can generally use multiple expander boards to support great numbers of sensor modules connected at a time (Fig. 6).

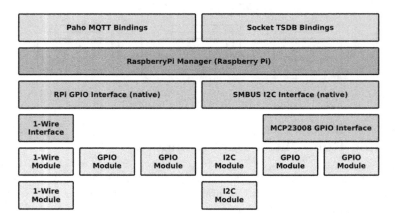

Fig. 6. Raspberry Pi computer extended interfaces

3.3 x86 Computer with IQRF Mesh Network

Following figure shows RoboPhery Manager service running on the x86 single board computer with USB connected to the IQRF coordinator with multiple sensor interfaces connected through a common mesh network. Individual mesh member implements an interface and can support multiple sensor modules (Fig. 7).

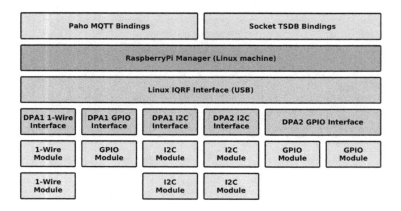

Fig. 7. x86 Computer IQRF Mesh interfaces

4 Conclusion

This article has shown the growing phenomena of Ambient Intelligence environment and what role it can play in daily life. We have introduced RoboPhery project for simple interaction layer for a wide variety of commonly used sensors and actuators. This layer is an integral part of any Ambient Intelligence Environment.

The main utility of the service list rests in the fact, that after little codebase changes can be operated on both computers and micro-controllers. It can reuse the sensor and actuator drivers on all platforms, thus greatly reducing the costs of migrations and reduces the vendor lock-in as you are free to use a lot of different devices available on the market. At the end the whole system is designed to be extremely modular a provides wide range of hardware interfaces and modules in the default setup and easy way to introduce any new interface or module.

The micro-controllers provide only a very limited computational power, operating memory, but it is enough to run the communication and sensoric operations. Modern single-board computers on the other hand can provide much more. They can communicate with external services, a provide enough power for other services in Ambient Intelligence stack as machine learning, time-series databases and event-driven automation. The further research is needed to explore possibilities of integrating the RoboPhery project with external services and expand the list of supported micro-controllers and devices.

Acknowledgement. This work and the contribution were also supported by project of specific science, Faculty of Informatics and Management, University of Hradec Kralove, Czech Republic.

References

1. Aarts, E., Encarnao, J.L.: True Visions: The Emergence of Ambient Intelligence. Springer, Heidelberg (2006)
2. Aarts, E., Harwig, R., Schuurmans, M.: Ambient Intelligence in The Invisible Future, pp. 235–250. McGraw-Hill Inc. (2002)
3. Arribas-Ayllon, M.: Ambient Intelligence: an innovation narrative, 16 April 2017. http://www.academia.edu/1080720/Ambient_Intelligence_an_innovation_narrative
4. Augusto, J.C., Nugent, C.D.: A new architecture for smart homes based on ADB and temporal reasoning. In: Proceedings of 2nd International Conference on Smart Homes and Health Telematic, ICOST 2004, pp. 106–113 (2005)
5. Bielikova, M., Krajcovic, T.: Ambient Intelligence within a Home Environment, 12 April 2017. https://www.ercim.eu/publication/Ercim_News/enw47/bielikova.html
6. Dunning, T., Friedman, E., Loukides, M.K., Demarest, R.: Time Series Databases: New Ways to Store and Access Data. O'Reilly Media, Sebastopol (2014)
7. Foster, I., Kesselman, C., Nick, J.M., Tuecke, S.: Grid services for distributed system integration. Computer **35**(6), 37–46 (2002)
8. Ha, Y., Sohn, J., Cho, Y., Yoon, H.: A robotic service framework supporting automated integration of ubiquitous sensors and devices. Inf. Sci. **177**, 657–679 (2007)

9. Horálek, J., Matyska, J., Stepan, J., Vancl, M., Cimler, R., Soběslav, V.: Lower layers of a cloud driven smart home system. In: Barbucha, D., Nguyen, N.T., Batubara, J. (eds.) New Trends in Intelligent Information and Database Systems. SCI, vol. 598, pp. 219–228. Springer, Cham (2015). doi:10.1007/978-3-319-16211-9_23

10. Huhns, M., Singh, M.P.: Service-oriented computing: key concepts and principles. IEEE Internet Comput. **9**(1), 75–81 (2005)

11. Lampkin, V., Tat Leong, W., Olivera, L., Rawat, S., Subrahmanyam, N., Xiang, R.: Building Smarter Planet Solutions with MQTT and IBM WebSphere MQ Telemetry. An IBM Redbooks Publication (2012)

12. Li, S., Xu, L.D., Zhao, S.: The internet of things: a survey. Inf. Syst. Front. **17**(2), 243–259 (2015)

13. Mani Chandy, K.: Event-Driven Applications: Costs, Benefits and Design Approaches (2006)

14. Papazoglou, M.P.: Service -oriented computing: concepts, characteristics and directions. In: Proceedings - 4th International Conference on Web Information Systems Engineering, WISE 2003, pp. 3–12 (2003)

15. Papazoglou, M.P., Van Den Heuvel, W.J.: Service oriented architectures: approaches, technologies and research issues. VLDB J. **16**(3), 389–415 (2007)

16. Preuveneers, D., Van den Bergh, J., Wagelaar, D., Rigole, P., Clarckx, T., Berbers, Y., Coninx, K., Jonckers, V., De Bosschere, K.: Towards an extensible context ontology for ambient intelligence. In: Ambient Intelligence: Second European Symposium, EUSAI 2004, pp. 148–159 (2004)

17. Rui, C., Yi-bin, H., Zhang-qin, H., Jian, H.: Modeling the ambient intelligence application system: concept, software, data, and network. IEEE Trans. Syst. Man Cybern. Part C (Appl. Rev.) **39**, 299–314 (2009)

18. Sobeslav, V., Horalek, J.: Communications and quality aspects of smart grid network design. In: Wong, W.E. (ed.) Proceedings of the 4th International Conference on Computer Engineering and Networks, vol. 355, pp. 1255–1262. Springer, Cham (2015). doi:10.1007/978-3-319-11104-9_143

19. Wu, D., Rosen, D.W., Wang, L., Schaefer, D.: Cloud-based design and manufacturing: a new paradigm in digital manufacturing and design innovation. CAD Comput. Aided Des. **59**, 1–14 (2015)

20. Zelkha, E., Epstein, B., Birrel, S., Dodsworth, C.: From devices to "Ambient Intelligence". Digital Living Room Conference (1988)

Lightweight Protocol for M2M Communication

Jan Stepan[1], Richard Cimler[1(✉)], Jan Matyska[1], David Sec[2],
and Ondrej Krejcar[1]

[1] Faculty of Informatics and Management,
Center for Basic and Applied Research (CBAR),
University of Hradec Kralove, Hradec Kralove, Czech Republic
{jan.stepan.3,richard.cimler,jan.matyska,ondrej.krejcar}@uhk.cz
[2] Faculty of Informatics and Management, University of Hradec Kralove,
Hradec Kralove, Czech Republic
david.sec@uhk.cz

Abstract. Increasing popularity of Internet of Things and Smart home
system raises new requirements for micro-controllers. There is a need
for affordable solutions with low power consumption and reasonably fast
communication transmission rate. This article deals with the communication protocols and introduces own protocol designed for machine to
machine communication. Different ways of data transfer between various
hardware layers are shown and compared. Every approach is discussed
with its benefits and downfalls. Benchmark of three selected types of communication using proposed protocol is performed and results are shown.

Keywords: Smart home system · Micro-controller · Communication
protocols · M2M

1 Introduction

New LOWPAN (Low power Wireless Personal Area Network) technologies for
machine to machine or Internet of things (IoT) communication are rapidly gain-
ing popularity [1]. Battery operation from small coin cell batteries is now available. On the other hand this brings limitation for data transmission. SigFox
radios are limited to 140 messages with maximum user data size of 12 bytes [2].
This requires to pay attention to data format [3]. LoRa has quite a low data rate
i.e. up to about 27 Kbps [4]. It can be used for applications requiring LoRaWAN
network. This technology is not ideal solution for real time applications requiring
low latency.

Another area of interest where data should be as small as possible are smart
home systems. Transferring large amount of data requires faster technologies.
Those might be more expensive and less reliable than industry standard solutions. It is also necessary to focus on data transfer capabilities for such reasons.
In this article we focus on the communication standards and discuss the solution which is used in the Smart home system - HAuSy (abbreviation for Home
AUtomation SYstem).

© Springer International Publishing AG 2017
N.T. Nguyen et al. (Eds.): ICCCI 2017, Part II, LNAI 10449, pp. 335–344, 2017.
DOI: 10.1007/978-3-319-67077-5_32

Traditional commercially available solutions for smart home systems are realized mostly as two layer architecture [5]. Those can be some sort of control unit, realized as custom circuit or derived from PC. Control unit communicates with wired or wireless modules with sensors, actuators or control elements. Control unit makes action based on module inputs or some user defined rules. Possibilities of rules customization vary from manufacturer to manufacturer. Two-layer approach is the simple from installation point of view but introduces huge security risks. The whole building might become uninhabitable or even dangerous when control unit has malfunction, even though the approaches for dealing with those issued are discussed in [6].

The paper is divided into 5 sections. In the following section communication protocols are discussed. In the third section HAuSy framework and its architecture are introduced. Own communication protocol design described there. In the fourth section test results are introduced and discussed. Last section is conclusion of this research.

2   Communication Protocols

One of the main challenges in IoT is to ensure communication between devices. In this case we speak about Machine to Machine Communication (M2M). This communication can be provided by metallic wires or various wireless technologies including mobile networks. Interesting survey on these technologies can be found in [4].

The traditional communication protocols such as RESTFUL services, MQTT or Web-sockets are mostly unsuitable due to extremely high communication overhead and should be replaced by other methods which can reduce overhead into a minimal level and help to reduce power consumption during the data transmission. In recent years, there were developed and implemented many communication protocols which can help to reduce data traffic in IoT like a CoAP (Constrained Application Protocol), XMPP (Extensible Messaging and Presence Protocol) or DDS (Document Delivery Service). Using this protocol message overhead could be reduced up to hundred percent in comparison with traditional HTTP REST protocols [7]. On the other hand these protocols could not be implemented into 8-bit micro-controllers with very limited hardware and computation capabilities. Due to this limitation we have decided to implement our own lightweight transmission protocol which brings low overhead and data reliability on between these simple devices. This protocol is implemented for communication in smart home system framework HAuSy.

3   HAuSy Framework

HAuSy is designed with different approach than traditional commercial solutions. Three-layer architecture is used, containing server, subsystems and nodes. Server is a generic replacement for control unit and Linux operating system is used. Java application embedded in web container is running there. Its goal is to provide API for UI applications, evaluate user defined rules, maintain security and fetch or send data to subsystems. Architecture of this approach is shown in Fig. 1.

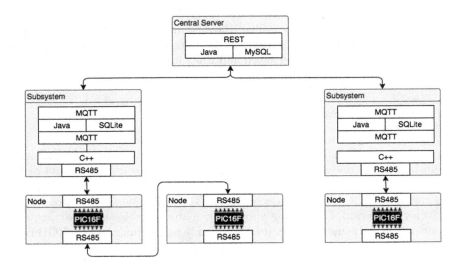

Fig. 1. Architecture of HAuSy solution

Subsystem is the major extension over traditional architectures. Single server present in home or cloud service communicates with multiple subsystem. These are installed in every major room in house and communicate with hardware endpoints called nodes. Subsystem seems to be redundant in correct mode, but when server has malfunction, subsystem is able to work with directly connected nodes. Subsystem is implemented on Raspberry Pi single board computers with option to easily migrate onto any available alternative. Two applications reside on them. One written in Java shares large amount of code base with server. It acts as gateway between C++ application and server as default. C++ application called controller then directly communicates with nodes. Whole architecture is described in detail in [8].

Nodes are the lowest, replacing modules from the two-layered solutions. Every node must contain micro-controller which works with peripherals. Subsystem and nodes communicate with full duplex industry standard RS485 [9] wired bus or with wireless Bluetooth Low Energy. Mission critical components as door locks, lights, switches should be connected with wires due to its better reliability. The rest of this paper only deals with wired nodes, wireless counterparts are described in [10]. Raspberry Pi uses internal UART with RS485 drivers. Nodes use identical drivers and awaits for commands from the subsystem. Micro-controller chosen for nodes implementation is 8 bit RISC from Microchip, namely 16F690 or any better pin compatible successors. ANSI C compiled with MicroC proprietary compiler was selected for their firmware development.

RS485 use bus topology where cable is routed from the subsystem to the first node, then from the first node to the second one and etc. Therefore the node shares one receiving channel from the subsystem where all of them get commands. Responses from all nodes into subsystem are send to the second channel. The node cannot send any data by itself because of these limitations. Data collision on bus might occur. The whole communication protocol is implemented as request-reply, where subsystem sends command to the specific node. The node has its unique identifier saved and when it is matched with that present in command, the response is send.

3.1 Node Properties

Every node in a system is defined with its type and channel count. The type specifies data handling, access type, data width and data count. Manipulation means whether the node is for reading (contains sensor or control input) or for writing (contains some sort of actuator). Access type is relevant for read-only nodes and specifies speed of data acquisition. That means as fast as possible or in a given time period. Data count shows how many variables node can acquire. The button has only one variable, its state. Data count equals to one. RGB light has conversely three variables, light intensity for each diode. Finally, data width shows how many bits carry information. The button has only two states, pressed and not pressed. Data width is equal to one. For example RGB light can have light intensity in percents ranging from 0 to 100 on each diode. 7 bits is therefore required to set its light state.

Communication protocol contains a message for dynamic address allocation, identification and data handling. Most frequent messages are:

REQUEST: The subsystem sends command to the specific node to send its current state. Own data contains a message identifier and two-byte address of node.

GET: The node responses to REQUEST message. The message contains ID for backwards verification and current data state.

SET: The subsystem sends new data state to the node. Output pins or some digital sensor state are changed in micro-controller. GET message is send as a response to subsystem for state verification.

3.2 Designed Communication Protocol

Particular protocol that is the using RS485 bus has been designed as a communication protocol between nodes and subsystem. RS485 only specifies voltage levels, cable structure and connection topology. Format of transmitted data depends on intended usage and it is possible to use some of industry standard formats or implement own one's. ProfiBUS and MODBUS are widely used in industry automation with mostly different requirements than HAuSy. Both protocol contains large amount of redundant and unneeded data which could slow down request-reply process.

Particular binary communication protocol with variable data length has been designed due to this reasons. Data frame packet structure and its requirements is shown on Fig. 2.

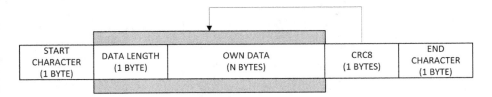

Fig. 2. Structure of data packet

First byte contains start number value 35 corresponding to ASCII char #. Following byte show information about own data length. Minimal permissible data length is one byte. Particular data follows next. Second to last is CRC8 checksum calculated from particular data with data length information included. The last byte contains end number value 36 corresponding to ASCII char $. Data packet is evaluated as valid only when all pieces of information fit. Further information about protocol design can be found in [11].

3.3   Transmission Methods

Nodes are using transmission rate of 230400 bauds per second. It is defined as standard 8 data bits, one start bit, one stop bit and no parity. 10 bits in total is required to send one data byte. Transmission time is reciprocal value of baud-rate. Conversion to microseconds and upper rounding tells that a single bit transmit time is approx 43 µs. Transmit time for the whole baud is therefore 430 µs. Precise external oscillator with frequency 11,0592 MHz is also used. This frequency ensures minimal deviance between real and required baud-rate. PIC16F690 need four oscillator ticks for instruction execution. Usable clocking frequency of micro-controller is therefore 2,7648 MHz. Time of the single instruction execution then takes approx 362 nS. PIC16F690 contains 35 instructions in total and their execution can take either one or two processor cycles. If all instructions in assembly produced from the C source code take two cycles (pessimistic scenario), it could be stated that node can evaluate at least 593 instructions instead of transmitting single byte of data over UART.

This statement means that it is required to put some thought about particular data content in communication protocol. The question is which way to send the new states for nodes in SET message or acquire states in GET messages to be more precise. There are several ways how to send data. The first one and far most simple is to take all node states as 32 bit integer values. This approach simplifies data handling on the subsystem. PIC MCU has 10 bit ADC inputs. Figure 3 shows how data from two-channel AD node are sent with the first approach.

Inefficiency of this scenario is obvious at first sight. The UART interface works with byte, therefore integer must be send as four bytes. Subsystem receives

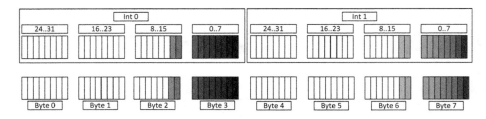

Fig. 3. Simplest approach to data transmission

those bytes and assembles integer back using bit shifts or helper structures. Only 10 bits are carrying information and whole 22 bits are send without any purpose. Transmission use whole bytes and thus two bytes are redundant. Data are sent as a big-endian where the first byte arrived is the most significant (MSB) and last is the least significant (LSB).

The second approach offers quickly implementable improvement and that is only to send necessary number of bytes. Algorithm of this procedure can be named as "byte by byte" approach (see Fig. 4). Data are send as a little endian so byte that arrives first contains the least significant bytes. They are also ordered as LSB inside. This approach eliminates two excessive bytes from the previous scenario.

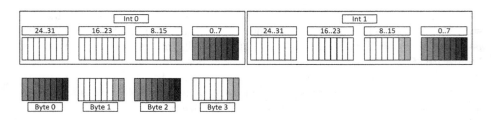

Fig. 4. Byte by byte transmission

The last presented approach is to use bit wise operators supported in any ANSI C/C++ compiler. It is possible to take any bit in byte and transform it into other variable with the help of binary expressions. Output of these transformations is shown in Fig. 5 and could be referred as "bit by bit" algorithm.

Most obvious differences between all of approaches emerge when the node contains single bit periphery or input (relay, button, switch, etc.). Information is carried by single bit only. Lets suppose node with eight channels. The simplest approach means transmission of 32 bytes, byte by byte needs to send 8 bytes and only single byte is required to send with bit by bit algorithm. But bit by bit algorithm requires many bit shift operations and their speed could be influenced by quality of compiler used. This requires testing before the final choice of algorithm can be made.

Fig. 5. Bit by bit transmission

4   Testing of Proposed Solution

Implementing test on a real existing hardware is the most obvious choice because testing the system as whole solution is necessary. Micro-controllers debugger is not able to monitor calculations in real time. Measured time could be distorted. The subsystem and node is needed to get the most accurate data. C++ application sharing codebase with HAuSy framework was created for the subsystem. It contains logic for node data acquiring in infinite loop. No data are sent to upper layers but processed messages count per second is stored. Therefore these values may be interpreted as messaging frequency with Hertz as unit. Information about connected node is also statically typed into applications. Application contains two selectable algorithms for data acquisition. The first one converts data into 32 bit integer bit after bit. The second one does it byte after byte. Therefore second algorithm can only process data from nodes with data width divisible by 8.

4.1   Types of Tested Nodes

Following nodes were implemented as micro-controllers firmware, all of them with eight channels.

- **Simulated digital input** The node sends simulated button state as one byte with all eight button state or as eight bytes with single data carrying bit.
- **Real digital input** Acts the same as the previous type but state is not simulated. This should show usage with real input connected to the nodes.
- **Simulated analog input** This type acts as eight 10 bit ADC inputs. Incremented dummy value is sent instead of sampling AD input. Values are sent either bit by bit (8 bytes total) or byte by byte (16 bytes total).
- **Real analog input** This node acts exactly the same as the previous one, but voltage on AD input is sent instead of dummy value.

4.2   Results

The subsystem uses GNU/Linux operating system which is not real time OS. Multiple system daemons such as systemd, cron, ssh server are running concurrently with the test application. Therefore the number of measured messages per second is not a constant but moves in certain ranges. Bit by bit algorithm allows to test all possible node implementations but byte by byte only those which send data aligned to whole bytes. Table 1 shows recorded results for all possible subsystem algorithm and node combination.

Table 1. Results of tests

Node firmware Subsystem algorithm	Bit by bit (Hz)	Byte by byte (Hz)
Digital input, simulated, 1 bit	942–946	NA
Digital input, real, 1 bit	937–940	NA
Digital input, simulated, 8 bit	605–611	607–617
Digital input, real, 8 bit	603–610	607–615
Analog input, simulated, 10 bit	513–524	NA
Analog input, real, 10 bit	400–406	NA
Analog input, simulated, 16 bit	463–467	463–465
Analog input, real, 16 bit	380–383	381–384

There are some results observable from the Table 1. Differences tend to be in a level of measurement error when comparing algorithms implemented on the subsystem (see Fig. 6a and b). If node simulates digital input and data state is send as a whole byte (otherwise it is not possible to directly compare subsystem's algorithms), bit by bit algorithm is capable of processing from 605 to 611 messages per second. Byte by byte algorithm is capable of processing from 611 to 617 messages per second. Very similar scenario happens when the nodes are loaded with firmware simulating analog inputs. Lower limit is exactly same at 463 messages per second. Upper limit is only three messages per second higher at bit by bit algorithm. These results could indicate that there is no significant difference and it does not matter which algorithm is chosen. But as it has been already told, byte by byte algorithm handles whole bytes only. When 1 bit and 8 bit digital node is compared (see Fig. 6c) or 10 bit and 16 bit analog node, (see Fig. 6d) results are very significant.

Bit by bit algorithm is capable of processing from 942 to 946 messages per second when node reads eight digital input. This range is more than 300 messages more against whole byte alignment (605 to 611 range). Data is simulated to minimize latency caused by micro-controller. Similar results occur when analog data are simulated. Bit by bit parses between 513 and 524 messages but two bytes alignment only from 463 to 467 messages per second. Gain of bit by bit is now not so significant. But in time critical application still offers more performance. There is an unambiguous conclusion telling that bit by bit algorithm is better approach of implementing communication protocol in performance constrained micro-controllers. If test were showing opposite results, it would probably indicate that C compiler used is producing very poor assembly code. Selecting other compiler or whole different micro-controllers should be the next step in such case.

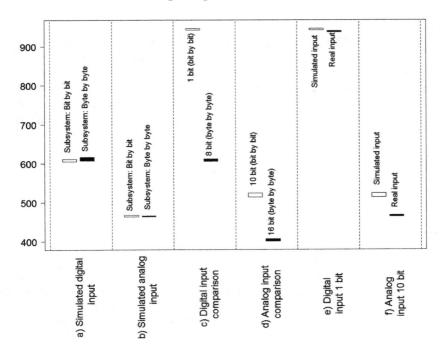

Fig. 6. Byte by byte transmission

The final step is to compare influence of reading actual digital or analog values as shown in the Fig. 6e and f. Charts contain only comparison when bit by bit algorithm is used. Messages minimum range drop from 942 to 937 and upper bound drops to 940 from 946 when simulated digital input is replaced with pin state reading. This indicates that pin reading is implemented very effective in MikroC. Sampling analog inputs instead of using dummy variables shows more significant drop from maximum of 524 messages to 406 messages. The main reason is that conversion of a single bit takes $1.6\,\mu s$. Sampling whole 10 bit input therefore takes $16\,\mu s$. Overhead caused by C language abstraction tends to be also little larger. But those processing rates are still higher in comparison of byte by byte approach.

5   Conclusion

This paper presents the framework for smart home automation and sensor networks called HAuSy. It's more resistant to breakdown due to three layer design. RS485 bus is used for communication between the subsystem and the nodes. Because the transmission rate is set only to 230400 bauds per second, it is necessary to design very effective approach for data transmission. New wireless IoT gateways like LoRa, SigFox or even Bluetooth Low Energy in most powerful mode have lower baudrates and data size limitations. Different approaches for sending data states must be tested. Three algorithms are presented. Naive approach with huge data redundancy, byte by byte which sends only necessary bytes needed and finally bit by bit which is sending only bits with the information needed. Results shown that bit by bit algorithm is much more effective even when it conducts large amount of bit shift operations needed to take data carrying information from byte variables. Because communication protocol is using only least necessary amount of other information needed to transmit data between nodes and subsystem, it can be said that it is the most effective way of transmission.

Acknowledgment. The support of Czech Science Foundation project GACR #15-11724S is gratefully acknowledged.

References

1. Botta, A., de Donato, W., Persico, V., Pescapé, A.: Integration of cloud computing and internet of things: a survey. Future Gener. Comput. Syst. **56**, 684–700 (2016)
2. Nolan, K.E., Guibene, W., Kelly, M.Y.: An evaluation of low power wide area network technologies for the internet of things. In: 2016 International Wireless Communications and Mobile Computing Conference (IWCMC) (2016)
3. Miorandi, D., Sicari, S., De Pellegrini, F., Chlamtac, I.: Internet of things: vision, applications and research challenges. Ad Hoc Netw. **10**(7), 1497–1516 (2012)
4. Ali, A., Shah, G.A., Farooq, M.O., Ghani, U.: Technologies and challenges in developing machine-to-machine applications: a survey. J. Netw. Comput. Appl. **83**, 124–139 (2017)
5. De Silva, L.C., Morikawa, C., Petra, I.M.: State of the art of smart homes. Eng. Appl. Artif. Intell. **25**(7), 1313–1321 (2012)
6. Naidu, S., Zafiriou, E., McAvoy, T.: Use of neural networks for sensor failure detection in a control system. IEEE Control Syst. Mag. **10**(3), 49–55 (1990)
7. Kuladinithi, K., Bergmann, O., Pötsch, T., Becker, M., Görg, C.: Implementation of CoAP and its application in transport logistics. In: Proceedings of IP+ SN, Chicago, IL, USA (2011)
8. Horálek, J., Matyska, J., Stepan, J., Vancl, M., Cimler, R., Soběslav, V.: Lower layers of a cloud driven smart home system. In: Barbucha, D., Nguyen, N.T., Batubara, J. (eds.) New Trends in Intelligent Information and Database Systems. SCI, vol. 598, pp. 219–228. Springer, Cham (2015). doi:10.1007/978-3-319-16211-9_23
9. Bayindir, R., Irmak, E., Colak, I., Bektas, A.: Development of a real time energy monitoring platform. Int. J. Electr. Power Energy Syst. **33**(1), 137–146 (2011)
10. Stepan, J., Cimler, R., Matyska, J.: Using IPv6 over Bluetooth Low Energy on Low Costs Platforms (2016). https://adibum.com/
11. Stepan, J., Matyska, J., Cimler, R., Horalek, J.: Low level communication protocol and hardware for wired sensor networks. J. Telecommun. Electron. Comput. Eng. (JTEC) **9**(2–4), 53–57 (2017)

Wildlife Presence Detection Using the Affordable Hardware Solution and an IR Movement Detector

Jan Stepan, Matej Danicek, Richard Cimler[(✉)], Jan Matyska,
and Ondrej Krejcar

Faculty of Informatics and Management, Center for Basic and Applied Research
(CBAR), University of Hradec Kralove, Hradec Kralove, Czech Republic
{jan.stepan.3,matej.danicek,richard.cimler,jan.matyska,
ondrej.krejcar}@uhk.cz

Abstract. Although there are many wildlife cameras in various price ranges available on the market, these may be too expensive for certain uses and may provide functionality that is not necessarily needed. This paper concentrates on guidelines for building a low cost device capable of storing times of movement detections and exposing simple API. This API can be accessed via Android counterpart application, that reads, stores and presents the data. Such technology may become useful for hunters trying to determine the times when they are most likely to encounter their prey at a certain location or similar situations when movement intensity based on time is the information of interest.

Keywords: Wildlife · Detection · Low-cost · Camera · Infrared · Movement · Android · BLE · Nordic

1 Introduction

There are many security/wildlife cameras available on the market varying in functionality, connectivity, quality, robustness and most importantly price. These are being frequently used for either amateur/semi-amateur hobby purposes or professional scientific animal herd tracking and count assessment [3] as well as a form of endangered species poacher protection [5]. Functionality of these devices varies from a simple camera capturing a picture upon movement detection to a sophisticated device equipped with full-HD camera, GSM module for virtually real-time image or video transmission and remote control capabilities. These devices can be quite expensive and although the price is warranted for many utilizations, there are many cases for which a much simpler and (most importantly) cheaper solution suffices plentifully.

One of the aforementioned utilizations is determining the times with highest probability of wildlife presence in a certain area, be it for the purposes of hunting, manual photo-shooting or simply observing of the animals. Another usage shall

© Springer International Publishing AG 2017
N.T. Nguyen et al. (Eds.): ICCCI 2017, Part II, LNAI 10449, pp. 345–354, 2017.
DOI: 10.1007/978-3-319-67077-5_33

be determining the most efficient times for cleaning publicly accessed areas. Similar scenario is for stars and elevator usage statistics.

The main features of presented device therefore include the price of the needed hardware equipment. Another feature is the ease of use, be it from the hardware point of view (e.g. using wireless technology to transfer data, so that the operator does not have to have any additional equipment such as cabling) or software point of view (such as having as simple and clear user interface as possible and options to store data in formats for easy parsing and analysis).

Another question in this research is the battery life. How long will the designed device be capable of operating with battery or a power bank as its power source is quite an important issue, since the intended usage for long term monitoring in somewhat remote outdoor locations prohibits frequent battery changing/recharging.

Aim of this research is to create a solution for a long term outdoor usage so the size and general layout must be taken into account. One of the problems in this area might prove to be having wireless radio turned on so power drain might be high. The only possible way seems to be a manual switch but considering device's possible not easily accessible location above ground, the switch might need to be on an extender cord which would pose another elements-proofing challenge.

2   Existing Solutions

As the previous chapter mentions, there are many commercially available wildlife (or intruder, for that matter) camera traps. This chapter provides an overview of the available commercial solutions including the ones at the lower and higher end of the price range and one simplistic and similar implementation using the Arduino platform.

2.1   ScoutGuard MG983G-30M

This camera trap is from the high price end of the spectrum with prices around $270, when ordered from an overseas retailer (and around $390 from a domestic distributor or directly from the manufacturer). This device is equipped with passive IR sensor for triggering photo or video recording with resolution up to 30 MPx. It also contains GSM SIM module for external triggers over Android application and 433 MHz interface for additional sensors. This device is an example of fully equipped professional photo trap and the price corresponds to the intended use; however, it is too expensive for the goals this project as aiming to fulfill.

2.2   König SAS-DVRODR11

One of the cheapest commercially available camera traps can be bought for around $100 from domestic retailers. It provides far fewer functions that the aforementioned MG983G, with the main difference being the absence of the GSM communications module and a limited photo/video resolution, namely 10 Mpx.

2.3   DYI Hobby Project

There are many available electronics hobby projects with similar purpose. For example this project uses a passive IR movement detector and an SD card module or shield for storing the data. Although the project builds on similar principles and approaches, there are several limitations that prevent it from being used in the intended way. Namely, the timestamps are saved in humanly readable format which would require parsing in order for the data to be presentable in a different way (e.g. graphs or charts). This, on its own, does not constitute a big of an obstacle; however, the timestamps are saved as relative values from the Arduino boot time. Another limitation is the absence of a data transfer interface other that the serial line (which would require difficult in-field cable manipulation) and the SD card (which would require in-field transferring of the card to the user's mobile device). Lastly, none of found solutions is not optimized for running over battery. They are using multiple modules with own voltage converters and their power consumption is tens of milliamperes. This means that even modern Li-Ion battery shall be discharged in matter of hours.

2.4   Conclusion of Related Projects

In general, although there is a variety of devices available, none of them provide the ideal solution for the intended purpose. The commercial devices could, in theory, be used with some additional parsing software (e.g. determining times of movement detection from filenames or their meta-data); however, their price is too high and a custom low-cost device is inevitable for the purposes of determining the most frequent times of animal presence. Summary of properties of existing devices and proposed solutions are summarized int Table 1.

Table 1. Overview of existing solutions

	Features	Price range	Battery life
MG983G-30M	FullHD camera, external triggers, remote oversight	270–390 USD	Months
SAS-DVRODR11	HD camera, external memory expansion	90–110 USD	Months
DYI projects	Various range of features	10–20 USD	1–30 h
Proposed solution	Only time stamps of movement	10–20 USD	At least a week

3   Proposed Solution

This paper focuses on designing and prototyping a device that uses passive infrared movement detection sensor to determine whether a warm-blooded animal is present. For cost and ease of development purposes an Nordic Semiconductor NRF52 development kit was chosen. This micro-controller board is based on the nRF52832, working at 16 Hz clock frequency and providing multiple digital and analog input/output pins. Most important thing is, that provides integrate Bluetooth Low Energy radio accessed over Nordic's SDK. It is successor to previous nRF58122 model, one of the lowest power requiring solution [2]. This setup provides plenty enough resources for the intended purpose, but since this model has only and 512 KBs of flash memory (which should not be frequently rewritten due to lifetime concerns), another type of storing the time-stamp/measurement values must be used. The cheapest and easiest way is using an SD card and storing the values on it.

As for the technology used for motion sensing, there were several candidates. The passive infrared detector was chosen because it operates by measuring the differential changes in the infrared (invisible) spectrum without actively transmitting any energy and is therefore the most energy-efficient and, compared to other technologies (microwave or ultrasonic which make use of the Doppler effect), the most impervious to false detections by inanimate objects with the same temperature as the environment (e.g. leaves moving in the wind).

The Bluetooth Low Energy (BLE) communication was selected because of its simplicity and ease of use. Another possibility would be using Wi-Fi or another protocol in the 2.4 GHz frequency range (such as ZigBee); however, establishing a BLE connection is much easier, much more straight-forward and user friendly than setting up a Wi-Fi connection, not to mention other protocols/technologies that are not generally implemented in Android (or other mobile) devices and very low power requirement on device's side [6].

Another concern is acquiring the current time in the NRF device. Best approach is to use built in real-time module build in NRF. It uses 32 kHz oscillator and adds only 1 μA to update time value every second. To achieve current real world time, initial configuration over Android must be performed before device deployment.

The Android application part of the solution is quite straight forward. Some way of storing the list of devices (and information about them, such as a user-given name, the MAC address of the Bluetooth radio and coordinates) as well as the timestamps from each of the devices is needed. Although a simple SQLite database provided "out-of-the-box" by Android API would have sufficed plentifully, in order to satisfy research and curiosity purposes an alternative data-storage library Realm was used.

Since the geographical coordinates of each Arduino device can be stored in the database, the Google Maps APIs can be used for initial position selection when saving a new sensor device (with the aid of the Android device's GPS module) as well as for showing an overview map with all the sensors' locations and names.

In order for the users to be able to easily determine the most probable time of animal encounter a chart needs to be available. Simply showing the times of movement detection on a time-line is not sufficient since it does not provide a clear view of the data. A graph containing cumulative movement detection counts on a per minute, per hour and per day basis needs to be shown in order for the trends to be clearly visible on the desired time level.

4 Implementation

The first step in prototyping and developing the project was choosing the hardware and implementing source code. As previously mentioned, NRF chip has been chosen as the main device because of its built in Bluetooth Low Energy radio, feature rich peripherals and a lot of available memory. Diagram of implemented solution is shown in Fig. 1.

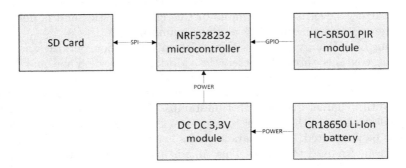

Fig. 1. Blocks of hardware architecture

4.1 Passive InfraRed Sensor

The HC-SR501 PIR motion sensor has been used. This sensor works with input voltage levels between 5 and 20 V because it contains internal power converter to 3.3 V. STM32 micro-controller required 3.3 V as well, therefore power converter was bypassed. PIR module requires only approximately 65 μA to operate.

This particular PIR provides "digital" values on the output (in the sense that either a 0 V level is outputted or 3.3 V level is outputted indicating a "high value") in two distinct modes: 1. Single trigger – low level on the output until a movement is detected, then high level for a customizable-length period (delay) 2. Repeat trigger – high level on the output is kept until the object leaves the sensor range. The single trigger mode is used in this project since the information about the length of the movement detection is not of particular interest. The delay is set to the lowest possible value, which, according to the data-sheet, should be 0.3 of a second, but in reality is approximately 2 or 3 s. Output pin from PIR sensor is connected directly into GPIO pin of NRF. Instead of reading pin state in loop, interrupt on rising edge is called.

The PIR is capable of operating in the temperatures between −15 and +70 °C with 110 ° angle, up to the distance of 7 m. This distance limit may prove insufficient in the real-world outdoor use and another PIR with better performance may need to be used.

4.2   Bluetooth Low Energy Service

Bluetooth Low Energy works in same bandwidth as older Bluetooth technology (now often referred as Bluetooth Classic), but it's architecture and protocols are all different [1,4]. Part of previous Bluetooth was RFCOMM specification that was allowing to abstract many layers and two radios were acting as a wired serial connection. New approach is used in BLE using Services, Characteristics and Roles. Role may be peripheral or central. Difference is that peripheral is advertising its capabilities and central scans for them and establishes a connection. Service can be for example standardized Battery Service from Bluetooth Special Interest Group. Its purpose is to provide battery level information of periphery running on battery. Battery Characteristic is provided, with notify handle. Battery Level is only variable in this characteristics and is defined as 8 bit integer with minimum of 0% and maximum of 100%.

Custom Low Energy profile with service called Motion Service was implemented. It contains three characteristics, Motion Request, Time and Motion Notify. After connection is established, central will write into Motion Request to initiate sending times of movement from periphery. Device then starts sending Motion Notify characteristics with timestamps of movements from the newest to the oldest along with counter to keep track of how many records are still needed to be transfered. Time characteristics is used when device is deployed to set time from Android to device. This time is afterwards incremented by NRF RTC module every second.

4.3   File and Time-Stamp Management

Part of Nordic SDK is library for file management on SD card. Card is able to work in two modes, SPI and SDIO. SPI is serial interface and is much slower, but speed is still sufficient for saving timestamps into file. PIR sensor is connected to GPIO pin of NRF and its set to wake up device on rising edge which occurs when sensor is triggered. Following steps are performed.

1. NRF is interrupted on movement
2. Time of event is added into in memory array
3. Count of events in memory is incremented
4. If number of records is larger than 32
   (a) Append records to RECORDS file on SD card
   (b) Overwrite records count in COUNT file

4.4   Android Application

Proposed system consists of HW solution as well as smart phone application which enables to operate with the detection system.

The MainActivity of the application consist of list of managed devices with the possibility of opening a chart containing the data from the given sensor by clicking on the device in the list. The title bar contains buttons for adding a new device to manage, connecting to a device and showing an overview map with markers for all the sensors.

Connecting to the Bluetooth device, sending request for data and receiving the values is done in a separate thread so the GUI does not become unresponsive.

For storing information about the managed devices and the time-stamp values, the Realm database was used as an alternative to the provided SQLite database. Each time-stamp is saved to Realm as an instance of a Timestamp wrapper class. Another approach could be storing the timestamps as an array of bytes (for example in a file) and although this approach may be less resource demanding, it was not taken due to more complicated future management (adding or removing values).

For showing the data in a chart, the MPAndroidChart library released under Apache 2.0 license was used. Although it provides many types of zoom-able and scalable charts, it does not provide a way of showing cumulative counts in minute/hour/day intervals. This cumulative calculation is done manually in the Histogram class with the use of gnu.trove lists and maps, that allow the use of primitive data types.

5   Testing and Assessment

Detection system has been tested indoors as well as outdoors. The testing proved that the system is functional and is possible to use it in the intended way. The price of the solution is fraction of the commercial solution price.

5.1   Indoors Testing

At first, the implementation was tested in a frequented indoors environment over the period of several days. Test criteria included the reliability of the Bluetooth connection establishing and data transfer, the reliability of individual movement detection and ease of determining trends in the movement detection times.

The movement detection proved to be reliable up to the declared range of 7 m with the correct positioning of the sensor being the key to successful movement detection at the edges of the detection zone. Movement data obtained during the several days of testing were plotted using the Android application with clearly visible times with higher amount of movement detections (during the day) as well as periods with no or minimal movement activity (during the night). See Fig. 2.

5.2   Outdoor Field Testing

The outdoor testing location was selected based of its proximity to water and animal trails presence (indicating movement of smaller animals such as foxes or cats).

During the tests 3000 mAh Li-Ion battery was used as a power source and even though the temperatures dropped to around $-14\,^{\circ}$C (and $-20\,^{\circ}$C respectively) the device was powered and fully functioning during the whole night.

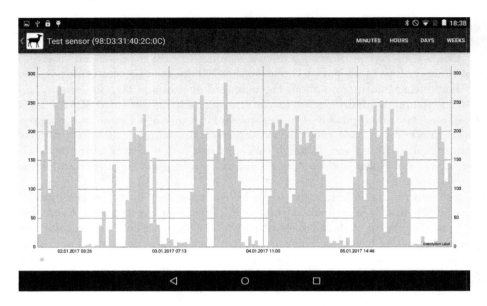

Fig. 2. Smart phone application. Detections per day.

The outdoor testing has shown that the device is capable of being used in the intended way and that the number of movement detections corresponds to the amount and positioning of animal footprints at the location, indicating the detection is indeed reliable. It has also shown that in a low movement environment a simple list of movement detection times may be informative to the user.

5.3   Battery Life

NRF chip is using only $5\,\mu A$ in IDLE mode. Advertising is set to repeat every 10 s and take 20 ms. Consumption rises to $240\,\mu A$ during this period. Unfortunately, devkit itself drains over than 20 mA, makes it very power demanding. Minimal evaluation kit from Raytac is used for real testing instead. PIR sensor is draining constant current of $65\,\mu A$ and SD card in IDLE takes $163\,\mu A$ and during writes $2240\,\mu A$. DC-DC circuit drains fixed current of $550\,\mu A$ and its efficiency at lower current drains is approximately 70%.

All of above means, that nRF58233 consumes $9\,\mu A$ in average. Total current without power DC-DC circuit is therefore $237\,\mu A$ and total drain from battery averages at $889\,\mu A$. Li-Ion battery used for powering circuit has capacity of 3000 mAH. It is not safe to discharge battery to less than 30% of its capacity, therefore life expectancy is over 2350 h (approx. 97 days).

5.4 Price of the Components

All prices are shown when only single piece is bought, discounts in high volumes tends to be more than 70%. Price of Raytac kit is 10USD, PIR sensor costs less than one dollar as well breakout board for SD card and DC-DC converter. SD card prices depends on capacity and starts from 4USD and same is applied for battery.

Difference between price and battery life time of existing solutions and proposed solution is show in Fig. 3.

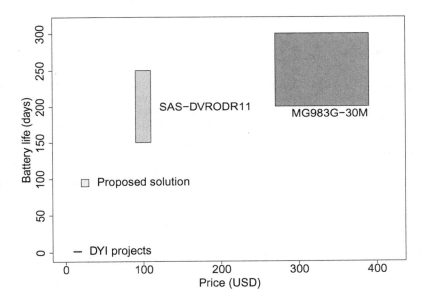

Fig. 3. Comparison of price and battery life

6 Summary and Future Advancements

Future development of this device may consist of performing more advanced power optimization on chip and tuning Bluetooth connection and advertising parameters. Second improvement might be brought by using custom Flash or eMMC storage instead of SD card and designing better custom DC-DC circuit. Third improvement is using small solar panel to charge up the battery. Last area of focus might be switching roles of device and Android application. Device might act as central and scans for phone nearby for a few seconds every few minutes. Influence to power consumption should be measured. Battery level information and Bluetooth service might be also added in the future. These improvements should move battery time over a year.

In the Android application, the ability to import the values directly from an SD card as well as some form of backup or data transference between multiple Android device may prove useful future functionality.

Another suggested functionality was providing the directional information of the movement using two or more movement detectors each pointing at a different area and the time difference between movement detections. This suggestion was dismissed after discussion in first development stage, because it increases difficulty of development and more sophisticated algorithms must be applied.

Acknowledgment. The support of Czech Science Foundation GACR project #15-11724S is gratefully acknowledged.

References

1. Al Kalaa, M.O., Balid, W., Bitar, N., Refai, H.H.: Evaluating Bluetooth low energy in realistic wireless environments. In: 2016 IEEE Wireless Communications and Networking Conference (WCNC), pp. 1–6. IEEE (2016)
2. Buccolini, L., Pierleoni, P., Conti, M.: Design and energetic analysis of a self-powered Bluetooth low energy speed sensor. In: 2016 IEEE 16th International Conference on Environment and Electrical Engineering (EEEIC) (2016)
3. Burton, A.C., Neilson, E., Moreira, D., Ladle, A., Steenweg, R., Fisher, J.T., Bayne, E., Boutin, S.: Review: wildlife camera trapping: a review and recommendations for linking surveys to ecological processes. J. Appl. Ecol. **52**(3), 675–685 (2015)
4. Gomez, C., Oller, J., Paradells, J.: Overview and evaluation of Bluetooth low energy: an emerging low-power wireless technology. Sensors **12**(9), 11734–11753 (2012)
5. Hossain, A.N.M., Barlow, A., Barlow, C.G., Lynam, A.J., Chakma, S., Savini, T.: Assessing the efficacy of camera trapping as a tool for increasing detection rates of wildlife crime in tropical protected areas. Biol. Conserv. **201**, 314–319 (2016)
6. Mackensen, E., Lai, M., Wendt, T.M.: Performance analysis of an Bluetooth low energy sensor system. In: 2012 IEEE 1st International Symposium on Wireless Systems (IDAACS-SWS) (2012)

Text Processing and Information Retrieval

Word Embeddings Versus LDA for Topic Assignment in Documents

Joanna Jędrzejowicz and Magdalena Zakrzewska[✉]

Faculty of Mathematics, Physics and Informatics, Institute of Informatics,
University of Gdansk, 80-308 Gdansk, Poland
{joanna.jedrzejowicz,magdalena.zakrzewska}@inf.ug.edu.pl

Abstract. Topic assignment for a corpus of documents is a task of natural language processing (NLP). One of the noted and well studied methods is Latent Dirichlet Allocation (LDA) where statistical methods are applied. On the other hand applying deep-learning paradigm proved useful for many NLP tasks such as classification [3], sentiment analysis [8], text summarization [11]. This paper compares the results of LDA method and application of representations provided by Word2Vec [5] which makes use of deep learning paradigm.

Keywords: NLP - topic assignment · Deep learning · LDA

1 Introduction

The Web is a very rich source of information which is quickly expanding and the number of documents is growing very fast. We are in need of better tools enhancing organization, browsing and understanding of huge amounts of data. The manual extraction of themes of scientific articles becomes an infeasible task and an automatic generation of themes is a challenge for machine learning algorithms and text mining in particular. Standard methods based on keywords identification proved useful but have their limitations.

This paper deals with topic modeling which is an automatic classification of a corpus of documents into themes. One of the well examined algorithms in this area is Latent Dirichlet Allocation (LDA). The idea of topic modeling using LDA is based on the following assumption. Though documents contain only words they are generated as mixtures of topics. Topics are latent variables which can be inferred from the distribution of words in documents, and this is the process reverse to documents generation. Topics identification in this case is an iterative process which starts with a random assignment of words in documents to topics. In the following steps of the algorithm these assignments are modified based on the information how prevalent are words across topics, and topics across documents.

Word2Vec [5] is a neural network implementation that allows to learn distributed representations of words. There are two main learning algorithms in Word2Vec: continuous bag of words and continuous skip gram. In our case it

© Springer International Publishing AG 2017
N.T. Nguyen et al. (Eds.): ICCCI 2017, Part II, LNAI 10449, pp. 357–366, 2017.
DOI: 10.1007/978-3-319-67077-5_34

was skip gram model which was trained with Google News dataset in English. We made use of Word2Vec implementation described in [7].

The remainder of this paper is organized as follows. Section 2 describes the methods and algorithms, that is Latent Dirichlet Allocation and the idea of Word2Vec with the application of distributed embeddings. Section 3 deals with computational experiments. Here the corpus of documents is described, the preprocessing performed as the first stage and the results of experiments. Finally, Sect. 4 summarizes conclusions and future work.

2   An Approach to Topic Assignment

2.1   Latent Dirichlet Allocation (LDA)

Topic models [1] are based on the following idea: documents are mixtures of topics and a topic is a probabilistic distribution of words. It assumes probabilistic procedure in which documents are generated. In the first step the number of topics is fixed as well as the distribution of topics over documents. Then each word in the document is generated by first picking a topic and then the word from the topic is chosen.

To reverse this process statistical methods can be used to recover the set of topics on which the documents generation was based. For this purpose the Collapsed Gibbs Sampling for Latent Dirichlet Allocation has been chosen. The Collapsed Gibbs Sampling Algorithm calculates the probability of a word associated with a topic in an iterative process, separately for each word assuming that topic assignment for other words is correct. In what follows let W stand for the set of words, T - for the set of topics, D - for the set of documents, $P(z_i = j | z_{-i}, w_i, d_c, *)$ is the conditional distribution where $z_i = j$ is token i assigned to topic j, z_{-i} is assignment of remaining tokens, w_i is a word that is being processed from document d_c, and $*$ describes remaining information such as words w_{-i} and documents d_{-c}, α is a number and describes Dirichlet parameter for the document-topic distribution, β is a number and describes Dirichlet parameter for the topic-word distribution.

$$p(z_i = j | z_{-i}, w_i, d_c, *) = \frac{C_{w_i j}^{WT} + \beta}{\sum_{w=1}^{W} C_{wj}^{WT} + W\beta} \cdot \frac{C_{d_c j}^{DT} + \alpha}{\sum_{t=1}^{T} C_{d_c t}^{DT} + T\alpha} \tag{1}$$

C^{WT} is a matrix such that $C_{w_i j}^{WT}$ counts how many times the word w_i was assigned to topic j without the current instance w_i and $C_{d_c j}^{DT}$ counts the number of times topic j was assigned to word from document d_c without the current instance w_i.

The Eq. (1) produces unnormalized probability. To calculate the actual probability of assigning word to topic j it must be divided by $|T|$.

The equation can be interpreted as follows:

$-\frac{C_{d_c j}^{DT} + \alpha}{\sum_{t=1}^{T} C_{d_c t}^{DT} + T\alpha}$ - counts the number of word occurrences in all documents to topic with taking Dirichlet parameter α into account. This part of equation describes the assignment of topic j to document d_c.

- $\dfrac{C_{w_ij}^{WT}+\beta}{\sum_{w=1}^{W} C_{wj}^{WT}+W\beta}$ - counts the number of word occurrences in specific document
  to topic with taking Dirichlet parameter β into account. This part of the
  equation describes assignment of a word w_i to topic j.
- parameters α and β were calculated by EM algorithm described by Nallapati
  et al. [6].

Parameter *iter* is the number of algorithm iterations that needs to be big enough
to achieve a stable result.

Algorithm 1. Collapsed Gibbs Sampling for Latent Dirichlet Allocation

---

Input: C - corpus of documents, J - number of topics, *iter* - number of iterations, α -
   Dirichlet topic distribution parameter, β - Dirichlet word distribution parameter,
Output: degree of membership of a word associated with a topic and topic being
   associated with a document
1: Count the the number of occurrences of each word in each document
2: Perform random assignment of word $\rightarrow$ topic and document $\rightarrow$ topic
3: **for** *iter* **do**
4:     **for all** document d_c **do**
5:         **for all** word $w_i \in d_c$ **do**
6:             decrement the degree of membership of word w_i to any topic and degree of
             any topic associated with document d_c
7:             **for all** topic j **do**
8:                 calculate the conditional distribution $P(z_i = j|z_{-i}, w_i, d_c, *)$ as in (1)
9:             **end for**
10:             normalize the probabilities
11:             from the probabilities calculated in step 8: draw topic k using polynomial
             distribution
12:             increase degree of membership for word-topic caused by assigning word w_i
             to topic k
13:             increase degree of membership for topic-document by assigning for docu-
             ment d_c topic k
14:         **end for**
15:     **end for**
16: **end for**

2.2 Distributed Representation Using Word2Vec

Another method of finding topics examined in the paper makes use of distributed representations which try to capture semantics of words by coding each word and its context in a real vector space embedding. They are
expected to be consistent with vector space operations such as sum or difference. Mikolov et al. [5] present an example which assumes that the output for
$v(\text{"}king\text{"}) - v(\text{"}man\text{"}) + v(\text{"}woman\text{"})$ will be close to $v(\text{"}queen\text{"})$.

The library Word2Vec developed by Google Inc allows to make use of distributed representations of words that were learned by a neural network implementation. Word2Vec uses skip-gram model to find representation of words to predict surrounding words in documents. For our tests we used the Word2Vec model [5] shared by Google company that was trained on 3 bilion of words. It was 300-dimensional space model. Then we used fuzzy c-means clustering to calculate the representation of topic-word tie.

Fuzzy c-means clustering is a method which allows one row of data to belong to two or more clusters. The method is based on minimization of the objective function

$$J_m(u,c) = \sum_{i=1}^{N} \sum_{j=1}^{noCl} u_{ij}^m dist(x_i, c_j)$$

where m is a fixed number greater than 1, N is the number of data rows, $noCl$ is the number of clusters, c_j is the center of the j-th cluster and u_{ij} is the degree of membership of the i-th data row x_i in cluster c_j.

To ilustrate how Word2Vec was combined with fuzzy c-means clustering we consider the following example.

Example 1. Let document 1 contain words "computer software year" and document 2 contains words "cell year sport car year". Table 1 gives the results of fuzzy c-means clustering of all the words from both documents if 5 topics (clusters) are assumed.

Table 1. Word topic assignment.

word	topic 1	topic 2	topic 3	topic 4	topic 5	topic assignment
computer	0,7357	0,1373	0,0195	0,0628	0,0447	topic 1
software	0,7013	0,1829	0,0952	0,0112	0,0094	topic 1
cell	0,3771	0,5962	0,0152	0,0026	0,0089	topic 2
year	0,1051	0,2013	0,6123	0,0264	0,0549	topic 3
sport	0,0152	0,0991	0,0262	0,7241	0,1354	topic 4
car	0,4163	0,0146	0,0251	0,0287	0,5153	topic 5

The greatest values (marked in gray) represent word-topic assignment. To assign documents to topics, for each document and each topic the number of words assigned to topic is divided by the number of words in document. The document-topic assignment is shown in Table 2.

Table 2. Document-topic assignment.

Document	Topic 1	Topic 2	Topic 3	Topic 4	Topic 5
Document 1	$\frac{2}{3} \approx 0,6667$	$\frac{0}{3} = 0$	$\frac{1}{3} \approx 0,3333$	$\frac{0}{3} = 0$	$\frac{0}{3} = 0$
Document 2	$\frac{0}{5} = 0$	$\frac{1}{5} = 0,2$	$\frac{2}{5} = 0,4$	$\frac{1}{5} = 0,2$	$\frac{1}{5} = 0,2$

The details of topic assignment using Word2Vec are given in Algorithm 2.

Algorithm 2. Topic assignment

Input: $\mathcal{C}$ - corpus of documents, $noCl$ - number of clusters, W - the set of all words in corpus $\mathcal{C}$, model - Word2Vec learned model,
Output: degree of membership of a topic being tied with a document
1: remove stop words from W
2: for all words in $w \in W$ obtain word vectors from the model
3: use fuzzy c-means algorithm to calculate topics-word assignment
4: calculate document-topic assignment by dividing the number of all words in document d assigned to topic t by the number of all words in document d

---

3 Computational Experiments

3.1 Datasets Used in Experiments

To compare the two approaches of topic modeling we conducted different experiments on publicly available documents corpus [2]. The corpus contains documents in English from different domains. For the experiments two data sets were created:

- the first set contains 15 documents about art, sport, science, traveling and animals, each group contains exactly 3 documents and is further referenced as Doc1,
- the second set is composed of 30 documents dealing with computer science and subjectively the following topics can be distinguished: work environment, software design, management, manipulation and employer. This set is referenced as Doc2.

For the purpose of experiments the following libraries were used: the implementation of LDA algorithm from "Topicmodels" package from R language [10], implementation of Word2Vec from gensim package [7] and implementation of fuzzy c-means from skfuzzy package [9] for python language.

We used several methods to validate the experiments:

- we performed cross validation for each document by calculating perplexity [12],
- we calculated semantic similarity for 50 most frequent words in each topic to check meaningfulness.
- we compared the probabilities of topic-document assignment.

3.2   Preprocessing Stage

The first stage of the algorithm is to prepare the corpus of documents. Often documents contain digits, dates, conjunctions, diacritic signs etc. that can be removed due to its meaninglessness in the process of assignment words to topics. Preprocessing makes use of the following procedures: tokenization, stop-words removal and stemming.

Tokenization divides data stream into tokens, which can be words, n-grams, sentences, etc. Moreover often in this process punctuation marks, brackets, dashes, are removed.

Stop-words removal is held to remove joining words like "and", "to", "or", "with", which have little value in text analytics. Because these words frequently appear in sentences and are meaningless in terms of finding topics they should be removed. Each natural language has its own stop list and [4] allows to make use of the list for English (as well as some other languages).

The purpose of stemming is to change multiple variations of words back into their root form. For example words "presented", "presentation" or "presenting" should all be replaced by the word "present". This process is based on the assumption that all these words will appear in the same topic. For the English language the most used are Porter's stemmer and Lancaster stemmer, both supplied by [4]. In the experiments WordNet Lemmatizer was also used. We used python NLTK library [4] implementations for WordNet Lemmatizer, Porter stemmer and Lancaster stemmer.

For different experiments some or all of the above procedures were applied.

3.3   Results of Experiments

In this section we compare LDA algorithm with Word2Vec algorithm by testing how well they assign topics to documents and words to topics for the above mentioned sets Doc1 and Doc2. The first set consists of 15 English documents about art, sport, science, traveling and animals, each group contains exactly 3 documents. The second set with 30 English documents can be subjectively determined with these as 5 general topics: work environment, software design, management, manipulation and employer.

Selection of Best Stemmer for LDA. Both algorithms require the removal of "stop-words" to get more accurate results. Moreover LDA algorithm requires the application of a stemmer.

Table 3 presents results obtained by LDA algorithm on documents preprocessed by tokenization and stemming by different algorithms. As we can see from (a) not using tokenization results in large amount of words without any meaning in the context of a topic. Even though WordNet Lemmatizer with ease determined the basic form of a word, words with the same meanings were found in topics. Both Porter stemmer and Lancaster stemmer did well with the different forms of the same words. The problem was in deciding what is the basis of each word.

Table 3. Comparison of topics for Doc1 after using: (a) no preprocessing, (b) stop-words removal and WordNet Lemmatizer, (c) stop-words removal and Porter Stemmer, (d) stop-words removal and Lancaster Stemmer.

a) without preprocessing					b) WordNet Lemmatizer				
topic 1	topic 2	topic 3	topic 4	topic 5	topic 1	topic 2	topic 3	topic 4	topic 5
the	the	the	and	the	animal	art	science	transport	sport
are	and	and	for	sports	form	painting	type	travel	game
animals	art	that	the	and	plant	form	human	human	physical
other	arts	which	transport	sport	cell	work	philosophy	people	include
have	has	science	with	such	energy	example	century	aircraft	participant
and	its	was	travel	are	phylum	style	modern	country	including
c) Porter Stemmer					d) Lancaster Stemmer				
topic 1	topic 2	topic 3	topic 4	topic 5	topic 1	topic 2	topic 3	topic 4	topic 5
anim	art	scienc	travel	sport	anim	art	sci	travel	sport
develop	paint	natur	transport	particip	develop	form	nat	transport	particip
organ	form	modern	human	includ	form	paint	typ	word	includ
year	artist	human	mean	physic	org	work	hum	includ	competit
cell	work	centuri	peopl	competit	cel	cre	modern	hum	gam
form	exampl	type	power	game	plant	styl	centuri	peopl	phys

Presented examples show that creating topics differs in each method. However we see that LDA properly created topics for documents by selecting words from each group. Because some words with the same meaning appear repeatedly in one topic we can observe that WordNet Lemmatizer performed worse. Both Porter stemmer and Lancaster stemmer gave similar results. Taking into account that Porter stemmer recognized easier the basic form of a word, we consider the most useful of all three considered.

Comparison of LDA and Word2Vec Algorithms. In the second step LDA and Word2Vec algorithms for the corpus Doc1 were compared. Table 4 presents perplexity for both algorithms and the results are similar. Probabilities for topics being assigned to document are given in Table 5.

Table 4. Comparison of perplexity for Doc1 after using: (a) LDA algorithm with stop-words removal and Porter Stemmer, (b) Word2Vec algorithm.

	Training set	Testing set
(a) LDA with Porter Stemmer	1072,7151	1495,6427
(b) Word2Vec	1041,6314	1465,7653

According to Table 5 the assignment of topics to documents was done correctly for both LDA and Word2Vec algorithms. The obtained probabilities for topic being assigned to document were much higher for matching topics (for example for documents about art they were matched with topics containing words "art", "paint" etc.). Other topics didn't get zero probability because in a given document words matching other documents occurred as well.

Table 5. Comparison of probabilities for Doc1 after using: (a) LDA algorithm with stop-words removal and Porter Stemmer, (b) Word2Vec algorithm.

document	a) LDA with Porter Stemmer					b) Word2Vec				
	topic 1	topic 2	topic 3	topic 4	topic 5	topic 1	topic 2	topic 3	topic 4	topic 5
animals doc 1	0,6015	0,0738	0,1144	0,1218	0,0885	0,6939	0,0173	0,1035	0,1279	0,0574
animals doc 2	0,7855	0,0410	0,0530	0,0578	0,0627	0,7953	0,0362	0,0273	0,0241	0,1171
animals doc 3	0,6955	0,0472	0,1024	0,0630	0,0919	0,7594	0,0691	0,1019	0,0279	0,0417
art doc 1	0,0357	0,5192	0,2816	0,0989	0,0646	0,0237	0,6049	0,1844	0,1072	0,0798
art doc 2	0,0317	0,7903	0,0980	0,0452	0,0347	0,0202	0,8145	0,1151	0,0415	0,0087
art doc 3	0,0748	0,5835	0,0723	0,1347	0,1347	0,0527	0,6457	0,0449	0,1531	0,1036
science doc 1	0,0482	0,0797	0,7338	0,0713	0,0671	0,0697	0,0363	0,7516	0,1063	0,0361
science doc 2	0,1502	0,1054	0,5527	0,1118	0,0799	0,1671	0,0821	0,5852	0,1291	0,0365
science doc 3	0,0585	0,0916	0,6972	0,0509	0,1018	0,0916	0,0082	0,7141	0,0683	0,1178
traveling doc 1	0,0774	0,0640	0,0741	0,6700	0,1145	0,0646	0,1041	0,1174	0,7016	0,0123
traveling doc 2	0,0981	0,0475	0,0633	0,7025	0,0886	0,1063	0,0705	0,0515	0,6816	0,0901
traveling doc 3	0,0772	0,1074	0,1208	0,6040	0,0906	0,0559	0,0951	0,1535	0,5892	0,1063
sport doc 1	0,0673	0,0427	0,0476	0,0591	0,7833	0,0026	0,0058	0,0126	0,1531	0,8259
sport doc 2	0,0808	0,1414	0,0758	0,0960	0,6061	0,2042	0,1849	0,0118	0,0139	0,5852
sport doc 3	0,0794	0,0631	0,0631	0,1075	0,6869	0,1010	0,0262	0,0742	0,1024	0,6962

Finally the experiment on Doc 2 with 30 documents, without the knowledge on correct assignment, was performed. These documents were divided into 5 topics. The differences between topics were difficult to define. Table 6 presents differences of perplexity.

Table 6. Comparison of perplexity for Doc2 after using: (a) LDA algorithm with stop-words removal and Porter Stemmer, (b) Word2Vec algorithm.

	Training set	Testing set
(a) LDA with Porter Stemmer	2183,7528	2699,1262
(b) Word2Vec	1473,5378	1527,1449

The Table 7 presents probabilities of document being assigned to topic. With dark gray color highest of them were distinguished and probabilities close to highest with light gray.

The results on perplexity were definitely better for Word2Vec than LDA model. Besides, the borders between topics were more clear for Word2Vec. The probabilities for 16.67% documents were close for the results of LDA, and for Word2Vec it was only 6,67%. Both algorithms did not provide satisfactory results for Doc 2. For LDA the most difficult task was the "TOGAF or not TOGAF" article about the possibilities of "The Open Group Architecture Framework". It was a typical technical text that should have been assigned to the first topic. Word2Vec algorithm did not process well "TheArchitectAndTheApparition" about the duties of a Software Architect. The algorithm Word2Vec coped better with the task.

For both algorithms, articles with similar topics turned out difficult to process. As presented in Table 7 all topics are related. Due to that fact the algorithms could not decide properly on the boundaries between them, which caused the words to duplicate.

Table 7. Comparison of probabilities for Doc2 after using: (a) LDA algorithm with stop-words removal and Porter Stemmer, (b) Word2Vec algorithm.

document	a) LDA with Porter Stemmer					b) Word2Vec				
	topic 1	topic 2	topic 3	topic 4	topic 5	topic 1	topic 2	topic 3	topic 4	topic 5
Beyond entities and relationships	0,1141	0,0472	0,0307	0,0767	0,7313	0,0431	0,0312	0,0014	0,1317	0,7926
Bigdata	0,5414	0,1049	0,1018	0,1049	0,1469	0,6923	0,0963	0,0525	0,0575	0,1014
Conditions over causes	0,3090	0,1146	0,1716	0,2895	0,1153	0,1736	0,3937	0,1942	0,1621	0,0764
Emergent design in enterprise IT	0,1620	0,0535	0,1122	0,6276	0,0447	0,1383	0,0264	0,1294	0,6947	0,0112
From information to knowledge	0,6197	0,1342	0,0829	0,0763	0,0868	0,7194	0,1699	0,0596	0,0149	0,0362
From the coalface	0,0997	0,4930	0,1784	0,0576	0,1713	0,0884	0,5368	0,2111	0,0719	0,0918
Heraclitus and Parmenides	0,2880	0,0975	0,0724	0,4387	0,1034	0,0746	0,0895	0,0991	0,5191	0,2177
Ironies of enterprise IT	0,2274	0,0879	0,4315	0,0980	0,1552	0,1097	0,2022	0,5001	0,0916	0,0964
Making sense of organizational change	0,1122	0,0553	0,6163	0,1545	0,0618	0,0401	0,0464	0,6171	0,1911	0,1053
Making sense of sensemaking	0,1070	0,1488	0,0679	0,1593	0,5170	0,0754	0,1713	0,0626	0,0172	0,6915
Objectivity and the ethical dimension of decision making	0,1315	0,3117	0,1364	0,2792	0,1412	0,0642	0,2016	0,4001	0,1615	0,1726
On the inherent ambiguities of managing projects	0,1925	0,5314	0,0823	0,1199	0,0739	0,1031	0,6618	0,0963	0,0621	0,0767
Organisational surprise	0,1533	0,1007	0,1396	0,1007	0,5057	0,1012	0,1063	0,1261	0,1609	0,5055
Professionals or politicians	0,6069	0,1038	0,0284	0,1014	0,1595	0,7953	0,1031	0,0628	0,0273	0,0115
Rituals in Information System Design	0,6549	0,0869	0,0594	0,0800	0,1188	0,6941	0,0916	0,1052	0,1082	0,0009
Routines and reality	0,1649	0,5232	0,1139	0,0930	0,1049	0,1963	0,5942	0,0385	0,0612	0,1098
Scapegoats and systems	0,1930	0,1591	0,4224	0,1069	0,1186	0,1053	0,0377	0,5732	0,1261	0,1577
Sherlock Holmes failed projects	0,0893	0,1037	0,0626	0,1263	0,6181	0,0183	0,1038	0,0089	0,2179	0,6511
Sherlock Holmes mgmt Fetis	0,2633	0,1482	0,1007	0,3640	0,1237	0,4016	0,1524	0,1971	0,1153	0,1336
Six heresies for BI	0,1627	0,1165	0,1331	0,4492	0,1386	0,1053	0,1126	0,1005	0,5725	0,1091
Six heresies for enterprise architecture	0,0808	0,0773	0,5923	0,1318	0,1178	0,1526	0,1019	0,6341	0,0572	0,0542
The architect and the apparition	0,1667	0,1923	0,2051	0,2692	0,1667	0,2721	0,0174	0,2814	0,1290	0,3001
The cloud and the grass	0,0680	0,0621	0,5888	0,1243	0,1568	0,0358	0,0345	0,6717	0,1052	0,1528
The consultants dilemma	0,1051	0,6462	0,0795	0,0872	0,0821	0,1001	0,7284	0,0353	0,0738	0,0624
The danger within	0,1370	0,0808	0,0521	0,4658	0,2644	0,1972	0,1007	0,1559	0,5035	0,0427
The dilemmas of enterprise IT	0,1115	0,6563	0,0529	0,0826	0,0967	0,0953	0,7293	0,0309	0,0618	0,0827
The essence of entrepreneurship	0,0555	0,0999	0,0435	0,7040	0,0971	0,0205	0,0924	0,0236	0,8263	0,0372
Three types of uncertainty	0,5966	0,1149	0,0953	0,1031	0,0901	0,7363	0,0953	0,0416	0,0837	0,0431
TOGAF or not TOGAF	0,1558	0,2078	0,1948	0,1948	0,2468	0,2142	0,2086	0,1256	0,2501	0,2015
Understanding flexibility	0,1490	0,1718	0,4631	0,0886	0,1275	0,2016	0,1052	0,5016	0,0915	0,0999

4 Conclusions

Main contribution of the paper is an attempt to compare the usefulness of representation provided by Word2Vec with the classical bag of words representation used in LDA. The drawback of both approaches applied in the examined algorithms is the necessity to specify the number of topics in case of Algorithm 1 and the number of clusters (corresponding to topics) in Algorithm 2. On the other hand the results achieved are quite encouraging for further research, specially in the area of Word2Vec since machine learning techniques other than clustering can be applied to Word2Vec representation of words. Another planned task is the investigation of multi-class learning algorithms applied to text categorization.

References

1. Blei, D.M., Ng, A.Y., Jordan, M.I.: Latent Dirichlet allocation. J. Mach. Learn. Res. **3**, 993–1022 (2003). http://www.jmlr.org/papers/v3/blei03a.html
2. Documents for tests (2016). http://hereticsconsulting.files.wordpress.com/2016/01/textmining.zip
3. Enríquez, F., Troyano, J.A., López-Solaz, T.: An approach to the use of word embeddings in an opinion classification task. Expert Syst. Appl. **66**, 1–6 (2016). http://dx.doi.org/10.1016/j.eswa.2016.09.005

4. Loper, E., Bird, S.: NLTK: the natural language toolkit. In: Proceedings of the ACL-02 Workshop on Effective Tools and Methodologies for Teaching Natural Language Processing and Computational Linguistics, ETMTNLP 2002, vol. 1, pp. 63–70. Association for Computational Linguistics, Stroudsburg (2002). http://dx.doi.org/10.3115/1118108.1118117

5. Mikolov, T., Sutskever, I., Chen, K., Corrado, G.S., Dean, J.: Distributed representations of words and phrases and their compositionality. In: Burges, C.J.C., Bottou, L., Ghahramani, Z., Weinberger, K.Q. (eds.) Advances in Neural Information Processing Systems 26: 27th Annual Conference on Neural Information Processing Systems 2013, Proceedings of a Meeting, 5–8 December 2013, Lake Tahoe, Nevada, USA, pp. 3111–3119 (2013). http://papers.nips.cc/book/advances-in-neural-information-processing-systems-26-2013

6. Nallapati, R., Cohen, W.W., Lafferty, J.D.: Parallelized variational EM for latent Dirichlet allocation: an experimental evaluation of speed and scalability. In: Workshops Proceedings of the 7th IEEE International Conference on Data Mining (ICDM 2007), 28–31 October 2007, Omaha, Nebraska, USA, pp. 349–354. IEEE Computer Society (2007). http://dx.doi.org/10.1109/ICDMW.2007.33

7. Řehůřek, R., Sojka, P.: Software framework for topic modelling with large corpora. In: Proceedings of the LREC 2010 Workshop on New Challenges for NLP Frameworksm, pp. 45–50. ELRA, Valletta, May 2010. http://is.muni.cz/publication/884893/en

8. Sakenovich, N.S., Zharmagambetov, A.S.: On one approach of solving sentiment analysis task for Kazakh and Russian languages using deep learning. In: Nguyen, N.-T., Manolopoulos, Y., Iliadis, L., Trawiński, B. (eds.) ICCCI 2016. LNCS (LNAI), vol. 9876, pp. 537–545. Springer, Cham (2016). doi:10.1007/978-3-319-45246-3_51

9. Skfuzzy: Fuzzy logic toolkit in python (2016). http://pythonhosted.org/scikit-fuzzy/

10. Topicmodels: Package for r (2016). https://cran.r-project.org/web/packages/topicmodels/

11. Yousefi-Azar, M., Hamey, L.: Text summarization using unsupervised deep learning. Expert Syst. Appl. **68**, 93–105 (2017). http://dx.doi.org/10.1016/j.eswa.2016.10.017

12. Zhang, W., Wang, J.: Prior-based dual additive latent Dirichlet allocation for user-item connected documents. In: Proceedings of the 24th International Conference on Artificial Intelligence, IJCAI 2015, pp. 1405–1411. AAAI Press (2015). http://dl.acm.org/citation.cfm?id=2832415.2832445

Development of a Sustainable Design Lexicon. Towards Understanding the Relationship Between Sentiments, Attitudes and Behaviours

Vargas Meza Xanat⬛ and Yamanaka Toshimasa$^{(\boxtimes)}$ ⬛

The University of Tsukuba, Tsukuba, Japan
ktdesignbox@kansei.tsukuba.ac.jp,
tyam@geijutsu.tsukuba.ac.jp

Abstract. Design education and practice have been deeply interlinked with industrialization throughout their history, but on recent years, global initiatives like the UN Sustainable Development Goals have challenged the conventional production and consumption models. Therefore, to understand the relationship between pro-environmental sentiments, attitudes and behaviours related to design, the present study proposes the development of a Sustainable Design lexicon. Through a combined method of semantic and sentiment analysis incorporating graphical symbols included in text based data, it is expected to uncover the psychological and contextual factors aiding the production and acceptance of Sustainable Design in developed and developing countries. The lexicon is expected to aid the development of an algorithm for video recommendations, which would improve creative people's learning experience of complex and biological related content.

Keywords: Sustainable design · Social networks · Semantic and sentiment analysis · Machine learning · Web systems development

1 Introduction

1.1 Why Sustainable Design?

Sustainable Design (also called Eco Design) seeks to incorporate responsibility for the systems that support life. Although it has been promoted for over a decade, its integration in the education and professional realms has not been fully accepted. This is partly because design has been deeply interlinked with consumerism, economical profit and technological advance, ignoring ecological and social well-being.

There also exist several challenges such as lack of trust in institutions, skepticism, and resistance among scholars, students and other related stakeholders. A change of focus from "clients" and "design objects", to "systems", "processes" and "relationships" is required as well, but such concepts are not easy to visualize.

Carter [6] suggests to find ways to explain and deal with the complexity of the designer role in the world, including the designer as educator and understanding the long-term effects of design decisions. This implies a change of the designers' perceptual field that could affect the design process. However, to understand how such a

© Springer International Publishing AG 2017
N.T. Nguyen et al. (Eds.): ICCCI 2017, Part II, LNAI 10449, pp. 367–375, 2017.
DOI: 10.1007/978-3-319-67077-5_35

change might occur, we first need to comprehend how people adopt and/or fail to adopt pro environmental behaviors.

1.2 Environmental Psychology and Designers

Nature and other nonhuman living entities are primarily handled as instruments by humans due to several psychological barriers. Gifford [13] identified 33, summarized as limited cognition, ideologies, social comparison, costs, discredence, perceived risk, and the limits of behaviour. Gutsell and Inzlicht [15] suggest that emotions like purity, sanctity and disgust might be mediating sustainable behaviour. However, to understand behavioral changes better, we could extend research on decision making, where the role of negative emotions seems to be particularly relevant. For example, sadness is related to attachment and longing for specific places [31].

Ahn [3] found that there were significant interaction effects of framing and emotion on risk choice, meaning that: (a) negative emotions made people more risk-averse in the face of gains; (b) anger made people more risk-averse and fear more risk-seeking in the face of lose; (c) sad people tended to be less sensitive to potential losses, more sensitive to potential gains and take risks; and (d) stressed or more cognitively loaded people tended to rely more on experience. Ahn's experiments support the assumption that conservative people might feel more encouraged towards pro environmental behaviours if messages are framed in the past, while most of sustainable related framing emphasises future related images (young animals and plants, children, etc.).

Promising results of behavioural change have been found with individualized social marketing approaches [2, 9]. Role modelling and feedback about others might also support pro-environmental behaviours [1, 32], partly because human attitudes towards nature are affected by the strength and stability of human relations [11]. However, such proposals emphasize positive emotions.

In the case of the design process, some studies provide evidence that students are keener to change than professionals [17, 37], while obstacles for behavioral change seem to be lack of time and evidence of effectiveness [23]. Nevertheless, the afore-mentioned researches have been conducted through self-reports mostly in developed countries.

1.3 Sentiment and Behavioural Analysis Through SNS

As Sustainable Design has been taught as a separate course, such knowledge has been widely diffused through internet. One of the most remarkable developments of the web on recent years are Social Networking Sites (SNS), which are services that store personal and relational data to facilitate communication and information sharing. In social networks, inbound groups wield more influence on attitudes, norms, behaviour and decisions than messages received from outsiders [8, 20]. Therefore, it is possible to study community behavior through social networks without relying on self-reports.

The link between everyday communication and behavioral and self-reported measures of personality, social behavior and cognitive styles has been explored before [34]. Correlations between human ratings and sentiment analysis software have been significant, meaning that algorithms are able to predict episodic feeling associated with human emotional responses to online content [25]. Hołyst et al. [16] findings on the interaction with mono emotional and poli emotional threads imply that: (a) people will tend to comment in longer conversations, (b) there will be more emotional attach in such conversations, and (c) the emotion reflected in new comments will tend to be similar to the more dominant one in the conversation.

In the case of online videos, weak positivity was found to be the most common sentiment in comments, corresponding to mild praise for the video, its author or topic; while negativity was associated with the densest discussions within comment sections and tended to be directed towards other commenters [36]. The positive effect of negativity on interaction has been confirmed in further studios about video social networks [39] and other online contexts [7].

In contrast with semantic analysis software, sentiment analysis software has been particularly focused on the analysis of emoticons, which tend to emphasize the emotion reflected in words. Some studies affirm that emoticon's meaning is universal [14], while others state the opposite [27]. However, the importance of context in the emoticons meaning is recognized to be relevant [18], although very few studies incorporate graphical representations other than human to the sentiment analysis [24].

1.4 Objective

It has been found that, at a global level and specifically in developing countries, there is a lack of in depth knowledge of attitudes and sentiment among sustainable designers, academics and other stakeholders. Pro environmental behaviours among the sustainable design community that involve the design process itself are largely unknown. Also, the study of negative emotions which might be more relevant to prompt behavioral changes is being neglected in favor of positive emotions, partly because it is assumed that designers are akin to innovation and future oriented. Therefore, this study proposes the development of a lexicon of sustainable design related words and graphic re-presentations to advance the answer of the following questions:

- Which are the behaviours, attitudes and feelings related to Sustainable Design expressed by the interested community?
- Which contextual factors are related to pro environmental behaviours, attitudes and feelings?

2 Methodology

2.1 Description of the Datasets

We selected some YouTube videos from a previous study on Sustainable Design related social networks [40]. English and Spanish were ta-ken in consideration to incorporate more developing countries. In the case of the English video dataset, comments from the most commented videos with at least 8,000 views and degree centrality equal or higher than one were extracted with YouTube Data Tools [30]. Due to the lower interaction shown in the Spanish video dataset, comments from the most commented videos with a degree centrality equal or higher than one were extracted.

Next, the comments were revised to discard other languages and spam. After cleaning, 13,978 comments from 163 English videos and 1,366 comments from 147 Spanish videos were considered for analysis. From this point onwards, English comments will be identified as "SDEN" and Spanish comments as "SDES".

2.2 Semantic Analysis

In order to aid the inference of relevant aspects of a text meaning in its context, semantic network analysis represents the content of a given text as a network of objects, which is then queried to answer a research question [38]. On the present study, TI [19] was employed to find out word frequency and co-occurrence of words. We will also employ PC-ACE [12] for topic modelling. Relevant words and topics will be classified under a basic scheme of subjects/nouns, adjectives/emotion related words, verbs, direct objects/topics, time/place related words, graphical symbols, and others. A second classification level will be based on the following Wordnet-Affect labels [33]: emotion, mood, trait, cognitive state, physical state, hedonic signal, emotion-eliciting situation, emotional response and behaviour. Semantic networks for the top frequent words, verbs, and emotional related words will be drawn with Gephi.

2.3 Sentiment Analysis

Sentiment analysis deals with the computational treatment of subjective expressions in written text, which are not easy to observe and verify. Combined, semantic and sentiment analysis can reveal holistic structures and dynamics of human expression in written content.

Sentiment can be measured in several ways, including polarization or orientation in binary or trinary scales; tipology or categorization such as happiness, fear and anger [21]; and dimensions such as valence or arousal [10]. Cyberemotions studies have mostly dealt with polarization scales. However, because we are interested in detecting negative emotions such as sadness and disgust, categorization will also be taken on account. Moreover, because arousal aids the prediction of the possibility of an action [26], the three types of emotional analysis will be conducted on the comments.

For the first type of sentiment analysis, SentiStrength [35] will be employed. For sentiment categorization, the syuzhet library for R [28] will be applied, which is based on Saif and Turney's [22] NRC Emotion lexicon. As for arousal, SentiStrenght employs mood setting, exclamation marks and repetitions of letters in a word to calculate it. However, a new dictionary for this software will be developed to detect arousal, based on the ANEW database [5, 29].

Once relevant keywords and sentiments related to pro environmental behaviours have been detected, we will generate matrices for keywords, sentiment and arousal scores for the comments. These matrices will be correlated through Quadratic Assignment Procedure (QAP) with the UCINET software [4].

2.4 Handling of Graphic Representations and Special Characters

The reason emoticons and other graphic representations are usually discarded from text analysis is that software can only interpret alphanumeric characters accurately. Therefore, identifiers were assigned to each special character and graphic representation to incorporate them to further analysis. For example, the emoticon "☺" was substituted with "gsgrin38", where gs stands for "graphic symbol" and "grin" for the action or most probable meaning shown in the emoticon.

Emotional polarity for the coded emoticons was assigned mostly based on the Sentistrength lexicons. For missing graphical representations, the Emoji Sentiment Ranking [30] was taken on account. Coding exceptions were emoticons expressing arousal and emoticons with repetition of elements (e.g. :(((). Another special case were the symbols "*" and "-", which stood for several meanings and therefore were coded in several different ways. Also, symbols contained in webpage links were considered irrelevant and thus were not coded.

We should be careful to correct for sample bias and data gaps. Because we are going to run Sentistrenght in comments from a relatively small community focused on a narrow topic, the lexicon will have to be modified to be context sensitive. Therefore, samples of 100 comments in Spanish and English will be separated from the databases to be coded by humans. The 6 coders will be chosen among design and/or psychology experts in order to detect words and their emotional valence in specific cases.

3 Preliminary Results and Discussion

Table 1 shows a sample of keyword and graphic symbol frequency in both datasets. The top 100 frequent words suggest topics around energy and architecture with the notions of system, community and sharing. SDEN showed more words related to change, future, technology, urban environments and costs in a male centered world. In contrast, SDES top words focused on communication, cordiality, political statements, natural materials, knowledge and a female architect. This, together with the videos content discussed in [40] suggest a prevalence of female Spanish speaking informal businesses and self-expression channels on internet. The attractiveness of flexible working hours, extra income and self-agency attracted women in the American continent to use internet as a promotion tool to create small world networks amid male backlash [41], but this case is especially true in Spanish speaking countries [42].

Representations for arousal (!) and emoticon representations of happiness were the most frequently found in both datasets. In second place, calculations and money related representations were found, suggesting a frequent mention of the economical barrier in the adoption of pro environmental behaviours. However, sadness and love related graphic representations were in third place of frequency in SDEN, while in the case of SDES, it was anger related graphic representations. Nevertheless, further analysis will clarify if such preliminary assumptions hold up.

Table 1. Top frequencies for emotional related words (2), verbs (3) and graphical symbols (6).

English word	Count	Category	Spanish word	Count	Category
Gsexc1 (!)	4692	6	Gsexc1 (!)	693	6
Not	3838	2	No	508	2
Like	2393	2	Gracias	257	2
No	1878	2	Si	247	2
Gsplus2 (+)	1651	6	Gsplus2 (+)	118	6
Need	1368	3	Br	117	2
Has	1310	3	Gsat2 (@)	105	6
Think	1291	3	Hacer	104	3
Know	1119	3	Bueno	78	3
Work	994	3	Hay	77	3
Good	937	2	Gshash2 (#)	72	6
Live	869	3	Contra	69	2
See	829	3	Barbarie	66	2
Br	816	2	Pronuncian	66	3
Say	733	3	Bien	58	2
Gsat2 (@)	722	6	Buena	58	2
Well	718	2	Excelente	56	2
Use	710	3	Gsat1 (@)	56	6
Great	701	2	Interesante	55	2
Look	694	3	Saber	55	3

4   Limitations and Future Work

Sentiment detection software has issues to detect irony. Moreover, Sentistrength has better performance when detecting positive emotions. Therefore, ambivalent and polisemic terms and graphical symbols detection and interpretation might not be very accurate. We are also creating the lexicon with romanic languages, which have semantic characteristics that might be incompatible with other widely used languages (e.g. Chinese).

The Sustainable Design Lexicon is expected to be the base to develop an algorithm for video recommendations. Matching interest level in Sustainable Design, and contextual and graphical cues might improve the personalization of the learning experience and thus enhance positive attitudes towards the environment. Creative people should approach complex content such as biomimicry, social inclusive design, etc.; with enthusiasm and curiosity instead of suspicion and a closed mind. It is also expected to eventually develop better methods to teach life related topics (e.g. ecology, biology, chemistry) to creative people.

In general terms, the present study is expected to advance the understanding of how text based communication expresses behaviour, attitudes and sentiments through the aforementioned combined methods.

Acknowledgements. The authors wish to thank Akira Uno, Constantine Pavlides, Mike Thelwall and Roberto Franzosi.

References

1. Abrahamse, W., Steg, L., Vlek, C., Rothengatter, T.: A review of intervention studies aimed at household energy conservation. J. Environ. Psychol. **25**(3), 273–291 (2005). doi:10.1016/j.jenvp.2005.08.002
2. Abrahamse, W., Steg, L., Vlek, C., Rothengatter, T.: The effect of tailored information, goal setting, and tailored feedback on household energy use, energy-related behaviors, and behavioral antecedents. J. Environ. Psychol. **27**(4), 265–276 (2007). doi:10.1016/j.jenvp.2007.08.002
3. Ahn, H.: Modeling and analysis of affective influences on human experience, prediction, decision making, and behavior. Ph.D. dissertation, Massachusetts Institute of Technology (2010). http://affect.media.mit.edu/pdfs/10.hyungil-phd.pdf
4. Borgatti, S.P., Everett, M.G., Freeman, L.C.: UCINET for windows: software for social network analysis. Analytic Technologies, Harvard (2002). https://sites.google.com/site/ucinetsoftware/home
5. Bradley, M.M., Lang, P.J.: Affective norms for English words (ANEW): instruction manual and affective ratings. Technical report C-1, The Center for Research in Psychophysiology, University of Florida (1999). http://www.uvm.edu/pdodds/teaching/courses/2009-08UVM-300/docs/others/everything/bradley1999a.pdf
6. Carter, C.M.: Sustainable consumption & sustainable design: moving sustainability theory towards design practice. Ph.D. dissertation, University of Texas (2013). https://repositories.lib.utexas.edu/handle/2152/22338
7. Chmiel, A., Sienkiewicz, J., Thelwall, M., Paltoglou, G., Buckley, K., Kappas, A., Hołyst, J.A.: Collective emotions online and their influence on community life. PLoS ONE **6**(7), e22207 (2011). http://journals.plos.org/plosone/article?id=10.1371/journal.pone.0022207
8. Cialdini, R.B.: Influence: science and practice (2001). https://www.academia.edu/570634/Influence_Science_and_practice
9. Daamen, D.D., Dancker, H., Staats, H., Wilke, A.M., Engelen, M.: Improving environmental behavior in companies: the effectiveness of tailored versus nontailored interventions. Environ. Behav. **33**(2), 229–248 (2001). doi:10.1177/00139160121972963
10. Dodds, P.S., Danforth, C.M.: Measuring the happiness of large-scale written expression: songs, blogs, and presidents. J. Happiness Stud. **11**(4), 441–456 (2010). doi:10.1007/s10902-009-9150-9
11. Easterlin, N.: A Biocultural Approach to Literary Theory and Interpretation. JHU Press, Baltimore (2012)
12. Franzosi, R.: PC-ACE (2016). https://pc-ace.com
13. Gifford, R.: 33 reasons why mankind isn't acting to stem global warming. PostMagazine (2015). http://www.scmp.com/magazines/post-magazine/article/1848858/33-reasons-why-mankind-isnt-acting-stem-global-warming
14. Gruzd, A.: Emotions in the twitterverse and implications for user interface design. AIS Trans. Hum. Comput. Interact. **5**(1), 42–56 (2013). http://aisel.aisnet.org/thci/vol5/iss1/4
15. Gutsell, J.N., Inzlicht, M.: A neuroaffective perspective on why people fail to live a sustainable lifestyle. In: Van Trijp, H.C. (ed.) Encouraging Sustainable Behavior: Psychology and the Environment, pp. 137–154. Psychology Press, New York (2013)

16. Hołyst, J.A., Chmiel, A., Sienkiewicz, J.: Detection and modeling of collective emotions in online data. In: Holyst, J.A. (ed.) Cyberemotions, pp. 137–158. Springer, Cham (2017). doi:10.1007/978-3-319-43639-5_8
17. Ji, E., Amor, M.C.: Bridging the gap between sustainable design education and application. Int. J. Des. Manag. Prof. Pract. **8**(3–4), 15–38 (2014)
18. Kelly, C.: Do you know what I mean > :(A linguistic study of the understanding of emoticons and emojis in text messages. Bachelor dissertation, Halmad University (2015). http://www.diva-portal.org/smash/record.jsf?pid=diva2%3A783789&dswid=-2179
19. Leydesdorff, L.: TI.exe for co-word analysis (2013). http://www.leydesdorff.net/software/ti/index.htm
20. Mckenzie-Mohr, D.: New ways to promote proenvironmental behavior: promoting sustainable behavior: an introduction to community-based social marketing. J. Soc. Issues **56**(3), 543–554 (2000). doi:10.1111/0022-4537.00183
21. Mishne, G.: Experiments with mood classification in blog posts. In: Argamon, S., Karlgren, J., Shanahan, J. (eds.) Proceedings of ACM SIGIR 2005 Workshop on Stylistic Analysis of Text for Information Access, vol. 19, pp. 321–327. Citeseer (2005). https://pdfs.semanticscholar.org/0982/8f26fd9bb7ef105538fa51a57456ae38e63e.pdf
22. Saif, M.M., Turney, P.D.: Crowdsourcing a word-emotion association lexicon. Comput. Intell. **29**(3), 436–465 (2013). doi:10.1111/j.1467-8640.2012.00460.x
23. Niedderer, K., Ludden, G., Clune, S., Lockton, D., Mackrill, J., Morris, A., Cain, R., et al.: Design for behaviour change as a driver for sustainable innovation: challenges and opportunities for implementation in the private and public sectors. Int. J. Des. **10**(2), 67–85 (2016)
24. Novak, P.K., Smailović, J., Sluban, B., Mozetič, I.: Sentiment of emojis. PLoS ONE **10**(12), e0144296 (2015). http://journals.plos.org/plosone/article?id=10.1371/journal.pone.0144296
25. Paltoglou, G., Theunis, M., Kappas, A., Thelwall, M.: Predicting emotional responses to long informal text. IEEE Trans. Affect. Comput. **4**(1), 106–115 (2013). doi:10.1109/T-AFFC.2012.26
26. Paltoglou, G., Thelwall, M.: Sensing social media: a range of approaches for sentiment analysis. In: Holyst, J.A. (ed.) Cyberemotions, pp. 97–117. Springer, Cham (2017). doi:10.1007/978-3-319-43639-5_6
27. Park, J., Barash, V., Fink, C., Cha, M.: Emoticon style: interpreting differences in emoticons across cultures. In: Seventh International AAAI Conference on Weblogs and Social Media, North America, June 2013
28. R Core Team: R: a language and environment for statistical computing. R Foundation for Statistical Computing, Vienna (2013). http://www.R-project.org
29. Redondo, J., Fraga, I., Padrón, I., Comesaña, M.: The Spanish adaptation of ANEW (affective norms for English words). Behav. Res. Methods **39**(3), 600–605 (2007). doi:10.3758/BF03193031
30. Rieder, B.: YouTube data tools. Computer software. Vers. 1.0. N.p (2015). https://tools.digitalmethods.net/netvizz/youtube
31. Scannell, L., Gifford, R.: Defining place attachment: a tripartite organizing framework. J. Environ. Psychol. **30**(1), 1–10 (2010). doi:10.1016/j.jenvp.2009.09.006
32. Schultz, P., Wesley, J.M., Nolan, R.B., Cialdini, N.J., Griskevicius, V.: The constructive, destructive, and reconstructive power of social norms. Psychol. Sci. **18**(5), 429–434 (2007). doi:10.1111/j.1467-9280.2007.01917.x
33. Strapparava, C., Valitutti, A.: WordNet affect: an affective extension of WordNet. In: Proceedings of the 4th International Conference on Language Resources and Evaluation, Lisbon, Portugal, vol. 4, pp. 1083–1086 (2004)

34. Tausczik, Y.R., Pennebaker, J.W.: The psychological meaning of words: LIWC and computerized text analysis methods. J. Lang. Soc. Psychol. **29**(1), 24–54 (2010). doi:10. 1177/0261927X09351676
35. Thelwall, M.: The heart and soul of the web? Sentiment strength detection in the social web with sentistrength. In: Holyst, J.A. (ed.) Cyberemotions, pp. 119–134. Springer, Cham (2017). doi:10.1007/978-3-319-43639-5_7
36. Thelwall, M., Sud, P., Vis, F.: Commenting on YouTube videos: from Guatemalan rock to el big bang. J. Am. Soc. Inf. Sci. Technol. **63**(3), 616–629 (2012). doi:10.1002/asi.21679
37. Ueda, E.S.: Industrial designers working towards an eco-innovation approach. Bull. Jpn. Soc. Sci. Des. **62**(1), 11–20 (2015). doi:10.11247/jssdj.62.1_11
38. Van Atteveldt, W.H.: Semantic network analysis: techniques for extracting, representing, and querying media content. Ph.D. dissertation, Vrije Universiteit (2008). http://dare.ubvu. vu.nl/bitstream/handle/1871/15964/title?sequence=2
39. Vargas-Meza, X., Yamanaka, T.: Local food & organic food: communication and sentiment in YouTube video networks. In: 6th International Kansei Engineering and Emotion Research Conference KEER, University of Leeds, Leeds, UK, 31 August–2 September 2016 (2016)
40. Vargas-Meza, X., Yamanaka, T.: Sustainable design in video social networks. In: Tsukuba Global Science Week 2016, Tsukuba, Japan (2016). http://jairo.nii.ac.jp/0025/00039976
41. Wallace, J.: Handmade 2.0: women, DIY networks and the cultural economy of craft. Ph.D. dissertation, Concordia University (2014). http://spectrum.library.concordia.ca/978912/
42. Zafra, R.: Ciberfeminismo y 'net-art', redes interconectadas. Mujeres en Red. El Periódico Feminista (2008). http://www.mujeresenred.net/spip.php?article

Analysing Cultural Events on Twitter

Brigitte Juanals[1](✉) and Jean-Luc Minel[2]

[1] IRSIC, Aix Marseille University, Marseille, France
Brigitte.Juanals@univ-amu.fr
[2] MoDyCo, University Paris Nanterre - CNRS, Nanterre, France
jean-luc.minel@u-paris10.fr

Abstract. In this paper, we first present a model to represent message flows and their contents on Twitter, then a model and an instrumented methodology to describe and analyze these flows and their distribution among the various stakeholders. The aim is to explore the engagement and interactions between different types of stakeholders. We apply our methodology and tools to the 12th edition of the cultural event "European Night of Museums" (NDM16).

Keywords: Circulation of information · Influence · Instrumented methodology · Social network · Twitter · European Night of Museums

1 Introduction

Since the 2000s social networks owned by American companies (Facebook, Twitter, Instagram, YouTube, etc.) have become very popular in the cultural field. As these social networks attract the general Internet audience, cultural institutions have incorporated into their communication strategies a strong presence in these digital spaces through the dissemination of contents (news, practical or cultural information, representations of works and associated information, etc.) and the development of interaction with their public. These evolutions go hand in hand with the development of cultural marketing. Lastly, the "eventualization" of culture, following the explosion of temporary exhibitions, has amplified a "shift of patrimonial institutions to the logic of working like streaming media". It is a fact that in France and in Europe, over the last decade, temporary cultural events have multiplied, in the form of public events or festivals - such as the "European Heritage Days", "European Night of Museums", "La Nuit de l'Archéologie", "Passeurs d'images". These events have now become recurrent, generating initiatives that contribute to cultural outreach in society as well as to the development of cultural tourism.

Our study is based on the 12th edition of the cultural event 'European Night of Museums' (NDM16). This event took place in the heritage institutions that were partners in the event and was extended to digital media on the website dedicated to the operation as well as on social networks (Twitter, Facebook, Instagram). The outline of this paper is the following. First, in Sect. 2, we will

© Springer International Publishing AG 2017
N.T. Nguyen et al. (Eds.): ICCCI 2017, Part II, LNAI 10449, pp. 376–385, 2017.
DOI: 10.1007/978-3-319-67077-5_36

present the specificities and the contribution of our approach and our methodology to collect and analyze flows of tweets. In Sect. 3, we describe the model designed then in Sect. 4 we discuss the results from quantitative and lexical analysis. Finally, we conclude in Sect. 5.

2   Related Work

A number of scientific studies have investigated how museums use social media, focusing on the communication and mediation policies and practices of museums and their uses of platforms (including Facebook, Twitter) for various purposes [6,14]. The aim is to move from a vertical logic of diffusion to the establishment of closer relationships based on audience interaction and participation. Nowadays, museums and cultural institutions use social media as a means to communicate and promote their cultural activities, as well as to interact and engage with their visitors, the main use of social media by museums remaining information and promotion of activities such as exhibition openings or events [6]. A qualitative study on the ways that museums use Twitter in this perspective shows that they choose to link resources, engage the public with new social media tools and favor a two-way form of communication [14]. Villaespesa Cantalapiedra [17] carried out fieldwork including a series of interviews with museum professionals which showed that the term 'engagement', 'can be interpreted in a variety of ways (. . .): From fostering inspiration and creativity in the user, originating a change of behavior, increasing the user's knowledge, receiving interaction from the user in the form of a like or a comment,' [17]. In particular, when Langa [10] studied the building of a relationship that forty-eight museums engaged on Twitter with online users, she showed that its primary use was as a marketing tool (public relations, events announcements, fact of the day, etc.) and that it led to a lesser engagement and a low audience participation. As mentioned in [7], tweet analysis has led to a large number of studies in many do-mains such as ideology prediction in Information Sciences [4], natural disaster anticipation in Emergency [15] and tracking epidemic [13] while work in Social Sciences and Digital Humanities has developed tweet classifications [16]. However, few studies aim at classifying tweets according to communication classes. An exception worth mentioning is the work presented in Lovejoy and Saxton [11] in which the authors (Twitter users) analyze the global behavior of nonprofit organizations on Twitter based on three communication classes: Information, Community and Action classes. Recently, several studies on tweet classification have been carried out in NLP [1,9] but to the best of our knowledge, only [3] has classified cultural institutional tweets in communication categories based on NLP techniques.

3   Designing a Methodology

3.1   Designing a Model

So far, scientific studies on the forms of digital communication engaged in by cultural institutions and audiences have analyzed the practices of institutions and audiences as well as their uses of digital platforms. Our contribution aims to design, in an inter-institutional cultural space, a model for the circulation of message flows on a social network platform, taking the case of Twitter. We chose an inter-institution space that corresponds both to a social network platform widely used by institutions and to a growing communication situation in the cultural field at the present time. This choice led us to focus on the category of cultural event programmed on a national scale. This enables us to explore how information circulates and is exchanged as well as the communication relations established between the different stakeholders present in an inter-institution space. Based on the analysis of message flows on the scale of the cultural event studied, we attempt to answer questions concerning the communicative practices of the stakeholders, such as for example, what is the current strategic usage of social media conducted by different types of stakeholders (not only museums or cultural institutions) during cultural events? We expect marketing and promotional messages to be present but we also inquire into initiatives fostering audience participation, providing cultural contents and favoring interaction with users. We will specifically examine the communication policies of cultural institutions during a cultural event on Twitter in order to assess whether they are part of their mission to democratize culture for a wide audience on this platform or whether it is rather a marketing campaign to promote a place or an event. A second question concerns the identification of passing accounts (see below), i.e. accounts that participate actively in the circulation of information during the cultural events studied. Finally, two periods of time were distinguished. A period before the event and a period during the event. This distinction is an empirical consequence of our studies on Museum Week 2014 et 2015 [3].

Our analysis focuses on Twitter messages (called tweets) sent by accounts of cultural institutions and by other institutional or non-institutional stakeholders who participated in a cultural event. A first step was to build a terminology to describe the objects studied according to three dimensions: the message, the stakeholders, and the forms of stakeholder participation. Concerning messages, we will call a message sent by a twitter account an "original tweet" and an original message sent by an account different from the issuing account a "retweet". The current Twitter API gives access to the original tweet (and its sending account) of a retweet. The generic term tweet includes "original tweet" and "retweet". Regarding stakeholder qualification, we distinguished Twitter accounts, accounts managed by institutions (called "organizational account", OA), and accounts managed by individuals (called "private account", PA). This distinction is based both on the official list of museums in France provided by the French open data website, and the description on the Twitter account provided by the Twitter API. The analysis of the description field is necessary because non-museum institutions such as the City of Paris (@Paris) participated in the Night of Museums. Analysis of the flows during the MuseumWeek 2014 and 2015, European Night of Museums 2016, Europeans Days Heritage 2016 events, led us to identify six attributes used to qualify accounts according to their modes of participation and one computed score was associated at each attribute (cf. Table 1). We used the terms "participant", "producer", "relayed", "relaying", "mentioned" and "passing". The attribute "participant" was assigned to an account if it produced at least one original tweet or retweet during of the two temporal periods of the event (before and during the event). The attribute "producer" was assigned to an account if it produced at least one original tweet. The attribute "relayed" was assigned to an account if at least one of its tweets was retweeted or quoted. The "relaying" attribute was assigned to an account if the account retweeted or quoted at least one tweet. The "mentioned" attribute was assigned to an account if its Twitter account name was mentioned at least once in a tweet. The attribute "passing" was assigned to an account if it was both "relayed" and "relaying". Our hypothesis is that an account with a high passing score is a key influential user who actively participates at the circulation ofinformation. Consequently, we computed for each account a passing score that is the product between the number of accounts that this account retweeted and the number of accounts that retweeted it. The value of this index is not significant in itself; it simply provides a means of comparing accounts. Note that these attributes were calculated irrespective of the number of followers. From these six attributes, it is possible, to compare the behavior of several accounts (see Sect. 4.2). Several quantitative analyses can be carried out. Quantitative analyses focus on the observation of flows and aim to identify accounts that contribute to the circulation of information during the cultural event. For each attribute (see above) a ranking of the accounts is computed, keeping only the first 10 or 15 accounts in this ranking. This make it possible to order the accounts that produce the largest number of original tweets, which are the most retweeted, and so on. In order to build a clasifier, a classification analysis of the contents of the messages was carried out in three stages. First, a team composed of two linguists and 10 community managers of cultural institutions designed a model, that is to say, determined the classes in which to categorize the tweets, and the features used to assign a tweet to a class. Four classes were identified (Encouraging participation, Interacting with the community, Promoting the event and informing about it, Sharing experience). The features selected were semio-linguistic (mostly lexical, but also including punctuation marks, emoticons), tweet-specific features (for example, the presence/absence of hashtags in tweets) as well as metadata such as the identity of the account. In the second stage, a classifier was built based on a corpus of 1000 tweets annotated by hand by cultural experts according to the categories defined in the previous step [3]. The classifier is based on the Naive Bayes and SVM models, with unanimous vote. In a third stage, the classifier was applied to the corpus of tweets to categorize all the tweets. Results of the thorough evaluation of the quality of the classifier carried out on the MuseumWeek 2015 campaign are detailed in [7]; the F-measure $F_{0.5}$ coefficient is 0.696.

Table 1. Score calculation method

Participant	0 (no participation in the considered period of time) or 1
Producer	Total amount of tweets and retweets
Relayed	Total number of accounts which relayed his/her original tweets
Relaying	Total number of accounts she/he relayed
Mentioned	Total number of mentions of the account in text tweets
Passing	Product of relayed score by relaying score

3.2   Implementing the Methodology

We implemented our methodology by building a workflow (cf. Fig. 1) based on the one hand, on open access tools like R[1] (statistical), TXM[2] (text mining), Gephi[3] (graphs visualization), Neo4j [4] (graphs mining), Scikit-learn[5] (machine learning), and Sphynx (faceted serach engine)[6], and on the other hand, we developed some scripts python. All data are stored in a NoSql database and the scripts are used to query this database and compute specific attributes.

4   Applying the Methodology on the European Night of Museums Event

The analyses of the European Night of Museums (NDM16) were carried in the structuring content of the organizational and media framework in order to understand what happened during this cultural event, going beyond the display of quantitative data communicated at the time of its closure, i.e. 3000 events organized in France and in Europe, more than 2 million visitors who participated in the European Night of Museums in France.

The data acquisition stage consisted in harvesting tweets with only one hashtag, the official event hashtag #NDM16. We developed a script Python, based on Twarc (http://github.com/docnow/twarc) module proposed by Ed Summers using the streaming option of Twitter Application Programming Interface (API). In this paper, we limit the analysis to tweets in French sent during the week preceding the event and the day of the event, that is to say from 14 May to 21 May 2016 midnight. The main figures are the following: 11 264 tweets of which 3 301 original tweets (29%), 7 963 retweets (71%) sent by 4 012 participants.

The specific figures are the following: more than half (56%) of organizational accounts and only 25% of private accounts produced original tweets. Participation in this event was largely limited to the action of retweeting (75% of tweets)

---

[1] https://www.r-project.org/.
[2] http://textometrie.ens-lyon.fr/.
[3] https://gephi.org/.
[4] https://neo4j.com/.
[5] http://scikit-learn.org/stable/.
[6] http://sphinxsearch.com/.

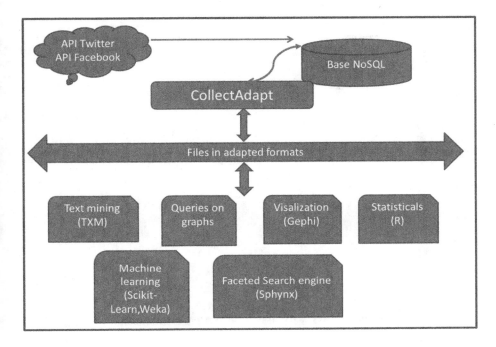

Fig. 1. Workflow and tools

the messages sent by the institutional partners. The findings of the supervised classification of original tweets sent during the event are shown in Table 2. It should be noted that some tweets have been categorized twice (that is why the sum of the total percentages is greater than 100%). This table shows that the organizational accounts tweets mainly (93%) serve to promote and inform about local events planned for the occasion. This result is consistent with the organizational framework of the cultural event, the success of which is partly linked to its attendance. There is also some interaction with the public but messages of engagement are very sparse. We applied the t-SNE algorithm [12] to get a global cartography of the networks (Fig. 1). Several communities are well distinguished. One community (in green) composed mainly of organizational accounts, which is more productive and interactive, and others, composed entirely of private accounts (in pink) which produced few tweets (see Sect. 4.1).

4.1   Key Influential Users

Passing Accounts. We analysed the 10 passing accounts that had a significant score (see Sect. 3). A remarkable point is the non-correlation between the passing score and the number of followers. The Pearson-correlation computed on the first fifteen passing accounts (NuitdesMusées excluded) is equal to 0.15875,

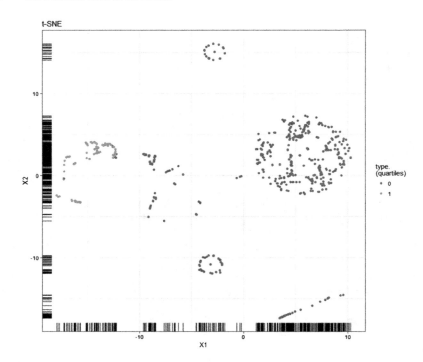

Fig. 2. t-SNE analysis

that is very low. Consequently, while the growth of followers of museums Twitter accounts is considered as a marketing target, the circulation of information during a cultural event is depending of key influential accounts which are not necessarily what is called 'big players' [5]. Some characteristics of these passing accounts must be pointed out. They are all organizational accounts with the exception of the account '@Mariette Escalier'. The owner of this private account is a professional in the field of cultural mediation. Ranked second after the organizational account of the event although she has less than 2 000 followers. In 5th place, the presence of the account of Alain Juppé, a French politician, mayor of the city of Bordeaux and former prime minister, shows that culture has become a political issue and plays an important role in marketing Cities and territories.

Great diversity was observed in the communicative practices of museum accounts. Some museums relay exclusively organizational accounts, while others relay only private accounts. We also noted that the passing accounts during the event are not the same as the passing accounts of the period preceding the event. Only the 'NuitdesMusées' and 'MinistèreCultureCom' accounts are common to both these periods (see [8] for more details).

In order to compare the 15 most active accounts, we performed a principal component analysis with 4 variables for each account: number of original tweets, mentioned score, relayed score and passing score (see Sect. 2). We did not take

Table 2. Supervised classification

Classes	Percentage
Encouraging participation	4%
Interacting with the community	19%
Promoting the event and informing about it	93%
Sharing experience	0%
Not classifified	2%

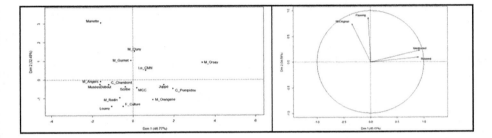

Fig. 3. Principal component analysis

into account the official account of the event (@NuitdesMusees) which due to its official position, presents specific characteristics. This analysis shows the positions occupied by the 15 most active stakeholders in the 2016 NDM (Fig. 1). Among them, the Musée d'Orsay is in a remarkable position, far on the right; it was mentioned and relayed many times because it organized an event (a jazz concert in the museum) which was very well attended. The success of this in situ event was relayed on Twitter which worked in this respect as a soundboard whereas the Louvre Museum, which organized conferences, was seldom mentioned or relayed despite its reputation. Similarly, the position of the private account Mariette Escalier (cultural mediator) occupies another remarkable position, in the left top corner; she is a passing account and she sent a lot of tweets, unlike most private accounts (Fig. 4).

Integration Score. The aggregated communities around the passing account are represented in Fig. 2. This spatial representation realized by applying the Louvain algorithm [2] makes possible to highlight several points. The private account of the politician Alain Juppé who sought to enhance his cultural program was largely relayed by the collective who supports him (Fig. 3), but Alain Juppé account (bottom right) is very isolated from the rest of the network. In order to support this, we calculated a score called integration score. This score is equal to the ratio between the cardinality of the set of accounts that have retweeted the passing account, among others, and the cardinality of the set of accounts that have only retweeted the passing account. Thus, if a collective of a passing account is totally isolated from the graph of all accounts, its integration score will be

Fig. 4. Main communities

equal to zero; Conversely, if a passing account is well integrated, in other words, the members of this group retweets other accounts, its integration score will be equal to 1. For exampe, Alain Juppé integration score is equal to 0.36 while the score of Mariette Escalier is equal to 0.8. Other communities are grouped around museum accounts (on the top left) and are partially connected to the rest of the network. Nevertheless, it must pointed out that different communities did not interact. In other words, the Musée d'Orsay and the Center Pompidou, two Parisian museums share few common accounts, at least for the Night of Museums event. On the contrary, the community of the private account @Mariette Escalier (on the top right) is much more immersed in the network.

5   Conclusion

We designed and implemented a methodology to analyze forms of engagement, participation and relationship between cultural institutions, organizations and audiences. Checking this methodology on the cultural event "European Night of Museums" 2016, we first applied the t-SNE algorithm to get a global cartography of the networks. We performed a correspondence analysis that shows the specificity of one private account and the role of two museums in Paris, the Muséee d'Orsay and the Centre Pompidou, which played a major role in disseminating information. We intend to dig deeper into the instrumented methodology: more specifically, we are working on the extension of our method in order to take into account an incremental analysis of graphs. This conceptual work brings together research issues on the production and circulation of information with data mining and visualization software. Data mining is seen as an heuristic iterative and incremental process.

References

1. Shou-De, L., et al. (eds.): 2nd Workshop on Natural Language Processing for Social Media. Association for Computational Linguistics and Dublin City University (2014)
2. Blondel, V., Guillaume, J.L., Lambiotte, R., Lefebvre, E.: Fast unfolding of communities in large networks. J. Stat. Mech. **2008**(10), P10008 (2008)
3. Courtin, A., Juanals, B., Minel, J., de Saint Léger, M.: A tool-based methodology to analyze social network interactions in cultural fields: the use case MuseumWeek. In: 6th International Conference on Social Informatics (SocInfo 2014), pp. 144–156 (2014)
4. Djemili, S., Longhi, J., Marinica, C., Kotzinos, D., Sarfat, G.E.: What does Twitter have to say about ideology? In: NLP 4 CMC: Natural Language Processing for Computer-Mediated Communication, pp. 16–25 (2014)
5. Espinos, A.: Museums on social media: analyzing growth through case studies. In: Museums and the Web (2016)
6. Fletcher, A., Lee, M.: Current social media uses and evaluations in American museums. Mus. Manag. Curatorship **27**, 505–521 (2012)
7. Foucault, N., Courtin, A.: Automatic classification of tweets for analyzing communication behavior museums. LREC **2016**, 3006–3013 (2017)
8. Juanals, B., Minel, J.: Information flow on digital social networks during a cultural event: methodology and analysis of the european night of museums 2016 on Twitter. In: SMS+Society Special Issue (2017)
9. Kothari, A., Magdy, W., Darwish, K., Mourad, A., Taei, A.: Detecting comments on news articles in microblogs. In: Kiciman, E., et al. (eds.) 7th International Conference on Web and Social Media (ICWSM). The AAAI Press (2013)
10. Langa, L.: Does Twitter help museums engage with visitors? In: iSchools IDEALS, pp. 484–495. University of Illinois, Aix-les-Thermes (2014)
11. Lovejoy, K., Saxton, G.D.: Information, community, and action: how nonprofit organizations use social media. J. Comput.-Mediated Commun. **17**(3), 337–353 (2012)
12. Van der Maaten, L., Hinton, G.: Visualizing high-dimensional data using t-SNE. J. Mach. Learn. Res. **9**, 2579–2605 (2008)
13. Missier, P., Romanovsky, A., Miu, T., Pal, A., Daniilakis, M., Garcia, A., Cedrim, D., da Silva Sousa, L.: Tracking dengue epidemics using twitter content classification and topic modelling. In: Casteleyn, S., Dolog, P., Pautasso, C. (eds.) ICWE 2016. LNCS, vol. 9881, pp. 80–92. Springer, Cham (2016). doi:10.1007/978-3-319-46963-8_7
14. Osterman, M., Thirunarayanan, M., Ferris, E., Pabon, L., Paul, N., Berger, R.: Museums and Twitter: an exploratory qualitative study of how museums use Twitter for audience development and engagement. J. Educ. Multimedia Hypermedia **21**(3), 241–255 (2012)
15. Sakaki, T., Okazaki, M., Matsuo, Y.: Tweet analysis for real-time event detection and earthquake, reporting system development. IEEE Trans. Knowl. Data Eng. **25**(4), 919–931 (2013)
16. Shiri, A., Rathi, D.: Twitter content categorisation: a public library perspective. J. Inf. Knowl. Manag. **12**(4), 1350035 (2013)
17. Villaespesa Cantalapiedra, H.: Measuring social media success: the value of the balanced scorecard as a tool for evaluation and strategic management in museum. Ph.D. thesis, University of Leicester (2015)

A Temporal-Causal Model for Spread of Messages in Disasters

Eric Fernandes de Mello Araújo$^{(\boxtimes)}$, Annelore Franke,
and Rukshar Wagid Hosain

VU Amsterdam, De Boelelaan 1105, 1081 Amsterdam, HV, The Netherlands
e.araujo@vu.nl, {a.franke,r.f.g.wagidhosain}@student.vu.nl

Abstract. In this paper we describe a temporal-causal model for the spread of messages in disaster situations based on emotion contagion and awareness works. An evaluation of the model has been done by simulation experiments and mathematical analysis. Parameter tuning was done based on two scenarios, including a credible message and a dubious message. The results are useful for the prediction of reactions during disasters, and can be extended to other applications that involve panic and supportive systems to assist people.

Keywords: Temporal-causal modeling · Disaster · Agent-based model

1 Introduction

The possibility of disaster and tragedy are a constant in everyone's lives. Apart from the problems caused by humans themselves (due to political decisions, errors, etc.), natural disasters also frequently occur in many places on Earth.

In 1953, a flood was caused by an extremely heavy storm in the Netherlands, England, Scotland and Belgium, called the North Sea Flood [1]. In the aftermath of this disastrous flood, a protective construction in the form of the Delta Works was created in the Netherlands in order to protect the country against similar natural disasters. Despite the protection of this new construction, a new flood is still a possibility. This was shown in a Dutch TV series called 'Als de dijken breken', or directly translated, 'If the dikes break'. This series raises the question of whether people are prepared for such a natural disaster [2]. What makes the scenario of 1953 different from nowadays is the current use of technology for communication. During the Twin Towers attack on the 9/11 in 2001 in New York, survivors made phone calls during the evacuation, most of which were not directed to emergency personnel, but to relatives, friends and family [3].

The spread of information in emergency situations can bring panic or can calm people down, like for relatives and friends that would remain in a stressful state in case of a lack of information about someone involved in a tragedy. In order to understand this scenario, we propose a temporal-causal model that considers how the act of sending a message could influence people's behavior through social contagion. Some of the questions that guided us are related to how

© Springer International Publishing AG 2017
N.T. Nguyen et al. (Eds.): ICCCI 2017, Part II, LNAI 10449, pp. 386–397, 2017.
DOI: 10.1007/978-3-319-67077-5_37

the information (message) is received. How people react regarding to the context, the sentimental and emotional charge of the message? Are they unable to perform any action, or intentionally not taking action when the message is not taken seriously? Does the source and the type of communication define if the message is serious or not?

In Sect. 2 we discuss the background theory. Section 3 discusses the temporal-causal network model in detail, with both a conceptual and a numerical representation. Section 4 is dedicated to the parameter tuning and datasets. In Sect. 5 are the simulation results in multiple scenarios, and the mathematical analysis. Lastly, Sect. 6 will be the discussion of the paper.

2 Social Contagion and Behavior in Disaster Situations

Modeling disaster situations is a huge challenge. Especially, because it is impossible to simulate realistic scenarios for this sort of event. Because of this, our model considers similar situations and combines different works that explain parts of the cognition of humans and presents some solid ground to build upon.

Blake et al. [3] studied the reactions of survivors of the World Trade Center attack on 9/11, in 2001. Over 20% of the survivors that participated in his research had made phone calls during the evacuation, and 75% of these calls were directed to relatives, friends and family, and not emergency personnel. The survivors wanted to inform their relatives about the situation, their whereabouts and warn them about what was going on. They also used the calls, text messages and emails (on Blackberry devices) to gather more information on the situation from outside during the evacuation process. In our model we consider the emergence of social media as a trend, and possibly an easy way to communicate with people during a disaster.

Paton [4] developed a model of disaster preparedness using the knowledge about the social cognitive preparation system. This model describes how people prepare for disaster situations that might occur in the future and how different factors can affect that process. The focus on disaster situations in the future is different from our approach as we want to look into the spontaneous occurrence of a disaster and how people respond here. Paton shows that there are clear indications that anxiety or fear can play a motivating or demotivating effect in preparing for a disaster.

Bosse et al. [5] propose a temporal-causal model for emotion contagion based on interaction between individuals. It shows how specific traits of people define how they affect each other. For our model we assume that the messages, used to communicate, carry some subjective content. This can be seen as the sentiment of the message. Furthermore, it carries other cues to which the cognitive system will pay attention on the attempt to unfold and understand, for instance, how serious the message is.

Thilakarathne and Treur [6] present a computational agent-based model to simulate emotional awareness and how this may affect the execution of an action. Thilakarathne and Treur [7] also introduce a neurologically inspired agent model which makes distinctions between prior and retrospective awareness of actions. Those concepts are used in our model, as a way of tracking awareness in our agents who will receive messages. Our model follows the Network-Oriented modeling approach, proposed by Treur [8].

3   Agent-Based Model

This section presents the designed temporal-causal model. The numerical representation for the connections in the network is also shown. This model represents a scenario where the person has a perception of the environment, and starts receiving messages from another person about a possible disaster happening. The internal states are based on the emotions and potential actions of the person. Figure 1 shows the conceptual representation of the temporal-causal model.

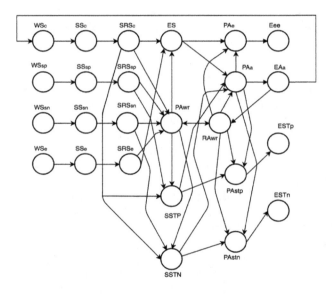

Fig. 1. Temporal causal network model

Table 1 describes the meaning of each state in the model. We based our model on the previous models by [6,7], but without the element of ownership. In addition, the model has been extended in the field of emotion and sentiment, both positive and negative. Despite the fact that we realize culture could be of some influence, we decided to not include this in the model.

The scenario for this model can be understood as someone who receives a message claiming that something bad is happening, and the person starts to investigate the environment to see if there is any cue that matches the message. It

Table 1. External and internal states of the model

Notation	Description
WS_c	World state context c
WS_{sp}	World state sentiment positive sp
WS_{sn}	World state sentiment negative sn
WS_e	World state emotion e
SS_c	Sensor state context c
SS_{sp}	Sensor state sentiment positive sp
SS_{sn}	Sensor state sentiment negative sn
SS_e	Sensor state emotion e
SRS_c	Sensory representation of context c
SRS_{sp}	Sensory representation of sentiment positive sp
SRS_{sn}	Sensory representation of sentiment negative sn
SRS_e	Sensory representation of emotion e
ES	Emotion state
$PAwr$	Prior-awareness state
$RAwr$	Retrospective-awareness state
$SSTP$	Sentiment state positive
$SSTN$	Sentiment state negative
PA_e	Preparation for action emotion e
PA_a	Preparation for action a
PA_{stp}	Preparation for action sentiment state positive stp
PA_{stn}	Preparation for action sentiment state positive stn
EE_e	Expressed emotion e
EA_a	Execution of action a
EST_p	Expressed sentiment state positive p
EST_n	Expressed sentiment state negative n

can be an alarm message about a storm approaching. In this case, the perceptions about the environment could come from taking a look outside of a window.

The model has four different world state inputs. The critical awareness of hazards [4] are represented as the world state *context* (the environmental context at the moment) WS_c. In order to include the anxiety factor, we added emotion (scariness, shock, excitement and shame, based on Ekman's basic emotions [9]), positive sentiment and negative sentiment, respectively as world states WS_e, WS_{sp} and WS_{sn}.

These external inputs are then sensed and lead to the sensor states SS_c, SS_{sp}, SS_{sn} and SS_e. Subsequently they proceed to the sensory representation states, SRS_c, SRS_{sp}, SRS_{sn} and SRS_e, which indicate how intense the stimuli is perceived by the person.

The emotional state ES is the current emotional state of the person, influenced by the *context* and *emotion* stimuli. The prior-awareness $PAwr$ is the awareness state of a person before they have executed any action. The $PAwr$ can then be suppressed by the retrospective-awareness, $RAwr$, when the person actually has executed an action which might have changed their awareness state. The two states for the sentiments are the positive, $SSTP$, and the negative state, $SSTN$. These two states are defined as the current sentiment state of the person, whether the stimuli had a positive or negative sentimental charge.

The model has four similar preparation states. The state preparation of a person to express an emotion, PA_e, leads to the external state expressed emotion EE_e. The state preparation for action PA_a prepares a person to execute an action, which leads to the external state execution of action, EA_a. Lastly, both states preparation for action sentiment state positive stp, PA_{stp} and preparation for action sentiment state negative stn, PA_{stn}. These two preparation states lead to the expressed sentiment state positive EST_p and expressed sentiment state negative EST_n.

Appendix A (www.cs.vu.nl/~efo600/iccci17/appendixA.pdf) shows all connections between the states within the model and where each of the combined functions were used. The temporal-causal behavior of the model is based on the methods proposed by [8]:

1. At each time point t each state Y in the model has a real number value in the interval $[0, 1]$, denoted by $Y(t)$.
2. At each time point t each state X connected to state Y has an *impact* on Y defined as $\mathbf{impact}_{X,Y}(t) = \omega_{X,Y} X(t)$ where $\omega_{X,Y}$ is the weight of the connection from X to Y.
3. The *aggregated impact* of multiple states X_i on Y at t is determined by a *combination function* $\mathbf{c}_\gamma(..)$ where X_i are the states connected to state Y.

$$\mathbf{aggimpact}_\gamma(t) = c_\gamma(\mathbf{impact}_{X_1,Y}(\mathbf{t}), \, ..., \, \mathbf{impact}_{X_k,Y}(\mathbf{t})) = \\ c_\gamma(\omega_{X_1,Y} X_1(t), ..., \omega_{X_k,Y} X_k(t)) \tag{1}$$

4. The effect of $\mathbf{aggimpact}_\gamma(\mathbf{t})$ on Y is exerted over time gradually, depending on *speed factor* η_γ

$$\mathbf{d}Y(t)/\mathbf{dt} = \eta_\gamma[\mathbf{aggimpact}_\gamma(t) \text{-} \ Y(t)] \tag{2}$$

5. Thus the following difference and differential equation for Y are obtained:

$$\mathbf{d}Y(t)/\mathbf{dt} = \eta_\gamma[\mathbf{c}_\gamma(\omega_{X_1,Y} X_1(t), ..., \omega_{X_k,Y} X_k(t)) - Y(t)] \tag{3}$$

The two combination functions used in our model were the identity and the advanced logistic (alogistic) functions. The identity function is $\mathbf{c}_\gamma(V) = \mathbf{id}(V) = V$, while Eq. 4 shows the advanced logistic function. The results of the simulations are shown in Sect. 5.

$$\mathbf{c}_\gamma(V_1, ...V_k) = \mathbf{alogistic}_{\sigma,\tau}(V_1, ...V_k) = \left(\frac{1}{1 + e^{\sigma(V_1 + ... + V_k - \tau)}} - \frac{1}{1 + e^{\sigma\tau}} \right) (1 + e^{-\sigma\tau}) \tag{4}$$

4   Data Generation and Parameter Tuning

This section describes which datasets were used to tune the parameters of the model. It is, furthermore, described how the parameters were tuned.

4.1   Experimental Datasets

Finding experimental data on the cognitive reactions during the spread of messages in disaster situations is difficult. The data concerning the messages is mostly missing or protected by messenger services. The information about cognitive states of people in the context of message receiving is difficult to obtain, due to limitations on the extraction of the data. Therefore, we created two different experimental datasets based on our understanding of the problem and on the literature. Both datasets contain information about a person receiving a message about a disastrous situation that might occur.

The first experimental dataset defines the course of all 25 states for a person receiving a message through the telephone with a tensed emotion and negative sentiment. This person is easily influenced by the sentiment and emotion of the message and believes that the message is true. In Fig. 2a the course of the 25 states in the first dataset is shown. Bosse et al. [5] state that the emotion of a person is influenced by the impact of the emotion of another person and the person's own belief. We assume that the impact of the emotion is big, because this person received a telephone call and this person is easily influenced. It is considered that the person will react spreading the message.

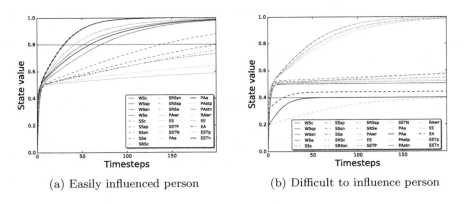

(a) Easily influenced person   (b) Difficult to influence person

Fig. 2. Experimental dataset

The second experimental dataset defines the course of all 25 states for a person that receives a message through a messenger service. The observed emotion of the sender is happy and positive. This person is influenced by the emotion and sentiment of the message, however, believes that the message is not accurate enough to spread. In Fig. 2b the course of the second dataset is shown. It can be seen that this person's emotion and sentiment also approach the observed emotion and sentiment. The reason for this is that the person believes that the sender is happy and has a positive sentiment and is, therefore, influenced by the happy and positive sender. Because the value of the expected action is below 0.5 and this person believes that the message is not accurate enough, it is decided that this person will not revert to the action of spreading the message.

4.2   Parameter Tuning Methods

To tune the model, optimal values for the speed factor values η, the steepness σ and the threshold τ in the advanced logistic formulas should be found. For reasons of simplicity, we have decided to use a speed factor value of 0.4 for each state, because we assume that the changes of the states within a person to deal with receiving a disastrous message does not occur quickly, however, not too slowly either.

The steepness and threshold of the advanced logistic function is more difficult to manually define. In this model there are 14 states that use the advanced logistic function. Thus, in total there are 28 steepness and threshold values that need to be tuned. The domain of the steepness and threshold values is assumed to be $[-\inf, \inf]$.

This is another reason why manually tuning the parameters is difficult. We chose the mean squared error (MSE) as the objective function that has to be minimized.

We used random search to tune the parameters. To make parameter tuning with this method more tractable, we tuned 2 parameters instead of 28, and we searched within a uniformly distributed domain of $[-e^{-5}, e^5]$. This method provided errors of approximately 0.10 on the first dataset and of approximately 0.09 on the second dataset. When using the corresponding steepness and threshold values (see Table 2) in the model, this gave us an acceptable simulation.

Table 2. The resulting MSE, steepness and threshold after tuning.

	MSE	σ	τ
Dataset 1	0.103	5.859	−5.945
Dataset 2	0.094	0.847	−5.565

Then, we performed random search with the modeling choice of tuning 28 parameters. This gave us an error of approximately 0.08, which is lower than tuning 2 parameters. However, when using the corresponding steepness and threshold values in the model, we got an abnormal simulation. We tried to decrease the domain to $[-e^{-2}, e^2]$. However, decreasing the domain did not make a difference.

5   Simulations and Results

In this section the simulation results are given. Three scenarios are simulated. For all simulations the steepness and threshold values from Table 2 are used for each of the 14 states that use the advanced logistic function. We used a step size of $\Delta t = 0.1$ for all simulations, and the speed factor value η is 0.4.

The value of all connection weights are 1.0, except for the connection weights of $(SRS_c, SSTP)$, $(SRS_c, SSTN)$, $(PAwr, SSTP)$, $(PAwr, SSTN)$, (PA_a, PA_{stp}), (PA_a, PA_{stn}). The values of these weights are 0.01, because $\omega_{SRS_{sp} \to SSTP}$ and $\omega_{SRS_{sn} \to SSTN}$ should weigh the heaviest to calculate the state values of SSTN and SSTP.

5.1   Scenario 1: Receiving a Tensed and Negative Phone Call

In this first scenario, a person receives a phone call from another person informing then that a disastrous situation might occur. The person observes that the sentiment of the message is negative, and that the emotion of the message is tensed. This person is easily influenced by the sentiment and emotion of the message, also because he/she received a telephone call, which is assumed to be a credible source. This person, therefore, believes more easily that the message is true.

The initial values of the input states can be found in Table 3. We have defined a tensed emotion to be 0.8 and a happy emotion to be 0.4. This is based on the valence and arousal model of [10]. In Fig. 3a all states are depicted for this person. In Fig. 3b only the input and output states are depicted.

Table 3. Initial values of the input states of scenarios 1 and 2

Input states	Initial values scenario 1	Initial values scenario 2
WS_c	0.0	0.0
WS_{sp}	0.0	1.0
WS_{sn}	1.0	0.0
WS_e	0.8	0.4

The output states EST_n, EA_a, EE_e and input state WS_c go up to around 0.9 due to the message's effect at the person. Since an action (i.e. spreading a message about a disastrous situation) increases awareness about the danger, the WS_c value increases.

5.2   Scenario 2: Receiving a Happy and Positive Text Message

In this scenario a person receives a message through a messenger service (textual) about a disastrous situation that might occur. This person, however, observes

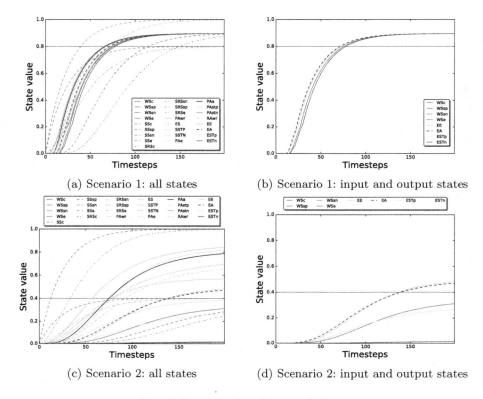

(a) Scenario 1: all states   (b) Scenario 1: input and output states

(c) Scenario 2: all states   (d) Scenario 2: input and output states

Fig. 3. Scenarios 1 and 2: simulations

that the message has a positive sentiment and a happy emotion. This person is influenced by the message, however, does not believe that it is true enough to spread the message as it is more difficult to determine the sentiment and emotion of a textual message.

The initial values of the input states can be found in Table 3. In Fig. 3c the simulation for this person is shown with all states. In Fig. 3d only the input and output states are shown for this person.

It can be seen that the EST_n is 0 over time, as the observed sentiment (WS_{sn}, WS_{sp}) of the message was positive. However, the EST_p state does not approximate to 1 throughout the simulation. This was expected since the connection weights that are important to define the EST_p are all 1.0. As expected, the E_e is higher than the WS_e, because this person observed a happy emotion and is influenced by it. The E_a is around 0.45. It is assumed that this person will not take action to spread the message, because this value is below 0.5.

5.3   Scenario 3: Outputs as Inputs

In this scenario the person from scenario 1 receives a telephone call informing that a disaster might occur and takes action by sending a textual message to the person from scenario 2.

Table 4. Initial values of the input states of the two persons in scenario 3

Input states	Initial values person 1	Initial values person 2
WS_c	0.0	person1 E_a
WS_{sp}	0.0	person1 EST_{sp}
WS_{sn}	1.0	person1 EST_{sn}
WS_e	0.8	person1 E_e

The initial values can be found in Table 4. In Fig. 4 the outputs of the first person and the outputs of the second person are shown. The outputs of the first person are used as inputs for the second person.

It can be seen that person 1 has a high EST_{sn}, and that person 2 also gets an EST_{sn} because of person 1. However, the EST_{sn} of person 2 is much lower than that of person 1. It is also shown that person 1 has a tensed emotion (i.e. $E_e = 0.8$), however, person 2 has a rather happy emotion (i.e. $E_e = 0.4$). It can, thus, be seen that person 2 is not influenced easily by the emotional state of person 1 as expected. The E_a of person 2 approximates a value of 0.4. Therefore, it is assumed that person 2 does not spread the message further.

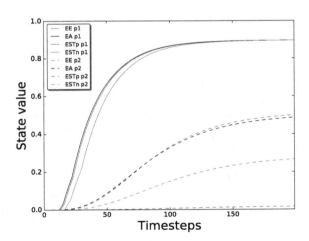

Fig. 4. Scenario 3: outputs of person 1 as inputs for person 2

5.4   Mathematical Analysis of the Model

To determine when the model is in equilibrium, we check when the states reach their stationary points. For instance, a state Y has a stationary point if $dY(t)/d(t) = 0$. The model is in equilibrium if every state has a stationary point at certain time t. Taking into account the difference and differential equations used in the model, the stationary point equation can be written as:

$$Y(t) = c_Y(\omega_{X_1,Y} X_1(t), ..., \omega_{X_k,Y} X_k(t)) \tag{5}$$

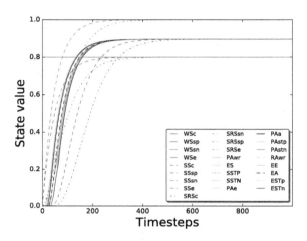

Fig. 5. Simulation for mathematical analysis

As an example we are going to determine the stationary points for the states SS_c, SS_{sn}, and SS_e. The verification method is the substitution of values in the stationary point equations. To determine the stationary points, the person from the simulation in scenario 1 (Fig. 3a) is used, however, with a longer simulation time and with a $\Delta t = 0.05$ (see Fig. 5).

The model was run until $Y(t + \Delta t) = Y$ holds. A stationary point for state SS_c was found at time point 531, with state value 0.8952. For state SS_e the stationary point is at time step 366 (state value of 0.7995). A stationary point for state SS_{sn} was found at time point 377 with state value 0.9995.

The connection states of SS_c, SS_e and SS_{sn} are respectively WS_c, WS_e and WS_{sn}. The state values of these connection states at the time points 531, 366 and 377 are, respectively, 0.8957, 0.8 and 1.0. The connection weights are all 1.0. When substituting these values in Eq. 5 we can see that the equation holds with an accuracy of $<10^{-2}$:

We found the stationary points for all states in the model. When taking into account that every state has to be stationary at time point t for the model to be in equilibrium, we can observe that the model is in equilibrium at time point 531 for the proposed set up.

6   Discussions and Future Works

In this paper a computational model is presented in order to model people's behavior on spreading messages in disaster situations. The model was designed as a temporal causal network model, following the approach of Treur [8], moreover, inspired and based on findings from previous research, as discussed in the background information, Sect. 2.

The proposed model can be a base for any type of disaster situation and can easily be extended in future research. Validation of this model is very difficult, if even possible, due to the lack of empirical data, because often observations during disaster situations are missing. Therefore, experimental data was created based on literature and experience to perform parameter tuning.

Within the scope of this paper we decided to incorporate different types of communication methods through the simulated scenarios. It could be interesting however to take those communication methods to the next level as well in order to learn why particular methods are more credible than others, or how messages spread more easily through some channels than others. Personality traits are somewhat incorporated in the simulation scenarios as well, however, for future research more traits should be explored. The same goes for culture and other possible influences.

Acknowledgments. E.F.M. Araujos funding is provided by the Science without Borders Program, through the CAPES (reference 13538-13-6).

References

1. Hall, A.: The north sea flood of 1953. Arcadia, (5) (2013)
2. Takken, W.: Wat doen we als de randstad onder water staat? NRC handelsblad (2016)
3. Blake, S., Galea, E., Westeng, H., Dixon, A.J.P.: An analysis of human behaviour during the WTC disaster of 11 based on published survivor accounts. In: Proceedings of the 3rd International Symposium on Human Behaviour in Fire, vol. 1, p. 3. 2004, September 2001
4. Paton, D.: Disaster preparedness: a social-cognitive perspective. Disaster Prev. Manag.: Int. J. **12**(3), 210–216 (2003)
5. Bosse, T., Hoogendoorn, M., Klein, M.C.A., Treur, J., Van Der Wal, C.N., Van Wissen, A.: Modelling collective decision making in groups and crowds: integrating social contagion and interacting emotions, beliefs and intentions. Auton. Agents Multi-Agent Syst. **27**(1), 52–84 (2013)
6. Thilakarathne, D.J., Treur, J.: Modelling the dynamics of emotional awareness, pp. 885–890 (2014)
7. Thilakarathne, D.J., Treur, J.: Modelling prior and retrospective awareness of actions, pp. 62–73 (2013)
8. Treur, J.: Network-Oriented Modeling and Its Conceptual Foundations. Springer International Publishing, Cham (2016). pp. 3–33
9. Ekman, P.: Basic Emotions. Handbook of Cognition and Emtion. Wiley, Hoboken (1999)
10. Valenza, G., Lanata, A., Scilingo, E.P.: The role of nonlinear dynamics in affective valence and arousal recognition. IEEE Trans. Affect. Comput. **3**(2), 237–249 (2012)

One Approach to the Description of Linguistic Uncertainties

Nikita Ogorodnikov[✉]

Moscow State University, Moscow, Russia
onm96@inbox.ru

Abstract. The basic concept of this paper is to interpret some uncertainties of our speech. It differs from traditional approach [1]. The main idea is to build new functions using the Fuzzy C-means Clustering Algorithm [4]. We will adhere to Zadeh's idea to represent a hedge as an operator acting on fuzzy sets and will consider four uncertainties: *much larger, much smaller, slightly smaller* and *slightly larger* (we will call them *modifiers*). These modifiers act on two fuzzy sets, so we have a linguistic variable with two values. The result of research is an algorithm, which determines the membership function of this variable's value after the modifier's action. Some properties, showing the conformity of the algorithm to the semantics of natural language and convergence in particular cases, are also described. The application of this work is the task of extracting from the database objects satisfying the semantics of the modifiers used.

Keywords: Fuzzy sets · Fuzzy clustering · Linguistic uncertainties · Linguistic variable

1 Introduction

In 1972 Lotfi Zadeh published an article "A Fuzzy-Set-Theoretic Interpretation of Linguistic Hedges" in the "Journal of Cybernetics". This article suggests an approach to deal with uncertainties of our speech using the Fuzzy Sets Theory methods. But nowadays one questions remains: does the formulated approach correspond to the natural semantics of the language? Ryjov [3] also researched this problem (in particular, modifiers *much larger* and *slightly smaller*) and he writes: "There are no general recommendations or theorems currently formulated"[1]. So, that became the reason for researching linguistic uncertainties and writing the present paper. Our problem is *to formulate a modification rule for the value of the linguistic variable*. In fact this paper consists of the algorithm description and properties, which show its conformity to the semantics of natural language and convergence in special cases. The algorithm (in particular cases) was tested on the database consisting of 1000 normally distributed numbers with expected value, equal to 10, and variance, equal to 1. The database was generated using a programming language *R* (the library *e-1071* was used for clustering), and the results of the algorithm working are described in Sect. 5.

---

[1] [3], p. 41.

© Springer International Publishing AG 2017
N.T. Nguyen et al. (Eds.): ICCCI 2017, Part II, LNAI 10449, pp. 398–406, 2017.
DOI: 10.1007/978-3-319-67077-5_38

2   Notations and Requirements

In the present paper some concepts of the Fuzzy Sets Theory are used. All of them are defined and discussed in [1], thus, further we rely on basic definitions and facts from this area.

Let us consider the linguistic variable and its term-set. The term-set consists of basic terms (for example, *large*, *small*) and their combinations with modifiers such as *much bigger*, *slightly smaller* etc. In the present paper we confine ourselves to modifiers *much larger*, *much smaller*, *slightly smaller* and *slightly larger* and assume that our linguistic variable has two basic terms; it will be conveniently to call them *large* and *small*. Generally speaking, we can use any qualitative adjectives; it depends on our task and context. Both these terms are described by membership functions, so, let us denote them μ_1 and μ_2 respectively. Also, let us denote some parts of the universal set U:

$U_1 = \{u \in U: \mu_1 (u) = 1\},$
$U_2 = \{u \in U: \mu_2 (u) = 1\},$
$U_{12} = \{u \in U: 0 < \mu_1 (u) < 1, 0 < \mu_2 (u) < 1\}.$

We will consider such membership functions that meet the requirements[2] (it will be used further):

(1) μ_1 does not decrease to the left of U_1, μ_2 does not increase to the right of U_2;
(2) $\forall u \in U \ \mu_1 (u) + \mu_2 (u) = 1.$

These functions can be constructed using a fuzzy clustering algorithm *c-means* (we use the algorithm with parameter m = 2 and Euclidean distance, denoted by ρ) and changing its results as follows:

Let us denote cluster's centers c_1 (cluster with membership function μ_1) and c_2 (cluster with membership function μ_2). From the formula for membership coefficients in [4], the object u_i, closest to the cluster's center c_j, has the largest grade of membership to this cluster. Let us put $\mu_j(u_i) = 1$, and for $k \neq j$ take $\mu_k(u_i) = 0$. Then we do this procedure for other cluster's center. So, $\exists u \in U: \mu_1(u) = 1$, $\mu_2(u) = 0$, and $\exists v \in U:$ $\mu_1(v) = 0$, $\mu_2(v) = 1$. After this $\forall w \in U: w > u$ we take $\mu_1(w) = 1$, $\mu_2(w) = 0$ and $\forall w \in U: w < u$ take $\mu_1(w) = 0$, $\mu_2(w) = 1$. We make these manipulations because degrees of membership must correspond to the semantics of terms *large* and *small* (or other qualitative adjectives).

3   The Main Algorithm: Description and Properties

Before formally setting the algorithm for changing the membership function of linguistic variable's value it will be useful to consider an example, which explains why this algorithm is exactly as described below.

---

[2] These requirements were taken from [2], pp. 54–57.

Suppose that we have the membership function for *large* objects and our aim is to get the membership function for *larger* objects. It is reasonable to assume that under the action of the modifier *larger* the new grades of memberships to the fuzzy set of *large* objects should be no more than the old ones. Thus, if the object was *large* with zero coefficient then it should also has a zero grade of membership to the fuzzy set of *larger* objects. It makes one think: in the new fuzzy set only objects, which were *large* with nonzero coefficients, must have nonzero grade of membership, and for other objects we should take the grade of membership equal to zero. But this consideration does not differ the action of *much larger* from *slightly larger*. So, let us change it based on the natural modifier's semantics.

Formal Description of the Algorithm and Some Requirements
To get the result of the modifier *much larger* action firstly let us discard from the universal set U all objects incoming in U_2. So we get the new universal set $\hat{U}$ and apply *c-means* to objects incoming in $\hat{U}$. Then it is necessary to change a new membership functions as it said in *Notations and Requirements*. As the result we have membership functions λ_1 and λ_2 defined on the universal set $\hat{U}$. $\forall u \in U_2$ let us take $\lambda_1(u) = 0$, and $\lambda_2(u) = 1$, and after that we have membership functions, defined on the U and satisfying the requirements (1) and (2). Similarly we define the action of the modifier *much smaller* by changing U_2 to U_1 and $\forall u \in U_1$ taking new functions as $v_1(u) = 1$ and $v_2(u) = 0$.

Now let us consider the modifier *slightly larger*. It is reasonable to require that under its action values of new degrees of membership to the set of *large* (*small*) objects should be no more (not less) than original grades of membership and not less (no more) than under the action of the modifier *much larger*. For this we discard *the left part* of U_2 from the universal set (i.e. objects $u \in U_2$ that are not larger than the median of U_2). Then we act similarly to the situation with the modifier *much larger*: $\forall u$ from *the left part* of U_2 we should take the membership function to the new set of *large* objects equal to 0, and the membership function to new the set of *small* objects equal to 1. For defining the modifier's *slightly smaller* action the same things should be done, but *the left part* of U_2 changes to *the right part* of U_1, which is also defined by the median, and $\forall u$ from *the right part* of U_1 we should take the membership function to the new set of *large* objects equal to 1, and the membership function to the new set of *small* objects equal to 0.

After the description of the algorithm is given, it is necessary to prove that it satisfies our requirements for changing the grades of membership (…under the action of the modifier *larger* the new grades of memberships to the fuzzy set of *large* objects should be no more than the old ones…).

Lemma 1. $\forall u \in U\ \lambda_1(u) \leq \mu_1(u)$, where μ_1 - the membership function of the fuzzy set containing *large* objects, λ_1 - the membership function of the new fuzzy set with *large* objects (after the action of the modifier *much larger*).

Proof
Following our notations, c_1 and c_2 are centers of clusters with *large* and *small* objects respectively, and let us denote centers of these clusters after the action of the modifier *much larger* as $\hat{c}_1$ and $\hat{c}_2$ respectively.

From this point and on, we argue on the assumption that cluster's centers shift to the right when we discard *the left part* of U_2 (and cluster's centers shift to the left when we discard *the right part* of U_1), and the more objects we discard, the more cluster's centers shift. In the present paper this is called *an assumption of cluster's centers shifts*, and this statement is clear, by natural reasons (because we apply *c-means* to the set of objects that are not less (no more) than original objects), but it would be better to prove this fact formally. So we consider that $\hat{c}_1 > c_1$ and $\hat{c}_2 > c_2$.

Suppose the opposite: $\exists u_0 \in U$, that $\lambda_1(u_0) > \mu_1(u_0)$. Now let us consider three possible cases:

(1) If $u_0 \geq c_1$, then $\mu_1(u_0) = 1$ (this follows from the construction of the membership functions in *Notations and Requirements*), and in this case $\lambda_1(u_0) \leq \mu_1(u_0)$.
(2) If $u_0 \leq c_2$, then $\lambda_1(u_0) = 0$, and then $0 = \lambda_1(u_0) \leq \mu_1(u_0)$.
(3) Now let $\hat{c}_2 < u_0 < c_1$; this means that

$$\mu_1(u_0) = \frac{1}{(1 + \rho(c_1, u_0)/\rho(c_2, u_0))}, \quad \lambda_1(u_0) = \frac{1}{(1 + \rho(\hat{c}_1, u_0)/\rho(\hat{c}_2, u_0))}.$$

This follows from the formula for membership coefficients in [4]; ρ is Euclidean distance, as it said in the previous chapter. By our assumption, $\lambda_1(u_0) > \mu_1(u_0)$, so we have:

$$\frac{\rho(\hat{c}_1, u_0)}{\rho(\hat{c}_2, u_0)} < \frac{\rho(c_1, u_0)}{\rho(c_2, u_0)},$$

$$\frac{(\rho(\hat{c}_1, u_0) \cdot \rho(c_2, u_0) - \rho(c_1, u_0) \cdot \rho(\hat{c}_2, u_0))}{(\rho(\hat{c}_2, u_0) \cdot \rho(c_2, u_0))} < 0,$$

$$\rho(c_2, u_0) \cdot (\rho(c_1, u_0) + \rho(c_1, \hat{c}_1)) - \rho(c_1, u_0) \cdot (\rho(c_2, u_0) - \rho(c_2, \hat{c}_2)) < 0,$$

$$\rho(c_2, u_0) \cdot \rho(c_1, \hat{c}_1) + \rho(c_1, u_0) \cdot \rho(c_2, \hat{c}_2) < 0.$$

Here we take advantage of the fact that $c_2 < \hat{c}_2 < u_0 < c_1 < \hat{c}_1$. Thus, we come to a contradiction, which completes the proof.

Lemma 1 *is proved*.

Corollary. $\lambda_1(u) = \mu_1(u) \Leftrightarrow u \leq c_2$ or $u \geq \hat{c}_1$.

Proof

If $u \leq c_2$, then $\lambda_1(u) = \mu_1(u) = 0$ (this follows from the construction of the membership functions for such objects), and if $u \geq \hat{c}_1$, then $\lambda_1(u) = \mu_1(u) = 1$.

Suppose, that $\lambda_1(u) = \mu_1(u)$, but $c_2 < u < \hat{c}_1$. The, if $u \in (c_2; \hat{c}_2]$, then $\mu_1(u) > 0$, $\lambda_1(u) = 0$. If $u \in [c_1; \hat{c}_1)$, then $\mu_1(u) = 1$, $\lambda_1(u) < 1$. If $u \in (\hat{c}_2; c_1)$, we proceed similarly to the third case of Lemma 1 and get $\rho(c_2, u_0) \cdot \rho(c_1, \hat{c}_1) + \rho(c_1, u_0) \cdot \rho(c_2, \hat{c}_2) = 0$. Anyway, we have a contradiction.

Corollary is proved.

Similarly to Lemma 1 can be proved.

Lemma 2. $\forall u \in U \; v_2(u) \leq \mu_2(u)$, where μ_2 - the membership function of the fuzzy set containing *small* objects, v_2 - the membership function of the new fuzzy set with *small* objects (after the action of the modifier *much smaller*).

From property (2) (in the previous chapter) it follows that $\forall u \in U \; v_1(u) \geq \mu_1(u)$ and $\lambda_2(u) \geq \mu_2(u)$, where v_1 - the membership function to the set of *large* objects after the action of the modifier *much smaller*, and λ_2 - the membership function to the set of *small* objects after the action of the modifier *much larger*.

Thus, we figured out with modifiers *much larger* and *much smaller*, and now let us introduce some notations:

λ_1' - the membership function to the set of *large* objects after the action of the modifier *slightly larger*;

λ_2' - the membership function to the set of *small* objects after the action of the modifier *slightly larger*;

v_2' - the membership function to the set of *small* objects after the action of the modifier *slightly smaller*;

v_1' - the membership function to the set of *large* objects after the action of the modifier *slightly smaller*.

Acting like in Lemma 1 and using property (2) we get the following result:

Lemma 3. $\forall u \in U \; \lambda_1(u) \leq \lambda_1'(u) \leq \mu_1(u)$, $\lambda_2(u) \geq \lambda_2'(u) \geq \mu_2(u)$ and $v_2(u) \leq$ (u) $\leq \mu_2(u)$, $v_1(u) \geq v_1'(u) \geq \mu_1(u)$.

It should be noticed that this relies on the *assumption of cluster's centers shifts* formulated in Lemma 1. Thus we get the algorithm for changing the membership function of linguistic variable's value. Thanks to the proved statements we can conclude that our algorithm looks reasonable, and now let us discuss its other important properties.

4   Convergence in Special Cases

The algorithm considered in the previous chapter can be used to select from the whole universal set those objects that correspond to the semantics of the expressions used as modifiers. Having received the database, it will output its *needed* records, for example, in decreasing order of membership coefficients. On this background, the question arises: can any *needed* records be found? In other words, can we guarantee that with the help of our algorithm the database user will either find what he needs, or will he be convinced that this does not exist? And to answer this question, let us remember, that if in U exists such u that $\mu_1(u) = 0$, then it is possible to construct $\lambda_1, \lambda_2, \lambda_1', \lambda_2'$, and if exists such u that $\mu_2(u) = 0$, then it is possible to construct v_1, v_2, v_1', v_2'. From the membership functions construction in the second chapter it follows that we always have at least one $u \in U$, that $\mu_1(u) = 0$ or $\mu_2(u) = 0$. So, we always implement the step of the algorithm. But this does not guarantee that the algorithm will finish its work for any user, because there is no *stop criterion*, in other words, it is necessary to specify a condition under which the algorithm will instead of re-building functions, issue a message about the inexpediency of further work. To establish such criterion, we consider the cases, and for further convenience we define several concepts.

Definition. Let us call the actions of *much larger* and *slightly larger* offsets to the right and the actions of *much smaller* and *slightly smaller* - offsets to the left.

Let us consider multiple offsets to the right. At each step, as it was shown above, we can construct the new membership functions, and, from the *assumption of cluster's centers shifts*, $\hat{c}_1 > c_1$, if the modifier *much larger* acts, and $\hat{c}_1' > c_1$, if the modifier *slightly larger* acts. But changing of U_1 depends on changing of the object, closest to $\hat{c}_1$ or $\hat{c}_1'$. Thus, we have two variants:

(1) The object, closest to $\hat{c}_1$ or $\hat{c}_1'$, shifted (i.e. now other object is closest), and in this case U_1 changes and we do the next step of the algorithm.
(2) If the object, closest to $\hat{c}_1$ or $\hat{c}_1'$, did not shift, then new U_1 has the same cardinality as the old one.

This gives an occasion to think: will not the number of objects in U_1 stabilize at some point?

Lemma 4. When the *assumption of cluster's centers shifts centers* is fulfilled, the number of elements in U_1 does not stabilize (i.e. it will decrease, and perhaps not strictly from time to time).

Proof
Let us consider the cluster's center c_2 and denote the object, closest to c_2, by u_2. Suppose that from time to time u_2 is the same object in the universal set U. From the membership function μ_2 construction, $u_2 \in U_2$ at each step of the algorithm. But since each time we discard at least one object from U_2, then after a finite number of steps we will reach u_2 and will discard it, so, after this number of steps another element will be considered as u_2. Thus, the element, closest to c_2, does not stabilize. But from this it follows that the element u_1, closest to c_1, also does not stabilize, otherwise it would turn out that $u_1 = u_2$, but this is impossible. So, it follows trivially that the number of objects in U_1 will not stabilize.
Lemma 4 *is proved.*

Theorem 1. For a finite number of offsets to the right, the set U_1 decreases to one object; for a finite number of offsets to the left, the set U_2 decreases to one object.

Proof
This statement for offsets to the right follows from the cases (1) and (2) considered above, and Lemma 4. For offsets to the left, it is sufficient to carry out the same arguments with the set U_2.
Theorem 1 *is proved.*

Thanks to Theorem 1, we can define the criterion for stopping the algorithm if only offsets to the right (only offsets to the left) are made: the algorithm terminates if the set U_1 (U_2) at the current step consists of one object. This condition looks reasonable from a logical point of view, because if there is only one reference *large* object, then there are no objects, *much* or *slightly larger* (or if there is only one reference *small* object, then there are no objects, *much* or *slightly smaller*). Thus, if the database user applies only offsets to the right (only offsets to the left), either he finds some objects he needs or will reach one object. And this object is either needed, or there are no such objects in the database.

5   Examples of the Algorithm Working

In Lemma 4 we talked about a non-strict decrease in the number of objects in U_1. In the course of researching the algorithm working with the particular database (see Sect. 1) this situation arose (Fig. 1).

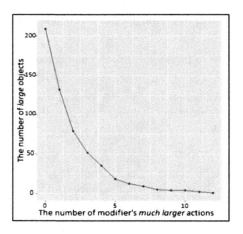

Fig. 1. The example of non-strict decrease in the cardinality of U_1.

After 10 applications of *much larger*, the cardinality of U_1 remained the same as after 9 applications - 4 objects. The Table 1 shows the results of the change in the cardinality of all sets U_1, U_2 and U_{12}:

Table 1. Changing in the cardinality of U_1, U_2 and U_{12}.

Iteration	Cardinality		
	U_1	U_2	U_{12}
0	209	199	592
1	132	407	461
2	79	596	325
3	52	769	179

(continued)

Table 1. (*continued*)

Iteration	Cardinality		
	U_1	U_2	U_{12}
4	35	832	133
5	18	890	92
6	12	932	56
7	9	962	29
8	5	976	19
9	4	984	12
10	4	990	6
11	2	992	6
12	1	995	4

This table illustrates the action of the modifier *much larger* on the particular database. These results were obtained using the programming language *R*, and, particularly, the library *e-1071* for clustering.

6 Results and Conclusions

As a result of the work done, the desired algorithm for changing the value of a linguistic variable under the action of modifiers was described; also some of its properties were proved, namely, we have obtained the relations between the new grades of membership (after the action of the modifier) and the old ones; the criterion of stopping in special cases also was determined. This criterion is based on the algorithm's convergence in such cases. It should be noticed that the results obtained in the present paper are in no way dependent on the particular database and can be applied to any universal set with a given order. The practical application of this theory is the task of extracting from the database those objects that satisfy the natural semantics of the modifiers used. In the future it is planned to expand the list of investigated modifiers (this, probably, can make our algorithm more personified and more corresponding to the natural language), to consider linguistic variables with a larger number of basic terms and to research what happens to the degree of fuzziness under the described algorithm.

Acknowledgements. My thanks are due to A.P. Ryjov for his insightful comments on earlier drafts and useful tips on the organization of my research.

References

1. Zadeh, L.A.: A fuzzy-set-theoretic interpretation of linguistic hedges. J. Cybern. **2**(3), 4–34 (1972)
2. Ryjov, A.P.: Elements of the theory of fuzzy sets and its applications. Moscow (2003). (in Russian)
3. Ryjov, A.P.: Some problems of optimization and personification of social networks. LAP Lambert Academic Publishing (2015). (in Russian)
4. Bezdek, J.C.: Pattern Recognition with Fuzzy Objective Function Algorithms. Plenum Press, New York (1981)

Complex Search Queries in the Corpus Management System

Damir Mukhamedshin[(✉)], Olga Nevzorova, and Aidar Khusainov

Institute of Applied Semiotics, Tatarstan Academy of Sciences, Kazan, Russia
damirmuh@gmail.com, onevzoro@gmail.com,
khusainov.aidar@gmail.com

Abstract. This article discusses the advanced features of the newly developed search engine of the "Tugan tel" corpus management system. This corpus consists of texts written in the Tatar language. The new features include executing complex queries with arbitrary logical formulas for direct and reverse search; executing complex queries using a thesaurus or word form/lemma list and extracting some types of named entities.

Complex queries enable to automatically extract and annotate semantic data from a corpus for linguistic applications. These options improve the search process and also enable to test the lexicon and collocations in the corpus.

1 Introduction

The electronic corpus as a collection of marked up texts is a rich source of research material. Corpus data processing is carried out on the basis of functions of a corpus management system, which include uploading data to the repository, as well as processing and output of search results within the corpus data. Texts of the electronic collection must have at least a grammatical markup of word forms. Corpus managers support direct search by the specified word form, search by lemma, as well as reverse search by a set of grammatical characteristics. Some developed corpora have syntactic and semantic annotations, alongside grammatical one. Syntactic annotation of corpus data requires a developed syntactic parser on the basis of the chosen formal syntactic model; semantic annotation needs a vast semantic classification of the lexicon. There are basically no systematic semantic dictionaries that would describe the lexicons of low-resource languages, which impedes a comprehensive semantic annotation of corpus data.

This paper presents new search functions that are being developed for the corpus manager system designed for the project "Tugan Tel electronic corpus of the Tatar language". At present the word forms of the Tatar corpus have only morphological markup generated by a morphological analyzer of Tatar. Semantic resources of the Tatar language are notably heterogeneous. There are various particular tightly-themed dictionaries in different subject spheres, an explanatory dictionary published in 2005, and diverse bilingual dictionaries; however, ideographic and extensive semantic dictionaries are missing. The absence of theoretical basis hampers development of semantic annotation models of the Tatar corpus data. New search queries enable to extract semantic data from the Tatar corpus which has neither syntactic nor semantic data markup. A new type of search query is a combination of properly standard direct

© Springer International Publishing AG 2017
N.T. Nguyen et al. (Eds.): ICCCI 2017, Part II, LNAI 10449, pp. 407–416, 2017.
DOI: 10.1007/978-3-319-67077-5_39

query (setting target words) and algorithms of processing contexts of search outputs. On the basis of the developed algorithms, we carried out a series of experiments. Thus basing on this type of search query one can extract practically any set of named entities from a corpus that is not marked up semantically, just by target search query.

The paper will present basic stages of search processing by the corpus management system for the Tatar language and will describe the search language and the formal methods of realizing search queries which are sent to the corpus data repository. A special attention will be paid to the new type of search based on identifying named entities, as well as to other complex search types that allow enhancing the capacity of data extraction from the corpus.

2 "Tugan Tel" Corpus Management System

The Tatar corpus management system (www.corpus.antat.ru) is developed at Institute of Applied Semiotics of the Tatarstan Academy of Sciences. The main functions of the corpus management system are searching for lexical units, making morphological and lexical searches, searching for syntactic units, n-gram searching based on grammar and others. The core of the system is the semantic model of data representation. The search is performed using common open source tools. We use MariaDB database management system and Redis data store. Our purpose is to design the corpus management system for supporting electronic corpora of Turkic languages. This line of research is developing very rapidly.

Among well-known electronic corpora projects for Turkic languages are the corpora of Turkish and Uyghur [1], Bashkir, Khakass, Kazakh (http://til.gov.kz), and Tuvan languages. "Tugan Tel" Tatar national corpus is a linguistic resource of modern literary Tatar. It comprises more than 100 million word forms, at the rate of November 2016. The corpus contains texts of various genres: fiction, media texts, official documents, textbooks, scientific papers etc. Each of the documents has a meta description [8]: author, title, publishing details, date of creation, genre etc. Texts included in the corpus are provided with morphological markup, i.e. information about part of speech and grammatical properties of the word form [9]. The morphological markup is carried out automatically on the basis of the module of two-tier morphological analysis of the Tatar language with the help of PC-KIMMO software tool.

2.1 Related Work

There are a lot of papers devoted to the development of corpus management systems. One of the most known system is Sketch Engine [3] corpus analysis tool (https://www.sketchengine.co.uk/). Having been conceived as downloadable software, it is nowadays a website with a multifunctional concordancer and many corpora for many languages which are ready to be explored through its many tools. The Sketch Engine (Czech Corpus [4]) has the following advantages: it supports arbitrary metadata of documents, uses its own query language (CQL – Corpus Query Language), supports reverse search, phrase search and semantic search using semantic annotations, allows one to view the statistics of the corpora and to select lists of words and n-grams by parameters; it also handles documents in different formats and uses NoSQL database. But the Sketch Engine has some limitations too. Among these are the following: it has complicated reverse search, processes search queries for a long time, has no speed optimizations, demonstrates DDoS vulnerability, and has a big search engine results page. There is a vast number of tools that process corpora in many different ways. Among them are Antconc [2], Wordsmith tools [5], the BonTen Corpus Concordance system [6] and others. A brief overview of the corpus management systems is given in [7].

The Russian National Corpus management system (www.ruscorpora.ru) [10] uses a ready kernel like "Yandex.Server". This system is proprietary and its full version is distributed on a commercial basis. The system enables to execute direct and reverse search queries, use semantic search functionality based on semantic annotations, logical expressions in morphological queries (AND, OR operators); it has no speed optimization for the morphological search, since it was developed as a system for direct search and search based on morphology (on lemmas) of the Russian language. The use of such systems requires fine-tuning the hardware and software environment to achieve the best results.

3   Arbitrary Morphological Formulas in Complex Queries

3.1   Building Search Queries

Typically, the user search query to the corpus management system is an ordered 7-tuple which includes the following elements: search type and mode, the lexical and morphological components, search direction, and the minimum and maximum intervals between the queries. A search query example is a tuple Q, which is displayed in (1).

In this search query Q_1 is the first tuple consisting of the following elements. *Lemma* is the search type (i.e. search by Lemma), *malay* is the lexical component (*malay (Tat)/boy*), *N|ADJ* is the morphological component (the formula that means "a noun or an adjective"), *right* is the search direction (i.e. to the right from the preceding tuple), *number 1* is the minimum interval from the previous tuple, considering the direction, *number 10* is the maximum interval from the previous tuple, considering the direction, *exact* is the search mode (i.e. exact search).

Q_2 is the second tuple of this query. It includes the following components. *Utyr* is the lexical component (*utyr (Tat)/sit*), *V|CONJ* is the morphological component (the morphological formula meaning "a verb or a conjunction"). Other elements duplicate those in Q_1 tuple.

Q_1 and Q_2 tuples together make Q tuple.

$$Q_1 = (\text{lemma, malay, N|ADJ, right, 1, 10, exact})$$
$$Q_2 = (\text{lemma, utyr, V|CONJ, right, 1, 10, exact}) \tag{1}$$
$$Q = (Q_1, Q_2)$$

Such elements as type, search mode, direction, and intervals are deterministic. No additional analysis is required to establish their values. However, we can determine the values of morphological and lexical components only in the course of post-processing.

Lexical Component of a Query. We can define the lexical component not only in the case of a separate word form or lemma (for instance *malay* (Tat)/boy). It can also be set as a part of a word structure (for example, *kita**) or as a rule with exceptions (for instance *malay-malaylar*). Search for lemmas is used in case it is required to find all word forms that contain lemma *malay*, excluding the word form *malaylar* (*malaylar* (Tat)/boys), in the latter case. A proper semantic interpretation of a query by the corpus management system is indispensable if we want to take into consideration this extended syntax of the lexical component.

The following pattern illustrates how the lexical component of a query is processed. At the beginning it is searched for by the system in the class of Word forms or the class of Lemmas, which depends on the search type. The elements that belong to the Word form class are 2-tuples in the form (2). Those of the Lemma class are 2-tuples in the form (3).

$$(\text{word form}, \text{identifier}) \tag{2}$$

$$(\text{lemma}, \text{identifier}) \tag{3}$$

For the first element of this 2-tuple, all parts of the lexical component must be separated by whitespaces. At this point search rules are formulated by the system for the results in the Parse Class set of elements. To illustrate this, for the query *mala** the search proceeds for all word forms that begin with *mala* (*malay, malaylar, malae*, etc.). These word forms constitute a W set of target forms. For instance, form (4) presents the rule for search under discussion. Here w is the "word form" component in a tuple of a Parse class element.

$$w \in W \tag{4}$$

The user is informed by the system about a syntax error in case if a significant, but not exclusive part of the lexical component is missing in the corresponding class.

Morphological Component of a Query. It is a major challenge to define the semantics of the morphological component of the search query. A complex morphological formula with morphological variables and logical operations can be used here (in particular, conjunction, disjunction, and negation). Thus the formula can constitute a conjunction of variables (5), a disjunction of variables (6), a negation of the disjunction of variables (7) or take any other logical form.

$$V, PRES, PASS \tag{5}$$

$$V|PRES|PASS \tag{6}$$

$$!(V|PRES|PASS) \tag{7}$$

$$!V|PRES, PASS \tag{8}$$

The morphological formulas in (5)–(8) can be described as follows:

(5): {V (verb) AND PRES (present tense) AND PASS (passive form)}. The system searches for a set of word forms which have all the above-mentioned morphological features.

(6): {V (verb) OR PRES (present tense) OR PASS (passive form)}. The system searches for a set of word forms with at least one from the above-mentioned morphological features.

(7): not {V (verb) OR PRES (present tense) OR PASS (passive form)}. The system searches for a set of word forms with no above-mentioned morphological features.

(8): {not (V (verb)) OR PRES (present tense) AND PASS (passive form)}. The system searches for a set of word forms which simultaneously do not have morphological feature V or have morphological features PRES and PASS.

Consequently, search queries are processed using various algorithms depending on the formula type.

3.2   Processing Search Queries

Figure 1 displays the general algorithm for processing the types of formulas that are presented in (5)–(7).

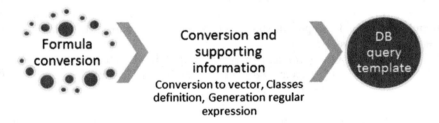

Fig. 1. General algorithm for processing arbitrary formulas.

The main algorithm steps include the following operations:

- The formula is converted to a tuple of morphological notations with the symbol of decomposition operation "," or "|", with any extra characters removed (tildes, brackets, exclamation marks and other signs).
- Information is converted and supported. (A) Converting morphological notations into a bit vector is necessary for using them in the query. (B) Classes for searching for the required data are defined by the algorithm on the basis of morphological notations. (C) The system generates a regular expression to mark the appropriate entities in the context.
- The database query is performed.

4 Performing Searches Using Thesauri or Lists of Words

Searching in a linguistic corpus using a thesaurus significantly expands the range of tasks performed by the search engine of the corpus management system. The search engine of the corpus management system "Tugan Tel" implements this functionality as a search by the list of word forms and/or lemmas. This approach allows one to search without restrictions within the existing database and to perform arbitrary searches.

Complex search queries using the list of word forms and/or lemmas support the same functionality as conventional searches in "Tugan Tel". For example, there is a possibility of adding to the search query a morphological formula that would apply to each item in the list. Also the search engine enables phrasal search using lists of word forms and/or lemmas.

The search query using lists of word forms and/or lemmas presents the same tuple of 7 components as Q_1 or Q_2 in (1), but the lexical (second) component is a list of word forms or lemmas divided between them by the vertical line «|», as shown in (9). This example shows a search through the list of two word forms: house ($yort$ − Tat), home (θy − Tat).

$$Q_1 = (\text{lemma}, yort|\theta y, \text{DIR, right, 1, 10, exact})$$
$$Q = (Q_1) \tag{9}$$

An example of phrasal search query using a thesaurus is shown in (10). The result of this query is an intersection of a multitude of contexts in which the first query component is found, and a multitude of contexts in which the second query component is found. The second component shows search query by lemma list: go (bar − Tat), leave (kit − Tat), pass (kil − Tat).

$$Q_1 = (\text{lemma}, yort|\theta y, \text{DIR, right, 1, 10, exact})$$
$$Q_1 = (\text{lemma}, bar|kit|kil, \text{""}, \text{both, 1, 10, exact}) \tag{10}$$
$$Q = (Q_1, Q_2)$$

Due to the database architecture of the search engine in "Tugan Tel" corpus management system it is required in average no more than 1 s to perform complex searches with the use of lists of word forms and/or lemmas.

5 Extracting Named Entities

Extracting named entities from corpus data allows, on the one hand, to directly retrieve the required data by query, and on the other hand, to test the corpus for containing particular information and to replenish it with documents that include the missing data. The algorithm of extraction of named entities proposed in this paper enables to obtain semantic samples for corpora that do not have semantic data markup. On the other hand, the algorithm has no restriction on the semantic types of extracted data, i.e. the semantic type is defined by the keyword in the query.

5.1   Describing an Algorithm of Extracting Named Entities

The algorithm for extracting named entities is based on the idea of comparing n-grams. The comparison is made within the entire corpus volume, thereby increasing the accuracy of the results.

The extraction process is iterative, the threshold number of iterations specified by the user. The first step presents sampling by the initial search query. The initial search query may be a query on the word form, lemma or phrase, or a search by morphological parameters. A list of bigrams and their frequency is collected across the sample. The bigrams which contain the results are advanced one position to the left or right (set by the user). The resulting list is sorted by frequency of bigrams in order from largest to smallest, to be cut to a predetermined covering index (for example, 95% of all results, this rate being set by the user). This result is used in the second iteration of the algorithm. Each bigram is searched for in the mode of phrasal search in the corpus. Search results are involved in composing a list of trigrams which are advanced one position to the left or right, and their frequency. The resulting list of trigrams is also sorted by frequency in order from largest to smallest, and is cut to a predetermined covering index.

The third and subsequent iterations (until the threshold number of iterations is reached or no match is found as a result of iterating) use the list of n-grams received from the previous iteration. The corpus is searched for each n-gram in the phrasal search mode, and a list of (n + 1)-grams is made up. The resulting list is then cut to a predetermined covering index and compared with the list of n-grams derived from the previous iteration. The comparison accuracy P is set by the user as a percentage. If n-gram frequency is less than P from the quantity of the found (n + 1)-gram, then the n-gram is considered the found named entity, otherwise the extraction proceeds. Thus, the final result will represent a list of the most stable n-grams of different lengths, including search results by the initial search query.

A request to retrieve named entities is an extension of Q-tuple, presented in (1). In addition to the search query, there are added components defining the threshold number of iterations to the left (L) and right (R), the covering index (C), and the accuracy of matching (P). A search example is presented in (11).

$$Q = (Q_1, Q_2, L, R, C, P) \tag{11}$$

5.2   Experiments

As part of the task of enhancing named entity search a number of experiments have been carried out. One of the most revealing of them was search for the names of ministries. The initial search query for the experiment was (12).

$$Q = ((\text{wordform}, ministrlygy, \text{""}, \text{right}, 1, 10, \text{exact}), 7, 0, 95, 80) \tag{12}$$

The result of this query is a list of 50 n-grams containing word form "*ministrlygy*" in the last position. The reference list of names of ministries presented on the Republic of Tatarstan government website (http://prav.tatarstan.ru/tat/ministries.htm), contains

17 items. 12 of 17 items were found in the corpus by means of the algorithm, so the results overlap is 70.6%. 5 items were not found in the corpus for the reasons described in Table 1. The remaining 33 n-grams are different spelling variants of names of ministries.

Table 1. List of unfound names of ministries.

Name	Reason
Urman huҗçalygy ministrlygy (Tat) – ministry of forestry	Overlap of the sequence of word forms with the sequence in another name «*huҗçalygy ministrlygy*» *(Tat) – ministry of property* and «*Transport həm yul huҗçalygy ministrlygy*» *(Tat) – ministry of transport and road management*
Yashlər eshlәre həm sport ministrlygy (Tat) – ministry of youth and sport	Corpus meanings not corresponding to the official name
Transport həm yul huҗçalygy ministrlygy (Tat) – ministry of transport and road management	Overlap of the sequence of word forms with the sequence in another name «*huҗçalygy ministrlygy*» *(Tat) – ministry of property* and «*Urman huҗçalygy ministrlygy*» *(Tat) – ministry of forestry*
Hezmət, halykny el bеlən təemin ity həm social yaklau ministrlygy (Tat) - ministry of labour, employment and social protection	Corpus meanings not corresponding to the official name
Ecologia həm tabigy baylyklar ministrlygy (Tat) – ministry of ecology and natural resources	Corpus meanings not corresponding to the official name

Another experiment was concerned with street names search. The search query for this experiment is (13).

$$Q = ((\text{wordform}, uramy, \text{""}, \text{right}, 1, 10, \text{exact}), 7, 0, 95, 80) \qquad (13)$$

The result of this query is a list of 600 n-grams containing word form "*uramy*" in the last position. We obtained the following results after manual data evaluation: 432 (72%) n-grams are street names, 72 (12%) n-grams are also street names, but require special character filtering, 96 (16%) n-grams are not street names for various reasons (for example, any sentences containing the word "*uramy*"; post addresses and others).

One more experiment was concerned with restaurant names search. The search query for the experiment is (14).

$$Q = ((\text{wordform}, restorany, \text{""}, \text{right}, 1, 10, \text{exact}), 7, 0, 95, 80) \qquad (14)$$

The result of this query is a list of 40 n-grams containing word form "*restorany*" in the last position. We obtained the following results after manual data evaluation: 7 (17.5%) n-grams are restaurant names, 5 (12.5%) n-grams are also restaurant names, but require special character filtering, 28 (70%) n-grams are not restaurant names (any sentences containing the word "*restorany*"). The main reason for obtaining such results is that restaurant names are often written in quotes, and the algorithm treats this symbol as a separate entity in the n-gram.

5.3 Temporal and Qualitative Indicators of Implementing a Query for Extracting Named Entities

The experiments showed that the time of implementing a query for extracting named entities depends on the number of found items and bigrams by the initial search query, and on indexes of covering and accuracy of comparison. All the experiments were executed on machine with following characteristics: 4 core Intel Core i7 2600 (2.6 GHz), 16 GB RAM (4 × 4 GB, 1333 Hz), SSD 120 GB, HDD 3 TB (3 × 1 TB, RAID 0). On the test machine Ubuntu Server 14.04 LTS is running. Table 2 shows the timing indicators of search implementation. Algorithm tests revealed dependence of the quality of the results on the number of results found in the first step of the algorithm. This is due to the fact that a smaller number of results increase the actual data coverage and the data which the algorithm works with, may initially include particular cases. More results in the first step suggest that at the first cutting of the bigram list, only those will remain that will be included in the final list of the extracted named entities. Thus it is only needed to find the left or the right border for this list.

Table 2. Temporal indexes of implementing searches for extraction of named entities

Search query	Quantity of found items	Quantity of found bigrams	Time elapsed
Q = ((wordform, ministrlygy, "", right, 1, 10, exact), 7, 0, 97, 80)	27746	68	127.37 s
Q = ((wordform, restorany, "", right, 1, 10, exact), 7, 0, 95, 80)	118	40	261.90 s
Q = ((wordform, uramy, "", right, 1, 10, exact), 3, 0, 95, 80)	9592	600	848.07 s

6   Conclusion

Many existing search engines of corpus managers have the capabilities of semantic search using a preliminary semantic text annotation. The presented types of complex searches allow for implementing semantic search in a corpus that has no semantic markup, as well as to automate the process of semantic annotation of corpus data. Extracted named entities will in the future enable to add a semantic markup to texts in Tatar, as well as to fine-tune the morphological analyzer, which will lead to a reduction in cases of morphological ambiguity in the corpus. One of the possible areas of their use is also classification and clustering corpus texts. Currently we are extensively exploring possibilities of filling the semantics of corpus texts metadata with the help of semi-automatic or automatic semantic annotation.

Common algorithms of implementing the expanded functionality of search engines of "Tugan Tel" corpus management system are applicable for the corpora of various national languages (not only for the Tatar corpus) without significant changes in the original code. The modules architecture guarantees full functionality for addressing various linguistic issues due to flexibility of search queries.

Acknowledgment. The reported study was funded by Russian Science Foundation (research project № 16-18-02074).

References

1. Aibaidulla, Y., Lua, K.T.: The development of tagged Uyghur corpus. In: Proceedings of PACLIC17, pp. 1–3 (2003)
2. Anthony, L.: AntConc: a learner and classroom friendly, multi-platform corpus analysis toolkit. In: Proceedings of IWLeL 2004: An Interactive Workshop on Language e-Learning, pp. 7–13 (2004)
3. Kilgarriff, A., Baisa, V., Bušta, J., Jakubíček, M., Kovář, V., Michelfeit, J., Suchomel, V.: The Sketch Engine: ten years on. Lexicography **1**(1), 7–36 (2014)
4. Křen, M.: Recent developments in the Czech National Corpus. In: Proceedings of the 3rd Workshop on Challenges in the Management of Large Corpora (CMLC-3), pp. 1–4 (2015)
5. Scott, M.: Wordsmith Tools. Oxford University Press, Oxford (1996)
6. Asahara, M., Maekawa, K., Imada, M., Kato, S., Konishi, H.: Archiving and analysing techniques of the ultra-large-scale web-based Corpus Project of NINJAL, Japan. Alexandria **25**(1–2), 129–148 (2014)
7. Kouklakis, G., Mikros, G., Markopoulos, G., Koutsis, I.: Corpus manager a tool for multilingual corpus analysis. In: Proceedings of Corpus Linguistics Conference 2007. http://www.birmingham.ac.uk/documents/college-artslaw/corpus/conference-archives/2007/244Paper.pdf
8. Nevzorova, O., Mukhamedshin, D., Kurmanbakiev, M.: Semantic aspects of metadata representation in corpus manager system. In: Open Semantic Technologies for Intelligent Systems (OSTIS-2016), pp. 371–376 (2016)
9. Suleymanov, D., Nevzorova, O., Gatiatullin, A., Gilmullin, R., Hakimov, B.: National corpus of the Tatar language "Tugan Tel": grammatical annotation and implementation. Proc. Soc. Behav. Sci. **95**, 68–74 (2013)
10. Zakharov, V.: Corpora of the Russian language. In: Habernal, I., Matoušek, V. (eds.) TSD 2013. LNCS, vol. 8082, pp. 1–13. Springer, Heidelberg (2013). doi:10.1007/978-3-642-40585-3_1

Entropy-Based Model for Estimating Veracity of Topics from Tweets

Jyotsna Paryani, Ashwin Kumar T.K., and K.M. George[✉]

Computer Science Department, Oklahoma State University, Stillwater, OK, USA
{paryani, tashwin, kmg}@cs.okstate.edu

Abstract. Micro-blogging sites like Twitter have gained tremendous growth and importance because these platforms allow users to share their experiences and opinions on various issues as they occur. Since tweets can cover a wide-range of domains many applications analyze them for knowledge extraction and prediction. As its popularity and size increase the veracity of the social media data itself becomes a concern. Applications processing social media data usually make the assumption that all information on social media are truthful and reliable. The integrity of data, data authenticity, trusted origin, trustworthiness are some of the aspects of trust-worthy data. This paper proposes an entropy-based model to estimate the veracity of topics in social media from truthful vantage point. Two existing big data veracity models namely, OTC model (Objectivity, Truthfulness, and Credibility) and DGS model (Diffusion, Geographic and Spam indices) are compared with the proposed model. The proposed model is a bag-of-words model based on keyword distribution, while OTC depends on word sentiment and DGS depends on tweet distribution and the content. For analysis, data from three domains (flu, food poisoning and politics) were used. Our experiments suggest that the approach followed for model definition impacts the resulting measures in ranking of topics, while all measures can place the topics in a veracity spectrum.

Keywords: Twitter · Micro-blog · Veracity · Information entropy

1 Introduction

Users in micro-blogging sites like Twitter can form a Social or an information network [1]. Since the data from the micro-blogging sites can be about multiple events/entities it has become an invaluable information source. Thereby there has been a consistent growth in the number of applications that analyze social media data for knowledge extraction and prediction [2]. In International Data Corporation report, it is estimated that "from now until 2020 the data will double every two years", therefore resulting in exabytes of data [3]. With this tremendous growth in information propagated via social media concerns about the trustworthiness of the data itself is inevitable. Data integrity, authenticity, source and trustworthiness are some of the aspects of data veracity [4]. According to IBM Big Data and Analytics Hub [5], 27% of the respondents were not sure about the accuracy of the data, and one in three decision-makers do not trust the information used for analyzing the data. In addition to this uncertainty volume, velocity

© Springer International Publishing AG 2017
N.T. Nguyen et al. (Eds.): ICCCI 2017, Part II, LNAI 10449, pp. 417–427, 2017.
DOI: 10.1007/978-3-319-67077-5_40

and dynamic nature of big data brings forth additional concerns about the trustworthiness of the data [4, 7, 8]. With the increase in size and variety of data, the data available for analysis should not be outdated or manipulated. Moreover, value of the data is hidden in jargon or linguistic which may result in the information not recorded in writing or it may mislead the recipient [6]. Data velocity can also affect the trustworthiness of the data because the application can remove or alter important pieces of data to improve the processing time. In this paper, we propose entropy as a veracity model of information in social media from the truthful vantage point. It depends on the bag-of-words definition of topics. The novelty of the proposed model is that it estimates the veracity of information from the blog data itself without relying on external data, as it may not be feasible to collect and analyze external information. There are some existing models that estimate information veracity in big data ecosystem. They are the Objectivity, Truthfulness, and Credibility (OTC) model [9] and the DGS measures of veracity [10]. OTC model partially depends on external data, and DGS measures use properties associated with the data and the distribution to estimate the veracity. The model proposed in this paper is depended only on the document content and topic definitions based on data domains.

Our contributions in this paper are as follows (1) A new model to measure veracity based on bag-of-words topic definition and entropy is proposed. The model is domain independent and applies to all types of microblog data. External verification is not part of the model. (2) An empirical analysis is presented based on Twitter data from different domains comparing the proposed model against two previously published models. (3) We have also shown that veracity measures can be applied as domain specific topic comparison measures. (4) As it is difficult to evaluate veracity as a binary outcome from data, we propose to view it as a spectrum and the models to determine the position of topics in the spectrum. As there is no benchmark for model validation, comparison of the models among themselves is an alternative. These models are applied to the topics that were identified in three Twitter datasets. The first dataset contains tweets about flu, the second dataset contains tweets about food poisoning, and the third contains tweets about politics. These topics were chosen based on their significance and data availability. Rest of the paper is organized as follows. Section 2, provides a review of related works. In Sect. 3 data collection and pre-processing methodologies are explained in detail along with the proposed model to estimate veracity. Section 4 analyzes the experimental results and in Sect. 5, conclusion and directions to future work are presented.

2   Review of Literature

The abundance and diversity of data has opened several new opportunities in big-data but on the same hand it has raised questions about blindly trusting data collection, pre-processing, storage, data quality, nodes which store data, cloud service providers and information sharing [11]. This paper deals with the trust factor of the data and is based on different works proposed earlier. In this section, the previous published researches that contribute to the model definition such as topic modeling, sentiment computation, information theory and veracity models are described in detail.

2.1   Topic Modeling

Topic modeling refers to a generative model that is used for analyzing large quantities of unlabeled data. A topic is the probability distribution over the collection of words (bag-of-word) and topic model is the statistical relationship between a group of observed and unknown random variables that specifies a probabilistic procedure to generate the topics. One of the most popular topics modeling technique is the Latent Dirichlet Allocation (LDA) [12]. LDA is a generative probabilistic model, which extracts the topics in the text, based on co-occurrence of words with the topic in the document. LSA [13] and PLSA [14, 15] are also similar models.

2.2   Information Measure

In the proposed model, information entropy measure is used to estimate veracity of topics. Shannon originally proposed Information entropy in [16] known as Shannon's mathematical theory of communication. Shannon's theory deals with information to be conveyed with three communication problems such as accuracy of the information to be transmitted; how precisely the meaning is transferred and from all the information transferred how much is selected from the set of messages. Equation 1 estimates entropy in which, n refers to the number of different outcomes. The range of entropy is $0 \leq \text{entropy} \leq \log(n)$. Maximum entropy (log n) occurs when all the probabilities have equal values i.e. $1/n$. Minimum entropy (0) occurs when one of the probabilities is 1 and rest is 0's [16].

$$H = -\sum_{i=1}^{n} p_i \log p_i \tag{1}$$

2.3   Big-Data Veracity Models

There are limited veracity models available for big-data in the literature, one such model is the big data veracity model: Objective, Truthful and Credible (OTC) [9]. Details of the OTC model are as follows. OTC is a three-dimensional model proposed by Lukoianova and Rubin [9]. The first dimension of the model is the objectivity/subjectivity of the data. Objectivity deals with facts, truth, reality and reliability [18]. The second dimension, truthfulness in the textual data can be determined by checking if there is any false belief or false conclusion in the text [20, 21], which can be verified with the help of deception test. The last dimension in the OTC model is the credibility of the data. Credibility deals with two qualities, trustworthiness and expertise. The credibility of the text can be calculated using Mutual Information between word, which consists of performing analysis on frequently occurring nouns and verbs with the trust and credibility [9] using online corpus COCA [22]. This model needs external information for computation. Another big data veracity model in the literature is the DGS model proposed in [10], which deals with evaluating the accuracy of data from the tweets themselves without reference to external sources. The objective of this model is to compute veracity measures without using external resources. The DGS model proposes three measures based on the spread of information in term of volume, geographic spread and repetition in the volume. It is based on the argument that the information with high volume and inflation rate spreads widely and could be questionable. All the measures are defined with tweets as the source of data. The three measures proposed in the DGS model are Topic Diffusion, Geographic Dispersion and Spam Index [10]. The first measure, Diffusion Index deals with how fast the information has spread through Twitter. It deals with the concept that fake information has spread faster than the truth. [17]. The second measure, Geographic Dispersion is used to measure the extent to which the information is spread geographically. Spam Index, which deals with the impact of repeated tweets by the same user. Repeated tweets can be viewed as inflating the diffusion. The measure is similar to spamming which propagates questionable information. OTC model proposed in [9] uses the content of tweets along with external sources to estimate the veracity of tweets. The DGS model uses the properties associated with the tweet to estimate its veracity but it doesn't use the tweet text itself. It is based on the Tweet distribution conditioned on the properties. The major characteristic of the proposed information entropy model is that it only depends on the key word distribution in the tweets. This will be very useful when any corroborating evidence is not readily available.

3 Methodology

This section contains a brief description of the data collection process, data preprocessing, implementation of the OTC model and the proposed evaluation measure. The tools used in this research are Apache Hadoop and Apache Flume.

3.1 Data Collection

Flu dataset consists of tweets streamed from the Twitter API during the time period 06/1/2015–11/30/2015. The size of this raw dataset is about 216 GB. In this dataset there were approximately 45,518,318 tweets. Food poisoning dataset consists of approximately 4,063,205 tweets streamed during the time period 11/1/2016–01/15/2017. The size of this raw dataset is about 18 GB. Tweets belonging to the politics dataset were collected during the time period 2/17/2017–3/17/2017. There were approximately 13,476,717 tweets in this dataset. Politics dataset was about 200 GB. Keywords used to collect the tweets related to flu, food poisoning and politics were obtained from [19, 23, 24] respectively.

3.2 OTC Model Implementation

Objectivity and Truthfulness were calculated using a text-processing library "*Text Blob*" in Python [8]. The sentiment property of "Text Blob" calculates the subjectivity and polarity of the given text [8]. Objectivity in the OTC model is calculated from subjectivity and varies from 0 to 1. Polarity calculated in "Text Blob" library provides a measure of truthfulness. Credibility is computed using mutual information (MI) between two words. The mutual information deals with co-occurrences between two words. Mutual Information is defined in Eq. 2. We compute MI using Eq. 3 where n1 and n2 are the number of occurrences of a pair of words.

$$MI = \frac{\text{probability of word1 } \& \text{ } word2}{\text{probability of word1} * \text{probability of } word2} \tag{2}$$

$$MI = \frac{n1 * \frac{2}{n^2} + n2 * \frac{2}{n^2}}{\frac{2n-1}{n^2}} \tag{3}$$

Finally, all three dimensions of the OTC model are combined into one veracity index, by normalizing the dimensions in the range of (0, 1) interval with 1 being maximum objectivity, truthfulness, credibility and 0 being minimum objectivity, truthfulness, credibility [9].

3.3 Entropy Measure Computation

Shannon's entropy can be used to measure the ambiguity in the information contained in a text. We interpret this ambiguity as the measure of veracity implicit in the tweets. The entropy is calculated based on Shannon's entropy formula in Eq. 1. In Eq. 1, where p_i represents the probability associated the ith keyword defining the topic. Keyword probabilities are computed by the formula $p_i = \frac{n_i}{N}$ where N is the total number of words obtained from related tweets after excluding stop words and other insignificant words and n_i is the number of occurrences of the ith key word. Lower the information entropy value higher the veracity. Higher the information entropy value, lower the veracity.

3.4 DGS Computation

The formulae given in [10] are used to compute the three components of the measure, namely Diffusion index, Geographic spread and Spam index which is given below:

$$\text{Diffusion Index} = (\#\text{Unique Users})/(\text{Total tweets}) \tag{4}$$

$$\text{Geographic Spread Index} = (\#\text{Unique Location})/(\text{Total tweets}) \tag{5}$$

$$\text{Spam Index} = \sum_{\text{over unique user}} \frac{1}{\text{unique user tweet count}} /(\text{Total tweets}) \tag{6}$$

4 Analysis of Results

This section contains a brief description of the topic extraction process, analyses of veracity measures OTC, DGS and information entropy model, comparison of these results and inferences are drawn from the comparison. For practical reasons, we adopted a manual topic extraction approach as opposed to applying another model such as LDA for the topic extraction.

4.1   Topic Extraction Process

Three topics each were defined for the Flu and food poisoning datasets based on key words collected from the Center for Disease Control (CDC). For flu, topics numbered 1 to 3 are symptoms, side effects, and treatments. The topics for the food poisoning dataset in order are Bacterial Foodborne germs, Viral Foodborne germs, and Parasitic Foodborne germs. Social Issues, Economic Issues, Domestic Issues, and International Issues are topics numbered 1 to 4 related to the politics data domain. The keywords defining the topics were downloaded from the website "ontheissues.org" [23]. The keyword frequencies of topics were computed from the tweets collected for each domain.

4.2   Computed Results for Models

This section tabulates the results obtained in tables. Table 1 shows the calculated entropy scores of topics of all the three datasets using Eq. 1. There are two approaches to the interpretation of these results. The first interpretation is to use the numbers as a measure of the degree of veracity of the tweets related to the topics as specified by the topic key words. In other words, what the proportion of the tweets is closely related to the topics. As the measures are somewhat farther away from zero, the interpretation could be that the tweets are not completely relevant. The second interpretation is to use the numbers as a measure of comparison of the topics. In other words, which topics are better defined by the keywords defining the topics and relevant to the domain. With this approach, notable exceptions are topic 3 of flu data set and topic 2 of politics data set.

Table 1. Entropy score of all datasets

Topics	Flu data	Food poisoning data	Politics data
Topic 1	0.4165	0.4041	0.3512
Topic 2	0.4126	0.4089	0.3266
Topic 3	0.3606	0.4098	0.3403
Topic 4	–	–	0.3456

Table 2 shows the computed average values of topics for the OTC model. The results in Table 2 can be interpreted similar to the entropy model. Based on the numbers, we can state that tweets do not relate to flu topic 1, food poisoning topic 2, and political data topic 3 as well as the other topics. The DGS model is a combination of three indices. Each index is viewed as an axis in three-dimensional space. Tables 3, 4, and 5 show the computed values of the indices of the topics for the three datasets. The tables show the distance between the computed DGS measure and (1, 1, 1). A topic closer to (1, 1, 1) is more trustworthy than another, which is farther. From Table 3 one can infer that the topic 1 is trustworthy in flu dataset.

From Table 4 one can infer that the topic 3 is more trustworthy than others in food poisoning dataset because it is closer to the point (1, 1, 1). From Table 5 one can infer that the topic 2 is less trustworthy in politics dataset.

Table 2. Average OTC score for all datasets

Topics	Flu data	Food poisoning data	Politics data
Topic 1	0.3547	0.4317	0.4429
Topic 2	0.4407	0.4270	0.4480
Topic 3	0.4473	0.4411	0.4332
Topic 4	–	–	0.4443

Table 3. DGS Score for flu dataset

Topics	Diffusion index	Geographic index	Spam index	Distance from (1, 1, 1)
Topic 1	0.32345	0.07823	1.0063e−13	1.62
Topic 2	0.08028	0.02046	2.8402e−13	1.72
Topic 3	0.19226	0.05098	1.1633e−13	1.69

Table 4. DGS score for food poisoning dataset

Topics	Diffusion index	Geographic index	Spam index	Distance from (1, 1, 1)
Topic 1	0.3536	0.01	1.79e−10	1.61
Topic 2	0.2272	0.0067	1.69e−11	1.68
Topic 3	0.4659	0.013	5.85e−11	1.51

Table 5. DGS score for politics dataset

Topics	Diffusion index	Geographic index	Spam index	Distance from (1, 1, 1)
Topic 1	0.1348	0.0237	8.31e−12	1.7133
Topic 2	0.007021	0.0006385	1.75e−09	1.732
Topic 3	0.1348	0.0237	8.17e−12	1.7133
Topic 4	0.1348	0.0237	8.80e−12	1.7133

The notion of trust needs to be interpreted based on the data domain. For example, in the case of food poisoning, trust implies that the data can be used to determine food poisoning outbreak with the high probability. On the other hand, in the case of politics, topic trust can be associated to positions of candidates. Trust in this case should be interpreted such that there is a high likelihood that the candidate is sincere in supporting the topic.

4.3   Comparison of Results

The ranking of topics in their order of veracity (Topic 3, Topic 2 and Topic 1) are same for the entropy and OTC models for the flu dataset. But the rankings of the DGS model differs from these two models. The ranking of topics in their order of trustworthiness (Topic 2, Topic 1 and Topic 3) is similar to the entropy and DGS model for the food poisoning dataset. But the rankings of the OTC model differ from these two models. For politics dataset, none of the models produce same rankings of topics. The OTC model and entropy model are based on word sentiment and word distribution, while the DGS model is more depended on the tweet count. This could be a possible explanation of their behavior. The analysis presented thus far is based on the collection of data during a time period. The comparison of results from these models did not yield a conclusive outcome in the case of ordering. However, they all place the topics closely in a spectrum of veracity from 0 to 1. In the next section we describe another approach by comparing the computed values of these models per-day basis by building a time series.

4.4   Time Series Analysis

The relationship between the models is calculated by performing analysis of variance (ANOVA) and by calculating the correlation coefficient. ANOVA deals with performing statistical hypothesis testing on sample data and testing the results from the null hypothesis. The test results are statistically significant if it is unlikely to have occurred by chance, that is, if the probability (p-value) is less than significance level and F-value is greater than F-critical, then it leads to rejection of the null hypothesis. The null hypothesis considered in this case is, means of all three models (groups) are same. Then the alternate hypothesis would be at least one of the mean is different from the mean of another model. The significance level considered is 0.05. The food poisoning dataset is used for this analysis.

In Table 6, all the p-values are less than 0.05 (significance level) and F-values are greater than 3.05 (F-critical) and so we reject the null hypothesis. So, there is a possibility that at least one mean of a particular model is different from the means of other models. To find the difference between the three models, correlation coefficient is

Table 6. Food poisoning topics p-value and F value

Topics	p-value	F value
Topic 1	1.1E−148	5691.78
Topic 2	6.9E−144	5261.38
Topic 3	2.9E−149	5792.25

Table 7. Correlation of measures for food poisoning data

Topic	Pairing of models	Correlation coefficient (r)	Relationship
1	OTC and DGS	0.002	Weak positive correlation
2	OTC and DGS	−0.057	Weak negative correlation
3	OTC and DGS	−0.146	Weak negative correlation
1	Entropy and DGS	0.103	Weak positive correlation
2	Entropy and DGS	0.057	Weak positive correlation
3	Entropy and DGS	−0.122	Weak negative correlation
1	Entropy and OTC	−0.5675154	Strong negative correlation
2	Entropy and OTC	−1	Strong negative correlation
3	Entropy and OTC	−0.49361	Strong negative correlation

calculated for each topic. Table 7 represents the correlation coefficient scores and the relationship between the models for Topic 1, Topic 2 and Topic 3 for food poisoning data. If the correlation coefficient (r) lies between ±0.5 and ±1; then it is said to be strong correlation, if r-value lies between ±0.3 and ±0.49, then it is said to be medium correlation and if r-value lies below ±0.29, then it is a weak correlation. There is no correlation if value is zero.

Figure 1 shows the daily variations of the three measures in one graph for each topic of the food poisoning dataset. Table 7 shows the correlations and the implied relationships. While the daily graphs demonstrate the agreement of the models on some days, the correlations show more disagreement than agreement. Repeating the time series analysis for all the other topics in the other datasets reveal similar outcomes.

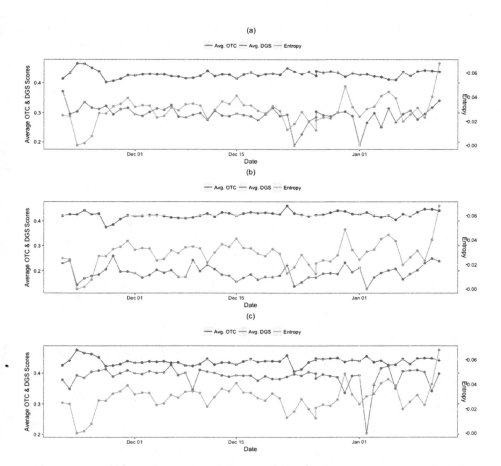

Fig. 1. Time series graphs for all topics in the food poisoning dataset. (a) Topic 1, (b) Topic 2, (c) Topic 3.

5   Conclusion

Micro-blogging sites like Twitter has become an important source of information where people post their real-time experiences and their opinions on various day-to-day issues which can be used to prediction, opinion-mining, and other purposes. Using untrustworthy information for analysis will yield results that will differ from the ground reality. In this paper, we proposed information entropy as a model to estimate veracity of topics (defined as bag-of-words) in micro-blog data. The measure of uncertainty property of the entropy is used as the model basis. This model is compared to the previously published models. The computed model results place the topics in a veracity spectrum in a consistent manner. Flu data available from CDC for the time period corresponding to Twitter flu data period are used for model validation. We interpret the tweets as indicator of flu. The data period analyzed is not considered flu season and the computed model values are not high. So, the data seems to validate the topic placement in the veracity spectrum by the models. No official data was available corresponding to the timeframe of the other datasets. Our analysis shows significant evidence that the model values are dependent on the approaches on which they are based in terms of topic discrimination. Therefore, for the models to be effective, they should be chosen based on the data domain and the meaning of veracity to the topics. However, in all cases, they can be used as measures to compare topics' veracity in a domain and to place them in a veracity spectrum.

References

1. Quan-Haase, A., Martin, K., McCay-Peet, L.: Networks of digital humanities scholars: the informational and social uses and gratifications of Twitter. Big Data Soc. **2**(1), 1–12 (2015)
2. Felt, M.: Social media and the social sciences: how researchers employ Big Data analytics. Big Data Soc. **3**(1), 2053951716645828 (2016)
3. Gantz, J., Reinsel, D.: The digital universe in 2020: Big Data, bigger digital shadows, and biggest growth in the far east. In: IDC iView: IDC Analyze the Future 2007, pp. 1–16 (2012)
4. Demchenko, Y., Ngo, C., de Laat, C., Membrey, P., Gordijenko, D.: Big Security for Big Data: addressing security challenges for the Big Data infrastructure. In: Jonker, W., Petković, M. (eds.) Workshop on Secure Data Management. LNCS, vol. 8425, pp. 76–94. Springer, Cham (2014). doi:10.1007/978-3-319-06811-4_13
5. The Four V's of Big Data, IBM Big Data and Analytics Hub. http://www.ibmbigdatahub.com/infographic/four-vsbig-data. Accessed 11 June 2017
6. Eembi, N.B.C., Ishak, I.B., Sidi, F., Affendey, L.S., Mamat, A.: A systematic review on the profiling of digital news portal for Big Data veracity. Procedia Comput. Sci. **72**, 390–397 (2015)
7. Yin, S., Kaynak, O.: Big Data for modern industry: challenges and trends [point of view]. Proc. IEEE **103**(2), 143–146 (2015)
8. TextBlob. https://pypi.python.org/pypi/textblob. Accessed 04 May 2017
9. Lukoianova, T., Rubin, V.L.: Veracity roadmap: is Big Data objective, truthful and credible? Adv. Classif. Res. Online **24**, 4–15 (2014)

10. Ashwin, K.T., Kammarpally, P., George, K.M.: Veracity of information in twitter data: a case study. In: International Conference on Big Data and Smart Computing BigComp, pp. 129–136. IEEE (2016)
11. Sänger, J., Richthammer, C., Hassan, S., Pernul, G.: Trust and Big Data: a roadmap for research. In: 25th International Workshop on Database and Expert Systems Applications DEXA, pp. 278–282. IEEE (2014)
12. Reed, C.: Latent Dirichlet allocation: towards a deeper understanding. http://obphio.us/pdfs/lda_tutorial.pdf. Accessed 04 May 2017
13. Landauer, T.K., Foltz, P.W., Laham, D.: An introduction to latent semantic analysis. Discourse Process. 25(2–3), 259–284 (1998)
14. Chuois, S.: Probabilistic latent semantic analysis. http://mlg.postech.ac.kr/~seungjin/courses/ml/handouts/handout06.pdf. Accessed 04 May 2017
15. Hofmann, T.: Probabilistic latent semantic indexing. In: Proceedings of the 22nd Annual International ACM SIGIR Conference on Research and Development in Information Retrieval, pp. 50–57. ACM, New York (1999)
16. Shannon, C.E.: A mathematical theory of communication. ACM SIGMOBILE Mob. Comput Commun. Rev. 5(1), 3–55 (2001)
17. Tapia, A.H., Moore, K.A., Johnson, N.J.: Beyond the trustworthy tweet: a deeper understanding of microblogged data use by disaster response and humanitarian relief organizations. In: Proceedings of the 10th International ISCRAM Conference, Baden-Baden, pp. 770–779 (2013)
18. Moser, P.K.: Philosophy After Objectivity: Making Sense in Perspective. Oxford University Press on Demand, New York (1993)
19. A-Z Index for Foodborne Illness: CDC (2016). https://www.cdc.gov/foodsafety/diseases/index.html. Accessed 04 May 2017
20. Buller, D.B., Burgoon, J.K.: Interpersonal deception theory. Commun. Theory 6(3), 203–242 (1996)
21. Zhou, L., Burgoon, J.K., Nunamaker, J.F., Twitchell, D.: Automating linguistics-based cues for detecting deception in text-based asynchronous computer-mediated communications. Group Decis. Negot. 13(1), 81–106 (2004)
22. Davies, M.: The Corpus of Contemporary American English (COCA): 400+ million words, 1990–2012. http://www.americancorpus.org. Accessed 04 May 2017
23. OnTheIssues Home page. http://www.ontheissues.org/default.htm. Accessed 04 May 2017
24. CDC: Influenza Flu. https://www.cdc.gov/flu/index.htm. Accessed 04 May 2017

On Some Approach to Integrating User Profiles in Document Retrieval System Using Bayesian Networks

Bernadetta Maleszka[(✉)]

Faculty of Computer Science and Management, Wrocław University of Science and Technology, Wybrzeze Wyspianskiego 27, 50-370 Wrocław, Poland
Bernadetta.Maleszka@pwr.edu.pl

Abstract. Information retrieval systems are more and more popular due to the information overload in the Internet. There are many problems that can cause this situation: user can not know his real information need when he submits a few words to the browser; these words can have many different meanings; user can expect different results depending on current context. To obtain satisfactory result, the retrieval system should save user profile and develop it. In this paper we propose a method for determining user profile based on content-based and collaborative filtering. Our approach uses Bayesian Networks to develop representative profile of the users' group. We have performed some experiments to evaluate the quality of proposed method.

Keywords: Document retrieval · Knowledge integration · Ontology-based user profile · Bayesian network

1 Introduction

Bayesian networks are very popular method for modeling uncertainty of some process, especially they are used to reason in uncertain conditions [6,7]. An example of environment with a high level of uncertainty is on-line document retrieval system: any time a new user can register to the system and a set of documents is continuously growing. In such a system we would like to model user interests to predict his preferences and recommend him new documents.

The most important issues in document retrieval system are connected with a user. The user would like to obtain relevant results using only a few terms in his queries – he can have a problem to formulate his information needs or simply he can not know specific terminology. On the other hand, his preferences are changing with time and the system should follow by his interests. To solve these problems, the system should gather information about users and analyze their activities to provide him better results. Other aspect of such systems are connected with a new user – the system does not know his preferences and the recommendations can be not satisfactory for the user. The system should

© Springer International Publishing AG 2017
N.T. Nguyen et al. (Eds.): ICCCI 2017, Part II, LNAI 10449, pp. 428–437, 2017.
DOI: 10.1007/978-3-319-67077-5_41

discover preferences of a new user based on only a few first queries or use collaborative filtering techniques to improve this process.

In this paper we consider the above-mentioned problem. An overall idea of our work is to develop a personalized document retrieval system that avoids "cold-start" problem with a new user. We join content-based (CB) and collaborative filtering (CF) methods to propose for a new user non-empty profile (built based on CF methods) and to update the profile using individual CB methods. In our previous works we have presented a method for determining a representative profile from the set of ontology-based users' profiles [9,11] and a method for determining user profile using Bayesian network [12]. In this paper we continue the research, focusing on method to develop a profile using Bayesian network approach. We consider a method for integrating profiles of a group of users with similar interests.

The paper is organized in the following way. In Sect. 2 we describe an idea of Bayesian networks and present some exemplary applications of these methods to discover user profile. Section 3 presents information about the model of documents set and user profile in our system. The developed method for determining user profile is described in Sect. 4. Section 5 shows the results of performed evaluations. In the last Sect. 6 we gather the main conclusions and future works.

2 Related Works

In this section we present an overview of methods for building user profile using Bayesian approaches. In our system we consider a set of terms which are connected to each other by some relations. To model the relations we use ontological nature of user profile. To build a user model we use some methods for developing Bayesian network which enable to discover connections between terms. Using statistical approach can enrich user profile with probabilistic measures that can be further developed for prediction aims.

2.1 Bayesian Network

A Bayesian network is a directed acyclic graph $D = (V, E)$, where V is a finite set of nodes and E is a finite set of directed edges between the nodes [2]. Each node $v \in V$ reflects a random variable X_v and is connected with its parent node $pa(v)$. It is also attached a local probability distribution, $p(x_v|x_{pa(v)})$. Let us use P for the set of local probability distributions for all variables in the network. A Bayesian network for a set of random variables X is then the pair (D, P).

When two nodes in the graph are not connected (there is no edge between them), it means that these nodes are independent. Then we can use the following formula to calculate the joint probability distribution:

$$p(x) = \prod_{v \in V} p(x_v|x_{pa(v)}).$$

We will also use the following theorem which was proved in [4]:

Given a selection variable X_s in a Bayesian network and a node X_i (other than X_s), such that X_i is not an ancestor of X_s, the conditional probability distribution of X_i given $pa(X_i)$ is the same in the general population and in the subpopulation induced by value x_s, i.e.,

$$Pr(x_i|pa(x_i), x_s) = Pr(x_i|pa(x_i))$$

2.2 Application of Bayesian Network in Information Retrieval Systems

In literature one can find two frequently used formulations of naive Bayes in context-based information retrieval. These are: the multi-variate Bernoulli and the multinomial model [14]. Both approaches share the following principles: each document has parametrized model:

$$P(d_i|\theta) = \sum_{j=1}^{|C|} P(c_j|\theta) \cdot P(d_i|c_j; \theta)$$

where c is a class which correspond to the parameters θ. The sum of total probability over all components determines the likelihood of a document d_i.

To calculate the probability we need to learn parameters θ based on training data. Using Bayes rule we can calculate the posterior probability of class membership for a test data.

$$P(c_j|d_i; \hat{\theta}) = \frac{(P(c_j|\hat{\theta}) \cdot P(d_i|c_j; \hat{\theta})}{P(d_i|\hat{\theta})}$$

where $\hat{\theta}$ is a set of estimated parameters of θ.

In paper [5] the main aim of building a probabilistic model is to estimate the probability that the structural characteristics of an element attract user to explore the content of this element and consider it as relevant. Such approach allows to determine context importance. The main advantage of naive Bayes classifier is a quick learning and low computational overhead. These are critical features for an online user modeling system [13]. The authors have proposed naive Bayes classifiers to improve predictive accuracy. Experiments have shown that the requirement of adaptation to the user's performance is also satisfied.

Naive Bayes has been shown to perform competitively with more complex algorithms and has become an increasingly popular algorithm in text classification applications [1]. Based on feedback from the user, the system automatically adapts to the user's preferences and interests using multi-variete Bernoulli event model:

$$P(d_i|c_j; \theta) = \prod_{t=1}^{|T|} (B_t \cdot P(w_t|c_j; \theta) + (1 - B_t) \cdot (1 - P(w_t|c_j; \theta)))$$

where T is a set of terms and B_t indicates if term t appears in document.

A very strong assumption in multi-variete Bernoulli model is that the terms can occur in document independently. Such an approach should not be implemented in a model with domain ontology and relations between the terms.

More complex application of Bayes approach can be found in [17], where probabilistic model is used to unify collaborative filtering and content-based filtering methods: based on content-based probabilistic models for each user's preferences, a system combines a society of users' preferences to predict an active user's preferences. The computational complexity of such solution is not a problem while the system does not require a global training stage and thus can incrementally incorporate new data. On the other hand, Bayesian networks are efficient in a problem of query processing techniques already developed for probabilistic networks and capture a more general class of probability distributions than the previously proposed probabilistic models [19]. Bayesian networks can be also combined with another machine learning techniques, e.g. with case-based reasoning [18].

A specific part of Bayesian approaches are hierarchical Bayesian networks that model tree-structured conditional probability distributions [8]. In document retrieval system with domain ontology, a hierarchical structure of network can be used in case of specification – generalization relation that is hierarchical relation. The power of hierarchical networks is as follows: a node may correspond to an aggregation of simpler types. A component of one node may itself represent a composite structure. This allows the representation of complex hierarchical domains [7]. An important advantage of hierarchical structure is that it reduces the number of parameters required to represent the distribution. Also, resultant hierarchical model is more interpretable than other models.

In this paper we would like to investigate the Bayesian network approach to develop a user profile and to integrate profiles of a group of users.

3   Personalized Document Retrieval System

In this paper we assume that the base of personalized document retrieval system is a library – a set of documents. The second aspect is a user that would like to retrieve some documents from the library. Each document is described by the set of terms coming from domain ontology. The system observes the user to find out user interests based on his search history and tries to recommend him some documents.

In subsections we present formal description of the system and its elements.

3.1   Model of Document

A library is a set of documents:

$$D = \{d_i : i = 1, 2, \ldots, n_d\} \tag{1}$$

where n_d is a number of documents and each document d_i is described by the set of weighted terms:

$$d_i = \{(t_j^i, w_j^i) : t_j^i \in T \wedge w_j^i \in \{0,1\}, j = 1, 2, \ldots, n_d^i\} \tag{2}$$

where t_j^i is index term coming from assumed set of terms T (domain ontology), w_j^i equals 1 if the term t_j^i occurs in index terms of this document and n_d^i is a number of index terms that describe document d_i.

We assume that domain ontology T is fixed: in concepts we have a term and its synonyms and we have a few relations defined between the terms: generalization – specification (relations: "is narrower" and "is broader") and relation "is correlated". An exemplary small part of the ontology is presented in Fig. 1.

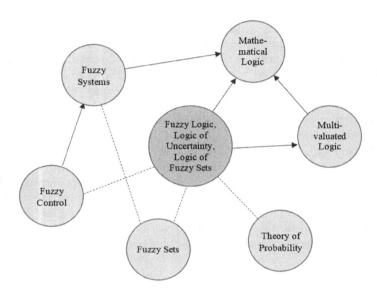

Fig. 1. An exemplary small part of the domain ontology (terms connected with term "Fuzzy Logic").

3.2   Model of User Profile

We assume that user profile has an ontological structure. The definition of ontology and its components are taken from [15].

Ontology-based user profile is defined as a triple:

$$O = (C, R, I) \tag{3}$$

where C is a finite set of concepts, R is a finite set of relations between concepts $R = \{r_1, r_2, ..., r_n\}$, $n \in N$ and $r_i \subset C \times C$ for $i \in \{1, n\}$ and I is a finite set of instances.

Developed user profile is based on domain ontology. Each concept contains the main term and its synonyms. Between selected concepts we have the following relations: "is narrower", "is broader" and "is correlated". A set of document that was relevant for a user are instances in our system.

In the user profile we define a probability table associated with each concept. We would like to determine the probability of document relevance for the user in condition of occurring a term t in his query. While there are many connections between the terms, we would like to model these probabilities using Bayesian network.

4 Method of Determining Profile for a New User

In this section we present a method for determining a profile for a new user of our system. We assume that we have a group of users with similar interests. It means that diameter of the profiles set is smaller than assumed threshold. We would like to develop a new profile which is a representative profile of the group. The representative profile will be assigned for a new user that is classified into this group (a method for a new user classification is out of scope of this paper – some methods were presented by author in previous papers [10] but using different structure of user profile).

Developing a profile for a new user can be considered in two cases: when the structure of network is fixed or not. Depending on the case, the method can differ in meaningful way.

In the first case, let us assume that we have a domain ontology O. We can say that we have fixed structure of the network, while the relations between concepts exist and do not change with time. In this approach the problem of developing a representative profile can be considered only on the level of probabilities tables' for each index term.

Let us use CPT for conditional probability table correlated with probability that document is relevant or not in condition of occurring particular terms $P(d_i|\tilde{T})$ where $\tilde{T}$ is a subset of terms' set T that contains terms which are in document description. Value of term's weight is equal to 1 if the term occurs and 0 if not.

Considering a set of CPTs: $\{CPT_1,\ CPT_2,\ \dots,\ CPT_{Tn}\}$, where Tn is a cardinality of users' set, the task is to determine table CPT^* which best represent the set of users' profiles. To obtain the representative CPT we can use some standard methods for determining a mean profile: calculating average value of each probability.

The second case is when we do not assume any model of relations between terms. A justification of using such an approach is as follows. When user is interacting with the system, it gathers information about his activities. Building a Bayesian network model based on the set of relevant and irrelevant document allows us to discover some hidden connections between terms that are not included in domain ontology. In this network, each concept is correlated with a conditional probability table.

When we consider a problem of integrating a set of users' profiles, we need to develop a method for merging Bayesian networks. As an input we have a set of users' profiles. Each profile was determined using Bayesian network building method based on the set of relevant and irrelevant documents for a particular user. As a result we would like to create a profile which best represents the group of the profiles' set. The task should be divided into two levels: structure of resultant profile and conditional probability table.

We propose a method to create integrated profile: determine Bayesian network based on the combined set of all documents that were judged by the users from a considered group.

In next section we present two exemplary profiles and the result of integrating procedure. We can consider some properties of this solution that enable us to develop more sophisticated method in the future.

5   Experimental Evaluation

To check the quality of proposed method for integrating users' profiles, we have performed some experimental evaluations.

The idea of the experiment is as follows. Let us assume that D means a set of relevant documents and D_{u1} – set of documents relevant for user U_1.

To perform experiments we have assume that documents are described by 5 terms (from t_1 to t_5). Each term can occur in document's description or not. We also assigned values to parameter "class": A for irrelevant document and B for relevant one. We have prepared an exemplary sets of documents which are relevant and irrelevant for both users. We have modeled a dependency between t_1, t_2 and t_5 (t_5 is a result of $t_1 \wedge t_2$) and between terms and relevance of document (document is relevant if at least two of terms from t_1 to t_4 occurs in document).

To obtain the results we have used Weka Software (Waikato Environment for Knowledge Analysis) [3,16]. An exemplary Bayesian networks are presented in Figs. 2 and 3. We present structures of network and conditional probability tables. In the graph we can interpret arrows as conditional dependence: in Fig. 2 we see arrow from node with term t_5 to node with term t_2 – it means that term t_2 is dependent on term t_5. A node with "class" is depended on terms t_1, t_2, t_3 and t_5. Node with term t_4 does not influence the result class. We can notice that Bayesian network has discovered assumed dependencies between terms and class nodes.

Each node in Bayesian network is connected with conditional probability table. An exemplary tables are presented for the node with "class" in Fig. 2 and 3. We can see that according to theorem which was proved in [4], node with "class" is not dependent on term t_4. In table from Fig. 2. we can see that probability of situation when document described by terms t_1, t_2, t_3 and t_5 is relevant is quite big – it is 0.964.

In our experiment we have two users with similar interests and in consequence with similar sets of relevant documents. We assume that 90% of relevant documents set of the first and second users are the same: $card(D_{u1} \cap D_{u2}) = 90\% \cdot card(D)$. Let us build user profile UP_1 and UP_2 for appropriately users U_1 and U_2 based on sets D_{u1} and D_{u2}. The results are presented in Figs. 2 and 3.

We can notice that the structures of both networks are similar but not the same. First, when comparing the structures, we see that dependencies between

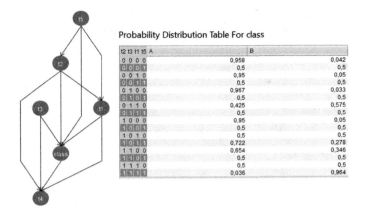

Probability Distribution Table For class

t2 t3 t1 t5 A	B	
0 0 0 0	0,958	0,042
0 0 0 1	0,5	0,5
0 0 1 0	0,95	0,05
0 0 1 1	0,5	0,5
0 1 0 0	0,967	0,033
0 1 0 1	0,5	0,5
0 1 1 0	0,425	0,575
0 1 1 1	0,5	0,5
1 0 0 0	0,95	0,05
1 0 0 1	0,5	0,5
1 0 1 0	0,5	0,5
1 0 1 1	0,722	0,278
1 1 0 0	0,654	0,346
1 1 0 1	0,5	0,5
1 1 1 0	0,5	0,5
1 1 1 1	0,036	0,964

Fig. 2. The result Bayesian networks for the first user.

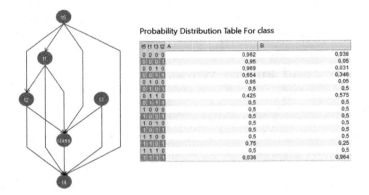

Probability Distribution Table For class

t5 t1 t3 t2 A	B	
0 0 0 0	0,962	0,038
0 0 0 1	0,95	0,05
0 0 1 0	0,969	0,031
0 0 1 1	0,654	0,346
0 1 0 0	0,95	0,05
0 1 0 1	0,5	0,5
0 1 1 0	0,425	0,575
0 1 1 1	0,5	0,5
1 0 0 0	0,5	0,5
1 0 0 1	0,5	0,5
1 0 1 0	0,5	0,5
1 0 1 1	0,5	0,5
1 1 0 0	0,5	0,5
1 1 0 1	0,75	0,25
1 1 1 0	0,5	0,5
1 1 1 1	0,036	0,964

Fig. 3. The result Bayesian networks for the second user.

t_1, t_2 and t_5 are quite different: in Fig. 2 node t_1 depends on t_2 and t_5, while in Fig. 3 term t_2 depends on t_1 and t_5. On the other hand, when we try to compare the conditional probability tables of these two cases, we can notice that the probabilities for the same values of t_1, t_2, t_3 and t_5 are very small, eg. for $t_1 = t_2 = t_3 = t_5 = 0$ the probability that such document will be relevant is equal to 0.958 in the first case and 0.962 for the second one.

The task of user profiles integration can be interpreted as merging these two Bayesian networks. The first idea of merging these networks is to build another Bayesian network based on the set of all relevant documents for both users. We supposed that the result should be close to previous networks while 90% of documents are the same. We do not assume any fixed structure of network. We present a result in Fig. 4.

On the structural level we can see that Bayesian network has discovered dependencies between terms t_1, t_2 and t_5. Also the node with term t_4 influences the node with "class". There are no dependencies between terms t_1, t_2, t_3 and t_4.

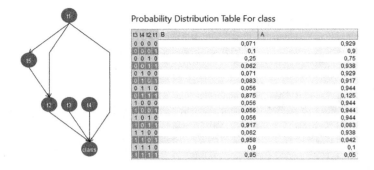

Probability Distribution Table For class

t3	t4	t2	t1	B	A
0	0	0	0	0,071	0,929
0	0	0	1	0,1	0,9
0	0	1	0	0,25	0,75
0	0	1	1	0,062	0,938
0	1	0	0	0,071	0,929
0	1	0	1	0,083	0,917
0	1	1	0	0,056	0,944
0	1	1	1	0,875	0,125
1	0	0	0	0,056	0,944
1	0	0	1	0,056	0,944
1	0	1	0	0,056	0,944
1	0	1	1	0,917	0,083
1	1	0	0	0,062	0,938
1	1	0	1	0,958	0,042
1	1	1	0	0,9	0,1
1	1	1	1	0,95	0,05

Fig. 4. Graphical representation of Bayesian network for integrated profiles.

Comparing the result Bayesian network with previous two networks built on a subset of the same documents' set, we see that the structure and conditional probability tables are different and it is not so obvious how the process of integration should be performed with unfixed structure of the network.

6   Summary and Future Works

In the paper we have investigate some approaches to building user profile based on Bayesian network methods. We are developing a personalized document retrieval system without the problem of so called "cold-start problem". A new user coming to our system is classified into proper group based on very first demographic data. A beginning profile is determined as a representative profile of the group. We have considered two approaches for combining knowledge from a group of profiles into single profile: with fixed or unrestricted structure of profile. Bayesian network techniques turned out to be very useful way to integrate knowledge about group of users.

In our further research we will develop more sophisticated method for integrating profiles and perform experimental evaluations using greater data sets.

Acknowledgments. This research was partially supported by Polish Ministry of Science and Higher Education.

References

1. Billsus, D., Pazzani, M.J.: A hybrid user model for news story classification. In: Kay, J. (ed.) Proceedings of the Seventh International Conference on User Modeling, vol. 407, pp. 99–108. Springer, Heidelberg (1999). doi:10.1007/978-3-7091-2490-1_10
2. Bottcher, S.G., Dethlefsen, C.: Learning Bayesian networks with R. In: Proceedings of the 3rd International Workshop on Distributed Statistical Computing, DSC 2003 (2003)

3. Bouckaert, R.R.: Bayesian Network Classifiers in Weka for Version 3-5-7. University of Waikato (2007)
4. Druzdzel, M.J., Diez, F.J.: Combining Knowledge from different sources in causal probabilistic models. J. Mach. Learn. Res. **4**, 295–316 (2003)
5. Dahak, F., Boughanem, M., Balla, A.: A probabilistic model to exploit user expectations in XML information retrieval. Inf. Process. Manag. **53**(1), 87–105 (2016)
6. Ferchichi, A., Boulila, W., Farah, I.R.: Towards an uncertainty reduction framework for land-cover change prediction using possibility theory. Vietnam J. Comput. Sci. **4**(3), 195–209 (2016)
7. Giftodimos, E., Flach, P.A.: Hierarchical Bayesian networks: a probabilistic reasoning model for structured domains. In: Proceedings of the ICML-2002 Workshop on Development of Representations, pp. 23–30 (2002)
8. DesJardins, M., Rathod, P., Getoor, L.: Learning structured Bayesian networks: combining abstraction hierarchies and tree-structured conditional probability tables. Comput. Intell. **24**(1), 1–22 (2008)
9. Maleszka, B.: A method for determining ontology-based user profile in document retrieval system. J. Intell. Fuzzy Syst. **32**, 1253–1263 (2017)
10. Mianowska, B., Nguyen, N.T.: A method for collaborative recommendation in document retrieval systems. In: Selamat, A., Nguyen, N.T., Haron, H. (eds.) ACIIDS 2013. LNCS, vol. 7803, pp. 168–177. Springer, Heidelberg (2013). doi:10.1007/978-3-642-36543-0_18
11. Maleszka, B.: A method for determining representative of ontology-based user profile in personalized document retrieval systems. In: Nguyen, N.T., Trawiński, B., Fujita, H., Hong, T.-P. (eds.) ACIIDS 2016. LNCS, vol. 9621, pp. 202–211. Springer, Heidelberg (2016). doi:10.1007/978-3-662-49381-6_20
12. Maleszka, B.: A method for user profile learning in document retrieval system using Bayesian network. In: Nguyen, N.T., Tojo, S., Nguyen, L.M., Trawiński, B. (eds.) ACIIDS 2017. LNCS, vol. 10191, pp. 269–277. Springer, Cham (2017). doi:10.1007/978-3-319-54472-4_26
13. Stern, M.K., Beck, J.E., Woolf, B.P.: Naive Bayes classifiers for user modeling. In: Proceedings of the Conference on User Modeling (1999)
14. Pazzani, M.J., Billsus, D.: Content-based recommendation systems. In: Brusilovsky, P., Kobsa, A., Nejdl, W. (eds.) The Adaptive Web. LNCS, vol. 4321, pp. 325–341. Springer, Heidelberg (2007). doi:10.1007/978-3-540-72079-9_10
15. Pietranik, M., Nguyen, N.T.: A multi-attribute based framework for ontology aligning. Neurocomputing **146**, 276–290 (2014)
16. Waikato Environment for Knowledge Analysis. Machine Learning Group at the University of Waikato. http://www.cs.waikato.ac.nz/ml/index.html. Accessed 10 Mar 2017
17. Yu, K., Schwaighofer, A., Tresp, V., Ma, W.-Y., Zhang, H.J.: Collaborative ensemble learning: combining collaborative and content-based information filtering via hierarchical Bayes. In: Proceedings of UAI 2003, pp. 616–623 (2003)
18. Schiaffino, S.N., Amandi, A.: User profiling with case-based reasoning and Bayesian networks. In: Proceedings of International Joint Conference IBERAMIA-SBIA, pp. 12–21 (2000)
19. Wong, S.K.M., Butz, C.J.: A Bayesian approach to user profiling in information retrieval. Technol. Lett. **4**(1), 50–56 (2000)

Analysis of Denoising Autoencoder Properties Through Misspelling Correction Task

Karol Draszawka and Julian Szymański[✉]

Department of Computer Systems Architecture,
Faculty of Electronic Telecommunications and Informatics,
Gdańsk University of Technology, Gdańsk, Poland
{kadr,julian.szymanski}@eti.pg.gda.pl

Abstract. The paper analyzes some properties of denoising autoencoders using the problem of misspellings correction as an exemplary task. We evaluate the capacity of the network in its classical feed-forward form. We also propose a modification to the output layer of the net, which we called multi-softmax. Experiments show that the model trained with this output layer outperforms traditional network both in learning time and accuracy. We test the influence of the noise introduced to training data on the learning speed and generalization quality. The proposed approach of evaluating various properties of autoencoders using misspellings correction task serves as an open framework for further experiments, e.g. incorporating other neural network topologies into an autoencoder setting.

Keywords: Autoencoder · Misspellings · Autoassociative memory

1 Introduction

Experiments presented in this paper aims at researching properties of autoencoders – neural network-based models of autoassociative memory [1]. To do this we need data that is scalable, easily obtainable, can be modified in such a way that allows to test particular properties of autoencoders, and corresponds to some popular, common sense task. The task of misspelling correction fulfills these requirements – it is widely used for supporting writers to ensure quality of their work [2]. The data can be represented in the form of large dictionaries of strings, that can be modified according to experiment requirements, e.g. by injecting a particular type of noise (typos).

Typical approach to complete the task of misspellings correction is to provide a priori dictionary and to select the entries that are the most probable given the distorted one presented on input. This requires the comparison between the input string and all entries in the dictionary using some similarity measure. This approach has a linear complexity $O(n)$, where n is the size of the dictionary. The quality of the results depends on the representation of dictionary entries and similarity measure used for string comparison [3,4]. It may be e.g.: Cosine,

© Springer International Publishing AG 2017
N.T. Nguyen et al. (Eds.): ICCCI 2017, Part II, LNAI 10449, pp. 438–447, 2017.
DOI: 10.1007/978-3-319-67077-5_42

Hamming, Levenshtein distances or their modifications [5]. Regardless of what similarity measure is used the complexity in that approach remains the same. One way to improve the efficiency is to used dedicated indexes [6] or hashing [7,8].

In this work we use misspellings correction as the task under which we examine properties of denoising autoencoders. Conducted experiments show that this task can be successfully completed using such neural networks. This solution provides high quality results, the $O(1)$ complexity and the possibility to tune the model for user preferences. Also representation of the dictionary is much more compact and saves memory.

2 Autoencoders as Typo Correctors

2.1 Denoising Autoencoders

As stated, for large scale dictionaries finding closest matches to a mistyped sequence of characters using some similarity measure may be inapplicable due to $O(n)$ complexity. On the other hand, existing techniques of scalability improvements introduce accuracy problems. For example, solutions based on Locality Sensitive Hashing [6], may decrease precision and/or recall of dictionary entries retrieval. The quality of results is strongly related to the similarity function that is employed – the more precise results we want to obtain the more sophisticated, and computationally expensive, function need to be used. In some circumstances, there is also a need for tuning of the similarity function so that it works better with a particular dictionary, e.g. to construct user personalized spelling corrections, when characteristic examples of his or her misspellings are provided.

Because of these drawbacks of traditional methods, we tested autoencoders as an alternative misspellings correctors. Autoencoder is an artificial neural network model, that tries to map a given input vector to itself (see Fig. 1). The computation is not trivial, because the information flowing from input to output goes first through encoding part (function f), which squeezes it into smaller number of a *bottleneck* layer (it may also be otherwise restricted, e.g. through sparse regularization), before it is recovered back with decoder function g. Autoencoders are trained by minimizing a loss function $L(\mathbf{x}, g(f(\mathbf{x})))$ that penalizes net output $g(f(\mathbf{x}))$ for being dissimilar from net input $\mathbf{x}$. For real valued input vectors L is typically chosen to be mean squared error (MSE), but for binary data it can also be binary cross-entropy (CE) loss [9]. When decoder is linear and L is the mean squared error, an undercomplete autoencoder learns to span the same subspace as Primary Component Analysis (PCA) [10]. Autoencoders with nonlinear f and g can learn more powerful nonlinear embeddings than PCA.

Typically, the aim of training such neural networks is to obtain a useful compact representation for a given data set, i.e. for the purpose of dimensionality reduction [11–13]. In these cases, we are not interested in the output of the decoder, but in the hidden layer, where internal representations of the input data are formed. However, autoencoders can also be used as noise reductors, for example like shown in [14] for enhancing speech signal, or in [15] for removing different types of noise from images. For such *denoising* autoencoders, training dataset consists of $(\hat{\mathbf{x}}, \mathbf{x})$ input-target data pairs, where $\hat{\mathbf{x}}$ is a corrupted version of the original $\mathbf{x}$ instance. The corrupted version is formed from original one during generation of training pairs.

2.2   Words Representation

In this context, misspellings correction can be seen as a denoising process, for which denoising autoencoders are perfectly suitable. All what has to be done is to transform sequence of characters into a fixed length feature vector, which is then directly applicable to neural networks input and targets. In this work, we used *one-hot* encoding of characters: let n be the number of symbols in the alphabet, then i-th symbol is associated with a vector $v(s^{(i)})$ of length n with all elements equal to 0 except i-th, where it has 1. Sequence of symbols is represented as a concatenation of all associated $v(s)$ vectors. For example, if alphabet $\mathcal{A} = \{$'a', 'b', 'c'$\}$, then $v($'a'$) = [1, 0, 0]$, $v($'b'$) = [0, 1, 0]$, $v($'c'$) = [0, 0, 1]$ and $v($'cab'$) = [0, 0, 1, 1, 0, 0, 0, 1, 0]$.

The reverse transformation, i.e. from binary vectors to strings of characters is analogical. However, vectors returned by a neural net will not be binary, but real valued, with all elements between 0 and 1. Thus, the returned vector is split into vectors representing each character, and then the position of the maximum element in each of the vectors determines decoded symbols. For example, if $\mathcal{A}$ is the same as in previous example, then vector $[0.780.210.01, 0.1, 0.89, 0.01, 0.0, 0.99, 0.1, 0.6, 0.37, 0.3]$ is decoded as string 'abba', because we have symbol vectors: $[0.780.210.01]$ with maximum at first position, $[0.1, 0.89, 0.01]$ with second element being maximal, $[0.0, 0.99, 0.1]$ also with max at second position, and $[0.6, 0.37, 0.3]$ with argmax equal to 1.

To obtain fixed-sized vectors for all input sequences, regardless of their length, space padding is used (space being added to alphabet). For the purpose of the experiments we assumed maximum length of the sequence, denoted as C, equal 16. All shorter expressions where padded to this size with spaces. In result, 576-element $(C \cdot |\mathcal{A}|)$ binary sparse vector represents a word that can be formed from 36-symbols alphabet not longer that 16 characters.

2.3   Output Layers and Loss Functions

The output of our model for typos correction is specific: it is interpreted as a concatenation of C probability distributions of characters from alphabet $\mathcal{A}$. Since this is a concatenation of distributions, and not a single probability distribution, we cannot apply a softmax output function to here. Instead, two possibilities were tested.

In the first one, the models is as seen in Fig. 1, i.e. with a normal dense layer of sigmoid units (with standard logistic activation function $\sigma(z) = 1/(1 + \exp(-z))$). This type of network models arbitrary binary outputs. The loss function L for a single training target $\mathbf{t}$ is a binary cross-entropy:

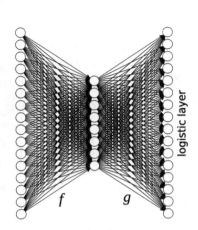

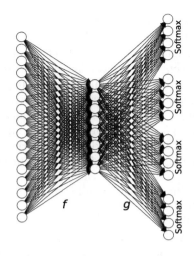

Fig. 1. Typical shallow autoencoder architecture: feed forward neural net with one *bottleneck* hidden layer. Sigmoid activation function at output.

Fig. 2. Autoencoder with multi-softmax output layer.

$$L(\mathbf{y}, \mathbf{t}) = - \sum_{i=0}^{|\mathcal{A}|C-1} \left[t_i \log(y_i) + (1 - t_i) \log(1 - y_i) \right], \tag{1}$$

where $\mathbf{y}$ is the output of the net.

The other one is presented in Fig. 2. This is a configuration, which we call *multi-softmax*, which is a concatenation of dense softmax layers each calculating normal softmax function $\sigma(\mathbf{z})_j = \exp(z_j)/\sum_{k=1}^{K} \exp(z_k)$. The loss function L for a single training target $\mathbf{t}$ for such an output layer is a sum of categorical cross-entropies, each corresponding to appropriate softmax:

$$L(\mathbf{y}, \mathbf{t}) = - \sum_{i=0}^{C-1} \sum_{j=0}^{|\mathcal{A}|} t_{i|\mathcal{A}|+j} \log(y_{i|\mathcal{A}|+j}). \tag{2}$$

This model has the same number of parameters as normal sigmoid layer, yet it enforces the output vector to be a valid concatenation of probability distributions.

3 Experiments

3.1 Experimental Setup

In all the experiments described below, we used our implementation of autoencoders in Keras [16] with Theano [17] back-end. We trained models for 1000 epochs using AdaDelta optimizer with default settings [18]. If not stated otherwise, training and testing data is based on a 1000 long Polish words dictionary (av. length of word is 13.02 characters).

Except an experiment that investigates generalization capabilities of presented models in detail (discussed further), all other experiments were conducted using the following strategy. Instead of creating fixed-sized sets of training and testing data, we use generators that create data on-the-fly during training and testing phases. During training, generator creates training pairs $(\mathbf{x}, \mathbf{x})$ simply taking a random word from a dictionary and transforming it into vector representation, or creates $(\hat{\mathbf{x}}, \mathbf{x})$ substituting one character of the word with a different random one from the alphabet. This way, any possible misspelling of a word (resulting from mistyped a single character) could be possibly generated. In one epoch, we give 100 such pairs per each dictionary entry. After each 10 epochs, models are evaluated using testing data generator: for each word in a dictionary, 100 one-character typos is generated. So, for 1000 words, 100000 tests are performed, and the percentage of times, when the net returns a correct word from typo is reported as test accuracy.

The number of units in the hidden layer is crucial for autoencoder to work properly. First experiment under these settings was conducted to selected the number of hidden units in the bottleneck layer. The results are shown in Table 1. It can be seen that more units give better results. However, this comes at the cost of complexity and computation time. Therefore we have chosen a shallow autoencoder of 300 hidden units as a good compromise between accuracy and complexity and all subsequent experiments use models with 300 neurons in the middle layer. We also tested deeper architectures (such as e.g. 576-1000-300-1000-576), but they performed similarity at much higher computational cost.

Table 1. The impact of the number of hidden units in the bottleneck layer of a shallow autoencoder. Each setup was trained and tested two times (dashed lines) and averaged (solid line).

Hidden layer size	10	30	100	300	400	576	700	1000
Average test accuracy [%]	0.13	36.2	67.3	71.2	71.1	71.9	72.6	73.5

3.2   Sigmoid vs Multi-softmax Output Layer

Providing the information to the network that each of the characters is coded as separate probability distribution significantly increases the speed of learning process. Our multi-softmax adjustment causes the network to perform much better than simple layer with sigmoid units, see Fig. 3. There, we present test accuracy results and number of epochs required to train the shallow autoencoders with 1 hidden layer of 300 units on word dictionaries containing 100 and 1000 elements.

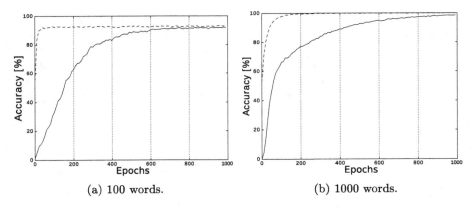

(a) 100 words.   (b) 1000 words.

Fig. 3. Sigmoid (solid line) and multi-softmax (dashed line) test accuracy for different sizes of training dictionary.

Multi-softmax output layer constantly outperforms normal sigmoid layer in both cases. In the case of 1000 words dictionary, multi-softmax enables autoencoder to achieve almost 100% test accuracy. Interestingly, performance on 100 word set is worse than on 1000 word set. This is because our 100 word dictionary contains very similar words (each begin with the same three letters), whereas 1000 word dictionary contains diverse entries and it is easier to find a correct word of misspelled one.

The time required to train the network is also very important factor to have practical applications of proposed approach. Training on low-end GPU, our 576-300-576 autoencoder took a couple of minutes for a dictionary of 100 words and about 2 h for 1000 words. We tested those models also on 10000 dictionary, where 1000 epochs took over 22 h, which was not enough to obtain a good test accuracy (69.3% for multi-softmax, 64.1% for sigmoid), showing that 576-300-576 models are too small for 10000 words.

3.3   Impact of Input Noise on the Model

Misspellings correction can be seen as a transformation of a noised input into an output without the noise. The type noise we use in the experiments is random character substitution, but other examples of misspellings, such as reversed order of neighboring chars or additional/missing chars, are possible as well. The question is: how much noise is needed for a denoising autoencoder to train it maximally effectively.

Table 2 shows the impact of the percentage of input training examples having typo. The case when all examples are without any single character changed is denoted as 0% case. On the other extreme, 100% case means that every time a word from the dictionary is presented to the model, it has a random character substitution at one random place. It can be seen that presenting distorted words at training significantly improves generalization of the autoencoder. This is expected, because the task is just to do this (at test time, all examples have one random character substitution). The results indicate that training an autoencoder on the correct data leads to very poor generalization. On the other hand, it is also beneficial to include in a training data an correct examples of words. The best result (93.3% test accuracy) was obtained when one 1 in a 100 examples was correct. Also 97% setting achieved better result than 100% distorted data.

Table 2. Impact of the percentage of distorted training input examples on model performance.

How often a training word contains a typo	0%	50%	67%	80%	90%	97%	99%	100%
Average test accuracy [%]	53.3	80.3	85.9	89.3	91.0	92.6	**93.3**	91.7

Another factor we evaluated was the level of distortion the network is still able to reconstruct. The assumption of the experiment presented here was that if we add more distorted data into the learning set the network can achieve better generalization properties. Positional representation of characters in the input vector means that there is a lot of independence among positions in the word. We tried to make use of this independence and tested whether more than one random substitutions can be made *in each* input example, although at test time examples still have only one substitution. The hope was that more 'denoising work' could be learned in a mini-batch processing than merely one random substitution in a random place per example.

In Table 3 we present results of the experiment showing how distortion level influence accuracy. The best performance was achieved when each training example contains 3 typos (test accuracy 97.3%). 5 typos per training word performs almost equally well. But 10 typos per word (when words are of 16 characters maximally) is too much – the words are distorted to such a degree that they cannot be reconstructed any more. This experiment was performed in two settings (same as in previous experiment), one in which 50% of examples is distorted and other are left unchanged, and second when 100% of examples is distorted. The second setting constantly outperforms the first one, and the results from this experiment conform previously drawn conclusions: it is better to show various forms of distorted data than to present many times (50% means each second epoch) the same correct version words.

Table 3. The impact of the number of character substitutions in every one distorted word presented to the model during training. Test accuracy [%].

Number of typos in each perturbed example	1	2	3	5	10
50% perturbed training examples	80.1	89.6	91.2	91.6	85.4
100% perturbed training examples	91.9	96.2	**97.3**	97.1	88.6

3.4   Generalization Abilities

In all experiments presented so far, training and testing data are generated online from two generators, which are sampling words from the same dictionary of words. Although typos made in training and testing examples are constantly different, it is possible that in some epoch t test generator will generate a typo exactly the same as previously created by train generator in epochs $\leq t$. In fact, the probability of such situation is monotonically growing during training, therefore, test accuracies will slowly converge to train accuracies. Although presented procedure is good for training misspellings correctors, it cannot reliably answer a question whether such systems can generalize to correct other typos than presented during training.

To answer this question we created four disjoint fixed-sized sets: one training and three testing data. The first one, denoted as L1, contains typos very similar to those from the training set: they are made in the same words and at the same positions as in the train dataset, only wrong character is different. The second test set, L2, holds typos made in the same words as examples from the training set, but at positions which are never altered in the training set (each word in the training set has one randomly chosen position in it, where it always has its correct character). The third test set, L3, has typos made in words not present in the training set dictionary at all.

Figure 4 shows accuracies of typos corrections for each of the datasets for 576-250-576 autoencoders with sigmoid and multi-softmax output layers. Correction unseen typos of L1 type is not a problem at all, the performance being essentially the same as for train data. Correction of L2 typos is only slightly harder for multi-softmax model, while sigmoid net struggles here, 500 epochs is not enough to obtain good accuracy, but increasing tendency indicates that it is not overfitting

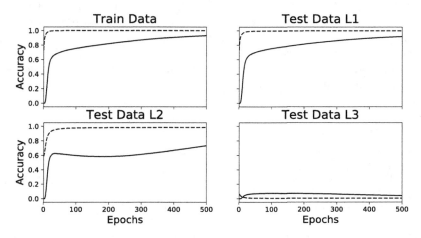

Fig. 4. Accuracy on train and three different test sets. Solid lines represent sigmoid model, dashed lines – multi-softmax model. Details describing datasets are in the text. Axes are shared between plots.

training data yet, and is able to generalize to L2 dataset. Neither model is able to correct words that they did not see during training. This is not a surprise, because the models are trained to associate their input to whole words from the dictionary used during training, and have no clue that out-of-vocabulary output is possible.

4   Summary and Further Works

In this paper we present results of the research on autoencoders for misspelling correction task. We argue that the observations given here regarding the properties of autoencoders' behavior are general and may be used for other applications that employ this type of networks. We shown that introduction of so-called *multi-softmax* layer significantly improves learning results. The results of the experiments shown that there is a finite capacity of the network. Developing it into larger topology is not a good way to extend the dictionary size because the time required to train the network significantly increases. Instead of enlarging the network, in future we plan to introduce hierarchical structure with root network at the top that will select subspace of the dictionary and indicates proper network to perform misspelling corrections in that subspace. This task can be performed using Self Organizing Map [19] that aggregate in each node the most similar entries from the dictionary. Selecting the size of the map allows us to configure the number of subnetworks used for a particular dictionary.

The framework we create during implementation of the experiments allows us to test other network topologies that may be applied as autoencoders. In future, instead of adding simple dense hidden layers we plan to test capacity of convolutional layers [20]. Also usage of deep residual networks [21] seems to be a good direction for extending the capacity and accuracy of the model. In the experiments presented here we used static vectors to represent the input strings. The other promising direction of research is to treat words as sequences where next character depends in some degree on the previous one. This naturally suggest recurrent networks [22], that can be incorporated to autoencoder model (as encoder as well as decoder part) and provide information on successive dependencies in training sequences.

References

1. Fausett, L.V.: Fundamentals of Neural Networks. Prentice-Hall, Upper Saddle River (1994)
2. Kukich, K.: Techniques for automatically correcting words in text. ACM Comput. Surv. (CSUR) **24**, 377–439 (1992)
3. Baeza-Yates, G., Navarro, R.: Faster approximate string matching. Algorithmica **23**, 127–158 (1999)
4. Szymański, J., Boiński, T.: Improvement of imperfect string matching based on asymmetric n-grams. In: Bădică, C., Nguyen, N.T., Brezovan, M. (eds.) ICCCI 2013. LNCS (LNAI), vol. 8083, pp. 306–315. Springer, Heidelberg (2013). doi:10. 1007/978-3-642-40495-5_31

5. Astrain, J.J., Garitagoitia, J.R., Villadangos, J.E., Fariña, F., Córdoba, A., de Mendıvil, J.G.: An imperfect string matching experience using deformed fuzzy automata. In: HIS, pp. 115–123 (2002)
6. Boguszewski, A., Szymański, J., Draszawka, K.: Towards increasing f-measure of approximate string matching in o(1) complexity. In: 2016 Federated Conference on Computer Science and Information Systems (FedCSIS), pp. 527–532 (2016)
7. Hantler, S.L., Laker, M.M., Lenchner, J., Milch, D.: Methods and apparatus for performing spelling corrections using one or more variant hash tables. US Patent App. 11/513,782 (2006)
8. Udupa, R., Kumar, S.: Hashing-based approaches to spelling correction of personal names. In: Proceedings of the 2010 Conference on Empirical Methods in Natural Language Processing, pp. 1256–1265. Association for Computational Linguistics (2010)
9. Rubinstein, R.Y., Kroese, D.P.: The Cross-entropy Method: A Unified Approach to Combinatorial Optimization, Monte-Carlo Simulation and Machine Learning. Springer Science & Business Media, Berlin (2013)
10. Jolliffe, I.: Principal Component Analysis. Wiley, Hoboken (2002)
11. Hinton, G.E., Salakhutdinov, R.R.: Reducing the dimensionality of data with neural networks. Science **313**, 504–507 (2006)
12. Bengio, Y., et al.: Learning deep architectures for AI. Found. Trends® Mach. Learn. **2**, 1–127 (2009)
13. Vincent, P., Larochelle, H., Lajoie, I., Bengio, Y., Manzagol, P.A.: Stacked denoising autoencoders: learning useful representations in a deep network with a local denoising criterion. J. Mach. Learn. Res. **11**, 3371–3408 (2010)
14. Lu, X., Tsao, Y., Matsuda, S., Hori, C.: Speech enhancement based on deep denoising autoencoder. In: Interspeech, pp. 436–440 (2013)
15. Agostinelli, F., Anderson, M.R., Lee, H.: Adaptive multi-column deep neural networks with application to robust image denoising. In: Advances in Neural Information Processing Systems, pp. 1493–1501 (2013)
16. Chollet, F.: Keras (2016). https://github.com/fchollet/keras
17. Theano Development Team: Theano: A Python framework for fast computation of mathematical expressions. arXiv e-prints abs/1605.02688 (2016)
18. Zeiler, M.D.: Adadelta: an adaptive learning rate method. arXiv preprint arXiv:1212.5701 (2012)
19. Kohonen, T.: The self-organizing map. Neurocomputing **21**, 1–6 (1998)
20. Kalchbrenner, N., Grefenstette, E., Blunsom, P.: A convolutional neural network for modelling sentences. arXiv preprint arXiv:1404.2188 (2014)
21. Zagoruyko, S., Komodakis, N.: Wide residual networks. arXiv preprint arXiv:1605.07146 (2016)
22. Botvinick, M.M., Plaut, D.C.: Short-term memory for serial order: a recurrent neural network model. Psychol. Rev. **113**, 201 (2006)

New Ontological Approach for Opinion Polarity Extraction from Twitter

Ammar Mars[1]([✉]), Sihem Hamem[2], and Mohamed Salah Gouider[1,3]

[1] ISG, SMART Lab, Université de Tunis, 2000 Le Bardo, Tunisia
ammar.mars@gmail.com
[2] ISG, Université de Gabes, Gabes, Tunisia
hamemsihem@gmail.com
[3] ESSECT, Université de Tunis, 1089 Montfleury, Tunisia
ms.gouider@yahoo.fr

Abstract. Since the past few years, we have been talking about opinion extraction, also known as opinion mining. It is a computational study of opinions and sentiments expressed in a text format. A lot of web resources contain user's opinions, e.g. social networks, micro blogging platforms, and Blogs. People frequently make their opinions available in these sources. It is important for a company to study the opinions of these customers in order to improve its services or the quality of these products. In this paper, we are interested in studying the opinions of users about a product and extracting their polarity (positive, negative or neutral), for example studying the opinion of users about the Nokia or Huawei brand. We collected data from Twitter because it is a rich data sources for opinion mining. We propose a new ontological approach able to classify the opinion of user's expressed in their tweets using Natural Language Processing (NLP) tools. This classification used a supervised Machine Learning Classifier: Support Vector Machine (SVM).

1 Introduction

In this we are interested by "Sentiment analysis" or also called "opinion mining" [8]. It aims to study of the people's opinions, sentiments, attitudes, and emotions on entities and their features (each product and their components). The task is very challenging and useful in decision making process. For few years, we have noticed an explosion of data caused by the increase of the use of the social networks like Facebook or Google Plus and Microblogging sites like Twitter. In this application, people always post real-time messages about their opinions on a variety topics, discuss current issues, complain and express positive sentiment for products they use in daily life. Consequently, it becomes a very important data sources of varied kind of information. The large amount of data available in these sources are unstructured and fuzzy. IBM Senior Vice President and Group Executive for the Software Group, cite that more than 80% of the world's data is unstructured that must be treated.

   In addition, the microblogging sites gives users the opportunity to express freely their opinions about anything, for instance Twitter. Many works are

© Springer International Publishing AG 2017
N.T. Nguyen et al. (Eds.): ICCCI 2017, Part II, LNAI 10449, pp. 448–458, 2017.
DOI: 10.1007/978-3-319-67077-5_43

mainly focused on analyzing opinions in reviews and classify reviews as positive or negative based on user sentiment. In this paper, we present the sentiment classification based on the combination of Natural Language Processing (NLP), sentiwordnet and Support Vector Machine (SVM) classifier.

The rest of this paper is as follows: Sect. 2 presents the related works, Sect. 3 defines the proposed ontological approach in combination with the supervised classifier, Sect. 4 shows experiments and results. We discuss in the last section we discuss the conclusion of this work and highlight some future work.

2   Related Work

One area of research has focused on sentiment based classification. Some of these works have focused on classifying the semantic orientation of conjoined adjectives [6] and the unsupervised learning algorithm for rating a review as thumb up or down based on the semantic orientation calculated using the PMI-IR algorithm [14]. The limitation of his work was the level of accuracy that depend on the domain of review, also the time of the treatment of queries and the difficulty of mixing the semantic orientation with other functionalities in a supervised classification algorithm. Besides et al. has adopted in [13] a bootstrapping[1] process to learn linguistic patterns of subjective expressions in order to classify the subjective expressions from objective ones. The process used a pattern extraction algorithm to learn subjective patterns. The learned patterns were then used to decide whether an expression was subjective or not. This proposed approach is used to identify the subjective sentences and achieve the accurate result around 90% during their tests. Furthermore, Agarwal et al. in [1] proposed a model based on a Support Vector Machine (SVM) for movie review documents. In the presented approach, features were extracted and, then data are converted to LIBSVM format and the SVM was applied for classifying reviews.

Moreover, Ghosh and Kar [5] used an unsupervised linguistic approach based on the SentiWordNet to calculate a sentiment score of each sentence. The results indicated that the SentiWordNet could be used as an important resource for sentiment classification tasks that would helps to measure the polarity of sentences based on the score of each word. Palanisamy et al. [10] utilised a simple lexicon based technique to extract sentiments from Twitter data. The lexicon consisted of positive, negative, negation and phrases. They used pre-processing steps such as: stemming, emoticons and hashtags detection. After that, the lexicon-based system would classify tweets as positive and negative based on the sentiment orientation of words. The drawback of this approach was that the dictionary edge was limited and it was so difficult to create a specific domain opinion word list. Other work we can cite, Mars et al. [9] develop a framework able to process a large amount of data and extract costumers opinions.

Other work was proposed by Vaitheeswaran and Arockiam [15] which a senti-lexical approach was employed to measure the sentiment on tweets using

---

[1] Refers to a problem setting in which one is given a small set of labeled data and a large set of unlabeled data, and the task is to induce a classifier.

lexicon-based approach. They used three dictionaries: Emoticon dictionary, Sentiment dictionary and Negation dictionary. The limitation of this work is that some of the positive emoticons are expressed with negative sentences or word that deceive the result and affect in the effectivness of the approach.

In addition, Zainuddin et al. [16] proposed a hybrid approach to analyze aspect-based Sentiments for tweets. The process began with a pre-processing phase and the Part of Speech (POS) was employed to identify nouns, verbs, adjectives and adverbs. This process is repeated for all sentences. The next process is calculating opinion words using Sentiwordnet. After that, the SVM was employed to determine the tweet-level classification and tweet aspect-based classification. The authors used two datasets and they found that SVM classifier is achieved 75.11% with Hate Crime Twitter Sentiment (HCTS) Dataset. Besides, they obtained 81.58% with Stanford Twitter Sentiment (STS). Thus, classifying opinions of different domains affect the effectivness of the results. It seems that the combination of text mining, Sentiwordnet and SVM is efficient to sentiment classification of tweets. The contribution lies in the use of a hybrid approach for twitter aspect-based sentiment analysis.

On the other hand, Kontopoulos et al. [7] propose an approach based on ontology. In the same way, Park and Baik [12] propose an approach based also on ontology. The system uses data about a particular topic from social networks such as Twitter. However, Duwairi and Qarqaz in [4] dealt with sentiment analysis in Arabic reviews using three machine learning classifiers: SVM, the naive bayes and the K-Nearest.

Besides, Pang et al. specified in [11], that learning techniques showed good results than linguistic ones; To make their comparisons, they based their experiments results on three supervised algorithms and they assured that SVM gave better results than Naive Bayes. Other work was proposed by Dalal and Zaveri [3] which they developed an opinion mining system for sentiment classification of users review using fuzzy functions applying on linguistic hedges (extremly, very, highly).

3   Our Approach

This work presents a new ontological approach for extracting customer opinions from Twitter, combined with machine learning and text mining tools. We used a lexical resource for opinion mining, that named the "SentiWordNet", that which classifies sentence as positive and negative based on the sentiment orientation of words. As we have mentioned, that we studying the opinions of user's opinions in social networks. Knowing the data collected is in a text format. In the following, we present our approach for extracting opinions from Twitter. It is composed by three tasks: Ontology Building, Tweets Collection and Tweets Analyze, that we will detail in this section:

3.1   Ontology Building

An ontology is an explicit conceptualization of a domain that provides knowledge about specific domains that are understandable by both developers and computers. In this work, we aim to develop an ontology which represents sentiment words. This Ontology improves the classification process of opinions. We aliment the ontology from the list of negative and positive words presented by Minqing Hu and Bing Liu [6]. The following table shows some words of this list: The ontology is automatically created from the list of presented words. It is a graph of positive and negative words of three categories: verbs, adverbs and adjectives. In the what follows, we present a fragment of the created ontology (Table 1 and Fig. 1).

Table 1. Fragment of positive and negative words list

Words	Polarity
Afraid, aggressive, agony, angry, batty, bad, bait, boil, buggy...	Negative
Accommodative, accomplish, achievement, admire, affable, alluring...	Positive

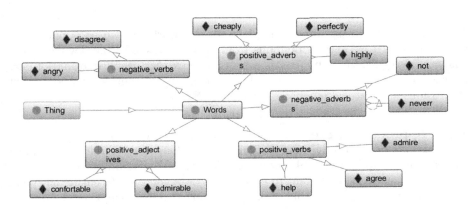

Fig. 1. Fragment of the ontology

After construction, we enrich the ontology such that a score is associated to each word of this ontology. Thus, we use the SentiWordNet library: It is a lexical resource for opinion mining that assigns to each word three sentiment scores: positivity, negativity, and objectivity. The SentiWordNet is described in details in the paper [2]. The final score of each tweet is calculated after considering the whole sentence structure and contextual information. In the following Table 2, there a presentation of a part of words and scores from the SentiWordNet.

Table 2. Fragment of the sentiwordnet

POS	ID	PosScore	NegScore	SynsetTerms	Gloss
a	00001740	0.125	0	able	"able to swim";
a	00002098	0	0.75	unable	"unable to get to town"
a	00003829	0.25	0	parturient giving birth	"a parturient heifer"
a	00005205	0.5	0	absolute perfect or complete or pure	"absolute loyalty"; "absolute silence";

3.2 Tweets Collection

Tweets collected from twitter were used as a data source. It is possible to extract tweets in a large scale from Twitter using the twitter public API that they provide. In this work, we will collect tweets about specific subjects. The products that we choose to analyze were: Nokia, Smartphone, Samsung, Iphone and Orange.

3.3 Tweets Analyze

This phase is the most important phase in our work. As we mentioned that in this approach we extract opinions from tweets collected from Twitter. Text analysis, also called text mining, is a process to extract useful information and knowledge from unstructured data. Most text mining systems are based on text expressions and natural language processing (NLP).

The Pre-processing Step. The data coming from Twitter contain s lots of noise, useless or additional information, special characters. These, which must be eliminated and keeping only meaningful information should be kept. For this reason, it was is very important to pre-process data collected to ensure the good quality of input. The main role of pre-processing task is to improve the quality of the input data.

Part of Speech Tagging. After the collection of the datasets, we assign part of speech[2]. It provides a way to "tag" the words in a string. That is, for each word, the "tagger" gets whether it's a noun, a verb, etc.

For example: **this is my new paper.** The output of tagger:

This/DT is/VBZ my/PRP new/JJ paper/NN

---

[2] It is a piece of software that reads text in some languages and assigns part of speech to each word (and other token) such as noun, verb, adjective, etc.

In this work, we choose to focus on tagging a specific gramatical classes. Our POS tagger will tag the 3 popular classes that occur in the languages of the world: verbs, adjectives and adverbs.

A Supervised Machine Learning Classifier: A Support Vector Machine. Support vector machine have been applied in order to classify sets of opinions. support vector machine have been employed in many text classification tasks due to their advantages: robustness in high dimensional space, most text categorization problems are linearly seperable. Also, SVM has achieved good results in opinion mining and this algorithm has overcome other machine learning techniques. SVM takes as input the ontology constructed from lexicons and the tweet collected from twitter and accordingly classifies the tweets to separate classes (Fig. 2).

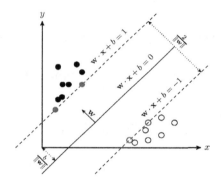

Fig. 2. The process of support vector machine

First of all, we will explain the SVM process that will be detailed in the following:

Step 1: Classification of Datasets: Our data are composed of n vectors X_i. Each X_i will also be associated with a value Y_i (Yi can take only the value 1 or -1). The more formal definition of an initial dataset is $D = f(X_i, Y_i) \mid X_i \in R$; $Y_i \in \{-1, 1\}$.

Step 2: Select Two Hyperplanes Separating The Data With No Points between Them:

$$H1 : w \times X_i + b \succeq +1 \text{ when } \quad Y_i = +1$$
$$H2 : w \times X_i + b \preceq -1 \text{ when } \quad Y_i = -1$$

d+ is the shortest distance to the closest positive point, d− the shortest distance to the closest negative point.

The margin of a separating hyperplane is d++d−. H1 and H2 are the planes. The points on the planes H1 and H2 are the tips of the Support Vectors. The plane H0 is the median in between where $w \times X_i + b = 0$, Where: w is a weight vector, x is the input vector and b is the bias.

Step 3: Maximize The Distance between The Two Hyperplanes: There are multiple hyperplanes that separate the data, but we have to choose a maximum margin: the maximum distance of $d+ + d-$. We consider the patterns from the class -1 and determine the hyperplane H1: The distance between origin and the hyperplane H1 is equal to $|-1 - b|/|w|$.

Similarly, the patterns from the class $+1$ and determine the hyperplane H2: The distance between origin and the hyperplane H2 is equal to $|+1 - b|/|w|$.

Based on the above considerations, the distance between hyperplanes (margin) H1 and H2 is $2/|w|$.

From these considerations, it follows that the identification of the optimum separation hyperplane is performed by maximizing $2/|w|$.

The problem of finding the optimum separation hyperplane is represented by the identification of (w, b) which satisfies:

$$w \times X_i + b \succeq +1 \text{ when } \quad Y_i = +1$$
$$w \times X_i + b \prec -1 \text{ when } \quad Y_i = -1 \text{ for which } |w| \text{ is minimum.}$$

Score Calculating. The classifier used two formula to calculate the score of each tweet.

$$F(tweet)pos = \frac{positivetweet}{Positivetweet + Negativetweet} + score(senti) \tag{1}$$

$$F(tweet)Neg = \frac{Negativetweet}{Positivetweet + Negativetweet} + score(senti) \tag{2}$$

Formulas (1) and (2) are detrminated to calculte the sentiment score of a tweet.

F tweet (pos) is the positive score of the tweet when considering the tweet positive.

F tweet (neg) is the negative score of the tweet when considering the tweet negative.

Score (senti) is the score of the word listed on the tweet retrieved from sentiwordnet.

We developed a classification according to the score calculated.

$$score = \begin{cases} If Ftweet(pos) >= 0.75 & \text{highly positif} \\ If 0.5 <= Ftweet(pos) <= 0.75 & \text{positif} \\ If Ftweet(pos) < 0.5 & \text{lowly positif} \\ If Ftweet(neg) >= 0.75 & \text{highly negatif} \\ If F0.5 <= tweet(neg) <= 0.75 & \text{negatif} \\ If Ftweet(neg) < 0.5 & \text{lowly negatif} \\ 0 & \text{objectif} \end{cases}$$

4   Experimentation and Results

In this section we test and validate our approach. We have realized this experimentation to measure the effectiveness of this approach. Following, we present the experimental study.

4.1   Mining Dataset of Joshi and Rose

To validate our approach, we need to test it with different datasets. We have selected the dataset of Joshi and Rose. It is a dataset of the opinion mining, released by Bing Liu's group from the University of Illinois, at Chicago. It consists of 200 review comments, each of 11 products presented in an XML format. This is an example of the review (Fig. 3):

```
<instance id="3" subpop="Canon-G3">
  <class cname="POS" />
  <text>in terms of the shots taken , this camera is insanely great !</text>
</instance>
```

Fig. 3. Fragment of the dataset

As it is displayed, the tag "<text>" presents the comment, the polarity (POS/NEG) is stored in the attribute "CNAME" of the "class" tag and the product name is stored in the "SUBPOP" attribute of the "instance" tag.

4.2   Evaluation of Our Approach

We measured the effectiveness of our approach by using the datasets mentioned in the previous section in comparaison with one related work (Akshat Bakliwal 2013) and our approach using 5 datasets for tweet-level sentiment classification in the first step. Therefore, our evaluation in this task is done using all datasets. The Table above synposis the results in precision and recall of all datasets (Table 3).

Table 3. Comparaison of results using precision and recall

Dataset	Akshat Bakliwal's method		Our approach	
	Precision	Recall	Precision	Recall
Cellular phone nokia	0.64	0.73	**0.81**	**0.78**
Digital camera nikon	0.54	0.73	**0.78**	**0.76**
Digital camera canon	0.63	0.80	0.67	0.82
DVD player	0.78	0.76	**0.812**	0.79
MP3 player	0.81	0.78	0.83	0.80

Table 4. Experimentation results

Datasets	Tweets	Positive	Negative	Unknown
Nokia	1970	820	1000	150
Smartphone	1966	612	1250	137
Samsung	6485	4235	2130	120
Iphone	7340	5210	2120	10
Orange	5465	2320	2900	245

Then, we collected a set of tweets from twitter and tested them with our proposed approach. The next table show the experimentation results (Table 4).

The next figure exposes the evaluation results of one related work used Naive Bayes and our approach using 5 datasets for tweet-level sentiment classification. Therefore, our evaluation is done to demonstrate the effectiveness of our classifier (SVM) in comparaison with other supervised machine learning classifier (Fig. 4).

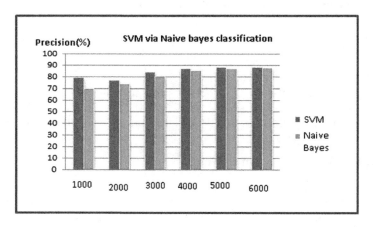

Fig. 4. Comparaison of our experiment result compared to related work used Naive bayes

5   Conclusion

In this work, we have proposed a method allows to extract customer opinion combined machine learning and text mining tools. First, we have developed an algorithm to identify opinion polarity. After that, we have tested and validated the effectiveness of this approach with datasets available in on the web in first time, and next we have used data collected from Twitter. The results show that the tweet classification using SVM is efficient. In order to improve the effectiveness of this approach, it seems that it is has been necessary it with other and large datasets. Another research axis will be to extract opinion polarity bot only about a product but the of a product's features.

References

1. Agarwal, B., et al.: One-class support vector machine for sentiment analysis of movie review documents. World Acad. Sci. Eng. Technol. Int. J. Comput. Electr. Autom. Control Inf. Eng. **9**(12), 2039–2042 (2015)
2. Baccianella, S., Esuli, A., Sebastiani, F.: SentiWordNet 3.0: an enhanced lexical resource for sentiment analysis and opinion mining. In: LREC, vol. 10, pp. 2200–2204 (2010)
3. Dalal, M.K., Zaveri, M.A.: Opinion mining from online user reviews using fuzzy linguistic hedges. Appl. Comput. Intell. Soft Comput. **2014**, 2 (2014)
4. Duwairi, R.M., Qarqaz, I.: Arabic sentiment analysis using supervised classification. In: 2014 International Conference on Future Internet of Things and Cloud (FiCloud), pp. 579–583. IEEE (2014)
5. Ghosh, M., Kar, A.: Unsupervised linguistic approach for sentiment classification from online reviews using SentiWordNet 3.0. Int. J. Eng. Res. Technol. **2** (2013). ESRSA Publications
6. Hatzivassiloglou, V., McKeown, K.R.: Predicting the semantic orientation of adjectives. In: Proceedings of the 35th Annual Meeting of the Association for Computational Linguistics and Eighth Conference of the European Chapter of the Association for Computational Linguistics, pp. 174–181. Association for Computational Linguistics (1997)
7. Kontopoulos, E., Berberidis, C., Dergiades, T., Bassiliades, N.: Ontology-based sentiment analysis of Twitter posts. Expert Syst. Appl. **40**(10), 4065–4074 (2013)
8. Liu, B.: Sentiment analysis and opinion mining. Synth. Lect. Hum. Lang. Technol. **5**(1), 1–167 (2012)
9. Mars, A., Gouider, M.S., Saïd, L.B.: A new big data framework for customer opinions polarity extraction. In: Kozielski, S., Mrozek, D., Kasprowski, P., Małysiak-Mrozek, B., Kostrzewa, D. (eds.) BDAS 2015–2016. CCIS, vol. 613, pp. 518–531. Springer, Cham (2016). doi:10.1007/978-3-319-34099-9_40
10. Palanisamy, P., Yadav, V., Elchuri, H.: Serendio: simple and practical lexicon based approach to sentiment analysis. In: Proceedings of Second Joint Conference on Lexical and Computational Semantics, pp. 543–548. Citeseer (2013)
11. Pang, B., Lee, L., Vaithyanathan, S.: Thumbs up?: sentiment classification using machine learning techniques. In: Proceedings of the ACL 2002 Conference on Empirical Methods in Natural Language Processing, vol. 10, pp. 79–86. Association for Computational Linguistics (2002)
12. Park, S.-M., Baik, D.-K.: Personal ontology-based sentiment analysis system for mobile devices. In: Proceedings of the International Conference on Semantic Web and Web Services (SWWS), the Steering Committee of the World Congress in Computer Science, Computer Engineering and Applied Computing (WorldComp), p. 42 (2013)
13. Riloff, E., Wiebe, J.: Learning extraction patterns for subjective expressions. In: Proceedings of the 2003 Conference on Empirical Methods in Natural Language Processing, pp. 105–112. Association for Computational Linguistics (2003)
14. Turney, P.D.: Thumbs up or thumbs down?: semantic orientation applied to unsupervised classification of reviews. In: Proceedings of the 40th Annual Meeting on Association for Computational Linguistics, pp. 417–424. Association for Computational Linguistics (2002)

15. Vaitheeswaran, G., Arockiam, L.: Hybrid based approach to enhance the accuracy of sentiment analysis on tweets. Int. J. Sci. Eng. Comput. Technol. **6**(6), 185 (2016)
16. Zainuddin, N., Selamat, A., Ibrahim, R.: Improving Twitter aspect-based sentiment analysis using hybrid approach. In: Nguyen, N.T., Trawiński, B., Fujita, H., Hong, T.-P. (eds.) ACIIDS 2016. LNCS, vol. 9621, pp. 151–160. Springer, Heidelberg (2016). doi:10.1007/978-3-662-49381-6_15

Study for Automatic Classification of Arabic Spoken Documents

Mohamed Labidi[1(✉)], Mohsen Maraoui[2], and Mounir Zrigui[1]

[1] Research Laboratory of Technologies of Information and Communication
and Electrical Engineering, Tunis, Tunisia
mohamedlabidi@protonmail.ch, mounir.zrigui@fsm.rnu.tn
[2] Computational Mathematics Laboratory, Monastir, Tunisia

Abstract. One of the important tasks in natural language processing is speech classification by domain. As shown in the literature, no prior studies have addressed this problem, specially the effect of using root N-grams and stem N-grams on Arabic speech classification performance. In this paper we describe a study for Arabic spoken documents classification, using the K-Nearest Neighbor, the Naive Bayes and the Support Vector Machine. We create a speech recognition system for the transcription of Arabic audio files. Then, we use four types of features: 1-gram, 2-gram and 3-gram word roots or stems as well as full words. The obtained results show that, compared to stem or word N-grams, the use of a 1-gram root as a feature provides greater classification performance for Arabic speech classification. It is that classification performance decreases whenever the number of N-grams increases. The data also exhibit that the support vector machine outperforms the Naïve Bayes and the k-nearest neighbor with 1 gram. Whenever the k-nearest neighbor is used, the 2-gram root achieves the best performance. The 3-gram root, on the other hand, achieves the best performance whenever the support vector machine was used.

Keywords: Arabic speech classification · Natural language processing · Artificial intelligence · N-gram · Stem · Root · KNN · SVM · Naïve Bayes

1 Introduction

Currently, the amount of spoken data available is so colossal that it has become difficult or impossible to ask a human being to perform a manual audio document classification task. It seems very interesting to count on a computer solution that allows assigning these audio files to a predefined set of categories in an automatic way. Audio classification has a various number of applications such as website classification, automatic indexing, etc.

The increase in spoken information today is highly noticeable even on the levels of certain languages like Arabic. The automatic classification of Arabic spoken documents appeared lately and only little work was interested in this field. The latter fact is due to the difficulty of the task and the complex morphological representation of the Arabic language as well as its orthographic variations linked to the phenomenon of agglutination of letters [22].

© Springer International Publishing AG 2017
N.T. Nguyen et al. (Eds.): ICCCI 2017, Part II, LNAI 10449, pp. 459–468, 2017.
DOI: 10.1007/978-3-319-67077-5_44

The impact of stemming on Arabic text classification was studied in [11]. The authors used two stemming approaches, namely, light stemming and word clusters to investigate the impact of stemming. They reported that the light stemming approach improved the accuracy of the classifier more than the other approach.

An approach for Arabic text classification was presented in [21] by applying stemming and without using stemming with three classifiers: the SVM, the decision trees and the NB. The results achieved 87.79% and 88.54% accuracy with the SVM and the NB, respectively, when stemming was used. On the other hand, the results without the application of stemming achieved lower accuracy of 84.49% and 86.35%.

An approach proposed in [3] based on the standard Naïve Bayes classifier with a rooting algorithm was employed for the dimensionality reduction of texts. The results reported an average accuracy of 62.23%, based on a corpus of 300 Arabic documents distributed over 10 categories, which were collected from several newspaper articles.

In [6] the authors made an experimental study to compare two approaches of dimensionality reduction using stemming and verified their effectiveness on Arabic text classification. Firstly, they applied the Latent Dirichlet Allocation (LDA) and Latent Semantic Indexing (LSI) for modelling a document set of Open Arabic Tunisian Corpus (OATC) containing 20,000 documents collected from Tunisian newspapers. They generated two matrices: the LDA (documents/topics) and the LSI (documents/topics). After that, they used the SVM for text classification. The classification results were evaluated by precision, recall and F-measure. The evaluation of classification results was performed on the OATC (70% training set and 30% testing set). Their experiments showed that the results of dimensionality reduction via the LDA outperformed the LSI in Arabic text classification.

The authors in [15] investigated the impact of using different indexing approaches (full-word, stem, and root) when classifying Arabic text. They used the NB with a stratified k-fold cross-validation on a corpus that consisted of 1000 Arabic documents. The overall results of this study demonstrated that the classifier achieved the maximum micro average accuracy of 99.36%, either by using the full-word form or the stem form.

The third type is N-grams: a sequence of N items from a given sequence of text. It can be deemed a window of length of N moves over the text. The content of this window is the N-gram. N-grams vary following the value that N takes: 1, 2, 3, and so on, and is called respectively: unigram, bigrams, trigrams and so on [16]. Two types can be distinguished: character N-grams and word N-grams.

The authors in [1] proposed to classify Arabic texts by combining N-grams with several similarity measures, including Dice, Manhattan, and Euclidean distances. The overall comparison results illustrated that combining the Dice measure with the trigram would give a better performance.

The effect of using n-grams was discussed in [12]. The authors utilized unigrams, 2-grams, 3-grams and 4-grams of full words with four feature selection methods, three representation schemas and three classifiers (SVM, KNN and NB). The main conclusion of their work was that the use of a full word was more efficient than the use of N-grams in terms of classification.

In general terms, implementing a classification system for spoken documents is similar to a text classification system. Generally, it undergoes several steps, namely collecting a representative dataset, dividing the sample into training and testing sets, extracting the features, selecting the representative features, representing the selected feature for the classifier, applying the algorithm, producing the classification model, applying the algorithm on testing data, and evaluating the performance of the classification model. The techniques used in each of the above steps affect the performance of the classification system of audio documents in various ways.

In this study, we investigate the classification of Arabic spoken documents. Furthermore, we study an effect that has not been studied in Arabic spoken documents classification and in Arabic text classification yet, which is the effect of root N-grams and stem N-grams.

Instead of single words, we make use of three feature types (1-grams, 2-grams, and 3-grams) of roots, stems and full words while using three different feature selection values and three classifiers.

This paper is organized as follows: Sect. 2 delineates our proposed approach and the released work. Section 3 exposits the realized experiments and the obtained results.

2   Proposed Approach

This section gives an overview of our proposed approach for the classification of Arabic spoken documents.

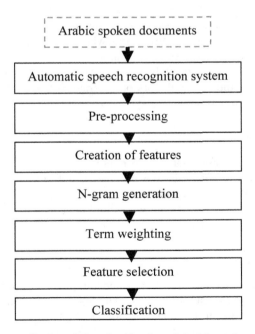

Fig. 1. Proposed approach for classification of Arabic spoken documents.

Figure 1 illustrates our approach for the classification of Arabic audio documents. First of all, we use our Arabic speech recognition system to transcribe the audio files. After that, a pre-processing phase removes stop words. Thirdly, a step to create different features (stem, root or full word) is performed. The fourth step is to generate N-grams (unigram, bigrams, and trigrams). The next step is term weighting, followed by feature selection and finally the classification using three different classifiers. All these steps will be well defined throughout the next sections.

2.1 Speech Recognition System

We use the CMU Sphinx tool to build our speech recognition system. The utilized data are described in the experimental setup (Sect. 3.1) and the obtained results are described in Sect. 3.2. The system gives us the transcriptions for the recognized speech files.

Like any other speech recognition system, our system built from an acoustic model followed by a language model. The acoustic model is used to describe the relationship between an audio signal and the phonemes. The model is trained by a set of audio files and their corresponding transcripts. It is constructed by taking audio recordings of speech, and their text transcriptions, and using the CMU Sphinx to create statistical representations of the phonemes that make up each word or phrase. We choose to make a phoneme-based acoustic model because it gives better results in Arabic speech recognition [14].

We realize our statistical language model utilizing the SRILM toolkit [22]. It is a probability distribution over sequences of words. Given such a W sequence, say of length m, it assigns a probability $P(W1, ..., Wm)$ to the whole sequence. In any speech recognition system, the computer essays to match sounds with word sequences. The language model provides context to distinguish between words and phrases that sound similar.

In the rest of our method we will work on the transcriptions of the audio files given by our speech recognition system.

2.2 Pre-processing

Data pre-processing is an essential step to remove useless data such as prepositions, pronouns, conjunctions words, etc. We use the list of stop words included in the Khoja's Stemmer [23]. The list contains 168 stop words. This stop word list can be considered to include the most frequent words in Arabic as it mainly encompasses the Arabic preposition and conjunction.

2.3 Creating Features

In this phase, we use three types of feature which are: full-word, stem, and root. For the root and stem extraction we opt for two tools. The first one is the Khoja's Stemmer. It is a root-based approach which uses the morphological analysis to extract the root of a given Arabic word according to vocalization variation and derivation pattern knowledge [23]. The Khoja's Stemmer attempts to find root for Arabic words based on predefined root lists and the morphological analysis. This approach produces abstract roots. The second one is the Alkhalil Morpho System [13], which is a morpho-syntactic parser of standard Arabic words. It is based on a very large set of Arabic morphological rules. The output of the analysis of each word contains the stem, its grammatical category, and its possible roots associated with corresponding patterns, proclitics and enclitics. We use the Alkhalil Morpho System to produce both roots and stems.

2.4   Generating N-grams

In this step, different types of N-grams are generated from the features created in the previous step. We use 1-gram, 2-grams and 3-grams of the root, the stem and the full word.

2.5   Term Weighting

There are many approaches for term weighting. In this work, a well-known approach called the TFI/DF is used [10]. As documents in the collection of text documents (transcriptions) vary in length (i.e., since the number of features in documents vary). The short ones might have less chance to be classified as relevant as long documents. Accordingly, the retrieval of any document must be done independently from its length. The latter goal can be reached by normalizing document vectors, hence allowing documents of all lengths to be retrieved.

2.6   Feature Selection

To reduce the size of the representation space and select the relevant feature, we choose the chi square χ^2 method. This choice stems from the fact that the latter (chi square χ^2 method) is considered among the most effective methods of feature selection for text classification [7]. We also opt for three different values to select the best ranked features; the values are 100, 500 and 1000.

2.7   Classification

Classification will be applied for each type of representation of our approach with three different numbers of features retained (100, 500 and 1000) with the chi square χ^2 method using the following three classifiers. The first one is the KNN classifier because it is considered efficient and provides good results in classification [9]. We choose the cosine similarity as it is among the most suitable for text classification [18] and a value of K = 13. The second one is the Naïve Bayes, thanks to its simplicity and speed, as it is widely used for classifying texts [2]. The third one is the SVM, which is considered one of the most powerful learning algorithms for text categorization [4]. The selection of the number of features (100, 500, 1000) returns to the work found in the literature concerning Arabic text classification.

3 Experiments

In this section we describe our experimental setup, the launched tests and the obtained results followed each time by a discussion.

3.1 Experimental Setup

Our acoustic model is built with the help of the CMU Sphinx [20]. We train it using 51 h of audio material for the modern standard Arabic, recorded by 41 native speakers. Each audio file is accompanied by its transcription. The audio files are converted to 16 kHz, 16 bits, mono speakers, and in an MS WAV format, as required by the Sphinx trainer. The phonetic dictionary has been similarly used by almost all researchers in the construction of Arabic speech recognition systems [24].

Our language model training corpora consist of around 200 million running full words including data from Ajdir Corpora, Tashkeela corpora [18], Abbas corpora [17], OSAC corpora [25] and collected corpora. Our statistical language model is constructed using the SRILM toolkit [22].

To evaluate our approach, our small audio corpus of 8 h is used for all our experiments.

We utilize the 2012 Arabic Newspapers dataset [19]. This dataset consists of 2,910 texts evenly divided into six classes: politics, economy, culture, sports, religion, and science and technology. The basic statistics of the dataset are illustrated in Table 1.

Table 1. Basic dataset statistic.

Topic	Politics	Economy	Culture	Sports	Religion	Sciences and technology
Number of words	379,632	363,858	407,080	376,832	367,253	312,814
Number of texts	474	534	464	534	430	474

3.2 Results and Discussion

We use the RapidMiner 7.2 [5] implementations of the classifiers SVM, KNN and NB to train and test the classification model. The evaluation of their performance is in terms of F-measure which combines both quality measures of precision and recall [2]. Table 2 presents the classification results using a full word in terms of F-measure, with three classifiers and three different values of features. The table also shows that the F-measure reaches its highest value (88.83%) with the SVM using 1-gram with 1,000 features and its lowest value (42.8%) with the NB using 3-grams and 100 features.

Table 3 provides the classification results using the Khoja's stemmer in terms of F-measure, with three classifiers and three different values of features. The table demonstrates that the F-measure reaches its highest value (90.39%) with the SVM using 1-gram with 1000 features, and the lowest one (49.37%) for the NB using 3-grams and 100 features.

Table 2. Classification results using full word.

Classifier	Grams	Number of features		
		100	*500*	*1000*
SVM	1	80.07%	86.74%	88.83%
	2	65.79%	76.52%	79.07%
	3	45.52%	55.22%	61.69%
KNN	1	78.43%	86.06%	87.46%
	2	65.53%	75.67%	79.02%
	3	43.31%	47.69%	50.89%
NB	1	72.21%	83.73%	85.57%
	2	67.18%	75.45%	77.27%
	3	42.8%	54.58%	62.5%

Table 3. Classification results using Khoja's stemmer.

Classifier	Grams	Number of features		
		100	*500*	*1000*
SVM	1	85.55%	87.93%	90.39%
	2	67.07%	79.07%	82.88%
	3	55.52%	62.7%	65.36%
KNN	1	83.23%	86.7%	87.32%
	2	69.79%	80.54%	84.12%
	3	50.27%	57.54%	60.64%
NB	1	80.53%	80.96%	82.87%
	2	65.08%	76.78%	80.01%
	3	49.37%	61.02%	64.3%

Table 4 gives the classification results using the roots of the Alkhalil Morpho System in terms of F-measure, with three classifiers and three different values of features. The table indicates that the F-measure reaches its highest value (89.71%) with SVM using 1-gram with 1000 features, and its lowest one (48.13%) for the KNN using 3-grams and 100 features.

Table 4. Classification results using roots of Alkhalil morpho system.

Classifier	Grams	Number of features		
		100	*500*	*1000*
SVM	1	85.92%	88.99%	89.71%
	2	66.1%	79.23%	82.5%
	3	53.67%	59.49%	63.58%
KNN	1	83.14%	86.35%	86.69%
	2	68.35%	80.24%	83.42%
	3	48.13%	55.04%	57.17%
NB	1	80.82%	81.38%	82.32%
	2	67.57%	76.67%	78.95%
	3	48.15%	58.94%	62.97%

Table 5 presents the classification results using the stems of the Alkhalil Morpho System in terms of F-measure, with three classifiers and three different values of features. The table shows that the F-measure reaches its highest (88.99%) with the SVM using 1-gram with 1000 features, and its lowest value (47.55%) for the KNN using 3-grams and 100 features.

Table 5. Classification results using stems of Alkhalil morpho system.

Classifier	Grams	Number of features		
		100	*500*	*1000*
SVM	1	83.44%	88.4%	88.99%
	2	65.12%	76.32%	79.77%
	3	51.76%	57.79%	63.36%
KNN	1	81.76%	86.85%	87.4%
	2	66.02%	76.94%	80.07%
	3	47.55%	51.24%	54.31%
NB	1	75.77%	84.09%	84.74%
	2	66.29%	75.31%	77.39%
	3	48.49%	56.69%	63.89%

Table 6. Comparison between the different classification results.

Grams	Number of features		
	100	500	1000
1	85.92% SVM, Roots Alkhalil	88.99% SVM, Roots Alkhalil	90.39% SVM, Khoja's Stemmer
2	69.79% KNN, Khoja's Stemmer	80.54% KNN, Khoja's Stemmer	84.05% KNN, Khoja's Stemmer
3	55.52% SVM, Khoja's Stemmer	62.7% SVM, Khoja's Stemmer	65.36% SVM, Khoja's Stemmer

Table 6 indicates that for unigrams, the SVM classifier using the roots of the Alkhalil Morpho System provides the best results for a number of features equal to 100 and 500 compared to other classification tests. The table also demonstrates that whenever the number of features is equal to 1000, the Khoja's Stemmer outperforms all other classification tests with a value equal to 90.39% of the F-measure. For bigrams, the KNN classifier using the Khoja's Stemmer is the most powerful among all others. For the trigrams, the SVM classifier for a number of features equal to 100, 500 and 1000 gives the highest result.

Tables 2, 3, 4 and 6 show that whenever the number of selected features increases, the results improve for all tests. The classification performance, however, decreases whenever the number of N-grams increases even when using the root or the stem as a feature.

4   Conclusion

The present work revolves around the classification of Arabic spoken documents. This domain has lately appeared and appealed the interest of only little work. The latter scarcity emanates from the complexity of the task and also the complexity of the morphological representation of the Arabic language. As a result, we have chosen a classification approach that encompasses several types of features and classifiers.

We have presented a new approach for the classification of Arabic audio documents, which combines the N-grams of three different feature types utilizing three classifiers tested with various feature selection values.

The bigrams and trigrams of full words, stems and roots established in our classification tests have certainly improved the results, but they are still insignificant in comparison with those achieved by unigrams.

References

1. Abbas, M., et al.: Evaluation of topic identification methods on Arabic corpora. JDIM **9**(5), 185–192 (2011)
2. Al-Badarneh, A., et al.: The impact of indexing approaches on Arabic text classification. J. Inf. Sci. **43**(2), 159–173 (2017)
3. Ali, M., et al.: Arabic phonetic dictionaries for speech recognition. J. Inf. Technol. Res. **2**(4), 67–80 (2009)
4. Aljlayl, M., Frieder, O.: On Arabic search: improving the retrieval effectiveness via a light stemming approach. In: Proceedings of 11th International Conference on Information and Knowledge Management, pp. 340–347. ACM (2002)
5. Al-Kabi, M., et al.: The effect of stemming on Arabic text classification: an empirical study. In: Information Retrieval Methods for Multidisciplinary Applications, p. 207 (2013)
6. Al-Molegi, A., et al.: Automatic learning of arabic text categorization. Int. J. Digit. Contents Appl. **2**(1), 1–16 (2015)
7. Al-Shalabi, R., Obeidat, R.: Improving KNN Arabic text classification with n-grams based document indexing. In: Proceedings of 6th International Conference on Informatics and Systems, Cairo, Egypt, pp. 108–112 (2008)
8. Al-Thubaity, A., Alhoshan, M., Hazzaa, I.: Using word n-grams as features in Arabic text classification. In: Lee, R. (ed.) Software Engineering, Artificial Intelligence, Networking and Parallel/Distributed Computing. SCI, vol. 569, pp. 35–43. Springer, Cham (2015). doi:10. 1007/978-3-319-10389-1_3
9. Ayadi, R., Maraoui, M., Zrigui, M.: LDA and LSI as a dimensionality reduction method in arabic document classification. In: Dregvaite, G., Damasevicius, R. (eds.) ICIST 2015. CCIS, vol. 538, pp. 491–502. Springer, Cham (2015). doi:10.1007/978-3-319-24770-0_42
10. Barigou, F.: Improving K-nearest neighbor efficiency for text categorization. Neural Netw. World **26**(1), 45 (2016)
11. Dai, P., et al.: A novel feature combination approach for spoken document classification with support vector machines. In: Proceedings of Multimedia Information Retrieval Workshop, pp. 1–5 (2003)
12. Duwairi, R., et al.: Feature reduction techniques for Arabic text categorization. J. Am. Soc. Inf. Sci. Technol. **60**(11), 2347–2352 (2009)

13. Harrag, F., et al.: Improving Arabic text categorization using decision trees. In: 2009 1st International Conference on Networked Digital Technologies, NDT 2009, pp. 110–115. IEEE (2009)
14. Khoja, S., Garside, R.: Stemming Arabic text. Computing Department, Lancaster University, Lancaster, UK (1999)
15. Lamere, P., et al.: Design of the CMU sphinx-4 decoder. In: INTERSPEECH (2003)
16. Lee, D.L., et al.: Document ranking and the vector-space model. IEEE Softw. **14**(2), 67–75 (1997)
17. Mesleh, A.M.: Support vector machines based Arabic language text classification system: feature selection comparative study. In: Sobh, T. (ed.) Advances in Computer and Information Sciences and Engineering, pp. 11–16. Springer, Dordrecht (2008). doi:10.1007/978-1-4020-8741-7_3
18. Noaman, H.M., et al.: Naive Bayes classifier based Arabic document categorization. In: 2010 7th International Conference on Informatics and Systems (INFOS), pp. 1–5. IEEE (2010)
19. Pilászy, I.: Text categorization and support vector machines. In: Proceedings of 6th International Symposium of Hungarian Researchers on Computational Intelligence (2005)
20. Qamar, A.M., et al.: Similarity learning for nearest neighbor classification. In: 2008 8th IEEE International Conference on Data Mining, ICDM 2008, pp. 983–988. IEEE (2008)
21. Saad, M.K., Ashour, W.: Arabic morphological tools for text mining. Corpora **18**, 19 (2010)
22. Schneider, K.-M.: Techniques for improving the performance of Naive Bayes for text classification. In: Gelbukh, A. (ed.) CICLing 2005. LNCS, vol. 3406, pp. 682–693. Springer, Heidelberg (2005). doi:10.1007/978-3-540-30586-6_76
23. Singh, S.R., et al.: Feature selection for text classification based on Gini coefficient of inequality. In: FSDM, vol. 10, pp. 76–85 (2010)
24. Stolcke, A., et al.: SRILM-an extensible language modeling toolkit. In: Interspeech, vol. 2002 (2002)
25. Zerrouki, T., Balla, A.: Tashkeela: novel corpus of Arabic vocalized texts, data for auto-diacritization systems. Data Brief **11**, 147–151 (2017)

"Come Together!": Interactions of Language Networks and Multilingual Communities on Twitter

Nabeel Albishry[1(✉)], Tom Crick[2], and Theo Tryfonas[1]

[1] Faculty of Engineering, University of Bristol, Bristol, UK
{n.albishry,theo.tryfonas}@bristol.ac.uk
[2] Department of Computing, Cardiff Metropolitan University, Cardiff, UK
tcrick@cardiffmet.ac.uk

Abstract. Emerging tools and methodologies are providing insight into the factors that promote the propagation of information in online social networks following significant activities, such as high-profile international social or societal events. This paper presents an extensible approach for analysing how different language communities engage and interact on the social networking platform Twitter via an analysis of the Eurovision Song Contest held in Stockholm, Sweden, in May 2016. By utilising language information from user profiles ($N = 1{,}226{,}959$) and status updates ($N = 7{,}926{,}746$) to identify and categorise communities, our approach is able to categorise these interactions, as well as construct network graphs to provide further insight on these multilingual communities. The results show that multilingualism is positively correlated with activity whilst negatively correlated with posting in the user's own language.

Keywords: Language networks · Multilingual communities · Community discovery · Network graphs · Social networks

1 Introduction

Despite the widespread use of Twitter globally – with 328 million monthly active users as of the first quarter of 2017 – little research has investigated the differences amongst users of various languages; there is a tendency to assume that the behaviours of English users generalise to other language users [1]. Language has featured as a facet of research on the geographies of Twitter networks [2,3], especially whether offline geography still matters in online social networks [4]. Linguistic-inspired studies have been performed on hashtags [5], as well as the volume and proportional of tweets in English and Arabic, as part of an analysis of the Arab Spring [6]. Nevertheless, language is clearly a vital component of affiliation and discourse on the web [7,8], with the creation and curation of

N. Albishry—This work has been supported by a doctoral research scholarship for Nabeel Albishry from King Abdulaziz University, Kingdom of Saudi Arabia.

N.T. Nguyen et al. (Eds.): ICCCI 2017, Part II, LNAI 10449, pp. 469–478, 2017.
DOI: 10.1007/978-3-319-67077-5_45

emerging multilingual networks and communities, representing well-established creative and cultural norms, including for minority languages such as Welsh [9], as well as investigations into the economics of linguistic diversity [10].

In the social network analysis domain, centrality measures such as degrees, betweenness, clustering coefficient, modularity and cliques have been used in various projects to measure influence or detect the emergence of new communities [11,12]. These measures provide the ability to assess network graphs that are constructed from collected data (for example, tweets). Selection of these centrality measures is dependent on the goal of the analysis; for example, the degree of a node helps to identify nodes with high number of connections within the network [13–15]. In a representation of a real-world network, this metric may help to identify highly connected persons, such as political leaders, sports stars or celebrities, who are potential "information spreaders" [16–18].

Clustering users in communities has been an important factor in social networking analysis, with a particular focus on clustering users based on their locations. However, for the sake of anonymity, many users tend not to disclose information about their identity, such as locations [19]; looking at Twitter, it has also been reported in the literature that geotagged tweets are generally low in number [20–22], the exponential growth in social media over the past decade has been joined by the rise of location as a central organising theme [23] of how users engage with online information services and, more importantly, with each other [24–26]. The work here examines the correlation between multilingualism of users and their associated activity.

The remainder of this paper is organised as follows: in Sect. 2, we introduce the methodology and key language themes; Sect. 4 presents the 2016 Eurovision Song Context case study, along with an analysis of the key data and results; Sect. 5 concludes the paper with a wider discussion and a summary of the potential application of our approach.

2   Methodology

The primary purpose of this study is to identify and define an extensible analytical approach for examining language uses, communities, and diversity on Twitter. The approach is based on network graphs and their properties, such as indegree, outdegree, and edge weights. Graphs are generated from language settings in users' profiles and those for statuses. First, we construct user graphs to analyse interactions and multilingualism at the level of individual users. Then, from the user graphs, we produce language communities graph that groups users based on common languages.

3   Language Entities

To generate the required graphs, we need three essential entities from each status; user ID, user profile language, and status language[1]. Those values can be

---

[1] For Twitter, *status* may also be referred to as *post*, or *tweet*.

extracted from *[status]['user']['id']*, *[status]['user']['lang']*, and *[status]['lang']*, respectively. It is important to note that the focus of this work is on the analytical approach, not necessarily the accuracy of language detection; therefore we assume that language of tweets are correctly identified. For profiles, users are expected to pick a language for their settings. Nevertheless, their language entity may show as the initial placeholder text "*Select Language...*" or a translated version that may provide information to the user's native language community.

3.1   Network Graphs

For this study, we need to generate two different graphs; one is based on individual users and their posting activity, while the other combines users into language communities. In the context of this study, all graphs must be directed to provide correct measurements, as demonstrated in Fig. 1.

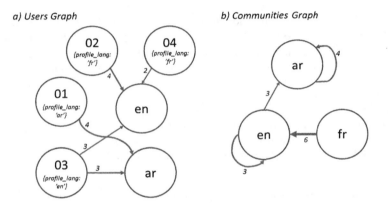

Fig. 1. Examples of simple models of language graphs

User Graph. This graph represents the core structure for our analysis. As shown in Fig. 1(a), nodes in this graph are of two types; users and posting language. Each posted tweet resulted in two nodes, one represents the user with profile language setting added to the node as the attribute '{*profile_lang:xx*}'. The other node represents language of the tweet. Edges link users with the posting languages they used, and their weight (thickness) measures the number of tweets that have been posted by the user (the starting node) in the target language (ending node). In the example above, the profile language setting for user '*03*' is '*en*', they posted three tweets in '*en*' and three in '*ar*' (Arabic). This graph will be referred to as the *user graph*.

Communities Graph. This second graph is derived from the user graph and has one type of node to represent language community, as shown in Fig. 1(b).

For each user node we generate one node from the '{*profile_lang:xx*}' attribute, and another node from the posting language to which it is connected. This resulted in combining all users of the same profile language into one node, with edge connecting to posting language and its weight measuring their activity. Theoretically, each tweet results in two language nodes, one for the user profile, and the other for language of the tweet. In our example above, users with '*fr*' (French) profiles have generated six tweets in '*en*'. In the case of '*ar*' node, we can see that users of the profile language as '*ar*' have posted four tweets in '*ar*' only – in graph terminology this is referred to as 'self-loop'; we will refer to this graph as the *communities graph*.

Throughout the paper, we refer to language communities in two ways; *profile community* to perceive the language as user profile settings, whereas *posting community* refers to the language as tweeting settings.

3.2   Measures

In this section, we will discuss how graph measures can be used to make deductions about users, associated community languages, posting language activity, and how different language communities are linked to each other. These measures and their interpretations, in the context of this study are as follows:

- *Indegree*: number of incoming edges;
- *Outdegree*: number of outgoing edges;
- *Edge Weight*: number of tweets on edge;
- *Weighted indegree*: total weights of incoming edges;
- *Weighted outdegree*: total weights of outgoing edges.

User Graph Properties. User nodes have *indegree* $= 0$, and posting languages have *outdegree* $= 0$; these two properties will be used to distinguish between nodes. Both outdegree of user nodes and indegree of posting languages must be greater than 0. The edge weight indicates the number of tweets associated with both end nodes. Referring to the example in Fig. 1(a), we can see that user '*03*' has indegree of 0 (user identifier), outdegree of 2 (number of languages he used), and weighted outdegree of 6 (total number of tweets posted). Also, in the same figure, we can see that for '*en*' posting language, it has outdegree of 0 (language nodes identifier), indegree of 3 (number of users posted in this language), and weighted indegree of 9 (total number of tweets posted); Table 1 presents main properties of this graph.

Communities Graph Properties. As discussed in Sect. 3.1, this graph is extracted from the user graph and contains one type of node: language community nodes. Nodes in this graph represent languages as profile language settings, posting language, or both. However, as the graph is directed, we can identify if a community node is for profile or posts by measuring the *indegree* and *outdegree* properties. Positive indegree implies posting language, and positive outdegree

Table 1. Node properties in user graph

	User	Language
Indegree	0	>0
Outdegree	>0	0
Edge weight	#Tweets	

Table 2. Node properties in communities graph

	Community node
Indegree	Posts
Outdegree	Profiles
Edge weight	#Tweets

indicates profile language settings. Figure 1(b) shows three language communities, two nodes appear as posting and profile nodes, while one node exists as a profile only node. The node '*ar*', for example, has outdegree of 1 and indegree of 2. In other words, at least one user has their profile language settings as '*ar*', and at least two users have posted in '*ar*'. In terms of edge weights, we can say that there are seven tweets posted in '*ar*' language, originated from two different profile language communities. For the '*fr*' node, we can see only outdegree, which means this language community exists as a profile-only node as no user posted in '*fr*'; these measures are summarised in Table 2.

4 Case Study and Discussion

In our case study, we explore the analysis of a dataset collected from the #Eurovision hashtag during the 2016 Eurovision Song Contest, based on the techniques presented in Sect. 2. Using the *user graph* and *communities graph*, we conduct analyses on multilingualism, activities and user behaviours in posting in different languages[2].

4.1 Case Study: 2016 Eurovision Song Contest

The 2016 Eurovision Song Contest[3] took place in May in Stockholm, Sweden, with the motto of *"Come Together!"*. There were 32 countries taking part, with two semi-finals taking place on 12 and 14 May, with 26 countries qualifying for the final on 16 May. This year's contest was perceived by many commentators to be tense and politically motivated, especially with Ukraine eventually winning the final [27]. Varying analyses see the contest as being influenced by political conflicts, friendships or cultural bias [28–31], with a range of news articles explicitly discussing the possibly biased results [32]. Twitter activity was very high throughout the event on the primary #Eurovision hashtag, with close to 8 million statuses, produced by nearly 1.25 million users.

The study focuses on original statuses (tweets) as the basic entity, as we wish to measure posting behaviour, not reactions. Preliminary analysis shows

---

[2] In this context, *different language* refers to tweet's language that is different to the user profile language settings.

[3] https://www.eurovision.tv/page/stockholm-2016/all-participants.

that they account for 48% of the total activity, of which 4% tweets with an *'unidentified'* language were eliminated. As for profiles, all users have chosen language preferences and no profile was found with the default language settings.

4.2 Multilingualism

The outdegree in the user graph shows the number of languages a user used; observing the outdegree of user nodes in the users graph revealed 20 groups of outdegree, ranging from 1 to 25. Figure 2 shows these groups, size of users and activities. Although 85% of users are monolingual, their activity accounts for 47% of all tweets. Additionally, while the average activity of users is five posts per user, monolingual users were the least active ones, scoring an average of two tweets per user. We found that 18% of tweets were in different languages, with a strong correlation between multilingualism and likelihood of using different languages.

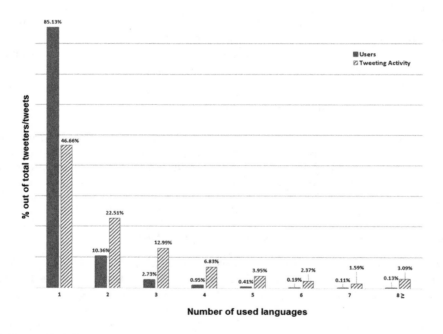

Fig. 2. Multilingual communities on #Eurovision and their associated activities.

We used the user graph to generate two communities graphs; the first will be used to explore language communities amongst monolingual users, while the other includes language communities for multilingual users only.

4.3 Monolingual Communities

This graph includes 63 language communities: 15 languages exist as profile-only and have not been used in any post, while 12 were used in posting but never show as a profile language. Moreover, about 13% of monolingual users used different languages in posts which form 10% of tweets in monolingual communities. Hence, strongest relationships exist as a self-loop, as discussed in Sect. 3.1.

To explore the relationships between language communities, we remove all self-loop edges from the graph. The resultant graph shows that monolingual users with '*en*' as profile language have posted in 47 other languages, causing 43% of tweeting activity, and that 48 other profile communities used '*en*' language in posting 43%. Also, we found that the strongest relationship (edge weight), 9% of activity, is when '*en*' profiles post in '*es*' (Spanish). A further interesting case to mention involve the '*el*' (Greek) and '*ru*' (Russian) languages. Although the number of profile communities that used '*ru*' is more than twice compared to the number of those that used '*el*', they were significantly lower in terms of activity.

4.4 Multilingual Communities

Although multilingual users form 15% of all users in the dataset, they generated 53% of tweeting activity. There are 48 language communities in this graph, 13 languages as profile-only, and 10 as posting languages. With self-loop edges excluded, activity in different languages measured 24% of multilingual users tweets. Also, we found that the strongest relations existed between the '*es*' profile community and the '*en*' posting language, which is the opposite to the monolingual case.

4.5 Visualisation

In Fig. 3, we present two communities graphs; the size of the node represents weighted indegree of community; how much a language was used in tweeting,

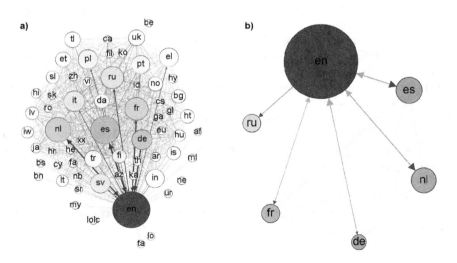

Fig. 3. Language communities graphs for `#Eurovision`.

and darkness reflects weighted outdegree; participation from users of language community. Edges link between *profile* and *posting* communities, and their thickness indicates the number of tweets posted.

Whilst Fig. 3(a) shows all language communities together, Fig. 3(b) presents a filtered graph. This filtered graph depicts relationships amongst language communities that scored high in weighted indegree and outdegree. Also, we eliminated users with activity lower than the overall average (five tweets/user), and generated the communities graph from the remainder.

5 Conclusions

This paper has presented an extensible approach for identifying interactions within language communities using a high-profile real-world case study – the 2016 Eurovision Song Contest – and its associated engagement and interactions on Twitter. This approach utilises network graph properties to explore the behaviour of monolingual and multilingual users. Surprisingly, even though monolingual users formed the largest proportion of users, they were less active than multilingual users. The results also confirmed that higher proportions of user multilingualism implies further distance from their profile language. In the profile community, large number of participants does not necessarily imply high language diversity, as a single post in other language is enough to take the community to a higher level of multilingualism. Therefore, filtering out those users with low activity would improve measurement accuracy. In a few cases, we witnessed users participating in a significant number of languages, up to 25 different languages. Such extreme cases may be interesting to investigate for possible spammer/false account detection or for sociolinguistics in more moderate cases (e.g. 2–5 languages).

The graph measures of users may be useful in confirming their association with language community, without the need to crawl their entire Twitter timeline. Although language settings for user profiles may indicate interface preference only, we found persistent activity in the same language across the users, especially for monolingual users.

A possible scenario for governments, politicians or campaigners would be to use this method to measure to what extent other languages are used within a profile community. It may also show how users associate themselves with one community in their profile while using other languages. Monitoring unusual activity for secondary languages may help to uncover important messages or opinions that could not be openly expressed, for a variety of reasons, to the rest of the profile community. This framework may also be extended to measure reactions via retweeting and replying using a variety of natural language processing and sentiment analysis techniques [33–35], to provide a different perspective for influence analysis.

References

1. Hong, L., Convertino, G., Chi, E.H.: Language matters in Twitter: a large scale study. In: Proceedings of the 5th International AAAI Conference on Weblogs and Social Media (2011)
2. Takhteyev, Y., Gruzd, A., Wellman, B.: Geography of Twitter networks. Soc. Netw. **34**(1), 73–81 (2012)
3. Magdy, A., Ghanem, T.M., Musleh, M., Mokbel, M.F.: Understanding language diversity in local Twitter communities. In: Proceedings of the 27th ACM Conference on Hypertext and Social Media, pp. 331–332 (2016)
4. Kulshrestha, J., Kooti, F., Nikravesh, A., Gummadi, K.P.: Geographic dissection of the Twitter network. In: Proceedings of the 6th International AAAI Conference on Weblogs and Social Media (2012)
5. Cunha, E., Magno, G., Comarela, G., Almeida, V., Gonçalves, M., Benevenuto, F.: Analyzing the dynamic evolution of hashtags on Twitter: a language-based approach. In: Proceedings of the Workshop on Languages in Social Media, pp. 58–65 (2011)
6. Bruns, A., Highfield, T., Burgess, J.: The arab spring and social media audiences: English and Arabic Twitter users and their networks. Am. Behav. Sci. **57**(7), 871–898 (2013)
7. Zappavigna, M., Martin, J.R.: Discourse of Twitter and Social Media: How We Use Language to Create Affiliation on the Web. Continnuum, New York (2012)
8. Zhuravleva, A., de Bot, K., Haug Hilton, N.: Using social media to measure language use. J. Multiling. Multicult. Dev. **37**(6), 601–614 (2015)
9. Gruffydd Jones, E., Uribe-Jongbloed, E. (eds.): Social Media and Minority Languages: Convergence and the Creative Industries. Multilingual Matters Ltd., Bristol (2013)
10. Gisnburgh, V., Weber, S.: How Many Languages Do We Need? The Economics of Linguistic Diversity. Princeton University Press, Princeton (2011)
11. Willis, A., Fisher, A., Lvov, I.: Mapping networks of influence: tracking Twitter conversations through time and space. Particip. J. Audience Reception Stud. **12**(1), 494–530 (2015)
12. Oatley, G., Crick, T.: Measuring UK crime gangs: a social network problem. Soc. Netw. Anal. Mining **5**(1), 1–16 (2015)
13. Borgatti, S.P., Everett, M.G.: Models of core/periphery structures. Soc. Netw. **21**(4), 375–395 (2000)
14. Rombach, M., Porter, M.A., Fowler, J.H., Mucha, P.J.: Core-Periphery Structure in Networks. SIAM J. Appl. Math. **74**(1), 167–190 (2014)
15. Liu, W., Pellegrini, M., Wang, X.: detecting communities based on network topology. Sci. Rep. **4**(5739) (2014). doi:10.1038/srep05739
16. Cha, M., Benevenuto, F., Haddadi, H., Gummadi, K.: The world of connections and information flow in Twitter. IEEE Trans. Syst. Man Cybern. **42**(4), 991–998 (2012)
17. Borge-Holthoefer, J., Rivero, A., Moreno, Y.: Locating privileged spreaders on an online social network. Phys. Rev. E **85**(066123) (2012). doi:10.1103/PhysRevE.85.066123
18. Zhang, J.X., Chen, D.B., Dong, Q., Zhao, Z.D.: Identifying a set of influential spreaders in complex networks. Sci. Rep. **6**(27823) (2016). doi:10.1038/srep27823
19. Kang, R., Brown, S., Kiesler, S.: Why do people seek anonymity on the internet?: informing policy and design. In: Proceedings SIGCHI Conference on Human Factors in Computing Systems, pp. 2657–2666 (2013)

20. Morstatter, F., Pfeffer, J., Liu, H., Carley, K.M.: Is the sample good enough? Comparing data from Twitter's streaming API with Twitter's firehose. In: Proceedings of 7th International AAAI Conference on Web and Social Media, pp. 400–408 (2013)

21. Tan, L., Ponnam, S., Gillham, P., Edwards, B., Johnson, E.: Analyzing the impact of social media on social movements: a computational study on Twitter and the Occupy Wall Street movement. In: Proceedings of IEEE/ACM International Conference on Advances in Social Networks Analysis and Mining (2013)

22. Kumar, S., Morstatter, F., Liu, H.: Twitter Data Analytics. Springer, Heidelberg (2014). doi:10.1007/978-1-4614-9372-3

23. Liang, Y., Caverlee, J., Cheng, Z., Kamath, K.Y.: How big is the crowd?: event and location based population modeling in social media. In: Proceedings of 24th ACM Conference on Hypertext and Social Media, pp. 99–108 (2013)

24. Cheng, Z., Caverlee, J., Lee, K.: You are where you tweet: a content-based approach to geo-locating Twitter users. In: Proceedings of 19th ACM Conference on Information and Knowledge Management, pp. 759–768 (2010)

25. Blamey, B., Crick, T., Oatley, G.: 'The first day of summer': parsing temporal expressions with distributed semantics. In: Bramer, M., Petridis, M. (eds.) Research and Development in Intelligent Systems XXX, pp. 389–402. Springer, Cham (2013). doi:10.1007/978-3-319-02621-3_29

26. Caverlee, J., Cheng, Z., Sui, D.Z., Yeswanth Kamath, K.: Towards geo-social intelligence: mining, analyzing, and leveraging geospatial footprints in social media. IEEE Data Eng. Bull. **36**(3), 33–41 (2013)

27. The Telegraph: Eurovision 2016: furious Russia demands boycott of Ukraine over Jamala's 'anti-Kremlin' song. http://www.telegraph.co.uk/news/2016/05/15/eurovision-2016-furious-russia-demands-boycott-of-ukraine-over-j. Accessed 01 Apr 2017

28. Ginsburgh, V., Noury, A.G.: The eurovision song contest. Is voting political or cultural? Eur. J. Polit. Econ. **24**(1), 41–52 (2008)

29. Charron, N.: Impartiality, friendship-networks and voting behavior: evidence from voting patterns in the Eurovision Song Contest. Soc. Netw. **35**(3), 484–497 (2013)

30. Blangiardo, M., Baio, G.: Evidence of bias in the Eurovision song contest: modelling the votes using Bayesian hierarchical models. J. Appl. Stat. **41**(10), 2312–2322 (2014)

31. Budzinski, O., Pannicke, J.: Culturally biased voting in the Eurovision Song Contest: do national contests differ? J. Cult. Econ. 1–36 (2016). https://link.springer.com/article/10.1007/s10824-016-9277-6

32. Kirk, A., Kempster, J., Franco, S.: Eurovision 2016: how does country bias affect the result? http://www.telegraph.co.uk/music/news/eurovision-2016-how-country-bias-affects-the-result. Accessed 31 Apr 2017

33. Oatley, G., Crick, T.: Changing faces: identifying complex behavioural profiles. In: Tryfonas, T., Askoxylakis, I. (eds.) HAS 2014. LNCS, vol. 8533, pp. 282–293. Springer, Cham (2014). doi:10.1007/978-3-319-07620-1_25

34. Sluban, B., Smailović, J., Battiston, S.: Mozetič I.: Sentiment leaning of influential communities in social networks. Comput. Soc. Netw. **2**(9), 1–21 (2015)

35. Mostafa, M., Crick, T., Calderon, A.C., Oatley, G.: Incorporating emotion and personality-based analysis in user-centered modelling. In: Bramer, M., Petridis, M. (eds.) Research and Development in Intelligent Systems XXXIII. LNCS, pp. 383–389. Springer, Cham (2016). doi:10.1007/978-3-319-47175-4_29

Bangla News Summarization

Anirudha Paul, Mir Tahsin Imtiaz, Asiful Haque Latif,
Muyeed Ahmed, Foysal Amin Adnan, Raiyan Khan, Ivan Kadery,
and Rashedur M. Rahman[(⊠)]

Department of Electrical and Computer Engineering, North South University,
Dhaka, Bangladesh
anirudhaprasun@gmail.com, asif.nobel@gmail.com,
akibl00095@gmail.com, adnanbd769@gmail.com,
raiyan.khan.106@gmail.com, ikadery@gmail.com,
{tahsin.imtiaz, rashedur.rahman}@northsouth.edu

Abstract. Document similarity calculation and summarization is a challenging task. Not many works have been done in this field for Bangla Language. Similarity calculation and summarization is more challenging for Bangla Language as Bangla grammar works differently than that of English. This paper proposes a way to calculate similarity between Bangla news and apply summarization on Bangla news documents taken from popular news portals by applying various data mining techniques as accurately as possible.

Keywords: Text summarization · Text mining · Document clustering · Data mining · Bangla News Summarization · DBSCAN

1 Introduction

Document clustering and summarization has a lot of applications. For example, document clustering has applications for web document clustering for search engines. Summarization is used to represent a document's information in fewer words accurately and completely so that a reader won't have to go through the entire document in order to get the main idea of the document. Although a lot of work has been done in English and other major languages, the work done for Bangla is still trivial and can be improved.

Bangla is a language spoken as first language by almost 210 million people all over the world. The amount of Bangla document in the internet is increasing exponentially day by day. The objective of this paper is to find a way to efficiently perform stemming to Bangla news documents collected from popular news portals in order to calculate the similarity between the documents and to perform summarization to Bangla news in such a way that the summarized document will represent the main document as accurately and completely as possible.

The main challenge in the field of text similarity calculation and summarization is the preprocessing of the documents. Since Bangla grammar is really complicated, the techniques used to stem English words do not accurately apply for Bangla words. In this paper, we have stemmed Bangla documents by firstly removing common stop words from the document and finally finding the longest common substring we can find from the dictionary for each word in the document.

© Springer International Publishing AG 2017
N.T. Nguyen et al. (Eds.): ICCCI 2017, Part II, LNAI 10449, pp. 479–488, 2017.
DOI: 10.1007/978-3-319-67077-5_46

2   Related Work

Uddin and Khan [1] have performed a survey on different techniques that have been applied in English or other languages for summarization and selected some of them in order to apply those techniques for Bangla. The summarization model they designed got the best result when they summarized the document down to 40%. Shaharia et al. [2] worked on summarizing resource-poor languages that are spoken mostly in east of India. These languages include Bangla, Assamese, Bodo etc. They introduced a rule based approach and used a dictionary of frequent Bangla words for the task of stemming. They succeeded to obtain 94% accuracy for Bangla and Assamese whereas 87% and 82% accuracy was obtained for Bishnupriya and Bodo respectively using this technique. Urmi et al. [3] designed a Bangla stemmer that is corpus based. It uses a 6-gram model and simple technique for threshold in order to identify the root. The next paper [4] used WordNet ontology for abstractive summary generation. Although their result was experimental, the summary that was generated was compressed well, easily understandable for humans and did not have any grammatical errors.

Baralis et al. [5] proposed an approach that can take a financial document collection and can summarize it though the documents are written in separate languages. The mining system they proposed has the capability of handling news written in separate languages and it can analyze the skill level of the users to cover specific parts in the summary and can also rank the summaries which are generated. In the next paper [6] weka-LibSVM classifier is used for multi-domain documents classification whereas term document matrix is used to represent the data.

3   System Overview

In this paper, Bangla news documents are collected from popular websites. The system is divided into two major parts. Part one consists of the process of finding similar documents using DBSCAN [9] after preprocessing. In part two, the system summarizes selected document depending on priority values assigned to the preprocessed sentences of that document. Both parts have preprocessing as a common step. The preprocessing part contains the task of deleting unnecessary words and stemming using a dictionary and custom word lists. In the part of finding similarity among the documents, cosine similarity [9] is used after the part of preprocessing. In part two, similar documents are taken and the target document which would be summarized is selected. The term frequency matrix is generated for each word occurring in the documents. Finally, sentences are given rank points in order to choose the top sentences.

3.1   Data Collection

For newspaper articles, we have designed a crawler using python script, which crawls data from popular newspaper websites in Bangladesh. We took all articles from last 7 days which was 993 articles in total. The crawled news articles were saved in an xls file where the columns contained the date of the news, source, category, headline and the news content respectively.

Table 1. Cleaning dictionary word list.

Word	Reason
কুড়ানি['কাগজ-'] তু	3rd Bracket enclosed redundant word and character
কুড়ানো, কুড়োনো	Comma separated same word
কুড়াল দ্র কুড়ুল	Redundant string 'দ্র'
কুড়ি '২০' তু কুড়ি	Numerical Values - '২০'

We got a word list of 98525 words from Kolkata: West Bengal Bangla Academy [7]. Some words in the complete word list contained some additional redundant character. Some examples are shown in Table 1. We got a list of 61790 words after cleaning the number and character containing words using Regular Expression [8]. This word list was used while stemming to find the largest common substrings for words in the news documents.

3.2   Preprocessing

In preprocessing, the unnecessary words (stop words) were deleted first and then the words were stemmed using dictionary lookup [2].

Before clustering, deleting stop words is needed because they occur very frequently and they are not the key points to determine the similarity of any two documents. The stemming needs to be used as a way to identify different representation of the same words as one. If two words have the same meaning but have different prefix or suffix depending on the tense and context, normal string matching algorithm will consider them as different words though they bear the same meaning. As a result, their Term Frequency and Inverse Document Frequency will be calculated incorrectly. Therefore, stemming is used to eliminate these additional prefix and suffix to find out the root word of all the words.

In Bangla documents, it is observed that most of the prepositions, conjunctions and some verbs and their variations used are not necessary while prioritizing words or sentences for finding similar documents or summarization process since these words are common and widely used everywhere. We generated a list of 49 prepositions and conjunctions and a list of 173 verbs and their variations. We got rid of those words from the documents before further processing. Table 2 shows such few words that we can get rid of before further processing.

In Bangla language, stemming words by cutting suffix or by considering prefix does not always give the correct result. Table 3 represents the comparison between prefix based result, suffix based result and result from word list look up.

The first example contains no additional suffix or prefix and is filtered perfectly by all three methods. The second and last example show the drawbacks of suffix and prefix stemming respectively. The letters "র" and "অ" work as common suffix and prefix respectively in Bangla and get removed although these are part of the original words in this case. Both the situations are handled perfectly by the Dictionary Lookup based stemming. In fact, this process gives 81% accuracy [2].

Table 2. Common removable verbs, prepositions and conjunctions examples

Verb	Prepositions/Conjunctions
করা (Do)	এবং (And)
হয় (is)	বরং (Rather)
বলা (Say)	কিন্তু (But)

Table 3. Comparison of stemming from different techniques.

Word	Intended Stem Result	Prefix Result	Suffix Result	Word list Lookup
ছাই	ছাই	ছাই	ছাই	ছাই
ময়ূরবিহার	ময়ূরবিহার	ময়ূরবিহার	ময়ূরবিহা	ময়ূরবিহার
অজগরটি	অজগর	জগরটি	অজগর	অজগর

3.3   Clustering

After stemming was done, we observed the cosine similarities [9] between the news documents. In DBSCAN [9], we set the minimum number of news articles to 3 in order to consider a group of articles as a cluster. By this we have tried to avoid fake or unreliable news. We also set the eps to 0.5. Which means news article whose similarity is between 0.5–1 will be in the same cluster. Figure 1 shows the correlation matrix which represents the similarity distance between the documents. Here, x-axis and y-axis represents news articles.

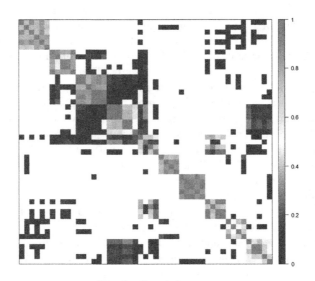

Fig. 1. Correlation map.

Cosine similarity calculates in terms of inner product space, how much two vectors are closed to each other. If two vectors are similar, the Cosine similarity value between

Table 4. Similarity calculation result on different thresholds.

Threshold	True positive	False positive	True negative	False negative
0.1	16	21	44	0
0.2	16	2	63	0
0.3	16	2	63	0
0.4	15	2	63	1
0.5	13	0	65	3
0.6	10	0	65	6
0.7	5	0	65	11
0.8	3	0	65	13
0.9	3	0	65	13

them will be close to +1 and if they are not similar, the value will be close to −1. But cosine similarity always works on numeric values; it cannot straightforward calculate the similarity between two strings or two documents. Therefore, each of the words in every news article is replaced with their TF-IDF values. Here each of the words in a news article is considered as a separate axis in the vector space. After this, if two news articles have the cosine similarity close to +1 this can be considered as similar. We tested various Cosine Similarity thresholds on several sample tests to find out the optimum Cosine Similarity value. Documents having cosine similarity above threshold are considered as similar match. Table 4 represents the accuracy for different threshold where we have calculated similarity between documents about a particular topic. In this table, 81 documents are considered where 16 documents are about the selected topic. The goal is to avoid false positive completely as it will mix up various news as the same. We decide to use the threshold value 0.5 for further analysis since this result shows that the threshold 0.5 gives us the optimal result.

3.4 Term Frequency Matrix for Document Summarization Process

After the preprocessing and finding similar documents, we select a particular list of similar news documents and generated term frequency matrix for all the words appearing in that list. This matrix tells us which word appears how many times in the documents in that particular list. The term frequency matrix looks like Table 5 after the system runs on similar documents.

3.5 Giving Priority Values to Sentence

For summarizing the selected document, we have decided to give the sentences priority values and select the top sentences having highest values among them. In order to do this, we implement the formula in (1).

$$S_{ik} = \frac{\sum w_i}{t_{ik}} \tag{1}$$

Table 5. Term frequency matrix.

Word	Document 1	Document 2	Document 3	Document 4
অধিনায়ক (Captain)	2	0	0	0
অনুশীলন (Practice)	1	1	1	0
অসুস্থ (Sick)	3	3	3	0
আগামী (Next)	1	2	1	0
আজ (Today)	1	2	1	0

Here,

S_{ik} = Score or priority value for k-th line in the i-th document
t_{ik} = Number of words in the k-th line of i-th document
w_i = i-th word score.

Method 1
In this method, we have used term frequency [12] to calculate w_i. Term frequency (TF) is calculated using (3). S_{ik} is then calculated using w_i calculated in (2).

$$w_i = TF \tag{2}$$

$$TF = \frac{x}{n} \tag{3}$$

Here,

x = number of occurrences of a word in all document
n = number of words in current document.

Method 2
In this method, we have summarized the article which has got the highest total score according to sentence scoring of method 1 and has shown it as the general summary for all articles in the cluster.

Method 3
In this method, we have used multiplication of term frequency and inverse document frequency. Inverse document frequency (IDF) [12] is calculated using (5). S_{ik} is then calculated using w_i calculated in (4).

$$w_i = TF * IDF \tag{4}$$

$$IDF = 1 + \log_e \frac{p}{q} \tag{5}$$

Here,

p = number of documents in the cluster
q = number of document in which this word exists.

Method 4
In this method, we have used the same process as method 2, but in this case, for sentence scoring we have used method 3.

3.6   Summarization

After all the sentences get a priority value, a threshold can be defined for the number of lines taken for the summarized document. We have decided to use the threshold value of 40% according to Uddin and Khan [1].

The following news is taken from The Daily Ittefaq [10].

"স্ত্রী হঠাৎ অসুস্থ হয়ে পড়ায় দেশের পথে রওনা দিয়েছেন মাশরাফি বিন মর্তুজা। ইংল্যান্ড এর সাসেক্সে যাওয়ার মাত্র দুই দিন অনুশীলনের পরে দেশে ফিরে আসছেন অধিনায়ক। জানা গেছে, বাংলাদেশের ওয়ানডে অধিনায়কের স্ত্রী সুমনা হক শনিবার হঠাৎ করে অসুস্থ হয়ে পড়েন। তাকে হাসপাতালে ভর্তি করা হয়। রবিবার বিকালে ঢাকায় পৌছানোর কথা থাকলে ফ্লাইট দেরি করায় রাতের বেলায় পৌছাবেন মাশরাফি। উল্লেখ্য, ইংল্যান্ডে প্রস্তুতি ক্যাম্প করতে ২৬ এপ্রিল রাতে ঢাকা ছেড়েছিল বাংলাদেশ দল। সোমবার সাসেক্সে প্রথম অনুশীলন ম্যাচ ডিউক অব নরফোক একাদশের সঙ্গে। ৫ মে আরও একটি অনুশীলন ম্যাচ খেলে ৭ মে আয়ারল্যান্ড যাবে দল। ১২ মে আয়ারল্যান্ডে ত্রিদেশীয় সিরিজের প্রথম ম্যাচ। তবে এক ম্যাচ নিষিদ্ধ থাকায় খেলতে পারবেন না মাশরাফি।"

The English translation of this article is as follows:

"Mashrafe is returning to his homeland due to sudden sickness of his wife. Just after two days of practice at Sussex, England, the captain is coming back. It has been informed that Sumona Haque, the wife of Bangladesh's one day captain has suddenly fallen sick on Saturday. She has been admitted to a hospital. Mashrafe was supposed to land of Sunday afternoon, but due to flight delay he has to reach Dhaka during night. It is noteworthy that Bangladesh team left Dhaka on the eve of 26th April to prepare for England camp. The first practice match is on Monday against Duke of Norfolk. After playing another practice match on 5th, the team will leave for Ireland on 7th. The first match of the tri nation series is on 12th May in Ireland. But due to a restriction, Mashrafe will not be able to play."

For this particular document, Table 6 shows the score for each sentence calculated using method 1.

Here is the result after summarization in 4 methods of the previous news article.

Method 1: Result

"স্ত্রী হঠাৎ অসুস্থ হয়ে পড়ায় দেশের পথে রওনা দিয়েছেন মাশরাফি বিন মর্তুজা। জানা গেছে, বাংলাদেশের ওয়ানডে অধিনায়কের স্ত্রী সুমনা হক শনিবার হঠাৎ করে অসুস্থ হয়ে পড়েন। উল্লেখ্য, ইংল্যান্ডে প্রস্তুতি ক্যাম্প করতে ২৬ এপ্রিল রাতে ঢাকা ছেড়েছিল বাংলাদেশ দল। ১২ মে আয়ারল্যান্ডে ত্রিদেশীয় সিরিজের প্রথম ম্যাচ।"

The English translation of this summarization is as follows:

"Mashrafe is returning to his homeland due to sudden sickness of his wife. It has been informed that Sumona Haque, the wife of Bangladesh's one day captain has suddenly fallen sick on

Saturday. It is noteworthy that Bangladesh team left Dhaka on the eve of 26th April to prepare for England camp. The first match of the tri nation series is on 12th May in Ireland."

Table 6. Priority values of sentences.

Sentence	Priority Value
স্ত্রী হঠাৎ অসুস্থ হয়ে পড়ায় দেশের পথে রওনা দিয়েছেন মাশরাফি বিন মর্তুজা	8.7
জানা গেছে, বাংলাদেশের ওয়ানডে অধিনায়কের স্ত্রী সুমনা হক শনিবার হঠাৎ করে অসুস্থ হয়ে পড়েন	6.34
উল্লেখ্য, ইংল্যান্ডে প্রস্তুতি ক্যাম্প করতে ২৬ এপ্রিল রাতে ঢাকা ছেড়েছিল বাংলাদেশ দল	5.69
১২ মে আয়ারল্যান্ডে ত্রিদেশীয় সিরিজের প্রথম ম্যাচ	5.375
তবে এক ম্যাচ নিষিদ্ধ থাকায় খেলতে পারবেন না মাশরাফি	5.33
ইংল্যান্ড এর সাসেক্সে যাওয়ার মাত্র দুই দিন অনুশীলনের পরে দেশে ফিরে আসছেন অধিনায়ক	5.29
৫ মে আরও একটি অনুশীলন ম্যাচ খেলে ৭ মে আয়ারল্যান্ড যাবে দল	4.86
রবিবার বিকালে ঢাকায় পৌছানোর কথা থাকলে ফ্লাইট দেরি করায় রাতের বেলায় পৌছাবেন মাশরাফি	4.64
তাকে হাসপাতালে ভর্তি করা হয়	3.39
সোমবার সাসেক্সে প্রথম অনুশীলন ম্যাচ ডিউক অব নরফোক একাদশের সঙ্গে।	3.23

Method 2: Result

"ইংল্যান্ডের ক্যাম্প ছেড়ে হঠাৎদেশে ফিরছেন টাইগার দলপতি মাশরাফি বিন মর্তুজা। তবে শুধু স্ত্রী সুমিই-ই নন, এই মুহূর্তে অসুস্থ মাশরাফির ছেলে সাহিল মুর্তজাও। তবে কোনো ইনজুরির কারণে নয়, মাশরাফির স্ত্রী অসুস্থ হয়ে পড়ায় দেশে ফিরছেন তিনি। জানা গেছে, মাশরাফির স্ত্রী সুমি অসুস্থ।"

The English translation of this summarization is as follows:

"Mashrafe is returning to his homeland due to sudden sickness of his wife. But not only his wife Sumona, his son Sahil Mortaza is sick as well. He is returning to the country because of the sickness of his wife, not because of his injury. It has been informed that wife of Mashrafe, Sumona is sick."

Method 3: Result

"স্ত্রী হঠাৎ অসুস্থ হয়ে পড়ায় দেশের পথে রওনা দিয়েছেন মাশরাফি বিন মর্তুজা। তবে এক ম্যাচ নিষিদ্ধ থাকায় সেই ম্যাচে খেলতে পারবেন না মাশরাফি। জানা গেছে, বাংলাদেশের ওয়ানডে অধিনায়কের স্ত্রী সুমনা হক শনিবার হঠাৎ করেই অসুস্থ হয়ে পড়েন। উল্লেখ্য, ইংল্যান্ডে প্রস্তুতি ক্যাম্প করতে ২৬ এপ্রিল রাতে ঢাকা ছেড়েছিল বাংলাদেশ দল।"

The English translation of this summarization is as follows:

"Mashrafe is returning to his homeland due to sudden sickness of his wife. But due to a restriction, Mashrafe will not be able to play. It has been informed that Sumona Haque, the wife of Bangladesh's one day captain has suddenly fallen sick on Saturday. It is noteworthy that Bangladesh team left Dhaka on the eve of 26th April to prepare for England camp."

Method 4: Result

"তবে শুধু স্ত্রী সুমিই-ই নন, এই মুহূর্তে অসুস্থ মাশরাফির ছেলে সাহিল মুর্তজাও। আগামী ১২-২৪ মে আয়ারল্যান্ডে অনুষ্ঠেয় ত্রিদেশীয় সিরিজ ও ১-১৮ জুন ইংল্যান্ডে অনুষ্ঠেয় চ্যাম্পিয়নস ট্রফিতে অংশ নিতে গত বুধবার দিবাগত রাতে ইংল্যান্ডের উদ্দেশে দেশ ছাড়ে বাংলাদেশ ক্রিকেট দল। তবে কোনো ইনজুরির কারণে নয়, মাশরাফির স্ত্রী অসুস্থ হয়ে পড়ায় দেশে ফিরছেন তিনি। একটি নির্ভরযোগ্য সূত্রে এ তথ্য জানা গেলেও মাশরাফির দেশে ফেরার ব্যাপারে এখন পর্যন্ত কোনো আনুষ্ঠানিক বিবৃতি দেয়নি বিসিবি।"

The English translation of this summarization is as follows:

"But not only his wife Sumona, his son Sahil mortaza is sick as well. Bangladesh team left Dhaka for England late night of last Wednesday in order to take part in the Ireland tri series in 12–24 May and the Champions Trophy tournament in 1–18 Jun. He is returning to the country because of the sickness of his wife, not because of his injury. This was not confirmed by BCB yet although this news is from a reliable source."

4   Method Evaluation

We have used ROUGE-N to evaluate the performance of the methods. ROUGE-N measures the similarity of N-gram units between human made summary and algorithm produced summary [11]. For the human made summary part, 10 journalists summarized the news articles of our dataset. We have checked the similarity of our algorithm produced summary against these human made summary using ROUGE-1, ROUGE-2, ROUGE-3. Table 7 shows the average performance of this comparison.

Table 7. Performance evaluation.

ROUGE-N	Metric	Method 1	Method 2	Method 3	Method 4
ROUGE-1	Avg_Recall	0.4219	0.3132	0.5354	0.3851
	Avg_Precision	0.4503	0.3145	0.4251	0.2976
	Avg_F-Measure	0.4313	0.3089	0.4689	0.3292
ROUGE-2	Avg_Recall	0.3515	0.2071	0.4391	0.2321
	Avg_Precision	0.3761	0.2077	0.3602	0.1897
	Avg_F-Measure	0.3597	0.2047	0.3929	0.2063
ROUGE-3	Avg_Recall	0.3202	0.1609	0.4079	0.1761
	Avg_Precision	0.3427	0.1638	0.3361	0.1491
	Avg_F-Measure	0.3277	0.1603	0.3662	0.1601

From this table, we can see that Method 1 gets the best precision score, it means the ratio of fetching relevant data in compare to irrelevant data is lower in this method. Method 3 gets the best recall score; this means it has better probability to retrieve correct and relevant data more while summarizing Bangla News Articles and this method also gets the highest F-measure score which also indicates if we consider both recall and precision, this method performs better than other methods.

5   Conclusion

In this paper, we have focused on avoiding incorrect stemming and mixing up of different news. The system also compared the obtained results with human produced summary.

In Bangla language, rule based stemming based on prefix elimination or suffix elimination does not work well since words in this language can have different types of variations. On the other hand, finding largest common substring by analyzing the dictionary for each word in the document has higher probability of selecting the right stem words.

The aggressive threshold selection to avoid false positive and unverified news makes the summary a valid one.

The performance evaluation against the human made summary shows us that TF-IDF based single document summarization as mentioned in Method 3 performs overall better than other methods. But if we want to generate a single summary for all the news in a single cluster, Method 4 also performs well without producing redundant summary.

References

1. Uddin, M.N., Khan, S.A.: A study on text summarization techniques and implement few of them for Bangla language. In: 2007 10th International Conference on Computer and Information Technology (2007). doi:10.1109/ICCITECHN.2007.4579374
2. Saharia, N., Sharma, U., Kalita, J.: Stemming resource-poor Indian languages. ACM Trans. Asian Lang. Inf. Process. 13(3), 1–26 (2014)
3. Urmi, T.T., Jammy, J.J., Ismail, S.: A corpus based unsupervised Bangla word stemming using N-gram language model. In: 2016 5th International Conference on Informatics, Electronics and Vision (ICIEV) (2016). doi:10.1109/ICIEV.2016.7760117
4. Dave, H., Jaswal, S.: Multiple text document summarization system using hybrid summarization technique. In: 2015 1st International Conference on Next Generation Computing Technologies (NGCT) (2015). doi:10.1109/NGCT.2015.7375231
5. Baralis, E., Cagliero, L., Cerquitelli, T.: Supporting stock trading in multiple foreign markets. In: Proceedings of 2nd International Workshop on Data Science for Macro-Modeling – DSMM 2016 (2016). doi:10.1145/2951894.2951897
6. Dsouza, K.J., Ansari, Z.A.: A novel data mining approach for multi variant text classification. In: 2015 IEEE International Conference on Cloud Computing in Emerging Markets (CCEM) (2015)
7. Bangla Word List [PDF] West Bengal Bangla Academy, Kolkata (n.d.)
8. List of Regular Expressions - Libreoffice Help. Help.libreoffice.org. N.p. (2017). Web: 5 May 2017
9. Tan, P., Steinbach, M., Kumar, V.: Introduction to Data Mining. Dorling Kindersley, Pearson, London (2015)
10. "আকস্মিক দেশের পথে মাশরাফি। খেলাধুলা । The Daily Ittefaq" Ittefaq.com.bd. N.p. (2017). Web: 3 May 2017
11. Ferreira, R., et al.: Assessing sentence scoring techniques for extractive text summarization. Expert Syst. Appl. 40(14), 5755–5764 (2013)
12. Ramos, J.: Using TF-IDF to determine word relevance in document queries. Technical report, Department of Computer Science, Rutgers University (2003)

Low Resource Language Processing

Combined Technology of Lexical Selection in Rule-Based Machine Translation

Ualsher Tukeyev[✉], Dina Amirova, Aidana Karibayeva,
Aida Sundetova, and Balzhan Abduali

Information Systems Department, Al-Farabi Kazakh National University,
Al-Farabi av., 71, 050040 Almaty, Kazakhstan
ualsher.tukeyev@gmail.com, amirovatdina@gmail.com,
a.s.karibayeva@gmail.com, sun27aida@gmail.com,
balzhanabdualy@gmail.com,
http://www.kaznu.kz

Abstract. This paper describes process of solving the task of lexical selection for English-Kazakh (and vice versa) machine translation system based on combined technology. Proposed combined technology is including the constraint grammar model and maximum entropy model for more effective solution of the problem of lexical selection for English-Kazakh (and vice-versa) language pair. Results are presented by comparing two technologies separately and together in Apertium English-Kazakh (and vice versa) system.

Keywords: Lexical selection · Combined technology · Machine translation · Source language · Target language · Sense · Ambiguity

1 Introduction

Machine translation has two main problems in text processing. First of it is lexical selection, which connected with problem of choosing corresponded translation by context. Second problem of a lexical words is order. Later consider the task of words order in sentence of target language. In word processing the main problem is lexical selection, which leads to disambiguation task.

The work of the translator's program is carried out in several stages. The development of algorithms that allow us to recreate the human ability to understand and choose the right meaning of a word is a difficult task. It consists in choosing the most suitable translation in the context under consideration from all possible translations. The solution of the problem involves finding the probable meanings of words, determining the relationships between these values and the context in which the words were used.

The scientific novelty of paper is in developing combined technology of lexical selection based on the constraint grammar model and maximum entropy model and them sequence applying to Kazakh language as source and target language in translation pairs Kazakh-English and English-Kazakh.

© Springer International Publishing AG 2017
N.T. Nguyen et al. (Eds.): ICCCI 2017, Part II, LNAI 10449, pp. 491–500, 2017.
DOI: 10.1007/978-3-319-67077-5_47

2   Related Works

Scientific research on a word sense disambiguation has long-term history. With a current of years the number of the proposed solutions and their efficiency steadily grew, but the task hasn't received the full decision yet. A large number of methods have been investigated: from the methods based on knowledge, rules, lexicographic sources, training with the teacher at corpus to training methods without teacher, clustering words on the basis of meaning. Among listed, today, training methods with the teacher have shown the best efficiency.

The problem of resolving ambiguity as a separate problem was formulated in the middle of 20th century, almost simultaneously with the advent of machine translation. Since that time, many methods of solving this problem have been developed, but it is still actual.

Tyers et al. in [1] uses the maximum entropy model for performing lexical selection in machine-based translation systems based on rules to English-Basque, English-Catalan. As a learning method, the method of teaching without the teacher (unsupervised methods) is used, which does not require an annotated corpus. The system uses the maximum entropy formalism for lexical selection as in Berger et al. [2] and Mareček et al. [3], but instead of counting the actual lexical selection event in the annotated corpus, they consider fractional occurrences of these events according to the model of the target language. Mareček et al. in [3] showed the model of the maximum entropy of translation in machine translation based on dependencies, which allows to develop a large number of features of functions in order to obtain more accurate translations. Tyers in [4] the general method of teaching rules for the module is described. Monolingual and bilingual corpora can be used for the method. For learning a monolingual corpus the method without the teacher (unsupervised method) is used. Also weighting method is described, based on the principle of maximum entropy. This method allows to take into account all the rules without having to choose between conflicting and overlapping rules.

Rule-based approaches in lexical selection not cover all possible translation of source language. Since it is not always possible to take into account all rules. In the statistical approach, lexical choice also does not guarantee for the correctness of the translation in context. Analyzing both approaches, we came to the conclusion about the development of technology combining the two approaches mentioned above with solutions to lexical selection problems for the Kazakh language in Apertium systems.

3   Combined Technology of Lexical Selection

The combined technology for solving the problem of lexical selection we propose consists of two models: the model of production rules (constraint grammar) and the maximum entropy models (Fig. 1).

In the process of translating a lexically ambiguous word, the system first solves the problem using the model of production rules. The model of production rules gives a good result, but does not completely solve the task. Because rules are not cover all cases of ambiguity of a certain word in any context. A word can have a different senses,

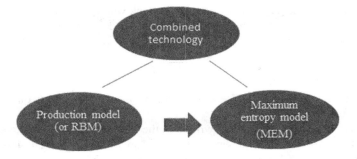

Fig. 1. The models of combined technology

and can be used with different other words in a context. And there can be many such combinations. And writing the rules can take a lot of time. Therefore, for the case when there is no written rule for some case, the selection of translation is solved by using data in semantic cube based on parallel corpora processed by statistical model. So, the ambiguous word is processed using the statistical part of the module.

The models of combined technology performed sequentially. Firstly is used the model of production rules and then maximum entropy model.

The technology of combined technology was introduced in the Apertium [5]. In the aperture, divide into two parts the lexical selection and lexical choice based on the semantic cube (Fig. 2).

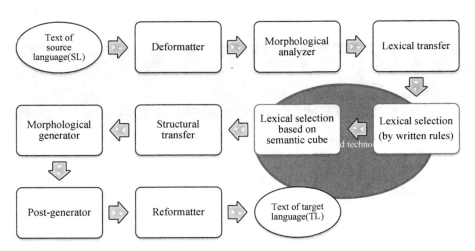

Fig. 2. The combined technology in Apertium work pipeline

3.1 Model of Production Rules

In the Apertium system, the lexical selection problem is solved by writing the rules manually in the lexical selection module using the Constraints Grammar, which has the form of product rules.

The model of product rules is a rule-based model, where knowledge is shown in the form "if (condition), then (action)". The product model can be represented in the following form:

$$i = \langle S, L, A \rightarrow B, T, \rangle$$

here S – case, L – translation, $A \rightarrow B$ – productivity kernel, T – postcondition of the production rule.

The rules of lexical selection are written in the file "apertium-eng-kaz.eng-kaz.lrx" for the English-Kazakh and in the file "apertium-eng-kaz.eng-kaz.lrx" for Kazakh-English pair of languages. All rules are written in XML format. The format of lexical rules is based on the technology of patterns. This module is used to select a translation when the meaning of ambiguous word refers to one part of the speech. For example, the word "year" is more often translated as "жыл"[zhyl], and the phrase "five years old" is translated as "бес жаста"[bes zhasta], that is, it indicates the age of a person. For this case the following rule is written:

```
<rule> <match lemma="year" tags="n.pl.*">
<select lemma="жас" tags="n.*"/>
</match> <match lemma="old" tags="adj.*"/></rule>
```

The meaning of the rule: if after the noun "year" is an adjective "old", then the translation "жас"[zhas] is chosen.

For example word "бас"[bas] has various translation (Fig. 3):

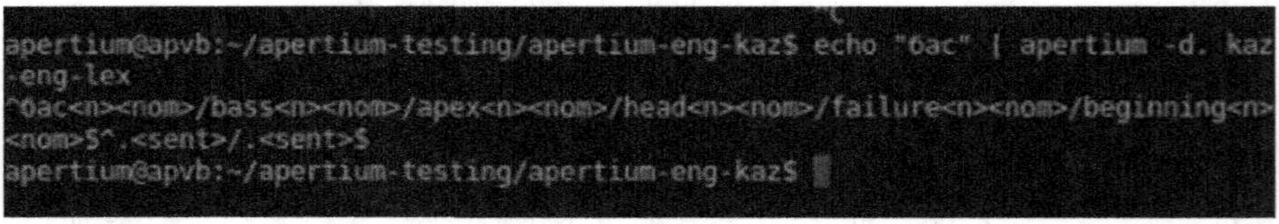

Fig. 3. The translations of word "Бас" in bilingual dictionary of English

The figures below show lexical units to word "бас(bas)". As you see given words have 4 translations.

```
<rule><match lemma= "адам" tags=n.*"/>
  <match lemma= "бас" tags="n.*">
<select lemma="head" tags="n.*"/></match></rule>
<rule>
  <match lemma="фильм" tags= "n.*"/>
  <match lemma= "бас" tags= "n.*"><select lemma= "beginning" tags= "n.*"/></match></rule>
```

The meaning of the rule: if word "бас"[bas] comes after word "адам"[adam], the translation will be as "head", whereas when it comes after with "фильм" [film] it translated as "beginning"

The model of production rules gives a good result, but does not completely solve the task. Because rules are not cover all cases of ambiguity of a certain word in any context. Kazakh-English and English-Kazakh language pairs in Apertium currently have 98 and 75 lexical selection rules respectively. A word can have a different senses, and can be used with different other words in a context; and there can be many such combinations. And writing the rules also depending of knowledge level of developer. The lexical selection rules are written to the various part of speech, namely to noun, verb, adjective, adverb, preposition and etc. Therefore, for the case when there is no written rule for some case, the selection of translation is solved by using statistics the maximum entropy model.

3.2   Maximum Entropy Model

Lexical selection maximum entropy model includes a set of binary functions and appropriate weights for each function [4]. The feature is defined as $h^s(t, c)$, where t is a translation and c – is a source language context for each source word s:

$$h^s(t, c) = \begin{cases} 1, & c \text{ the value of t under the condition } c \\ 0, & other \end{cases} \tag{1}$$

During the learning process each function is assigned a weight λ^s, and combining the weights as in Eq. (2) gives the probability of a translation t for word s in context c.

$$p_s(t|c) = \frac{1}{Z} exp \sum_{k=1}^{n_F} \lambda_k^s h_k^s(t, c) \tag{2}$$

where Z – is a normalizing constant, n_F- numbers of features for s. Thus, the most probable translation can be found using Eq. (3)

$$\hat{t} = \arg \max_{t \in T_s} p_s(t|c) = \arg \max_{t \in T_s} \sum_{k=1}^{n_F} \lambda_k^s h_k^s(t, c) \tag{3}$$

where T_s – all possible translations for s source word, λ- weight of known translation.

It is important to note that the rules for the function $h^s(t, c)$ will be different depending on the language pair. Consider an example: 'Mark is a bass player. He fried the bass'. In these sentences the word *bass* is ambiguous. In the first sentence, the sentence is translated into Kazakh as a "*bass guitar*", and in the second as a type of fish "*алабуға*"[alabuga]. Then the function has the form

$$h^s(t, c) = \begin{cases} 1, & if \ t =' \ бас - гитара' \ and \ 'player' \ after \ bass \\ & 0, \ in \ other \ cases \end{cases}$$

In combined technology the maximum entropy model is realized through the construction of a semantic cube.

4   Algorithms of Semantic Cube

The algorithm for implementation consists of the following steps:

Step 1. Create a frequency list of words.

First, to build a cube, we create a frequency list and a list of ambiguous words. The frequency list is a list of the most often met words in the corpus.

Step 2. Create a list of ambiguous words.

After composing the frequency list, it is necessary to find among them ambiguous ones, namely, the lexical ambiguity is taken into account, that is, when possible translations of the required word belong to one part of the speech (Table 1).

Table 1. Example of ambiguous words.

Ambiguous words	Part of speech	Sense 1	Sense 2	Sense 3
String	Noun	жол	жіп	ішек
Order	Noun	рет	жарлық	орден
Part	Noun	бөлік	партия	дене
Small	Adjective	кішкентай	ұсақ	шағын
Thing	Noun	зат	нәрсе	дүние
Discover	Verb	байқау	ашу	табу
Information	Noun	ақпарат	хабар	мәлімет

Step 3. Prepare bilingual parallel corpus.

Next, we prepare a bilingual corpus. Each statistic machine translation consist the greater number of text array. These arrays named as corpora. The word "prepare" means that "unnecessary" words or stop words are deleted, which have a special effect in calculating probabilities, as well as choosing the sense of the word. Such words, for example, articles like *the, a, an*, numbers, punctuation marks, etc. It is also necessary to bring all the words from the context to the basics, that is, to the initial form of the word. This completes the preparation of the corpus.

Step 4. "Training" stage. Semantic cube building.

The result of applying the maximum entropy model to solving the problem of lexical selection is the construction of a "semantic cube". At first the statistical system passes a stage of "training" at which statistical data on the translation of separate words and phrases from source language on target language are taken. The maximum entropy model is trained on a pre-prepared parallel bilingual corpus. On the basis of this stage the cube is formed. The cube represents a three-dimensional set of tables. The tables contain the senses of ambiguous words, the context, the meanings of the probabilities of the senses (translations) of ambiguous words that depend on words from the context.

There two variables can be denote as a linear releationship:

$$amb_{word} : t_i \rightarrow f_i \in C$$

Choosing or finding word's translation basically depends on two main variables - translation and context.

The semantic cube model determines the translation by weight of word's context in parallel corpora. The model gives information about frequency and probability to determined words and his neighbourhood words. The frequency of a translation is determined by the number of occurrences of the features, and when the known features are repeated, the number increases to 1.

The probability of translation calculated by next formula:

$$P_{amb\_word}\left(v_{f_{ij}}|f_n\right) = \frac{v_{f_{ij}}}{\sum_{i=1}^{n} f_n} \tag{4}$$

where, $v_{f_{ij}}$ is the frequence of a word with a certain translation, $\sum_{i=1}^{n} f_n$ – total amount of features in corpora for a particular translation.

The found probability gives the weight of a specific word translation.

The model of semantic cube chooses the translation from the corresponding set of tables of multivalued words maximizing their frequency and probability:

$$\tilde{t} = argmax_{t \in C} \sum_{i=1}^{n} P_{amb_{word}}\left(t_i|f_n\right) \tag{5}$$

The advantages of this model it works with parallel corpora.

Step 5. "Testing" stage.

Next is the "testing" stage which is made on another separate text, consisted of sentences, in which there are ambiguous words. During the translation process, this system calculates the most likely translation of the source sentence based on the data obtained during training. By comparing probabilities, as a translation for ambiguous word is selected the sense which has higher probability. It should be noted that the larger the volume of the corpus, the higher the quality of translation.

There are three outputs of the result for this technology. In the first case, when the system recognizes that there is an (lexical) ambiguous word in the sentence, then the word goes into the lexical selection module. First, the ambiguity is solved by using the model of product rules. If there is a written rule in the file for the required ambiguous word, then the translation for the word is selected based on this rule. In the second case, if there is no such rule, then a transition to the next stage occurs, where the problem is solved using the maximum entropy model. In this case, the translation for the required word will be that sense, which probability is higher. In the third case, when the translation cannot be found using the product rules model, or the maximum entropy model, as the default translation selects the meaning that is first specified in the bilingual dictionary.

Compared with other works of Berger et al. [2] and Dell Pietra et al. [6], our model differs in that the whole sentence is used as a context. Tyers et al. [1] use monolingual corpora; we use bilingual parallel corpora and a frequency list of ambiguous words with certain senses. Tyers uses a monolingual corpus of the source language and a statistical model of the target language. In his model the volume of context for an ambiguous word is four words, two on each side. We have the context of all the words included in the sentence, with the ambiguous word.

In the below figures the results of combined technology are presented (Figs. 4, 5):

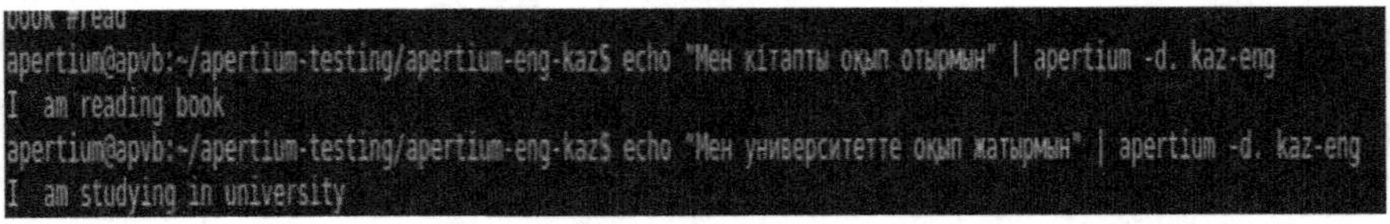

Fig. 4. The results to word "оқу"[oku] with combined technology.

At the Fig. 5 Kazakh word "күн"[kun] depending of context is translated as "sun".

Fig. 5. The results to word "күн"[kun] with combined technology

At the Fig. 6 Kazakh word "күн"[kun] depending of context is translated as "day".

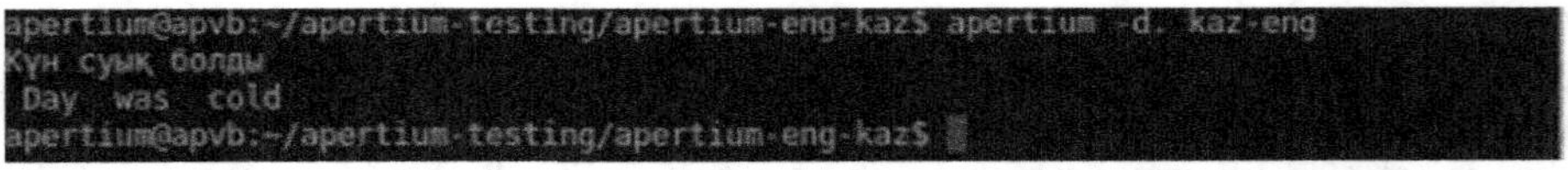

Fig. 6. The translation's results to word "күн"[kun] with combined technology

In Fig. 7 is given examples for English-Kazakh language pair. The word "plant" is ambiguous word can be translated into Kazakh as "зауыт"[zauyt] and "өсімдіктер"[osimdikter]. In both situations combined technology model chose right senses, for the first "зауыт"[zauyt] and for the second "өсімдіктер"[osimdikter].

For realization of the combined technology the rules and parallel corpora are used to determine the context. By using corpora for training and testing we distinguish context for ambiguous word. To resolving the task of lexical selection we collect and use parallel corpora taken from different sources such as texts in electronic form from various famous literary novels, fairy tales, open Internet resources, news portals.

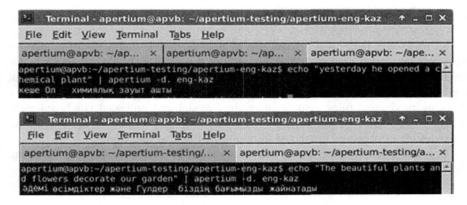

Fig. 7. The translation results for word "plant" with combined technology

To train the part of semantic cube of combined technology we use following corpora:

- corpus_lab_iis, that collected with books, fairy tales.
- Akorda collected from the official website of the President of the Republic of Kazakhstan [7]
- Egov collected from Electronic Government of the Republic of Kazakhstan [8]

  We have collected the English-Kazakh parallel corpus of $\sim 30\ 000$ sentences.

5   Experiments Results

In experiment for determining the translation were estimated in modes of checking productive rules, maximum entropy model and proposed combined technology in two direction of translation: Kazakh-English and English-Kazakh. Results are given in Table 2.

Table 2. Results of experiments on corpora

Language pair	Total number of sentences	Productive rule model, (%)	Maximum entropy model, (%)	Combined technology, (%)
Kazakh-English	26078	52	72	87
English-Kazakh	26078	48	65	78

By the results we can see that the proposed combined technology works better.

6 Conclusion and Future Work

In this paper was performed a combined technology uniting the model of productive rules and maximum entropy model for solving the problem of lexical selection for English-Kazakh and Kazakh-English language pairs. Also parallel bilingual English-Kazakh corpus has been developed with ~ 30 000 sentences. Experiments of checking productive rules, maximum entropy model and proposed combined technology in two direction of translation: Kazakh-English and English-Kazakh shows better results of proposed combined technology of lexical selection.

In future work we plan to increase the volume of parallel corpora for receiving more exactly results in solving of lexical selection.

References

1. Tyers, F.M., Sánchez-Martınez, F., Forcada, M.L.: Unsupervised training of maximum-entropy models for lexical selection in rule-based machine translation. In: Proceedings of the 18th Annual Conference of the European Association for Machine Translation (EAMT 2015). Antalya, Turkey, pp. 145–153 (2015)
2. Berger, A., Pietra, S.D., Pietra, V.D.: A maximum entropy approach to natural language processing. Comput. Linguist. 22(1), 39–71 (1996)
3. Mareček, D., Popel, M., Z Žabokrtský, Z.: Maximum entropy translation model in dependency-based MT framework. In: Proceedings of the Joint 5th Workshop on Statistical Machine Translation and MetricsMATR, pp. 201–206, Uppsala, Sweden, 15–16 July 2010
4. Tyers, F.M.: Feasible lexical selection for rule-based machine translation. Ph.D. thesis – Universitat d'Alicante, May 2013, 110 p.
5. https://www.apertium.org/
6. Della Pietra, S., Della Pietra, V., Lafferty, J.: Inducing features of random fields. IEEE Trans. Pattern Anal. Mach. Intell. 19(4), 1–13 (1997)
7. Akorda. www.akorda.kz
8. Public services and information online. www.egov.kz

New Kazakh Parallel Text Corpora
with On-line Access

Zhandos Zhumanov[✉], Aigerim Madiyeva, and Diana Rakhimova

Laboratory of Intelligent Information Systems,
Al-Farabi Kazakh National University, Almaty, Kazakhstan
z.zhake@gmail.com, {rockinfuture_7, di.diva}@mail.ru

Abstract. This paper presents a new parallel resource – text corpora – for Kazakh language with on-line access. We describe 3 different approaches to collecting parallel text and how much data we managed to collect using them, parallel Kazakh-English text corpora collected from various sources and aligned on sentence level, and web accessible corpus management system that was set up using open source tools – corpus manager Mantee and web GUI KonText. As a result of our work we present working web-accessible corpus management system to work with collected corpora.

Keywords: Parallel text corpora · Kazakh language · Corpus management system

1 Introduction

Linguistic text corpora are large collections of text used for different language studies. They are used in linguistics and other fields that deal with language studies as an object of study or as a resource. Text corpora are needed for almost any language study since they are basically a representation of the language itself. In computer linguistics text corpora are used for various parsing, machine translation, speech recognition, etc. As shown in [1] text corpora can be classified by many categories:

- Size: small (to 1 million words); medium (from 1 to 10 million words); large (more than 10 million words).
- Number of Text Languages: monolingual, bilingual, multilingual.
- Language of Texts: English, German, Ukrainian, etc.
- Mode: spoken; written; mixed.
- Nature of Data: general, specialized (dialect, idiolect, sociolect, etc.).
- Nature of Application: research, illustrative, learner, translation, aligned comparable, parallel, reference.
- Dynamism: dynamic (monitor), static.
- Temporal characteristic: diachronic, synchronic.
- Authorship: one author, two and more.
- Annotation: unannotated, annotated (morphologically, semantically, syntactically, prosodically, etc.).
- Access: free, commercial, closed.

© Springer International Publishing AG 2017
N.T. Nguyen et al. (Eds.): ICCCI 2017, Part II, LNAI 10449, pp. 501–508, 2017.
DOI: 10.1007/978-3-319-67077-5_48

There are quite a lot of text corpora for different languages. Among them are:

- American National Corpus;
- British National Corpus;
- Brown Corpus;
- Russian National Corpus;
- Europarl Corpus;
- EUR-Lex corpus.

There are number of text corpora for the Kazakh language. But all of them are monolingual:

- Almaty Corpus of Kazakh;
- Kazakh text corpora on Sketch Engine;
- Open-Source-Kazakh-Corpus;
- Kaz Corpus [2].

At the moment there is not a lot of parallel data that involves Kazakh language. Also the data is presented in raw format, usually it is a plain text files with one sentence on each line. Files in different languages are aligned on sentence level.

Parallel text corpora are very useful for comparative studies of all kinds. But they are also much more difficult to gather. Since so little ready to use parallel text corpora exist for Kazakh language there is a clear need to collect them somehow. That is the first task of this research.

When collected, the text corpora have to be accessed and worked with. Raw plain text formats are good for computers, but not for humans. Some system has to be put in place to facilitate the text corpora. Setting up such system is the second task of this research.

Following sections are dedicated to more detailed description of: Sect. 2 – the process and approaches of collecting parallel text corpora we have used; Sect. 3 – analysis of text corpora we have gathered; Sect. 4 – description of corpus management setup; Sect. 5 – conclusion and future work.

2   Process of Parallel Text Corpora Collection

In order to collect parallel text corpora we used 3 different approaches:

1. finding all significant ready to use aligned parallel texts;
2. using bitextor tool for crawling websites that contain same texts in several languages and aligning them;
3. using scrips for crawling texts from websites that contain same texts in several languages and using InterText tool with integrated hunalign tool for aligning them.

Approaches 2 and 3 seem to be similar to each other, but they have produced different amount of results which is described below.

There are not many places to find ready to use parallel text corpora that have Kazakh as one of the languages. In fact, there is one such place - the OPUS project [3].

There were some parallel Kazakh-English texts collected from Tatoeba and OpenSubtitles. That gave us 4480 aligned sentences.

Another ready to use resource is the Bible. It has been repeatedly translated into many languages. The Kazakh is also among them. There were several translations prepared by several organizations. The most resent one is called "New World Translation of the Christian Greek Scriptures" published on different media by Jehovah's Witnesses. Despite the nature of the organization it is turned out to be a great parallel resource since the text of the book has strictly numbered chapters and verses across all translations. That provided us Kazakh-English parallel text of 32358 sentences.

Bitextor is a free open source application for collection of translation memories from multilingual websites. [4] The application downloads all HTML files from a website, then pre-processes them into a consistent format and applies a set of heuristics to select the file pairs that contain the same text in two different languages (bitexts). Using LibTagAligner library translation memories in TMX format are created from these parallel texts. The library uses HTML-tags and length of the text segments for alignment. After cleaning the resulting translation memory from TMX format tags, we receive a parallel corpus with sentences in different languages aligned with each other.

We have run bitextor for following websites: http://www.kaznu.kz, http://www.bolashak.gov.kz, http://www.enu.kz, http://www.kazpost.kz, http://www.archeolog.kz, http://e-history.kz, http://inform.kz, http://egov.kz, http://primeminister.kz, http://tengrinews.kz and etc. (Fig. 1).

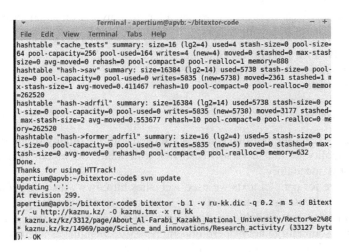

Fig. 1. An example of running bitextor for www.kaznu.kz

As a result of bitextor's work from each site we obtained *.tmx file with the following format (Fig. 2):

In this format, tag <tu> includes a pair of aligned segments (in this case - sentences); tag <tuv> - separate sentences in two languages; tag <prop> - HTML file addresses from which these sentences have been extracted; tag <seg> - sentences

```
-<tmx version="1.4">
    <header adminlang="en" srclang="ru" o-tmf="PlainText" creationtool="bitextor" creationtoolversion="4.0"
    datatype="PlainText" segtype="sentence" creationdate="20151017T180048" o-encoding="utf-8"> </header>
 -<body>
    -<tu tuid="1" datatype="Text">
      -<tuv xml:lang="ru">
          <prop type="source-document">Bitextor/esep.kz/rus/showin/article/1964.html</prop>
          <seg>Счетный комитет - Структурные подразделения</seg>
      </tuv>
      -<tuv xml:lang="kk">
          <prop type="source-document">Bitextor/esep.kz/kaz/showin/article/1964.html</prop>
          <seg>Есеп комитеті - Құрылымдық бөлімшелер</seg>
      </tuv>
    </tu>
    -<tu tuid="2" datatype="Text">
      -<tuv xml:lang="ru">
          <prop type="source-document">Bitextor/esep.kz/rus/showin/article/1964.html</prop>
          -<seg>
              Трудовую деятельность начал в 1977 году экономистом-аналитиком в Опытном хозяйстве Казахской
              машиноиспытательной станции. С октября 1978 года по октябрь 1979 года - экономист совхоза
              «Алатау».
          </seg>
      </tuv>
```

Fig. 2. A format of obtained parallel corpus for Kazakh-Russian language pair

themselves. In such *.tmx file sentence in one language corresponds to the sentence in another language. It should be noted that comparison quality depends on the website. Thus, we receive a file with parallel texts.

During cleaning of TMX files recurring segments, erroneous and meaningless sentence pairs were deleted. After removal of tags, we received Kazakh-English parallel corpus with 5 925 sentences.

The third approach is partially automated but also involves manual checking of the results. It consists of following stages:

- crawling parallel texts on the internet;
- cleaning and formatting of gathered texts;
- sentence splitting;
- sentence alignment;
- manual checking.

All stages except the last one can be automated. But the quality of the parallel text will affect quality of the tasks to be solved with them. So in our opinion human involvement is mandatory.

As a source for parallel texts we used web-sites http://www.akorda.kz/ and https://www.ted.com/. Texts from the first one were collected using scripts links.pl and extract_text.pl that are available in Apertium project's repository using the following link: https://sourceforge.net/p/apertium/svn/59905/tree/languages/apertium-kaz/texts/akorda/. Texts from the second site were collected manually.

After cleaning, formatting and sentence splitting we had two lists of sentences in two languages that were translations of each other but the sentences themselves were not aligned due to various translation reasons. To align them we used hunalign tool [5]. Hunalign has remarkably high quality: we got 6–8% of incorrectly aligned sentences out of unaligned lists mentioned above. But low percentage still meant that we had 2000–3000 alignment mistakes. It is quite many and that is why manual checking was due. Parallel text alignment editor called InterText was used for that (Fig. 3) [6].

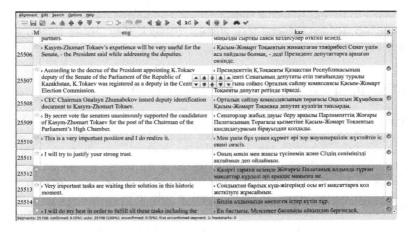

Fig. 3. InterText parallel text alignment editor

Approach described above resulted in two text corpora:

1. Akorda – 24 148 aligned sentences.
2. TED – 6 120 aligned sentences.

It seems to be logical that we should have tried to use bitextor on the Akorda and TED sites. And so we did. But for some unknown reasons bitextor did not produce any results or produced very few (under 100) aligned sentences when we tried to apply it with different settings.

All the raw text corpora described in this section are available on: https://drive. google.com/drive/folders/0B3f-xwS1hRdDM2VpZXRVblRRUmM.

3   Analysis of Collected Text Corpora

Information about all the text corpora that we have gathered is provided in the Table 1.

Table 1. Description of gathered text corpora

#	Method	Corpus	# of sentence pairs	# of words kaz	# of words eng
1	Ready to use	OPUS	4 480	19 892	27 839
2	Ready to use	New World Bible	32 358	548 258	824 398
3	Bitextor	Lab IIS	5 925	112 658	157 313
4	Semi-manual	Akorda	24 148	341 154	456 689
5	Semi-manual	TED	6 120	54 965	79 320
		TOTAL:	73 031	1 076 927	1 545 559

According to the classification shown earlier we gathered medium sized, bilingual, Kazakh-English, written, general text, aligned, parallel, static, many author, unannotated, free text corpora.

4 Setting up a Corpus Management System

Our second task – setting up a system to work with parallel text corpora – has been achieved with the help of an open-source tool called Manatee. [7] It is employed as the main corpus management tool for several large text corpora, including the Czech National Corpus, and we plan to use this system as the basis to maintain the parallel text corpora for Kazakh-English and Kazakh-Russian language pairs. Manatee is able to deal with extremely large text corpora and is able to provide a platform for computing a wide range of lexical statistics. It has such features as text preparation, concordancing, meta-data management, tokenization, efficient corpus storage, corpus annotation, computation of statistics and it is language and tag- set/annotation independent. Moreover, Manatee also functions as a corpus management server, which

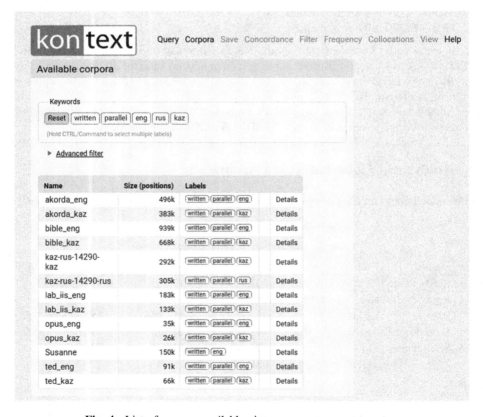

Fig. 4. List of corpora available via corpus management system

satisfies the condition of the web-accessibility of the corpus manager. To access its features we will use the GUI called KonText. It is a fully featured corpus query interface for the Manatee corpus search engine. It started as an extension of the Bonito 2.68 web interface but now is gradually becoming more independent. It is maintained by the Institute of the Czech National Corpus and the source code of the project is available at the URL: https://github.com/czcorpus/kontext/. All the key features of the Bonito 2.98.3, primarily a support for parallel text corpora, is present in the current version of the interface.

At first we've tried to implement the system that combines Manatee and Bonito into a corpus management tool called NoSketch Engine. We installed and locally hosted it during the testing period but since KonText has a lot of new features, more enhanced user interface and improved code documentation, we decided that it will be more reasonable to switch from Bonito to KonText. KonText comes with default plug-ins (located in lib/plugins directory), which provide a complete, working set of replaceable components needed to run it with all the features enabled but for the time being we have made it work with only the basic functionality such as search, concordances and frequency analysis, our main focus being the ability to manage parallel texts.

For now the system is hosted on Google App Engine virtual machine as the service provides flexible and inexpensive platform to experiment with different setups. It can be reached at http://104.197.218.108/corpora/corplist. Figure 4 shows a page from the system with list of all available corpora.

5 Conclusion

We have presented new parallel text corpora for Kazakh language with on-line access. Using 3 different approaches to collecting parallel data we have gathered medium sized, bilingual, Kazakh-English, written, general text, aligned, parallel, static, many author, unannotated, free text corpora. The corpora are available in raw text format along with web accessible corpus management system that is based on corpus manager Mantee and web GUI KonText.

For future we plan to continue work on the corpus and corpus manager. One direction of our efforts will cover collecting and possibly creating more parallel data including Kazakh and other languages. Another direction will cover implementation of all the other functionalities available in KonText and/or extend it even further at the next stages of our work.

References

1. Sereda, I.: Approaches to corpora classification in modern corpus linguistics (2012)
2. Makhambetov, O., Makazhanov, A., Yessenbayev, Z., Matkarimov, B., Sabyrgaliyev, I., Sharafudinov, A.: Assembling the Kazakh language corpus. In: EMNLP, pp. 1022–1031, October 2013

3. Tiedemann, J., Nygaard, L.: OPUS-an open source parallel corpus. In: Proceedings of the 13th Nordic Conference on Computational Linguistics (NODALIDA). University of Iceland, Reykjavik (2003)
4. Esplá-Gomis, M., Forcada, M.L.: Bitextor, a free/open-source software to harvest translation memories from multilingual websites. In: Proceedings of MT Summit XII, Ottawa, Canada. Association for Machine Translation in the Americas (2009)
5. Varga, D., Németh, L., Halácsy, P., Kornai, A., Trón, V., Nagy, V.: Parallel corpora for medium density languages In: Proceedings of the RANLP 2005, pp 590–596 (2005)
6. Vondřička, P.: Aligning parallel texts with InterText. In: Calzolari, N., et al. (eds.) Proceedings of the Ninth International Conference on Language Resources and Evaluation (LREC 2014), pp. 1875–1879. European Language Resources Association (ELRA) (2014)
7. Rychlý, P.: Manatee/bonito-a modular corpus manager. In: 1st Workshop on Recent Advances in Slavonic Natural Language Processing, pp. 65–70, December 2007

Design and Development of Media-Corpus
of the Kazakh Language

Madina Mansurova[✉], Gulmira Madiyeva, Sanzhar Aubakirov,
Zhantemir Yermekov, and Yermek Alimzhanov

Al-Farabi Kazakh National University, Almaty, Kazakhstan
mansurova0l@mail.ru, gbmadiyeva.kz@gmail.com,
c0rp.aubakirov@gmail.com,
yermekovzhantemir@gmail.com, aermek81@gmail.com

Abstract. The aim of this work was design and development of a media-corpus of the Kazakh language. The media-corpus is hosted by the al-Farabi Kazakh National University and serves linguists as an empirical basis for research on contemporary written Kazakh. The information system for media-corpus was built on the basis of component software architecture. To make the processes of collection, storage and analysis of media-texts in the Kazakh language automatic, four components of the information system were designed and developed. The text files are saved in XML format. At the stage of analysis such tasks as text normalization, removing stop words, adding metadata and morphological analysis are performed. The morphological analyzer receives an input of a plain text, and at the output gives the text in XML format, which is further convenient to work with as it is easily converted to JSON format. The XML format is defined using XML Schema Definition (XSD). XSD allows to convert data into any other format, which simplifies the data exchange between the systems. For the case of incomplete morphological markup and the presence of homonymy, a special interface to perform manual markup is developed.

Keywords: Media-corpus · Corpus linguistics · Morphological parsing

1 Introduction

A natural language corpus is an invaluable tool that reduces costs of technical work on the study of language phenomena and gives the possibility to find reference information within few minutes. The Kazakh language corpus is not a single technical support of linguistic research, it is a modern innovation tool, reference and information base on the Kazakh language that helps answer many questions facing home and foreign researchers, students, consumers who study the Kazakh language.

The concept "corpus" is identified by many researchers with the concept "a set of texts or language units" and this does not give us a necessary theoretical-methodological basis to consider a corpus as not only a universal phenomenon, i.e. possessing a definite set of characterological properties and features inherent to different types, styles of any advanced language, but also as a phenomenon of an idioethnic order determined by the peculiarities of national mentality imprinted in the national conceptual picture of the

© Springer International Publishing AG 2017
N.T. Nguyen et al. (Eds.): ICCCI 2017, Part II, LNAI 10449, pp. 509–518, 2017.
DOI: 10.1007/978-3-319-67077-5_49

world. This problem was solved for many well studied languages of the world (British English, American English, Armenian, French, Polish and others).

A special place in modern corpus linguistics is occupied by media-corpora. A media-corpus of a language is an information-reference system of a data-base of marked up media-texts in Kazakh in an electronic form. The base of media-texts includes solid sampled news texts issued in mass media. There is no doubt that a media-corpus is a quite valuable source for collection, analysis of any news information for a wide range of consumers who can search by different grounds (key words, rubrics, themes, etc.). It also can be a training tool for would-be specialists-journalists, reviewers, politicians, specialists in any media-sphere and others.

The aim of the work described in this paper was to implement conversion of documents from the web into a media-corpus. The data are collected from 44 web sites with Kazakh language content, including 10 emergency portals, 11 news portals, 13 educational portals, and 10 entertainment resources. The media-corpus of the Kazakh language developed by the authors will be available within a public web portal. It will become a new tool for research, analysis, study and teaching of the Kazakh language, intended for a wide range of consumers in the domestic and world arena.

The structure of the work is as follows. Section 2 presents the review of related works dealing with the corpus linguistics for Turkic languages. Section 3 describes the design of the component architecture and development of a media-corpus on its basis. Section 4 describes the data format being proposed by us allowing to do away with the existing drawbacks of the text corpus development. Section 5 presents the morphological analyzer and post processing after analyzer. In Sect. 6, we formulate conclusions and present a plan for further investigations.

2 Related Work

Within the framework of the Kazakh linguistics and applied linguistics, in Kazakhstan, of special interest is the study and development of the National Corpus of the Kazakh language due to insufficient development of the problems in this field. Despite the achievements in this field (the attempt to compile a corpus with the necessary markup, the presence of scientific investigations in the form of monographs, theses, text-books on the styles of the Kazakh language, works of a comparative character analyzing the differences in colloquial and literary languages, studies on its separate aspects), limits of investigations do not go beyond the frames of traditional linguistics, this restricting the attempts on development of the corpus or reducing them to mechanistic detection of lexical, phonetic and other differences of the Kazakh language.

The Kazakh language refers to the class of agglutinative languages and together with Uzbek, Kyrgyz, Bashkir, Tatar, Azerbaijani, Turkish and other languages forms a Turkic linguistic family. Agglutinative languages are characterized by a consecutive addition of suffixes or endings bearing a grammatical meaning to an unchangeable root or stem having a lexical meaning.

The sequence order of affixes in agglutinative languages is strict. For example, for nouns first a suffix is added to the stem and then the ending of the plural followed by a possessive ending, then comes a case ending and finally the ending of the conjugation form (which is only added to animate nouns) [1, 2].

The Turkic corpus linguistics has begun to develop intensively only since 1990, therefore projects of creation of public corpora of Turkic languages are especially urgent. Today there is a small amount of representative corpora in Turkic languages such as:

1. The Turkish national corpus with 50 million word usages which is the balanced and representative case of modern Turkish. It consists of samples of text data in a wide variety of genres covering the period of 20 years (1990–2009) [3].
2. Bashkir poetic corpus with more than 1.8 million word usages. It is the world's second poetic corpus. Its feature is that the corpus consists of works of the Bashkir poets of 20th and the beginning of the 21st century [4].
3. The written corpus of the Tatar language with more than 116 million word usages with the number of various word forms – about 1.5 million [5].

The experience in the development of the corpora of Turkic languages has positively influenced development of that of the Kazakh language. The problem of creating a National corpus of the Kazakh language is still actual. Among the really existing and functioning developments of the corpus of the Kazakh language, one should mention the Corpus of the Kazakh language placed on the portal of the state language of the Committee on languages of the Ministry of culture and information of the Republic of Kazakhstan [6], the Corpus of the Kazakh language created by the workers of National laboratory of Astana (NLA) of L. Gumilev Eurasian University [7]. One should mention the so-called English-Kazakh parallel corpus compiled by T.E. Kaldybekov on the basis of legal texts [8] as well as the Kazakh national corpus [9] which is another attempt of creating a full-fledged corpus of the Kazakh language and can be considered to be one of the first corpora which does not refer to accessible open corpora. There are attempts to create a new resource of the Kazakh language with a linguistic annotation – this is a research of scientists from Nazarbaev University [10], most likely, a tool of grammatical markups which is in the state of development, a joint project of G. Altynbek and W. Xiao-long who worked out a corpus of the Kazakh language at Xinjiang University (2010). As T.F. Kaldybekov notes, there is no information on the latter corpus [8]. Almaty corpus of the Kazakh language is close to a polyfunctional corpus [11].

3   Component Software Architecture

The component software architecture is a programming paradigm based heavily on the concept of a "component" [12, 13]. The component architecture means creation of a system consisting of components as multiply used nodes. This allows to distinguish the development of a separate component from the development of a system in a whole, this making it possible for different commands to develop and support components independent of each other.

The information system built on the basis of the component architecture meets the following requirements [14]:

- It works autonomously, does not require an operator;
- The system components do not depend on one another;
- All the system components are horizontally scaled;
- The system is fault-tolerant, resistant to reboots and shutdowns.

And each component has the following four properties: the possibility to be repeatedly used, interchangeability, extensibility, composability [14]. In view of the above arguments, the information system presented in this work was built on the basis of component software architecture.

To make the processes of collection, storage and analysis of media-texts in the Kazakh language automatic, an information system was designed and realized. This system consists of four components:

1. A component of information collection;
2. A component of data storage;
3. A component of data analysis;
4. A component of data visualization.

The information system component architecture is shown in Fig. 1.

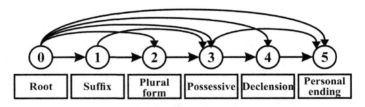

Fig. 1. Rules of affixes attachment for nouns

The use of queues allows the system to be easily scaled and fault-tolerant (Fig. 2).

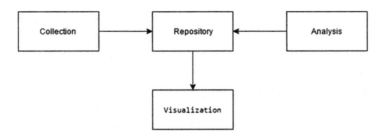

Fig. 2. The component architecture of the information system

In Fig. 3, the system architecture is presented in the form of the UML diagram. Descriptions of work of each component are given below.

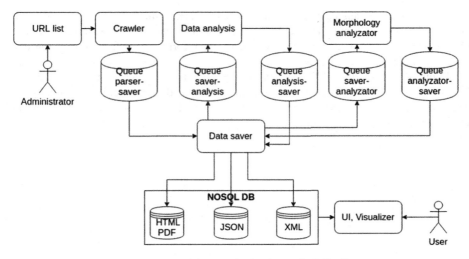

Fig. 3. System architecture in the form of UML diagram

The component of information collection. The task of the information collection component is to continuously monitor sites and download new information. The component is realized using technologies Jsoap, OpenMQ and Java EE.

The component of data storage. The task of the data storage component is to provide a stable and quick access to the data store. NoSQL data base MongoDB 3 is used as a data store.

The component of data analysis. The task of the data analysis component is preprocessing of accumulated data. Preprocessing is regarded as a process of text normalization, clearing from stop words, adding metadata. Also, a morphological analysis and marking up of texts are performed at this stage. The component is realized using technologies Apache Lucene, Apache Spark, Apache Hadoop and MPJ.

The component of data visualization. The task of the data visualization component is to provide a user with a handy search tool for accumulated data and search result display. The component subtasks are: acceleration of the search process, improvement of the search results and relevancy, visualization of the search results. The component was realized using technologies HTML5 and Elasticsearch, a platform for data processing built on the basis of Apache Lucene.

4   Data Storage Format

To store corpus files, the eXtensible Markup Language (XML) language is used. The collection component extracts the relevant text from the HTML code of the pages using the Jsoap library. Then the data are processed in the data analysis component. At this stage, a morphological analysis of texts in the Kazakh language is performed. The morphological analyzer receives an input of a plain text, and at the output gives the text in XML format, which is further convenient to work with, for example, it is easily converted to JSON format. The XML format is defined using XML Schema Definition (XSD). XSD allows to convert data into any other format, which simplifies the data exchange between the systems.

Figure 4 shows the XSD schema of the document, which is a file-based data storage format. XSD makes it possible to exchange data conveniently. It helps to exchange the marked texts processed by a morphological analyzer of the Kazakh language with other components.

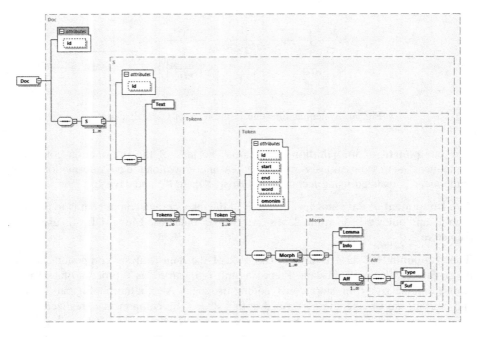

Fig. 4. Schema of XML-document

Figure 5 shows execution of a word markup in case if a morphological analyzer encounters homonymy. Inside the tag "Token" there is an attribute "omonim" which indicates whether this word has a homonym or not. In case if there is a homonym, inside the tag there will be two or more tags "Morph", this indicating that the word has two or more variants of morphological analysis.

5   Morphological Parsing and Post Processing

The corpus contains a special markup that is additional information on the properties of composing it texts. Markup is the main characteristic of the corpus; it differentiates the corpus from simple collections (or "libraries") of texts. The richer and more diverse the markup, the higher is the scientific value of the corpus.

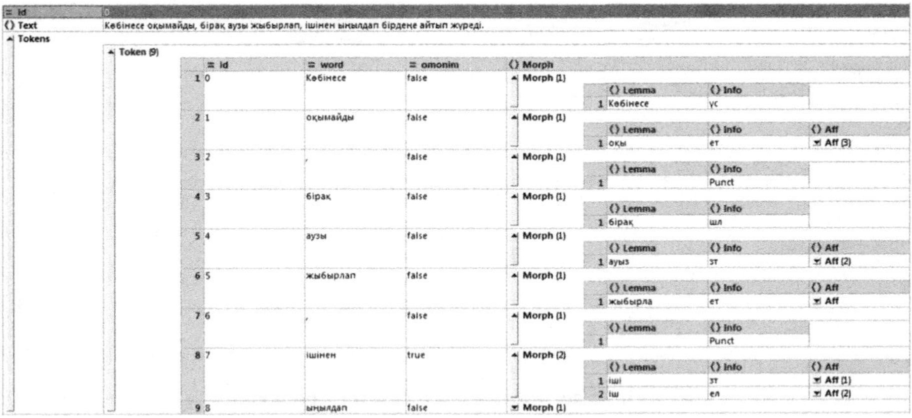

Fig. 5. Schema of text markup

The work of a morphological analyzer is as follows. At its input, we have an array of words, punctuation marks and numbers marked from the input text at the stage of lexical analysis, with lexical characteristics [15]. For every word, the analyzer performs the search for words in the dictionary loaded into memory. All the stems the word being analyzed can begin with are looked for. If a stem in turn satisfies this condition, a line containing all possible affixes for this stem is extracted from the dictionary of affixes. Each affix from this line is added by turns to the stem and the result is compared with the word being analyzed. In case of their exact coincidence, a new record is introduced into the list of the search results: according to the ordinal number of the affix in the line of affixes, variable morphological parameters of the word are determined (for example, for a noun – the number and case), and by the lexical information of this stem its constant parameters (noun, verb, adj, ...) are determined. If, as a result of such search, not a single successful variant is found, the user is requested to enter a new stem into the dictionary. In case he refuses to do it, performance of the morphological analysis is stopped. If the new word is introduced into the dictionary, the procedure of searching is repeated. At the output of the morphological analyzer, we have a list of lemmas (a normal form of the word) + affix + morphological characteristics of the word (part of speech, case, number).

Thus, the results of the morphological analysis in total are presented in the form of a dynamic array. The number of its elements is equal to the number of its lexemes in the sentence. The elements of the array are other arrays each of which retains all possible interpretations of its lexemes, homonyms. A dictionary of word stems, a dictionary of geographic names, a dictionary of names, and a dictionary of affix conjunctions are used as initial lexical materials.

In the case of incomplete morphological markup and the presence of homonymy, the expert linguist can perform a complete manual markup for a specific word. For manual processing of texts a special interface is developed.

Figure 6 shows an interface for manual removal of homonymy, which allows to choose the option offered by the morphological analyzer or to execute a new

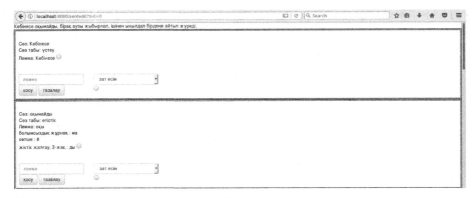

Fig. 6. Interface for manual removal of homonymy

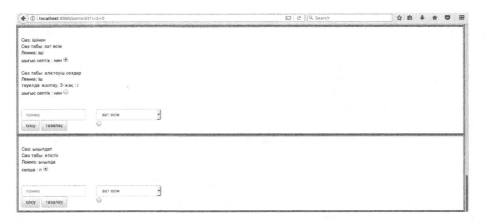

Fig. 7. Example of the word with homonymy

markup. When pressing the button "қосу" (kosu), there appears the possibility to choose a part of speech for the word being analyzed, to add an affix or several affixes corresponding to the given part of speech. When pressing the button "тазалау" (tazalau), it is possible to clear away all affixes in case of error and start addition once again. In case of a manual removal of homonymy, all affixes are added and a manual layout switch is chosen (Fig. 7).

6   Conclusion

The work proposes the design and development of a media-corpus of the Kazakh language. The authors described conversion of the documents from the web into a media-corpus. The media-corpus is hosted by the al-Farabi Kazakh National University and serves linguists as an empirical basis for research on contemporary written Kazakh. The text files are saved in XML format. At the stage of analysis such tasks as text normalization, removing stop words, adding metadata and morphological analysis are performed. For the case of incomplete morphological markup and the presence of homonymy, a special interface to perform manual markup is developed.

The developed media-corpus of the Kazakh language will allow to:

- Provide an open access to everyone;
- Search by morphological parameters;
- Use the corpus for solution of problems of Natural language Processing;
- Make a frequency analysis of texts;
- Teach the language using translations of words.

Acknowledgments. This work was supported in part under grant of Foundation of Ministry of Education and Science of the Republic of Kazakhstan "Development of intellectual high-performance information-analytical search system of processing of semi-structured data" (2015–2017).

References

1. Bekmanova, G.T.: Some approaches to the problems of automatic word changes and morphological analysis in the Kazakh language. Bulletin of the East Kazakhstan State Technical University Named by D. Serikbayev, vol. 1, pp. 192–197 (2009) (In Russian)
2. Zhubanov, A.H.: Basic principles of formalization of the Kazakh text content. Almaty (2002) (In Russian)
3. Turkish National Corpus. http://www.tnc.org.tr/index.php/en/
4. Bashkir poetic corpus. http://web-corpora.net/bashcorpus/search/?interface_language=ru
5. Written corpus of the Tatar language. http://corpus.tatar/
6. Portal of the state language of the Committee on languages of the Ministry of culture and information of the Republic of Kazakhstan. http://til.gov.kz/wps/portal/!ut/p/
7. Corpus of the Kazakh language created by the workers of National laboratory of Astana of L. Gumilev Eurasian University. http://kazcorpus.kz/klcweb/en/
8. Kaldybekov, T.E.: The Anglo-Kazakh parallel corpus for statistical machine translation. J. Young Sci. **6**, 92–95 (2014). (In Russian)
9. Portal of a state language of the Republic of Kazakhstan. http://dawhois.com/www/til.gov.kz.html
10. Makazhanov, O.A., Makhambetov, O.E., et al.: Development of morphological, syntactic and lexical sets of tags for tagging of texts in Kazakh. Philol. Cult. **2**(36), 37–39. Kazan University, Kazan (2014) (In Russian)
11. Almaty corpus of the Kazakh language. http://web-corpora.net/KazakhCorpus/search/?interface_language=ru
12. Szyperski, C.: Component Software: Beyond Object Oriented Programming. Addison-Wesley Professional, Reading (1997)
13. Aubakirov, S.S., Akhmed-Zaki, D.Z., Trigo, P.S.: News classification using apache Lucene. KazNU Bull. Math. Mech. Comput. Sci. Ser. **3**(91), 59–65 (2016)

14. Bass, L., Clements, P., Kazman, R.: Software Architecture in Practice (SEI Series in Software Engineering), 3rd edn. Addison Wesley, Boston (2012)
15. Azarova, I.V.: Morphological markup of the texts in Russian, using the formal grammar AGFL. Department of mathematical linguistics of St. Petersburg State University. http://www.dialog-21.ru/Archive/2003/AzarovaAFGL.htm

Morphological Analysis System of the Tatar Language

Rinat Gilmullin[(✉)] and Ramil Gataullin

Institute of Applied Semiotics, Tatarstan Academy of Sciences, Kazan, Russia
rinatgilmullin@gmail.com

Abstract. This paper presents the description of the morphological analysis system for the Tatar Language based on a two-level morphology model. The morphological system is used for grammatical annotation of the Tatar national corpus. This paper shows the results of evaluation of completeness of the system using statistical information that was obtained from the corpus data and describes the ways to improve this system.

Keywords: Morphological analysis system · HFST · Alphabet · Phonological rules · The Tatar language · Tatar National Corpus

1 Introduction

Morphological analysis is challenging, especially for the tasks related to natural language processing for agglutinative languages. This is the case of Turkic languages – such as Turkish, Tatar, Kazakh, Kyrgyz, Crimean Tatar, Turkmen, Uygur and others – that are characterized by rich and complex agglutinating morphology.

There are many works that describe models and systems of morphological analysis for Turkic languages [1–6]. Most of the systems (including those for the Tatar language) are based on a rule-based morphological model with the finite state transducer. This choice is logical due to the peculiarity of Turkic languages that are characterized by high degree of regularity and almost automatic morphology. Unfortunately, when describing the rules for the Tatar language, the authors encountered a large number of violations in morphology caused both by non-assimilated borrowings, mainly from Russian and Arabic, and by non-effectiveness of symbols in the Cyrillic alphabet. Moreover, works [7–9] on morphological analysis systems for the Tatar language lack data on the use of these systems and experimental results for their evaluation. In this article, the authors will try to fill this gap.

This paper presents a description of the current version of the morphological analysis system of the Tatar language that is reimplemented using the Helsinki Finite-State Transducer Technology (HFST) [10], so that it corresponds to the requirements for modern morphological analysis systems.

The completeness of this system was assessed on the basis of statistical information that was obtained upon morphological analysis of "Tugan Tel" Tatar National Corpus data [11]. This paper presents our latest results on increasing the completeness of this morphological model.

© Springer International Publishing AG 2017
N.T. Nguyen et al. (Eds.): ICCCI 2017, Part II, LNAI 10449, pp. 519–528, 2017.
DOI: 10.1007/978-3-319-67077-5_50

We plan to use the methodology and results of this work for the grammatical annotation of the Socio-Political Corpus of the Tatar language [12].

The paper has the following structure: Section 2 describes linguistic features of the Tatar language; Section 3 gives an overview of our morphological system; Section 4 shows statistical information on the Tatar language that was obtained from the corpus data and discusses the way to improve the model of our morphological system; Section 5 concludes the paper.

2   The Tatar Language

The Tatar language belongs to the Turkic group that forms the subfamily of Altaic languages. The Tatar language is spoken in West-central Russia (in the Volga region) and southern parts of Siberia. The number of Tatars in Russia in 2010 was 5.31 million people [13].

Different dialects of Tatar can be identified: Western, Kazan (Middle) and Eastern. In 2013, the existing language classifications [14, 15] described the Tatar language as an under-resourced language. However, recent results in machine translation [16, 17], speech analysis and synthesis [18], as well as the Tatar Nacional Corpus [11] can change this situation.

3   Tatar Morpfological Analysis System

One of the first versions of the morphological analysis system for the Tatar language was created on the basis of a two-level morphology model using the PC-KIMMO tool [7]. This tool had some technical limitations, therefore currently the system uses HFST in order to improve its technical parameters.

The main technical parameters of the systems are presented in Table 1.

Table 1. Technical parameters of the systems

Technical parameters	PC-KIMMO	HFST
Morphotactical rules	49	49
Phonological rules	42	55
Vocabulary	26,500	28,000
Processing speed	∼500–1000 (words per second)	∼5000–20,000 (words per second)
Coverage (recall)	80–85%	93–95%
Alphabet	Latin	Cyrillic
Programming language	C/C++ and Delphi	C/C++ and Python
Has API	No	Yes
Interface as	Desktop application	Web application

The Cyrillic alphabet of the Tatar was adopted in 1939 and consists of 39 characters: 12 vowels and 28 consonants.

The rule for vowel harmony in the Tatar language states that in one word the vowels share certain characteristics, such as being front or back. It is called the law of synharmonism and it is a basic linguistic feature of nearly all Turkic languages.

As shown in Table 1, 13 rules were added to the current version of the morphological module. This is primarily due to the implementation of orthography rules for Cyrillic where е, ю and я symbols represent different sounds in the language depending on the harmony of the vowels in the word. For example, *e: каен* - [kayın] (birch), *кибет* - [kibet] (store), *тиен* - [tiyen] (squirrel); *ю: аю* - [ayu] (bear), *юкә* - [yukə] (linden-tree); *я: ял* - [yal] (rest), *яшел* - [yəshel] (green). To describe these symbols, additional rules were created in the file with rules of the morphological analysis system. Moreover, many violations in morphology arise from not assimilated borrowings from Russian and Arabic, for example, words without vowel harmony: *бэла* (trouble) instead of *бэлә*, *кабинет+ы* (his/her/their room) instead of *кабинет+е*, *әдәбият+ы* (literature) instead of *әдәбият+е*. To take into account these features of the language, we created additional exceptional rules and a dictionary of exceptional words.

Although officially the Tatar language uses the Cyrillic alphabet, before it used the Arabic and Latin scripts. We should mention here that the most effective in terms of describing and processing of the Tatar language in information technology is the Latin graphics.

Two main language issues, namely Rule Components (alphabet and phonological rules) and Lexicon (stems, morphemes and morphotactic rules), can be dealt with XFST. First, we compile the Rules via encoding a finite-state network. Secondly, we build a finite-state transducer for the Lexicon. Then, we unite the newly formed network and the transducer into a single final network that covers all morphological aspects of the language, such as morphemes, derivations, inflections, alternations and geminations [19].

3.1 Rule Components of the Morphological Analysis System of Tatar Language

In XFST, Rule Components consist of auxiliary words: Alphabet, Sets, Rule.

Alphabet а б в г д е ё ж з и й к л м н о п р с т у ф х ц ч ш щ ъ ы ь э ю я ә ө ү ж ң h Л:л Л:н Д:д Д:т Д:н А:я А:а А:ә а:ы А:ы ә:0 А:0 Ы:ы Ы:е Ы:0 ы:е ы:0 С:с С:0 Г:г Г:к Г:н Г:0 Й:й Й:а Й:ә Й:и Й:я У:у У:ү У:в У:ю ь:0 й:0 к:г п:б 0:ы е:0 е:и а:я ә:я у:ю ү:ю

Sets – declared sets;
CS = h х б ц ж д ф г ж к л м н ң п р с ч ш щ т й з ъ ь в Л Д С Г Й – set of all consonants;
NASAL = м ң н – nasalization;
VOWEL = э а е ы и о ө ү у ә я ю Ы А – set of all vowels;
BACKV = а ы о у – set of all back (hard) vowels;
FRONTV = э ә и ө ү е – set of all front (soft) vowels;
ZVONKCS = р н ң ж л м й з – set of voiced consonants (special case)

ZVOPARE = б д г ж л р й з м н ң – set of all voiced consonants;

GLUPARE = п т с к ф ш һ ч щ – set of all unvoiced consonants;

GLUHCS = х б ц ч д ф г һ к п с ш т – set of voiced and unvoiced consonants (special case);

WUY = у ү ю

The alphabet contains 39 symbols from the modern Cyrillic Tatar alphabet, as well as 43 pairs of correspondences that the Rules uses to describe their morphological manifestations in the language.

Some capital letters such as А, Ы, С, Л, Д, Г, Й, У are defined on the intermediate level and they are invisible for the user. These representations are used for substitution, such as А is for а, я, ə or 0, and C is for с and 0. Here 0 stands for an empty character. Symbols are used to denote affixes and their manifestations in the form of corresponding context dependent allomorphs.

Thus, Table 2 contains affixes with an A-symbol and their allomorphs. It can be observed that some affixes can contain more than one capital letter, and for each capital symbol – called a lexical symbol – a separate rule is used to describe its corresponding manifestation – called a surface symbol.

We introduced 55 rules for full description of the morphological model of the Tatar language that are described in detail in [7, 20]. 12 rules out of 55 deal with exceptional situations, which appear due to imperfections in Tatar orthography and the presence of non-assimilated borrowings that violate the law of synharmonism. All these rules are implemented with XFST tools [19]. Let us consider some of them. For example, for the A lexical symbol that is represented by affixes in Table 2 and has a surface manifestation as a, the rule in XFST tools notation [19] might be:

"rule #1"
A:a=> :BACKV (CS) (:0*) (CS)_;

The rule states that the symbol A can have an a manifestation if it can be preceded by consonant letters or an empty symbol occurring zero and more times, and before it there may be another consonant letter necessarily preceded by any letter from the set of back vowels.

This rule could be enough to observe the law of synharmonism, provided that there are no factors for its violation:

- the use of the imperfect Cyrillic alphabet with the е, ю, я symbols that may denote neither back nor front vowels;
- many borrowings, mainly from Russian and Arabic, that remain unassimilated due to ineffective academic grammar of the modern Tatar language. For such words, we created a dictionary of exceptions with a special label that signals the emergence of such kind of context and the impossibility of applying regular rules.

To implement the morphology model taking into account all these factors of rule violations, we introduced corresponding changes in form of different exceptional contexts. Thus, "rule # 1" upon transformation received the following cumbersome form:

Table 2. Affixes with an A-symbol and their allomorphs

Lexical symbol	Grammatical type	Surface symbol
-КАлА	Modality	-кала, -гала, -кәлә, -гәлә
-ЫргА	Infinitive	-ырга, -ергә, -рга, -ргә
-мАк	Infinitive	-мак, -мәк
-мА	Negation	-ма, -мә
-сА	Condition	-са, -сә
-КАн	Past categorical	-кан, -ган, -кән, -гән
-АчАк	Future categorical	-ачак, -әчәк, -ячак, -ячәк
-АУ		-ау, -әү
-КА	Directive case	-га, -ка, -гә, -кә, -а, -ә
-ДА	Locative case	-да, -дә, -та, -тә, -нда, -ндә
-ДАн	Ablative case	-дан, -дән, -тан, -тән, -ннан, -ннән
-ДАгЫ	Possessiveness	-дагы, -дәге, -тагы, -тәге, -ндагы, -ндәге
-чА	Comparative	-ча, -чә

A:a=> :BACKV (e) (CS) (%>:0) (CS) (:0*) (CS)_;
[#ю|BACKV й:0 %>:0 У:ю] %>:0 CS_;
:BACKV (й:0) %>:0 :я (%>:0) CS_;
Ы:ы %>:0 0:н CS_;
BACKV й:0 %>:0 С:0 Ы:е %>:0 Г:н_;
:BACKV (CS) (%>:0) (Ы:0) (CS) %>:0 (CS)_;
:BACKV %>:0 Ы:0 (CS) (%>:0) (CS*) _; !абайла+ЫргА
:BACKV (й:0) (%>:0) :e (CS) (%>:0) (CS) (:0*) (CS)_;
[a:я|у:ю|ы:е] (e) (CS) (%>:0) (CS) (:0*) (CS)_;
[a:я| :BACKV|у:ю|ы:е] %>:0 Ы:0 (CS) %>:0 Г:0_;!дөньяма
[a:я|у:ю|ы:е] %>:0 Ы:0 (CS)* (%>:0) (CS)*_;!дөньямда
BACKV й:0 %>:0 Ы:е (CS*) (%>:0) (CS) _;!тоелма

4   Test and Evaluation

The morphology analysis system was tested on the data of the "Tugan Tel" Tatar National Corpus that was developed in the Institute of Applied Semiotics of the Tatarstan Academy of Sciences and the Kazan Federal University [11]. The volume of the corpus is 63,508,127 lexical units. Table 3 contains the linguistic data that was received upon corpus testing.

The research was conducted on a part of the "Tugan Tel" corpus with approximately 63 million lexical units, including words, punctuation marks, numbers, words in Latin, etc. 67.30% of all the corpus is represented by words. Our morphological analyzer recognizes 90.37% of them (60.83% of all the corpus) and does not 9.62% (6.48% of all the corpus). Some of the lexical units are tagged as "Error", which means that they consist of Cyrillic letters, Latin letters, numbers and some other symbols at the

Table 3. Distribution of lexical units of the Tatar National Corpus "Tugan Tel" according to their type

Lexical unit	Tags	Unique words	%	All words use	%
Words		375,298	40.55	38,629,659	60.8
Punctuation marks	Type2	138	0.01	11,134,395	17.5
End of sentence	Type1	4	0.00	5,305,378	8.3
Not Recognized	NR	482,338	52.12	4,113,428	6.4
Number, Latin, Letter, Signs, etc.	Type3	18,720	2.02	2,872,947	4.5
Error	Error	49,022	5.30	1,452,320	2.3
Overall		925,520	100	63,508,127	100

same time, so they cannot be recognized as words of Tatar language. Such words take 2.29% of the whole corpus.

The part-of-speech distribution is given in Table 4.

Table 4. Part-of-speech distribution in the Tatar National Corpus "Tugan Tel"

POS	POS Tag	Unique words	%	All words use	%
Noun	N	137,530	55.67	11,331,955	40.11
Verb	V	83,916	33.97	6,007,993	21.27
Pronoun	PN	1398	0.57	2,411,290	8.54
Adjective	Adj	12,234	4.95	2,187,637	7.74
Particle	PART	73	0.03	1,813,949	6.42
Proper name	PROP	9893	4.00	1,231,996	4.36
Postposition	POST	61	0.02	890,237	3.15
Adverb	Adv	569	0.23	787,836	2.79
Conjugation	CNJ	30	0.01	785,302	2.78
Number	Num	1162	0.47	426,047	1.51
Modal word	MOD	66	0.03	311,207	1.10
Interjection	INTRJ	116	0.05	63,781	0.23
Imitational word	IMIT	15	0.01	2141	0.01

As expected, most (77.65%) of the words are nouns (40.11%), verbs (21.27%), pronouns (8.54%) or adjectives (7.74%). Note that these calculations do not include ambiguous words, which take 26.86% of all recognized words.

The distribution of ambiguous words and the number of possible parsing options is given in Table 5.

Ambiguous words take 26.86% of all recognized words. Most of them (19.78%) have only two parsing options.

Table 5. Distribution of ambiguous words according to the number of possible variants

Number of variants	Unique words	%	All words use	%
1 (not ambiguous)	247,063	65.83	28,251,371	73.13
2	94,699	25.23	7,640,618	19.78
3	14,074	3.75	1,690,613	4.38
4	15,292	4.07	889,574	2.30
5	603	0.16	52,147	0.13
6	3078	0.82	80,256	0.21
7	21	0.01	7842	0.02
8	274	0.07	10,182	0.03
9	123	0.03	272	0.00
10	42	0.01	6726	0.02
11	1	0.00	1	0.00
12	27	0.01	54	0.00
15	1	0.00	3	0.00

Table 6 illustrates the distribution of words according to the length of the affixal chain, whereas Table 7 describes the distribution according to the number of not recognized words that is useful for system evaluation.

Table 6. Distribution of words according to the length of the affixal chain

Affixes count	Unique words	%	All words use	%
0	16,162	6.5417	13,555,659	47.98
1	73,860	29.8952	9,845,456	34.85
2	97,728	39.5559	3,913,178	13.85
3	47,928	19.3991	862,454	3.05
4	9883	4.0002	70,470	0.25
5	1438	0.5820	4057	0.01
6	63	0.0255	96	0.00
7	1	0.0004	1	0.00

Upon the analysis of 1000 most frequent not recognized words, we discovered that the majority (41.60% of the analyzed words) of these words were not included into the analyzer's dictionary; so even if these words have regular forms, the analyzer cannot recognize them. Mostly (24.60% of the analyzed words) these words are represented by proper names (e.g. Татнефть (Tatneft), Метшин (Metshin)); others are rarely used words and abbreviations (e.g. агросәнәгать (agroindustry), мчс (Emergencies Ministry)) or loanwords (e.g. брифинг (briefing), телесериал (TV series)). The solution is to add these words into the dictionary.

Another reason why we did not recognize words in our corpus is the incompleteness of the rules in the analyzer. There are several (21.50% of analyzed words) exceptional situations with unassimilated loanwords word forming (e.g. кит (whale), кабинет (office), компьютер (computer)). Also, some (10.20% of analyzed words)

Table 7. Number of not recognized words

Reasons	Unique words	%	All words use	%
Word stem is not in the dictionary				
Proper names	246	24.6	324,544	24.2
Other	170	17.0	234,276	17.4
Incompleteness of the rule				
Russian or European origin of the word	190	19.0	257,597	19.2
Tatar origin of the word	102	10.2	173,778	12.9
Arabic, Persian origin of the word	25	2.5	41,683	3.1
Mistakes or non-Tatar words	267	26.7	296,097	22.0
Overall	1000	100	1,342,108	100

cases are explained by the use of the Cyrillic alphabet. For example, the majority of exceptional cases appear due to the presence of complex letters, such as е, ю, я (e.g. баюын (their getting rich), ел (year), юл (way)). The possible solution in this case is to fix some rules and, subsequently, fix the dictionary according to them. Other words (26.70% of analyzed words) are mistakes or non-Tatar words. In these cases, there is no need for further improvements.

5   Conclusions

In this paper we present a Morphological analysis system of Tatar Language. The Tatar National Corpus was used to evaluate the completeness of the system.

The best indicator of the completeness of the system was obtained when the rules for exceptional situations with morphology violations were added; it raised from 85% to 95%.

Among the remaining 5% of unrecognized wordforms, 1000 most frequent words were selected and analyzed. 41.6% of these words were not recognized because of their absence in the system's dictionary. Basically, these are words from specialized dictionaries of names and last names, geographical names, thematic dictionaries, etc. 24.6% are proper names, 21.5% are borrowings from Russian and Arabic that do not obey the law of synharmonism. The presence of the remaining 12.3% is explained by the use of the Cyrillic alphabet, namely е, ю, я symbols, which do not observe the law of synharmonism. This problem can be solved by adding exceptional words to the dictionary, and partly by correcting the rules.

In the future we plan to improve the completeness of the system taking into account the tests results.

At present, the Morphological analysis system is used not only for morphological annotation of the Tatar language corpus, but also in the Yandex system of Russian-Tatar machine translation [21]. The service is also available at: http://tatmorphan.pythonanywhere.com/. The implemented system may be used also for the grammatical annotation of domain specific corpora of the Tatar language, for example, the new Tatar Socio-Political Corpus [12].

Acknowledgements. The reported study was funded by Russian Science Foundation, research Project № 16-18-02074.

References

1. Oflazer, K.: Two-level description of Turkish morphology. Lit. Linguist. Comput. **9**(2), 137–148 (1994)
2. Altintas K., Cicekli I.: A morphological analyzer for Crimean Tatar. In: Proceedings of the 10th Turkish Symposium on Artificial Intelligence and Neural Networks (TAINN'2001), pp. 180–189 (2001)
3. Çöltekin, Ç.: A set of open source tools for Turkish natural language processing. In: Calzolari, N. et al. (eds.) Proceedings of the Ninth International Conference on Language Resources and Evaluation (LREC 2014), pp. 1079–1086 (2014)
4. Kessikbayeva, G., Cicekli, I.: Rule based morphological analyzer of Kazakh language. In: Proceedings of the 2014 Joint Meeting of SIGMORPHON and SIGFSM, pp. 46–54. ACL, Baltimore, June 2014
5. Tantug, C., Adali, E., Oflazer, K.: Computer analysis of the Turkmen language morphology. In: 5th International Conference on NLP (FinTAL 2006), Turku, pp. 186–193 (2006)
6. Orhun, M., Tantug, C., Adali, E.: Rule based analysis of the Uyghur Nouns. Int. J. Asian Lang. Proc. **19**(1), 33–44 (2009)
7. Suleymanov, D.S., Gilmullin, R.A.: Dvukhurovnevoye opisaniye morfologii tatarskogo yazyka [Two-level description of the Tatar language morphology]. In: Proceedings of "Language Semantics and Image of the World" International Scientific Conference, vol. 2, pp. 65–67. Kazan State University, Kazan (1997, in Russian)
8. Gökgöz, E., et al.: Two-level Qazan Tatar morphology. In: Proceedings of the 1-st International Conference on Foreign Language Teaching and Applied Linguistics, Sarajevo, 5–7 May 2011, pp. 428–432 (2011)
9. Davliyeva, A.R.: An investigation of Kazan Tatar morphology. Doctoral Dissertation, San Diego State University (2011)
10. HFST. https://kitwiki.csc.fi/twiki/bin/view/KitWiki/HfstHome
11. Tatar Nacional Corpus. http://tugantel.tatar/?lang=en
12. Socio-Political Corpus of the Tatar language. http://tugantel.tatar/corpus/op/
13. Lewis, M.P., Simons, G.F., Fennig, C.D. (eds.): Ethnologue: Languages of the World, 19th edn. SIL International, Dallas. Online version: http://www.ethnologue.com (2016)
14. Berment, V.: Méthodes pour informatiser des langues et des groups de langues peu dotées. Ph.D. Thesis, Joseph Fourier University, Grenoble I (2004)
15. Krauwer, S.: The basic language resource kit (BLARK) as the first milestone for the language resources roadmap. In: Proceedings of International Workshop Speech and Computer SPEECOM, Moscow, pp. 8–15 (2003)
16. Yandex Translate: Online version: https://translate.yandex.com/translator/Russian-Tatar
17. Suleymnov, D., Gatiatullin, A., Gilmullin, R.: Lexicograficheskaya baza dannykh dlya system mashinnogo perevoda blizkorodstvennykh yazykov. In: Proceedings of Third International Conference «Informatizatciya obschestva», pp. 585–587. Astana, Kazakhstan (2012)
18. Khusainov, A.F., Suleymanov, D.S.: Language identification system for the tatar language. In: Železný M., Habernal I., Ronzhin A. (eds.) SPECOM 2013. LNCS, vol. 8113, pp. 203–210. Springer, Cham (2013). doi:10.1007/978-3-319-01931-4_27
19. Beesley, R.K., Karttunen, L.: Finite State Morphology. CSLI Publications, Stanford (2003)

20. Gilmullin, R.: Matematicheskoye modelirovaniye v mnogoyazykovykh sistemakh obrabotki dannykh na osnove avtomatov konechnykh sostoyaniy, pp. 48–94. Ph.D. Thesis, Kazan (2009)
21. Sokolov, A., Egorov, A., Gubanov, S., Khrystich, D., Shmatova, M., Galinskaya, I., Baytin, A.: Eksperimental'naya versiya tatarsko-russkogo statisticheskogo mashinnogo perevoda. In: Proceedings of the International Conference "Turkic Language Processing: Turklang-2015", pp. 67–76. Academy of Science of the Republic of Tatarstan Press, Kazan (2015)

Context-Based Rules for Grammatical Disambiguation in the Tatar Language

Ramil Gataullin[1], Bulat Khakimov[1,2]([✉]) [iD],
Dzhavdet Suleymanov[1,2] [iD], and Rinat Gilmullin[1]

[1] Institute of Applied Semiotics, TAS, Kazan, Russia
bulat.khakeem@gmail.com
[2] Kazan (Volga Region) Federal University, Kazan, Russia

Abstract. The paper is dedicated to the problem of grammatical ambiguity in the Tatar National Corpus and describes the methodology and software used for automation of the disambiguation process. Grammatical ambiguity is widely represented in agglutinative languages like Turkic or Finno-Ugric. Disambiguation in the corpus is based on the context-oriented classification of ambiguity types which has been carried out on corpus data in the Tatar language for the first time. In this study the corpus is used as a source for the research and at the same time as a destination for implementing the results. The grammatical ambiguity types are detected automatically using the finite-state morphological analyzer and then classified. In order to build up the grammatically disambiguated subcorpus, a special software module was developed. It searches for ambiguous tokens in the corpus, collects statistical information and allows creating and implementing the formal context-based disambiguation rules for different ambiguity types.

Keywords: Disambiguation · Grammatical homonymy · Context-based rules · Linguistic software · Turkic languages · Corpus linguistics

1 Introduction

The problem of grammatical disambiguation is very relevant for the modern computer and corpus linguistics, especially in relation to such morphologically rich languages as the Turkic group that forms a subfamily of Altaic languages. For example, the Tatar language is spoken in the far part of Eastern Europe (in the Volga region of Russia) and southern parts of Western Siberia. The number of speakers in Russia was 5.31 mln in 2010.

The Tatar National Corpus named "Tugan Tel" [1] was developed by the "Applied semiotics" Research Institute of the Tatarstan Academy of Sciences and the Kazan Federal University in Russia [2]. It's general concept is presented in [3]. For the automated morphological annotation a finite-state morphological analyzer is used [4]. The grammatical tagset for corpus annotation of Tatar word forms is continuously improved and aimed to be more relevant to the Tatar language peculiarities [5]. On the basis of statistical corpus data, the contextual constraints for certain grammatical

© Springer International Publishing AG 2017
N.T. Nguyen et al. (Eds.): ICCCI 2017, Part II, LNAI 10449, pp. 529–537, 2017.
DOI: 10.1007/978-3-319-67077-5_51

homonyms in the Tatar language were investigated for the purposes of the grammatical disambiguation.

The grammatical annotation model of the Tatar National Corpus is oriented at presenting all of the grammatical forms and meanings to the maximum extent. It was carried out using a special tool which was originally created on the basis of the PC-KIMMO two-level morphology model [4]. Currently, the system is redeveloped using Helsinki Finite-State Transducer Technology (HFST) [6], which improved technical parameters of the system (see Table 1).

Table 1. Technical parameters of the Tatar morphology modules

Technical parameters	PC-KIMMO	HFST
Number of morpho-tactical rules	49	49
Number of phonological rules	42	42
Vocabulary amount (words)	26500	28000
Processing speed (words per second)	~500–1000	~5000–20000
Coverage (recall)	80–85%	93–95%
Alphabet	Latin	Cyrillic
Programming language	C/C++ and Delphi	C/C++ and Python
API	No	Yes
Interface	Desktop application	Web application

2 Related Works

Grammatical homonymy in the Tatar language has been studied by native linguists for many years. Starting from works of Kurbatov [7] and Salimgarayeva [8], those studies were based mostly on the descriptive and qualitative approaches. One of the last works in this field is Salakhova's monograph on suffix homonymy [9]. As for automated disambiguation, there were attempts to use both rule-based and statistical approach [10]. However, working with relatively low-resourced language forces researchers to implement rule-based disambiguation first. Among the related works we can mention Brill's unsupervised learning algorithm for automatically training a rule-based part-of-speech tagger [11]. Yuret and Ture [12] presented a rule-based model for morphological disambiguation of Turkish using supervised training. Nevzorova et al. [13] investigated the issues of context-based part-of-speech disambiguation for Russian and presented the method of calculation of disambiguating contexts. Until today, there are no completed studies and accuracy evaluation results on automated disambiguation for Tatar. Our approach is to determine the most frequent types of part-of-speech and affix ambiguity and develop context-based rules for them. It helps to cover more linguistic data with the same effort and furthermore to start using statistical models on a partially disambiguated corpus.

3 Detection of Grammatical Ambiguity Types and Relevancy Evaluation

From the limited subcorpus of the Tatar National Corpus, the frequency of ambiguous tokens was obtained (see Table 2). For this purpose, the subcorpus was tagged using the HFST-based morphological analyzer tool which was mentioned above. We considered the tokens with more than one parsing option as ambiguous and exported a list of such examples. The total volume of the subcorpus was 21,940,452 tokens, and the share of ambiguous tokens amounted to 25.75%.

Table 2. Statistics of ambiguous tokens in the sample subcorpus

Parsing options	Number in subcorpus	Share in subcorpus (%)
2	4282108	19.51
3	1045392	4.76
4	296547	1.35
5+	26773	0.12
Ambiguous tokens in total	5650820	25.75

Based on this statistics, from the list of 500 most frequent types of automatically detected ambiguous word forms 150 types with two parsing options was extracted to assess their relevancy and to remove the redundant types of ambiguity.

The analysis showed that there are irrelevant examples among automatically detected grammatical ambiguity types. We found that sometimes additional parsing options appear due to the errors of the morphological analyzer. There are two main reasons which cause such errors: the redundancy in the vocabulary of lexemes and incorrect morphotactical and phonological rules. Those redundancies and incorrect rules are iteratively removed from the model of grammatical corpus annotation as they are identified during the analysis.

There is another case of "irrelevant ambiguity" which we conventionally interpret in that way despite it is not caused by the errors of the analyzer tool. Sometimes one of the parsing options corresponds to the Tatar morphology model correctly, but almost not used in the language. However, they unreasonably increase the ambiguity index of the text. We think that parsing options which are used in less than 1% cases, may be considered irrelevant and removed automatically.

As our previous studies show, those irrelevant types of grammatical ambiguity share about 8.5% of alternative parsing options in the sample subcorpus, and cover 2.1% of the corpus texts [10].

4 Frequent Types of Grammatical Ambiguity in the Tatar Language

From the list of the ambiguity types which were found linguistically relevant, we collected the most frequent types with only two parsing options. Then the following classification of automatically detected subtypes was proposed (Table 3).

Table 3. Frequent ambiguity types with two parsing options

ID	First homonym	Second homonym	Regular
1	Noun	Pronoun	–
2	Verb	Noun/adjective	+
3	Pronoun	Numeral	
4	Noun	Adjective	+
5	Postposition	Noun/Numeral	–
6	Noun	Adverb	–
7	Adjective	Noun + attributive affix	+
8	Noun/adjective	Noun + possessive affix	+
9	Adjective	Noun + directive case	–
10	Adjective	Verb	+
11	Verb	Verb	+
12	Adjective	Adverb	+
13	Pronoun	Pronoun + locative case	+
14	Noun	Adjective + affix chA	+
15	Pronoun	Noun	–

The types marked as "Regular" in the Table 3 are related to regularly derived grammatical forms. Context-based disambiguation rules for those types use morpho-syntactical characteristics of the grammatical forms which define the disambiguating context. Exceptions are types 1, 3, 5, 6, 9 and 15.

Type #1 includes the word *ul* ('he/son'). Each alternative has a certain disambiguating context with specific morphological and syntactical constraints.

Type #3 also includes only one word *ber* ('one'). It has two POS functions: the numeral 'one' and the indefinite pronoun with the meaning close to the indefinite article. These two variants also have their own context templates.

There are four subtypes in Type #5. They are respectively represented by one word form. It may be a postposition: *öçen* ('for'), *turında* ('about') and *buyınça* ('on'), or a pronoun *tege* ('that'). Alternative variants for these lexemes are nouns in certain forms. The context rules for this type are determined by the grammatical characteristics of ambiguous words and the syntactic functions of the corresponding postpositions.

Types #6 and #9 include the words *bik* ('very/bolt') and *başka* ('other/head + DIR'), respectively.

Type #15 is also an example of such part-of speech ambiguity; it refers to the word *bez* ('we/awl').

Also there is another specific case of verb ambiguity. It is caused by the multifunctionality of voice affixes in Tatar verbs. According to the statistical study of corpus data, the total number of such cases is 408346 tokens (which is 1.8% of the total volume of texts and 7.2% of all the alternative parses). The V/V + REFL subtype is the most frequent subtype among them. This subtype is characterized by the fact that the same verb form may be used in two meanings. The first meaning refers to a whole lexeme, and it is included in the vocabulary of the morphological analyzer separately. The second meaning is the reflexive voice form of another verb. For example, *ezlänergä* ('to research/to be searched'), *alınırğa* ('to take on/to be taken'). Disambiguation of this type is quite a difficult task and in many cases, in addition to grammatical features, requires consideration of the semantic characteristics of the disambiguating context.

5 Methodology of Disambiguation

The agglutinative characteristics of the Tatar grammar determine that the word forms are derived by adding derivational and inflectional affixes to the stem. Each grammatical meaning is usually expressed by an individual affix.

According to statistical data from the Tatar National Corpus, functional (part-of-speech) and affixal disambiguation is the most important and first-to-do task in grammatical disambiguation for the Tatar language.

The academic book of Tatar grammar [14, 15] represents the main formal grammatical models of Tatar word combinations (15 basic and 80 particular types) indicating the dependencies, as well as grammatical, and in some cases, lexical semantics. These models can serve as a basis for determining the disambiguation contexts. A certain strictness of the agglutinative syntactic structure allows expecting the detection of clear context restrictions.

The rule-based approach for POS and affix disambiguation in Tatar, as our previous studies have shown [10, 16], is an extremely time-consuming, requires a thorough linguistic investigation of each ambiguity type. We apply a method of contextual grammatical disambiguation which involves several stages:

1. Classification of the functional (POS) and affix (homoforms) ambiguity types;
2. Investigation of a minimum disambiguation context for each type. Then it is needed to build a set of disambiguating contexts with minimum complexity of recognition. In algorithmic writing, this requirement is expressed by the following rule: if to a functional homonym X having type T1 or T2, a rule from the set of disambiguating contexts is applicable, the homonym type X is determined by the applied rule, or an alternative type is attributed;
3. Building a management structure of a generalized rule ensuring maximum recognition accuracy.

For each type of POS or affix ambiguity, a generalized rule for disambiguation of this type is developed. The generalized rule is an ordered set of rules, written in a special formal language. Each rule within a set works with the certain disambiguation context.

Let's consider the functional (part-of-speech) disambiguation of the following type, which is represented with one ambiguous word *soň*:

Adv|N+Sg+Nom|PART|POST, where
Adv means Adverb ('late')
N+Sg+Nom means Noun in singular and nominative case ('end')
PART means Particle (emphatic, non-translatable)
POST means Postposition ('after')

This type of ambiguity is resolved with the following rule, which includes multiple sub-rules. The order of sub-rules reflects their hierarchy. See Fig. 1.

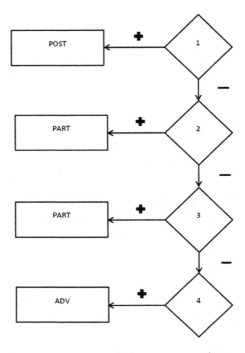

Fig. 1. Management structure of the context-based rule Adv|N+Sg+Nom|PART|POST

Sub-rule 1. If there is "ABL" (ablative case) morpheme in the interval of −1 from current ambiguous word form, and there are no interrogative morphemes in the same interval, the result should be POST.

Sub-rule 2. If there is an interrogative morpheme in the interval of −1 from current ambiguous word form, the result should be PART.

Sub-rule 3. If there are no particles *ğına, uq, bit, tugel,* or any verbs in the interval of +1 from current ambiguous word form, the result should be PART.

Sub-rule 4. In any other context, the result should be ADV.

Note: the **N + Sg + Nom** option is not represented in the language use, so this ambiguity type never resolved with that option.

The accuracy evaluation for this rule showed not less than 95%. It was calculated on 10,000 sentences randomly selected from a 21-million word tagged corpus.

6   Software Tool for Building Context Rules of Grammatical Disambiguation

Though construction of context-based disambiguation rules is mostly a manual process, a special supporting software tool was developed. It helps to create, edit and test the context-based rules based on the Tatar texts [16] (Fig. 2).

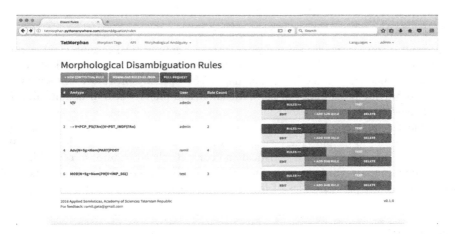

Fig. 2. Main page of contextual rules development tool's GUI

This software tool can be used not only separately, but also in combination with the probabilistic and statistical methods. The second part of the toolkit is described in [16]. The database of context-based rules is used for grammatical disambiguation in Tatar texts. It was the first experience in Tatar linguistics when a special toolkit taking the particularities of the Tatar language into account was developed. Obviously, such a tool also increases the effectiveness of the research work of a linguist (Fig. 3).

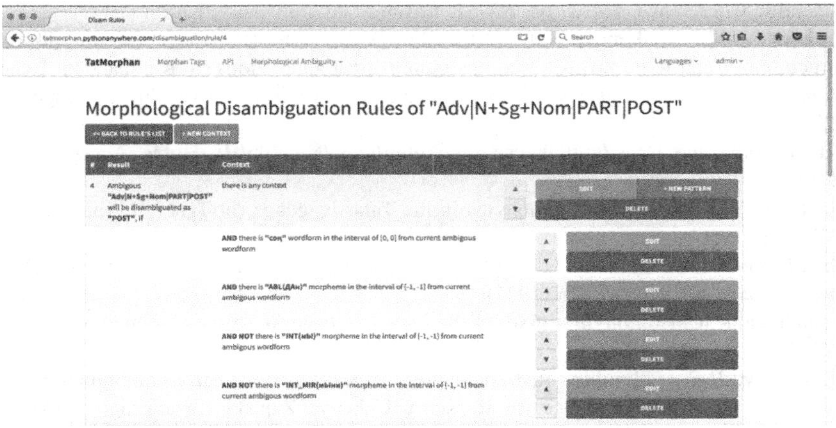

Fig. 3. Contextual rule page

We also developed a web application that makes working with the texts being prepared for the Tatar National Corpus more convenient and provides possibilities for statistical research [16]. This application facilitates the process of annotating of the Tatar National Corpus (including manual elimination of ambiguity).

In this software module, the corpus extension and morphological annotation possibilities are implemented, as well as support for manual grammatical disambiguation.

7 Conclusion

As a result of our research, a set of context-based grammatical disambiguation rules based on the formal context-oriented homonyms classification and statistical corpus data was formulated for the Tatar language. The special software tools which we developed provide possibilities of partially automated grammatical disambiguation for the linguistic corpus resources.

In the future research, we plan to continue the investigation of disambiguating contexts in order to develop the context-based grammatical disambiguation rules. We suppose that the analysis of the statistical and context-related features of other kinds of linguistic ambiguity will help us to find more effective approaches to disambiguation with regard to the particularities of the Tatar language.

References

1. «Tugan Tel» Tatar National Corpus Homepage. http://tugantel.tatar/?lang=en. 05 June 2017
2. Suleymanov, D.S., Nevzorova, O.A., Gatiatullin, A.R., Gilmullin, R.A., Khakimov, B.E.: National corpus of the Tatar language "Tugan Tel": grammatical annotation and implementation. Procedia Soc. Behav. Sci. **95**, 68–74 (2013)
3. Suleymanov, D.S., Khakimov, B.E., Gilmullin, R.A.: Corpus of Tatar: conception and linguistic aspects (in Russian). Philol. Cult. **4**(26), 211–216 (2011)
4. Suleymanov, D.S., Gilmullin, R.A.: Two-level description of the Tatar morphology (in Russian). In: Proceedings of "Language Semantics and Image of the World" International Scientific Conference, vol. 2, pp. 65–67. Kazan State University, Kazan (1997)
5. Galieva, A.M., Khakimov, B.E., Gatiatullin, A.R.: A Metalanguage for describing the structure of Tatar word forms for corpus grammatical annotations (in Russian). In: Uchenye Zapiski Kazanskogo Universiteta, vol. 155(5), pp. 287–296. Seriya Gumanitarnye Nauki (2013)
6. HFST Homepage. https://kitwiki.csc.fi/twiki/bin/view/KitWiki/HfstHome. Accessed 20 Apr 2017
7. Kurbatov, K.: Grammatical homonyms in the Tatar language (in Tatar). J. Tatar Lang. Lit. 307–311 (1959)
8. Salimgarayeva, B.: Homonyms in modern Tatar language: abstract of dissertation (in Tatar). Bashkir State University, Ufa (1971)
9. Salakhova, R.R.: Homonym suffixes of the Tatar language (in Russian). Gumanitarya, Kazan (2007)
10. Khakimov, B.E., Gilmullin, R.A., Gataullin, R.R.: Grammatical disambiguation in the corpus of the Tatar Language (in Russian). Uchenye Zapiski Kazanskogo Universiteta. Seriya Gumanitarnye Nauki **156**(5), 236–244 (2014)

11. Brill, E.: Unsupervised learning of disambiguation rules for part of speech tagging. In: Proceedings of the Third Workshop on Very Large Corpora, vol. 30, pp. 1–13. Association for Computational Linguistics, Somerset (1995)
12. Yuret, D., Ture, F.: Learning morphological disambiguation rules for Turkish. In: Proceedings of the Human Language Technology Conference of the North American Chapter of the ACL, pp. 328–334. ACL, New York (2006)
13. Nevzorova, O.A., Zinkina, Y., Pyatkin, N.: Resolution of functional homonymy in the Russian language based on context rules (in Russian). In: Proceedings of "Dialog'2005" International Conference, pp. 198–202. Nauka, Moscow (2005)
14. Tatar Grammar: Morphology (in Russian), vol. 2. Tatar Publishing Company, Kazan (1993)
15. Tatar Grammar: Morphology (in Tatar), vol. 2. Insan, Moscow. Fiker, Kazan (2002)
16. Gataullin, R.R., Gilmullin, R.A.: Web interface for removing morphological ambiguity in the corpus of the Tatar language (in Russian). In: Open Semantic Technologies for Intelligent Systems OSTIS-2015 Proceedings of IV International Scientific and Technical Conference, pp. 451–454. BSUIR, Minsk (2015)

Computer Vision Techniques

Evaluation of Gama Analysis Results Significance Within Verification of Radiation IMRT Plans in Radiotherapy

Jan Kubicek[1](✉), Iveta Bryjova[1], Kamila Faltynova[1],
Marek Penhaker[1], Martin Augustynek[1], and Petra Maresova[2]

[1] FEECS, VSB–Technical University of Ostrava,
K450, 17. listopadu 15, 708 33 Poruba, Ostrava, Czech Republic
{jan.kubicek,iveta.bryjova,kamila.faltynova.st,
marek.penhaker,martin.augustynek}@vsb.cz
[2] Faculty of Informatics and Management, University of Hradec Kralove,
Rokitanskeho 62, 5000 Hradec Králové, Czech Republic
petra.maresova@uhk.cz

Abstract. The Gamma analysis is a method, which currently represents a standard for results verification of patient IMRT (Intensity modulated Radiation Therapy) plans. It is just a numerical concept, which within standard usage contains neither relevant clinical information nor geometrical parameters of a given area. The main aim of the paper is an application development, which would be able to assess and predict of the Gamma analysis significance from the portal dosimetry on the base of the information from the planning system about irradiated volume and critical organs from the BEV (Beam's Eye View) projection. Consequently, a frequency of unsatisfactory results in matrixes of Gamma indexes will be tracked (Gamma index is greater than 1). The last part of the analysis deals with an investigation of a geometrical coherence of these points coordinates with particular pairs of laminations of multilevel collimator and their position against BEV projection.

Keywords: Radiotherapy · IMRT · Dosimetry · Gamma analysis

1 Introduction

Against the conformal radiotherapy, where laminations define the irradiation field, IMRT laminations are in move. By this movement, they modulate radiation beam intensity. Therefore, it is possible to decrease fluency in this part of the field where a critical organ is. On the base of this fact, the critical organs are much more saved in comparison with the conformal radiotherapy. Two transversal slices (Fig. 1) for prostate irradiation are shown. It seems that in the IMRT technique lines imagining dose distribution as much as possible surrounds of PTV (planning target volumes), and less oppress of the bladder and rectum then it is on CRT (clinical target volume). If we want to compare of dose load of both techniques we should take into account, and not just one slice. This fact is shown by DVH (dose volume histogram). It seems that

© Springer International Publishing AG 2017
N.T. Nguyen et al. (Eds.): ICCCI 2017, Part II, LNAI 10449, pp. 541–548, 2017.
DOI: 10.1007/978-3-319-67077-5_52

doses, which the critical organs obtain during IMRT (indicated by triangles), are significantly lower [1–4].

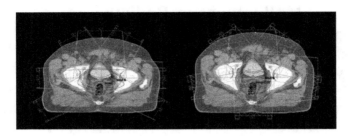

Fig. 1. Transverse CRT slices (left) and IMRT planes (right)

The IMRT technique can be realized in two modes, segmented or dynamical. In the segmented mode (step and shoot), the radiation field is divided into several subfields from a given gantry angle. These subfields have different lamination position (MLC). The radiation is released in the phase when laminations (LMC) stop moving themselves. In the dynamical mode (sliding window), radiation also goes along lamination direction [5–7].

The IMRT radiation plan is determined on the base of the calculation of beam intensity modulation (beam dose profile) from given gantry angles. This method is called the inverse planning [8–11].

From a view of the clinical practice it would be worth developing application intended for a prediction of the Gamma analysis significance from the portal dosimetry dosimetry on the base of the information from the planning system about irradiated volume and critical organs from the BEV (Beam's Eye View) projection. A second consequent issue deals with a determination and tracking of unsatisfactory results in matrixes of the Gamma indexes.

2 Application for Evaluation of Gamma Analysis

The main applicable result of our research is an application intended for evaluation of the Gamma analysis results performed during prostate IMRT plans verification. For this purpose, 30 reliable patients have been selected who treated in 2010-2015 with the prostate cancer on the radiotherapeutic and oncologic clinic of University hospital of Kralovske Vinohrady. All those patients have been treated by 3D CRT in the fractionator regime: 25 fractions with 1.8 Gy per one fraction, and consequently the boost technique has been applied from five radiation in the fractionator regime: 15 fractions with 2 Gy per one fraction.

Figure 2 is generated from the Eclipse planning system 8.6. From this system, BEV projections have been exported from each irradiation angle (35°, 85°, 180°, 275° and 325°). These projections contain point's coordinates, and their intersections indicate contours of the critical organs (bladder and rectum) and the target planning volume (PTV). For each respective angle matrix is generated carrying information about the Gamma analysis.

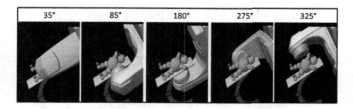

Fig. 2. Patient position and gantry of linear accelerator during prostate irradiation by IMRT technique

For all irradiation angles, the aggregate matrixes have been created, where information about critical organs and PTV position is saved of all patients involved to the study. These matrixes express occurrence frequency of OaR (organ-at-risk) and PTV in a given point. Points which do not meet Gamma analysis criterions, and which lay out of target structures were projected on the aggregate matrixes (Fig. 3).

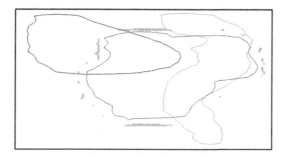

Fig. 3. Exported structures and points not meeting the Gamma analysis criterions in SW MATLAB

Since main target of the analysis is evaluation of the Gamma analysis, we should take into account critical the organ position and PTV. This task is done by creation of two weight matrixes, which were multiplied by matrix containing original results of the Gamma analysis. Point not meeting Gamma analysis criterions is given bigger weight then point lying out of the target structures.

2.1   Processing of Exported Data

For each patient, five graphs are generated reflecting projection of the target volume, bladder and rectum from a given angle of rotation arm of accelerator to perpendicular plane to beam axis going through the center. We can observe (Fig. 4) great percentage of unsatisfactory points lay on the PTV boundary or the critical organs. It is caused by the fact that in those spots there is the higher dose gradient. It would be worth making a rim 5 mm around projection of the critical points, and if point with the Gamma>1 laid in this rim, it would be distinguished from other points laying out of the mentioned volumes, and the point would be assessed as potentially dangerous.

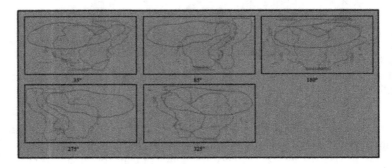

Fig. 4. Critical organs (bladder is blue and rectum is violet), PTV (red) and points which do not meet of Gamma analysis (green) from all radiation angles (Color figure online)

Table 1 shows amount of the points which do not meet the Gamma analysis for all patients from five radiation angles lay in OaR or PTV, and how many points lay out of the RoI. It seems that 50% of points lay out of the area of the interest, 30% lay inside the PTV. It is supposed that it is caused by higher dose gradient, which is on the boundary of the target planning volume. Rest of 20% points lay in the critical organs (9% in the bladder and 11% in rectum).

Table 1. Localization of points not meeting the Gamma analysis criterions

	Number of not meeting points				
	Bladder	Rectum	PTV	OUT	Overall
Number	3948	4910	13035	21955	43848
%	9	11	30	50	100

2.2   Aggregate Matrix of Occurrence OaR and PTV

In order to assess of an area where the critical organ or PTV is located with certain accuracy, the aggregate matrix is generated. This matrix is given by a sum of the BEV projections for a given OaR and PTV in a given radiation angle for all the patients. Next projections of the tracked volumes is consecutively added to this matrix. By this fact, prediction accuracy of the area is systematically increased in which tracked volume projection (target volume or critical organ) from given angle occur (Fig. 5).

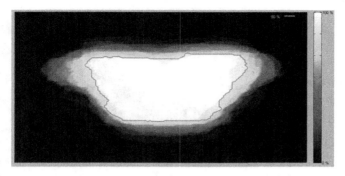

Fig. 5. Aggregate matrix PTV for the arm of the accelerator angle 180°. Area in which on 90% patients involved to the study occurred target volume is indicated by green line (Color figure online)

3   Data Evaluation

On the base of an assessment of the target volume projection positions and critical organs for individual patients, but also in the aggregate matrix, we can state that during the prostate radiotherapy the assumption dealing with a certain anatomical uniformity among individual patients with the prostate cancer has been proved. We show results for radiation angle 35° and 85°. Those results were consequently proved by radiotherapy clinicians. The following results reflect aggregation matrix of the target areas (bladder) for different radiation angles (35° and 85°).

3.1   Results for Radiation Angle 35°

The aggregate matrixes are represented by color scale of individual critical organs occurrence (bladder and rectum) and PTV. Area in which of 20% patients occurred given structure is indicated by blue color. Also, there are indicated points not meeting the Gamma analysis (green color indicates points, which belong to the area in which of 20% occurred a given structure, points occur out of this area are indicated by light blue color) (Figs. 6, 7 and 8).

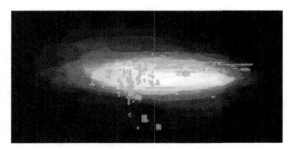

Fig. 6. Aggregate matrix of bladder occurrence and points not meeting the Gamma analysis criterions for given patient they occurred in the bladder for radiation angle 35°. The blue line indicates the area in which 20% cases occurred the bladder (Color figure online)

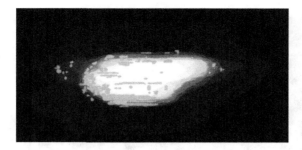

Fig. 7. Aggregate matrix of PTV occurrence and points not meeting the Gamma analysis criterions for given patient they occurred in the PTV for radiation angle 35°. The blue line indicates the area in which 20% cases occurred the PTV (Color figure online)

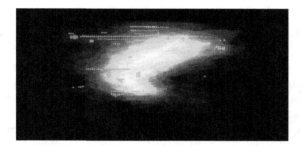

Fig. 8. Aggregate matrix of rectum occurrence and points not meeting the Gamma analysis criterions for given patient they occurred in the rectum for radiation angle 35°. The blue line indicates the area in which 20% cases occurred the rectum (Color figure online)

3.2   Results for Radiation Angle 85°

On the base of the results, it is apparent that choice of the boundaries of the area for projection of the bladder from angle 85° by the contour tracing the area with more that 20% bladder occurrence frequency is sufficient (Figs. 9, 10 and 11).

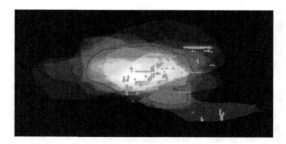

Fig. 9. Aggregate matrix of bladder occurrence and points not meeting the Gamma analysis criterions for given patient they occurred in the bladder for radiation angle 85°. The blue line indicates the area in which 20% cases occurred the bladder (Color figure online)

Fig. 10. Aggregate matrix of PTV occurrence and points not meeting the Gamma analysis criterions for given patient they occurred in the PTV for radiation angle 85°. The blue line indicates the area in which 20% cases occurred the PTV (Color figure online)

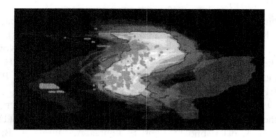

Fig. 11. Aggregate matrix of rectum occurrence, and points not meeting the Gamma analysis criterions for given patient they occurred in the rectum for radiation angle 85°. The blue line indicates the area in which 20% cases occurred the rectum (Color figure online)

4   Conclusion

The main result of the analysis is an application completing the Gamma analysis result about refining records in regards to assumed projection position of critical organs and target volume in the prostate cancer. On the base of the evaluation and comparison of all output records, it goes to a notification on a possible problem of the critical organ occurrence, and also by the assumption when the general Gamma analysis meets of selected tolerances. Contrarily, if there is an unsatisfactory Gamma analysis for greater percentage of pixels, the application help analyzing extend to witch those pixels can interfere to areas with assumed occurrence of the critical organs. The whole procedure has been proved on the clinical data, and it can be assumed that with successive completing BEV projections for further patients, the prediction accuracy will be increased, and by this way, sensitivity and specificity will increase as well.

Acknowledgment. The work and the contributions were supported by the project SV4506631/2101 'Biomedicínské inženýrské systémy XII'. This study was supported by the research project The Czech Science Foundation (GACR) No. 17-03037S, Investment evaluation of medical device development.

References

1. Cooperberg, M.R.: Re: 10-year outcomes after monitoring, surgery, or radiotherapy for localized prostate cancer. Eur. Urol. **71**(3), 492–493 (2017)
2. Miksys, N., Vigneault, E., Martin, A.-G., Beaulieu, L., Thomson, R.M.: Large-scale retrospective Monte Carlo dosimetric study for permanent implant prostate brachytherapy. Int. J. Radiat. Oncol. Biol. Phys. **97**(3), 606–615 (2017)
3. Maspero, M., Seevinck, P.R., Schubert, G., Hoesl, M.A.U., Van Asselen, B., Viergever, M.A., Lagendijk, J.J.W., Meijer, G.J., Van Den Berg, C.A.T.: Quantification of confounding factors in MRI-based dose calculations as applied to prostate IMRT. Phys. Med. Biol. **62**(3), 948–965 (2017)
4. Kim, K.H., Lee, S., Shim, J.B., Chang, K.H., Cao, Y., Choi, S.W., Jeon, S.H., Yang, D.S., Yoon, W.S., Park, Y.J., Kim, C.Y.: Predictive modelling analysis for development of a radiotherapy decision support system in prostate cancer: a preliminary study. J. Radiother. Pract. **16**, 161–170 (2017)
5. McIntosh, C., Purdie, T.G.: Voxel-based dose prediction with multi-patient atlas selection for automated radiotherapy treatment planning. Phys. Med. Biol. **62**(2), 415–431 (2017)
6. Haehnle, J., Süss, P., Landry, G., Teichert, K., Hille, L., Hofmaier, J., Nowak, D., Kamp, F., Reiner, M., Thieke, C., Ganswindt, U., Belka, C., Parodi, K., Küfer, K.-H., Kurz, C.: A novel method for interactive multi-objective dose-guided patient positioning. Phys. Med. Biol. **62**(1), 165–185 (2017)
7. Yildirim, B.A., Onal, C., Dolek, Y.: Is it essential to use fiducial markers during cone-beam CT-based radiotherapy for prostate cancer patients? Jpn. J. Radiol. **35**(1), 3–9 (2017)
8. Ingrosso, G., Carosi, A., di Cristino, D., Ponti, E., Lancia, A., Murgia, A., Bruni, C., Morelli, P., Pietrasanta, F., Santoni, R.: Volumetric image-guided highly conformal radiotherapy of the prostate bed: toxicity analysis. Rep. Pract. Oncol. Radiother. **22**(1), 64–70 (2017)
9. Kubicek, J., Penhaker, M.: Guidelines for modelling BED in simultaneous radiotherapy. IFMBE Proc. **43**, 271–274 (2014)
10. Kubicek, J., Penhaker, M., Feltl, D., Cvek, J.: Guidelines for modelling BED in simultaneous radiotherapy of two volumes: tpv1 and tpv2. In: Proceedings of SAMI 2013 - IEEE 11th International Symposium on Applied Machine Intelligence and Informatics, pp. 131–135 (2013). Article no. 6480960
11. Penhaker, M., Novakova, M., Knybel, J., Kubicek, J., Grepl, J., Kasik, V., Zapletal, T.: Breathing movement analysis for adjustment of radiotherapy planning. Stud. Comput. Intell. **710**, 105–116 (2017)

Shape Classification Using Combined Features

Laksono Kurnianggoro[✉], Wahyono, Alexander Filonenko,
and Kang-Hyun Jo

The Graduate School of Electrical Engineering,
University of Ulsan, Ulsan 44610, Korea
{laksono,wahyono,alexander}@islab.ulsan.ac.kr, acejo@ulsan.ac.kr

Abstract. Shape classification is an active research field due to its usefulness. In this work, hand crafted shape descriptors are combined with features extracted using convolutional neural network to do the classification task. Extensive experiments were performed on public data sets to reveal the performance of the proposed method compared to the other state of the arts shape classification methods.

Keywords: Shape classification · Machine learning · Neural network

1 Introduction

Shape classification is an active research field and has many applications. Various methods were proposed in the past few years including bag of contour fragments [26], common base triangle area [9], and centroid-based tree structure [24].

In this work, the centroid-based tree structure features are combined together with feature extracted by a deep neural network to perform the classification task of various shape.

The paper is organized as follows. The proposed pipeline is presented in the next section, followed by experiments as well as discussion of the results. Finally, the summary and conclusion of the paper is presented in the last section.

2 Related Works

There are various shape analysis methods available nowadays [19,23,27,29]. In the past, most of the methods are focusing on feature engineering such as shape context [5] or shape signatures [4]. Recently, machine learning are utilized in shape analysis methods such as support vector machine in [26] and deep learning in [1,2,18].

In the earlier years of shape analysis development, region-based feature extraction are popular. There are various methods utilizing region information such moments [8,10,12,22] rectangularity ratio, hole area ratio, convexity and solidity. The rectangularity ratio is obtained by comparing the size of a shape to its bounding box region. Meanwhile for hole area ratio, it is obtained by dividing

© Springer International Publishing AG 2017
N.T. Nguyen et al. (Eds.): ICCCI 2017, Part II, LNAI 10449, pp. 549–557, 2017.
DOI: 10.1007/978-3-319-67077-5_53

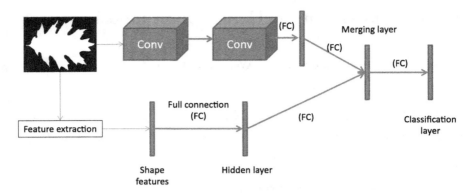

Fig. 1. The proposed hybrid deep network which allows multiple types of inputs. Bold connection represents the internal connections within the network which are trainable. (Color figure online)

the size of hole in a shape to the size of the corresponding shape. Convexity and solidity are parameters related to convex hull of a shape. They are represent the ratio between perimeter of convex hull to the shape perimeter and ratio between size of a shape to the size of the convex hull, respectively.

The contour-based shape descriptor such as shape context [5] is quite powerful for shape classification. This descriptor encodes the contour information from a given shape. Basically, several control points are extracted from the object contour. Using each of them, the number of points nearby are counted and stored in a circular histogram.

The recently published paper proposed bag of contour fragments [26] verified the usefulness of shape context. It utilize shape context features for each fragments of contour. To make the algorithm more robust, it employ a dictionary method to convert the features using a learned codebook. The transformed features from several sub region of the shape are then concatenated as the final features. For classification, the support vector machine (SVM) is utilized.

Another type of shape descriptor is the skeleton segmentation-based descriptors [15,28]. Part-to-part analysis is performed in [28] to conduct the shape matching process. In this case, the parts are generated by skeleton-based segmentation. From segments that represent protruding parts of the silhouette, various descriptors are then extracted. In the other hands, [15] proposed three kinds of descriptors including local tangent function, area descriptor, and regional skeleton descriptor.

The usage of deep learning for shape classification is demonstrated in [1,2]. There are several network architectures were tested in [1] perform the classification task. In total, there are three types of CNN architectures are considered where the input image is set as 32×32 binary image and the output is set as the classification probability.

Alongside the feature extraction and shape analysis algorithms, there are various standard dataset were published to asses the quality of a shape analysis method. The most commonly utilized dataset are including MPEG-7 [13], Kimia's [20], animal [3], ETH80 [5], Swedish leaf [21], and 100 plants leave [17] dataset.

Fig. 2. Illustration of the region partitioning, each color represents child index of the shape. In this case, two levels of partitioning is performed, where each parent shape is divided into 4 parts. (Color figure online)

3   The Proposed Method

The machine learning system proposed in this work is shown in Fig. 1. The system accepts two kinds of inputs including raw image data and hand crafted feature.

The convolutional block acts as feature encoder. It will converts the raw input image into features based on the training procedure. Those features are then processed by merging layer which accept two kind of features as shown in the Fig. 1. At the end of the network, a classification layer is utilized to perform the final task.

In this work, the centroid based tree structure features are utilized as the handcrafted feature. It employs multiple levels of region partitioning on the shape (Fig. 2). In the first level, the root shape, which represents a chicken, is divided into four parts. In the second level of partitioning, each child is being partitioned again into four sub-regions. Hence, in this case, two levels of partitioning is performed to the root shape where each shape are partitioned into four parts (denoted as centrees-4-2).

Feature descriptors are extracted for each sub-shapes. In total, there are eight types of descriptor used in this work, including normal angle, elongatedness, normalized centroid distance, occupation ratio, compactness, perimeter ratio, eccentricity, and inertia axis ratio.

The normal angle is defined as an angle spanned from the parent's major axis ($v_{p,2}$) to a line connecting the center of sub-shape to its parent's centroid ($l_{p,c}$). This value can be calculated using dot product rule of $v_{p,2}$ and $l_{p,c}$ as shown in (1).

$$f_1 = acos \left(\frac{l_{p,c}^T v_{p,2}}{\|l_{p,c}\|} \right) \tag{1}$$

The elongatedness is used as the second feature, as shown in (2), it is defined as a minimum ratio between width and height of shape bounding box.

$$f_2 = min \left(\frac{w_c}{h_c}, \frac{h_c}{w_c} \right) \tag{2}$$

A distance form the origin point of child shape (O_c) to the origin point of its parent (O_p) is defined as the normalized centroid distance (3). Since the shape of an object might appear at different scale, it is important to make this feature become scale invariant. In this case, the distance is normalized by the maximum value of the parent's bounding box side length.

$$f_3 = \frac{\|O_p - O_c\|}{max \, (w_p, h_p)} \tag{3}$$

The occupancy ratio is defined as the fourth feature as shown in (4).

$$f_4 = \frac{S_c}{w_c h_c} \tag{4}$$

The compactness of the sub-shape relative to its parent is defined as the fifth feature. As shown in (5), it is defined as a ratio between the size of a child region to the size of its parent.

$$f_5 = \frac{S_c}{S_p} \tag{5}$$

The perimeter ratio also utilized in this work. It is represent the ratio between the perimeter length (p) of a shape to its bounding box perimeter as shown in (6).

$$f_6 = \frac{2 \, (w_c + h_c)}{p_c} \tag{6}$$

Table 1. Detailed parameters of the proposed network

Input	Attributes	Layer 1		Layer 2		Layer 3	Layer 4	Layer 5
Image (96 × 96)	Type	Conv	Pool	Conv	Pool	Flaten	FC-Merge	Classifier
	Filter size	5 × 5	2 × 2	5 × 5	2 × 2	441	541	100
	Stride	1	2	1	2	1	1	1
	#filters	8	8	32	32	1	100	8
	Output size	92 × 92	46 × 46	42 × 42	21 × 21	441	100	8
	Activation	ReLU	N/A	ReLU	N/A	ReLU	ReLU	Softmax
Features (8)	Type	FC		N/A		N/A	Merged	Merged
	Filter size	100						
	Stride	1						
	#filters	100						
	Output size	100						
	Activation	ReLU						

The eccentricity (7) and inertia-axis ratio (8) are utilized. Both of them are defined based on the ratio of Eigenvalues.

$$f_7 = \frac{\lambda_c, 1}{\lambda_c, 2} \tag{7}$$

$$f_8 = \frac{\lambda_c, 2}{\lambda_p, 2} \tag{8}$$

3.1 Invariant Analysis of the Handcrated Features

A robust shape descriptor is expected to be invariant from various transformations such as rotation, scaling, and translation. It means that the extracted descriptors should be stable or in other words if the value is changed under some transformation, it should not affect the descriptor value significantly. The invariant properties of the handcrafted features are summarized as follows:

- **Translation invariance** is achieved using an absolute reference of a fixed location. In this case, the features are calculated with respect to a fixed location in the shape which is the center point or centroid of the shape. Thus, the features are expected to be rotation invariant.
- **Rotation invariant** means that the feature will not change significantly due to any rotation transformation. In this case, the rotation invariance is achieved by utilizing principal axes of the shape. It should be noted that the principal axis itself is rotation invariant. Hence, all parameters that extracted relative to the principal axes are also inherited the rotation invariant properties.
- **Scale invariance** is achieved using the normalization strategy. As explained earlier, several features are calculated based on ratio including compactness, eccentricity, inertia-axis ratio, occupancy ratio, and perimeter ratio. Therefore, the scale invariant properties can be achieved.

4   Experiments

In the experiment, the convolutional block is constructed with $5 \times 5 \times 8$ kernel for the fist block and $5 \times 5 \times 32$ kernel for the second block. In other words, there are 8 feature channels in first convolutional block and 32 features channels in the second convolutional block. In the both convolutinal block, rectified linear unit activation function and 2×2 max-pooling layer are utilized. Meanwhile, 100 neurons are set for all of the fully connected hidden layer while the last classification layer is set as same size with the number of class in the dataset. Detailed information of the proposed architecture is summarized in Table 1.

Fig. 3. Categories in ETH80 dataset, each column are examples from apple, car, cow, cup, dog, horse, pear, and tomato class. The objects may looks similar in several categories such as apple, pear, and tomato.

Table 2. Performance comparison on ETH-80 dataset

Algorithm	Accuracy
Color histogram [14]	64.86
PCA gray [14]	82.99
PCA mask [14]	83.41
SC + DP [14]	86.4
IDSC + DP [16]	88.11
IDSC + morphological strategy [11]	88.04
Height function [25]	88.72
robust symbolic [6]	90.28
kernel-edit [7]	91.33
BCF [26]	91.49
DeepNet + centrees-4-2	**97.56**

To train the deep network, RMSProp algorithm and back-propagation is used. The network is trained within 100 training cycles with label preserving data augmentation in order to increase the number of the dataset.

The experiments were carried out using ETH-80[1] dataset. This dataset contains 80 objects from 8 categories. To be specific, there are ten different objects for each category where for each object there are 41 color images taken from different viewpoints. In total there are 3280 color images accompanied with their corresponding contour. The experiment in this work only use the contour data where several examples are shown in Fig. 3.

In this dataset, the leave one object out evaluation is utilized, this evaluation strategy is also utilized in [26]. Specifically, one object is used for testing in each round and leaves the other as the training data. This strategy is suitable for deep network structure since a big amount of training data is provided.

The evaluation result is presented in Table 2. As expected, since there are a lot of training data were provided, the performance of the proposed method achieve the best result compared to other available methods.

5   Conclusion

A shape classification method using features extracted by deep neural network combined with 8 centroid-based tree structure features is presented. The proposed method has been evaluated in public dataset and the result shows it superiority compared to the other available method with 97.56% of classification accuracy. As the future work, more shape features such as shape context,

---

[1] https://www.mpi-inf.mpg.de/departments/computer-vision-and-multimodal-comp uting/research/object-recognition-and-scene-understanding/analyzing-appearance-and-contour-based-methods-for-object-categorization/.

common base triangle area, and other feature will be incorporated to determine the best possible combination for the proposed method.

Acknowledgement. This work was supported by the National Research Foundation of Korea (NRF) Grant funded by the Korean Government (2016R1D1A1A02937579).

References

1. Atabay, H.A.: Binary shape classification using convolutional neural networks. IIOAB J. **7**(5), 332–336 (2016)
2. Atabay, H.A.: A convolutional neural network with a new architecture applied on leaf classification. IIOAB J. **7**(5), 226–331 (2016)
3. Bai, X., Liu, W., Tu, Z.: Integrating contour and skeleton for shape classification. In: 2009 IEEE 12th International Conference on Computer Vision Workshops (ICCV Workshops), pp. 360–367. IEEE (2009)
4. Beghin, T., Cope, J.S., Remagnino, P., Barman, S.: Shape and texture based plant leaf classification. In: Blanc-Talon, J., Bone, D., Philips, W., Popescu, D., Scheunders, P. (eds.) ACIVS 2010 Part II. LNCS, vol. 6475, pp. 345–353. Springer, Heidelberg (2010). doi:10.1007/978-3-642-17691-3_32
5. Belongie, S., Malik, J., Puzicha, J.: Shape matching and object recognition using shape contexts. IEEE Trans. Pattern Anal. Mach. Intell. **24**(4), 509–522 (2002)
6. Daliri, M.R., Torre, V.: Robust symbolic representation for shape recognition and retrieval. Pattern Recogn. **41**(5), 1782–1798 (2008)
7. Daliri, M.R., Torre, V.: Shape recognition based on kernel-edit distance. Comput. Vis. Image Underst. **114**(10), 1097–1103 (2010)
8. Flusser, J., Suk, T.: Pattern recognition by affine moment invariants. Pattern Recogn. **26**(1), 167–174 (1993)
9. Hu, D., Huang, W., Yang, J., Shang, L., Zhu, Z.: Shape matching and object recognition using common base triangle area. IET Comput. Vis. **9**(5), 769–778 (2015)
10. Hu, M.K.: Visual pattern recognition by moment invariants. IRE Trans. Inf. Theory **8**(2), 179–187 (1962)
11. Hu, R.X., Jia, W., Zhao, Y., Gui, J.: Perceptually motivated morphological strategies for shape retrieval. Pattern Recogn. **45**(9), 3222–3230 (2012)
12. Kejia, W., Honggang, Z., Lunshao, C., Ping, Z., et al.: A comparative study of moment-based shape descriptors for product image retrieval. In: 2011 International Conference on Image Analysis and Signal Processing (IASP), pp. 355–359. IEEE (2011)
13. Latecki, L.J., Lakamper, R., Eckhardt, T.: Shape descriptors for non-rigid shapes with a single closed contour. In: Proceedings of IEEE Conference on Computer Vision and Pattern Recognition, vol. 1, pp. 424–429. IEEE (2000)
14. Leibe, B., Schiele, B.: Analyzing appearance and contour based methods for object categorization. In: Proceedings of 2003 IEEE Computer Society Conference on Computer Vision and Pattern Recognition, vol. 2, pp. II–409. IEEE (2003)
15. Lin, C., Pun, C.M.: Shape classification using hybrid regional and global descriptor. Int. J. Mach. Learn. Comput. **4**(1), 68 (2014)
16. Ling, H., Jacobs, D.W.: Shape classification using the inner-distance. IEEE Trans. Pattern Anal. Mach. Intell. **29**(2) 286–299 (2007)

17. Mallah, C., Cope, J., Orwell, J.: Plant leaf classification using probabilistic integration of shape, texture and margin features. Signal Process. Pattern Recogn. Appl. **5**, 1 (2013)

18. Nguyen, T.T.N., Le, T.-L., Vu, H., Nguyen, H.-H., Hoang, V.-S.: A combination of deep learning and hand-designed feature for plant identification based on leaf and flower images. In: Król, D., Nguyen, N.T., Shirai, K. (eds.) ACIIDS 2017. SCI, vol. 710, pp. 223–233. Springer, Cham (2017). doi:10.1007/978-3-319-56660-3_20

19. Pavlidis, T.: A review of algorithms for shape analysis. Comput. Graph. Image Process. **7**(2), 243–258 (1978)

20. Sebastian, T.B., Kimia, B.B.: Curves vs. skeletons in object recognition. Signal Process. **85**(2), 247–263 (2005)

21. Söderkvist, O.: Computer vision classification of leaves from swedish trees (2001)

22. Teague, M.R.: Image analysis via the general theory of moments. JOSA **70**(8), 920–930 (1980)

23. Terrades, O.R., Tabbone, S., Valveny, E.: A review of shape descriptors for document analysis. In: Ninth International Conference on Document Analysis and Recognition, ICDAR 2007, vol. 1, pp. 227–231. IEEE (2007)

24. Wahyono, Kurnianggoro, L., Yang, Y., Jo, K.H.: A similarity-based approach for shape classification using region decomposition. In: Huang, D.S., Jo, K.H., et al. (eds.) ICIC 2016. Lecture Notes in Computer Science, pp. 279–289. Springer, Cham (2016)

25. Wang, J., Bai, X., You, X., Liu, W., Latecki, L.J.: Shape matching and classification using height functions. Pattern Recogn. Lett. **33**(2), 134–143 (2012)

26. Wang, X., Feng, B., Bai, X., Liu, W., Latecki, L.J.: Bag of contour fragments for robust shape classification. Pattern Recogn. **47**(6), 2116–2125 (2014)

27. Yang, M., Kpalma, K., Ronsin, J.: A survey of shape feature extraction techniques (2008)

28. Yasseen, Z., Verroust-Blondet, A., Nasri, A.: Shape matching by part alignment using extended chordal axis transform. Pattern Recogn. **57**, 115–135 (2016)

29. Zhang, D., Lu, G.: Review of shape representation and description techniques. Pattern Recogn. **37**(1), 1–19 (2004)

Smoke Detection on Video Sequences Using Convolutional and Recurrent Neural Networks

Alexander Filonenko, Laksono Kurnianggoro, and Kang-Hyun Jo[✉]

Graduate School of Electrical Engineering,
University of Ulsan, Ulsan, Republic of Korea
{alexander,laksono}@islab.ulsan.ac.kr, acejo@ulsan.ac.kr

Abstract. The combination of a convolutional neural network (CNN) and recurrent neural network (RNN) is proposed to detect the smoke in space and time domains. CNN part automatically builds the low-level features, and RNN part finds the relation between the features in different frames of the same event. For this work, the new dataset was constructed with at least 64 sequential frames for each set giving the network ability to analyze the behavior of the smoke for at least 2 s. While being not too deep thus allowing fast processing, the proposed network outperformed state of the art deep CNNs which do not consider the change of the object in time.

Keywords: Smoke detection · Convolutional neural network · Recurrent neural network

1 Introduction

Smoke detection on videos is not a trivial task for an artificial system due to a great variety of its shapes, movement speed and direction, color tones, and density. Deriving appropriate low-level features and combining them into robust detectors usually works under very strict scenarios for which they were carefully chosen leading to low detection performance in slightly different conditions. If the number of required features is significant, then researchers apply automated machine learning methods such as AdaBoost and support vector machine (SVM) to find a good separation of classes.

The next level of detection is the derivation of features and their role in the describing an object by applying convolutional neural networks (CNN) which take raw input frames, generate corresponding features and adjust weights for accurate detection.

While modern CNNs use deep levels of convolution, they do not seek for new features in a time domain. Smoke has a dynamic nature, and it changes its shape. Recurrent neural networks (RNN) can reinforce this knowledge for a set of sequential frames.

Recent novelties in hardware and parallel implementation on graphics processing units (GPU) allows using neural networks for image and video

© Springer International Publishing AG 2017
N.T. Nguyen et al. (Eds.): ICCCI 2017, Part II, LNAI 10449, pp. 558–566, 2017.
DOI: 10.1007/978-3-319-67077-5_54

processing tasks rapidly. This paper proposes using a combination of CNN and RNN for smoke detection on video sequences. The results were compared to state of the art deep CNNs that utilize a single frame as input thus not using smoke dynamics.

2 Related Works

Methods for detection of smoke on videos have a long history since this task is important in providing safety to nature and people. Many existing approaches consider that the cameras are fixed, and therefore smoke can be separated from a background. The merit of this approach is the reduced false positive rate with a cost of missing very slow or nearly transparent smoke.

Lee et al. in [1] analyzed spatial and temporal domains and achieved a detection rate of 83.5%. Torabnezhad et al. in [2] fused visual and thermal data to detect smoke regions using the feature that smoke was not visible by the infrared camera. Chen et al. [3] applied the wavelet transform to detect smoke using frequency characteristics. Maruta et al. in [4] used improved local binary patterns that they applied to the AdaBoost algorithm. Researchers develop specialized hardware to deal with smoke detection task, e.g. Rashmi and Nirmala in [5] created field-programmable gate array which applies a fuzzy neural network.

A number of research papers on utilizing deep neural networks for smoke detection are a few where authors chose one of the networks and did not explain the choice.

Xu et al. utilized AlexNet to detect artificially rendered smoke on videos [7] where they reported maximum 94.7% accuracy. The performance of detection of the real smoke scenarios is a crucial one because it can help catching the real danger situations.

Frizzi et al. in [8] built their own CNN very similar to well-known LeNet-5 with an increased number of feature maps in convolution layers. For tests, authors used real smoke scenarios and achieved 97.9% accuracy which is higher than the performance of traditional machine learning approaches.

The network proposed by Tao et al. in [9] consists of 5 convolutional layers followed by three fully-connected layers which implement AlexNet for two classes and a single GPU. Authors used open dataset and reported the accuracy of their method making it possible direct comparison of state-of-the-art CNNs to the network proposed in [9].

In this work, we evaluate the accuracy of the state of the art CNN for smoke detection task. The most recent and effective networks on ImageNet dataset are the Inception-V4 and Xception.

2.1 Inception-V4

Szegedy et al. in their work on GoogLeNet [10] introduced a new approach to building the network where layers are not only stacked upon each other like in traditional networks, but sets of layers are run in parallel.

The network consists of a set of such a parallel modules named inception modules. With this architecture, authors could increase a total number of layers by keeping number of parameters from growing fast. By replacing 5×5 connection via two 3 × 3 layers and then replacing 3 × 3 convolution via 3 × 1 with three output units decreased processing requirements. The latest version of "pure" Inception module, Inception-V4, was described in [11] which has a simplified architecture and more inception modules that lead to higher performance.

2.2 Xception

Chollet in [12] revised the idea of Inception modules and offered to use depthwise separable convolutions by maximizing the number of towers in a module as shown in Fig. 1. This "extreme" version of Inception module named Xception had the same number of parameters with Inception-V3 and outperformed it on ImageNet dataset.

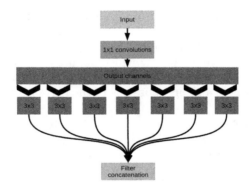

Fig. 1. The Xception module.

3 Smoke Detection

Smoke detection network consists of a combination of convolutional and recurrent parts as shown in Fig. 2.

The input to the network comprises a set of sequential color frames. Each frame is fed to a separate convolutional layer. All the convolutional layers share their weights. As the activation function, a rectified linear unit (ReLU) is placed at the output of convolutional layers. Convolutional part (CONV in Fig. 2) of the proposed network is shown in Table 1 and consists of the convolution layer with 64 features map with 5 by 5 receptive field followed by max pooling with stride 2. Next three layers represent the context module by checking wider area via a set of dilated convolutions. Context module is necessary for smoke detection since smoke is captured by a camera as a blob which tends to cover larger areas.

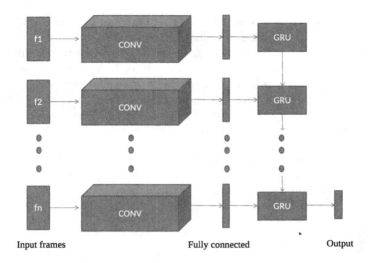

Fig. 2. Proposed network

Table 1. Convolutional part of the proposed network

Layer	1	2	3	4
Filter size	5×5	3×3	3×3	3×3
Dilation (width, height)	-	(1,2)	(2,1)	(2,2)
# Feature maps	64	64	64	64
Non-linearity	ReLU	ReLU	ReLU	ReLU
Resizing	Max pooling	-	-	Max pooling

All dilated convolution layers are zero-padded to prevent loss of resolution. The next step of the network is a fully connected layer which consists of 64 neurons. This layer is needed to feed the results to the recurrent part of the network. Results were refined by 0.2 dropout.

In the next step, the relation between frames in the time domain is found by the gated recurrent units (GRU) [13] that have a similar performance to long short-term memory (LSTM) while requiring fewer parameters. The choice of the best number of frames in a sequence is discussed in the experimental part.

4   Experiments

Experiments were conducted on the following hardware setup: Intel Core i7-2700K working at 4.6 GHz, 32 GB RAM working at 2133 MHz, Nvidia Titan X Pascal with 12 GB RAM working at stock frequency. The networks were implemented in Keras [14] and ran on Ubuntu 16.04.

Existing smoke detection datasets don't contain a sequence of frames, or they offer a small variety of the scenes, especially night time videos. For this task we created our dataset[1] which consists of at least 64 frames for each set. The total number of samples in the dataset is shown in Table 2.

Smoke appearance varies from scene to scene according to the different distance to the camera, time of day, a direction of air flow. For example, in Fig. 3 smoke has similar structure in the last two rows while showing two opposite color temperatures. In the sixth row, smoke does not always cover the whole image which reveals the factory structures. Smoke in first and seconds rows appears in the wildfire. Samples contain parts of trees and dry grass which may affect the performance of the detection. The brightness of the sparse smoke is low when recorded at night as it can be seen in the fifth row of Fig. 3. Instability in color, brightness, and edges may lead to the low performance of algorithms which depend on these features only.

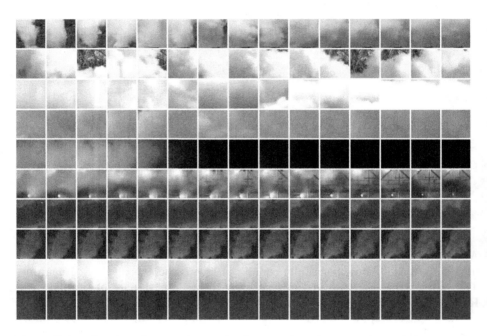

Fig. 3. Example of smoke samples in the dataset. Samples in each row belong to the same set retrieved each ten frames (Color figure online).

Some negative samples are illustrated in Fig. 4. One cannot rely only on background subtraction to filter out non-smoke regions because camera may start changing its pose as it shown in rows two, four, and eight. Strong wind may cause jitter as in the first row. Moving objects add difficulty to scenes as portrayed in rows three, seven, and ten.

---

[1] https://github.com/filonenkoa/smoke-detection-cnn-rnn.

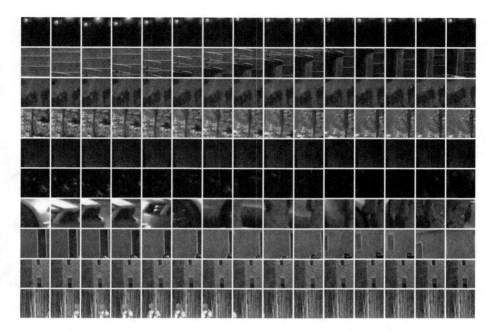

Fig. 4. Example of negative samples in the dataset. Samples in each row belong to the same set retrieved each ten frames.

80% of sets were chosen for training, and 20% of sets were chosen for testing randomly. The number of epochs was 200; the number of batches was set to 16 to prevent memory overflow. For the optimization, RMSprop[2] was used which divides the learning rate for weight by a running average of magnitudes of recent gradients for that weight. Four tests were run for each CNN. Learning rate was set to 0.001.

To find the optimal number of frames which should be considered in the detection, multiple tests were conducted with the proposed network. The proposed network was tested with and without context module to show if knowledge of the wider area is necessary. The results for the latter case are illustrated in Fig. 5. The maximum accuracy of 91.41% was achieved with 61 frames in a sequence. Before 61 frames in a sequence, it is difficult to see the trend how the accuracy is changing, but for 61 and 63 frames it is clear that the performance is improved. It worth noting that the maximum accuracy on training data was 97.47% which is much higher than results on testing data. Context module was added then to the network and results shown improvement in accuracy (Fig. 6) to 96.25% with 42 frames in a sequence. Four metrics were computed to evaluate the work of the network: number of true positives (TP), true negatives (TN), false positives (FP), and false negatives (FN). The term "accuracy" in this paper is TP+TN/(TP + TN + FP + FN). The ability of the network to choose correct

---

[2] http://www.cs.toronto.edu/~tijmen/csc321/slides/lecture_slides_lec6.pdf.

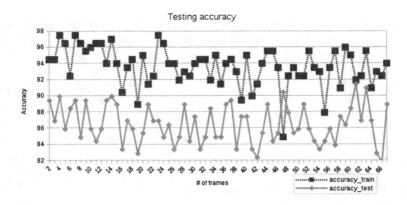

Fig. 5. Testing results without context module

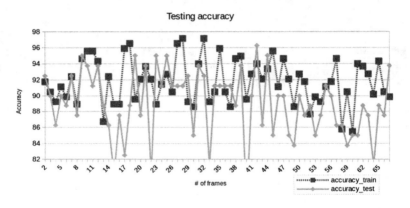

Fig. 6. Testing results with context module

answers and filter out wrong ones is represented by precision and recall. Figure 7 shows how the quality of the proposed network changes with the number of input frames.

To compare to the state of the art CNN, the Inception-V4 network was trained on the smoke dataset with same parameters, but without any connection between frames and with increased frame size to 299 × 299. The accuracy of Inception-V4 was 58.75% and 59.18% on test data and training data respectively. Xception network could not get good generalization and resulted with 41.25% test accuracy and 41.82% train accuracy.

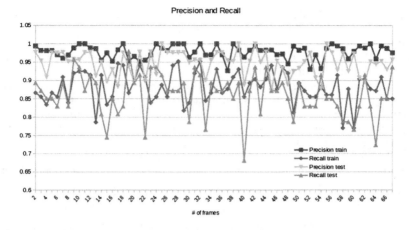

Fig. 7. Precision and recall result for the range 2 to 67 frames in a sequence

Table 2. Dataset

Type	# of sets	# of frames
Smoke	162	61,950
Non-smoke	234	39,018

5   Conclusion

The combination of CNN with a context module and RNN was presented in this paper that could successfully catch the behavior of smoke in a sequence of frames. Our network outperforms the state of the art Inception-V4 and Xception when a number of samples are much lower than those in ImageNet dataset and shape and color tone of the object varies much.

In the future work, we will go deeper in a combination of CNN and RNN giving the opportunity for a network to built connections in the high level of abstraction.

Acknowledgments. This work was supported by the National Research Foundation of Korea (NRF) Grant funded by the Korean Government (2016R1D1A1A02937579).

References

1. Lee, C.-Y., Lin, C.-T., Hong, C.-T., Su, M.-T.: Smoke detection using spatial and temporal analyses. Int. J. Innov. Comput. **8**(6), 1–23 (2012)
2. Torabnezhad, M., Aghagolzadeh, A., Seyedarabi, H.: Visible and IR image fusion algorithm for short range smoke detection. In: Proceedings of ICRoM, Tehran (2013)

3. Chen, J., Wang, Y., Tian, Y., Huang, T.: Wavelet based smoke detection method with RGB contrast-image and shape constrain. In: Proceedings of VCIP, Kuching (2013)
4. Maruta, H., Iida, Y., Kurokawa, F.: Anisotropic LBP descriptors for robust smoke detection. In: Proceedings of IECON, Vienna (2013)
5. Rashmi, G.P., Nirmala, L.: FPGA based FNN for accidental fire alarming system in a smart room. Int. J. Adv. Res. Comput. Commun. Eng. **3**(6), 6902–6906 (2014)
6. Hinton, G.E., Srivastava, N., Krizhevsky, A., Sutskever, I., Salakhutdinov, R.: Improving neural networks by preventing co-adaptation of feature detectors, Computing Research Repository (2012). http://arxiv.org/abs/1207.0580
7. Xu, G., Zhang, Y., Zhang, Q., Lin, G., Wang, J.: Deep domain adaptation based video smoke detection using synthetic smoke images, fire safety journal, under review. https://arxiv.org/abs/1703.10729
8. Frizzi, S., Kaabi, R., Bouchouicha, M., Ginoux, J.M., Moreau, E., Fnaiech, F.: Convolutional neural network for video fire and smoke detection. In: 42nd Annual Conference of the IEEE Industrial Electronics Society, IECON 2016, Florence, pp. 877–882 (2016). doi:10.1109/IECON.2016.7793196
9. Tao, C., Zhang, J., Wang, P.: Smoke detection based on deep convolutional neural networks. In: 2016 International Conference on Industrial Informatics - Computing Technology, Intelligent Technology, Industrial Information Integration (ICIICII), pp. 150–153, (2016). doi:10.1109/ICIICII.2016.0045
10. Szegedy, C., Liu, W., Jia, Y., Sermanet, P., Reed, S.E., Anguelov, D., Erhan, D., Vanhoucke, V., Rabinovich, A.: Going deeper with convolutions, computing research repository (2014). http://arxiv.org/abs/1409.4842
11. Szegedy, C., Ioffe, S., Vanhoucke, V., Alemi, A.A.: Inception-v4, inception-ResNet and the impact of residual connections on learning. In: AAAI 2017, pp. 4278–4284. AAAI Press (2017)
12. Chollet, F.: Xception: deep learning with depthwise separable convolutions, computing research repository (2017). http://arxiv.org/abs/1610.02357
13. Cho, K., van Merrienboer, B., Bahdanau, D., Bengio, Y.: On the properties of neural machine translation: encoder-decoder approaches, computing research repository (2014). http://arxiv.org/abs/1409.1259
14. Chollet, F.: Keras (2015). https://github.com/fchollet/keras

Intelligent Processing of Multimedia in Web Systems

Improved Partitioned Shadow Volumes Method of Real-Time Rendering Using Balanced Trees

Kazimierz Choroś[(✉)] and Tomasz Suder

Faculty of Computer Science and Management,
Wrocław University of Science and Technology,
Wyb. Wyspiańskiego 27, 50-370 Wrocław, Poland
kazimierz.choros@pwr.edu.pl

Abstract. Shadows are probably one of the most important factors of the 3D scene. They enhance realism and comprehension of rendered images. They are usually generated with some variation of shadow volumes or shadow mapping, although recently a new algorithm called Partitioned Shadow Volumes appeared. The paper presents an improved version of the Partitioned Shadow Volumes algorithm, dedicated for the static scenes, that optimizes the speed of shadowing by using binary space partitioning heuristics to generate more optimal ternary object partitioning tree. There is an obvious trade-off between ternary object partitioning tree building time and the rendering time, but the tests have shown that the proposed modification can significantly speed up the rendering time of scenes when the light position and shadow caster model do not frequently change.

Keywords: Real-time rendering · 3D graphics · Partitioned Shadow Volumes · Shadow mapping · Precomputed lighting · Static shadows · Binary space partitioning · Rendering time

1 Introduction

An effective real-time rendering is all the time a great challenge in computer science. It is very important to ensure the best realism of images in photorealistic 3D architectural visualizations, computer games, virtual worlds, or movies. But on the other hand, computer systems demand rendering algorithm enabling the realistic images be performed in real-time. In [1, 2] it has been discussed how the speed of real-time rendering depends on properties of rendered objects and scenes and on theirs properties such as the number of triangles in a 3D model, the scale of the object, and the number of moving light sources. It is often noticed [3] that for heterogeneous scenes there is no single rendering algorithm capable to generate scenes of all types sufficiently fast and with the same high image quality. The long list of different journal and conference papers as well as books can be find in one of the recent publications [4].

The main processes of real-time rendering are shadow volumes and shadow mapping. At the beginning 3D graphics rendering algorithms were not suitable for real-time rendering, but they became more and more accessible along with the

© Springer International Publishing AG 2017
N.T. Nguyen et al. (Eds.): ICCCI 2017, Part II, LNAI 10449, pp. 569–578, 2017.
DOI: 10.1007/978-3-319-67077-5_55

development of modern GPU cards. Shadow volumes became widely used in real-time rendering after the appearance of the stencil buffer-based implementation. Such an implementation was good for rendering of simple scenes in the early days of modern graphic processing units, but nowadays is clearly obsolete and unsuitable for today's requirements. On the other hand, shadow mapping is prone to many issues related to shadow texture filtering and memory requirement. Developers often have to deal with such problems like Peter Panning or shadow acne.

In this paper an improved method of rendering based on Partitioned Shadow Volumes (PSV) using balanced trees is presented. The method optimizes the speed of shadowing by using the Binary Space Partitioning (BSP) heuristics to generate the Ternary Object Partitioning (TOP) tree.

The paper is structured as follows. The next section briefly describes related work on real-time rendering. The rendering method of Partitioned Shadow Volumes is shortly presented in the third section. The disadvantages of this method is discussed in the fourth section. Whereas, the fifth section presents the proposed improved method of Partitioned Shadow Volumes (bPSV) using balanced trees and BSP heuristics. Then the test results showing the influence of the balance of ternary object partitioning trees on the maximum stack size and mainly on the rendering speedup are presented in the sixth section. The final conclusions are discussed in the last section.

2   Related Work

There are two major trends in the real-time shadows rendering methods: shadow volumes and shadow mapping [5].

Shadow volumes proposed many years ago [6] are the more mature but nowadays also partially obsolete approach for rendering real-time shadows. The main idea of shadow volumes is to generate extra geometry that will describe shadow that is cast by the object. The shadow geometry is generated by extruding the object edges in the opposite direction from the light. The basic approach requires extrusion of all triangle edges in the model, alongside with front and back caps addition, but a lot of work has been done to improve it with various silhouette-detection algorithms and mesh connectivity caching. Unfortunately, most of those algorithms are not suitable for meshes with open seams or "dangling triangles". The most common shadow volume algorithms are carried out in two passes. In the first pass the object silhouette geometry is rendered to the stencil buffer and then, in the second pass, the stencil buffer data is used to determine whether a given point is in shadow. The pixel visible from the eye is shadowed if it is covered by more front-facing shadow volume polygons than back-facing shadow volume polys. Due to the geometrical nature of this approach the generated shadow edges are hard and are not suitable for real time effects like blurring. Also the texture-based transparency (alpha tested textures) are not casting a proper shadows with this approach, because the entire algorithm is working at the geometry level.

Shadow mapping also proposed many years ago [7] is still the most common approach for generic real-time shadows. The shadow mapping techniques relies on the image representation of the shadow. The texture containing shadow information is called "shadow map". This texture-based approach has both advantages and disadvantages. The main advantage is that various soft-shadow effects are relatively easy to archive by simple blurring and sampling the image. The main disadvantage is that the hard (precise) shadows are not possible to achieve with this approach. Also shadows covering larger areas are harder to deal with, but there are techniques dedicated specially for them, like Cascaded Shadow Maps (CSM) [8]. Also, the shadow mapping approach suffers from issues like Peter Panning and shadow acne, which sometimes are hard to tune and overcome. The shadow mapping approach is performed in two passes: first pass is the shadow rendering to the texture, which is done from the light point of view. This way the distance of each pixel to light is stored in shadow map. Second pass is using shadow map to compare each pixel distance to light with the distance stored in shadow map in order to determine whether given point is shadowed.

Although shadow mapping is well known for years many improvements and modifications have been proposed all the time, also nowadays new approaches are presenting. In the Light Space Partitioned Shadow Maps (LSPSM) algorithm [9] the main idea is to create non-overlapped partitions for shadow maps in light space. The number of these partitions depends on the number of light and view directions.

There was also a Shadow Volume Binary Space Partition (SVBSP) approach [10], but it was abandoned due to the overheat. This early approach did not use the silhouette detection. All the shadow casting polygons were filtered down the BSP tree in a front-to-back order from the light. Of course, the polygon splitting was required due to the nature of the BSP tree [11]. The shadow volume planes was used for splitting, causing several fragments for every polygon. The fragment that was located inside at least one of such shadow volumes was considered shadowed. Unfortunately, the polygon splitting which was required by this approach was prone to various numerical issues, memory troubles, and geometrical degeneracies. The memory requirement was hard to predict and the requirement of polygon splitting operation made it unsuitable for real-time shadowing.

The SVBSP has a certain similarities to the Partitioned Shadow Volume algorithm. The binary space partition approach is a common point, but the PSV is using a new TOP tree structure while the SVBSP is using a classic BSP tree.

3 Rendering Method of Partitioned Shadow Volumes

The Partitioned Shadow Volumes – PSV [12] is a new algorithm for rendering hard shadows (shadows based on model geometry). The PSV is working in a different way than classic shadow volumes and is free of common shadow volumes drawbacks, like fill rate and geometrical limitations. The PSV is using a new data structure, the ternary object partitioning tree in order to check if a given point is shadowed. Each TOP tree node is defined by a plane equation and three sets of children:

- front child, containing all the geometry in the front of the node plane (in the positive half-space defined by the node plane)
- back child, containing all the geometry in the back of the node plane (in the negative half-space defined by the node plane)

- intersection/cross child, containing all the remaining geometry (geometry intersecting the plane).

The classic BSP approach had only two children: front and back, so, the intersecting geometry should be divided, and in fact increases computational cost and leads to many numerical issues and errors. The TOP tree solves this problem by simple storing the intersecting geometry in the third, so-called "intersection" child. The Partitioned Shadow Volume algorithm is performed in two passes:

- first pass, the TOP tree generation is run in the compute shader on the GPU. This step inserts all of the shadow casting triangles in the random order into the new data structure called TOP tree. The cost of the single triangle insertion into TOP tree is given as O (log n), where the entire TOP tree computation is given as O (n log n). The TOP tree memory footprint is easily predictable and given as O (n).
- second pass, the TOP tree shadowing (rendering) is run in the fragment shader. It's preferred to be run in the deferred pipeline but it works with forward lighting as well. For each pixel the constructed previously TOP tree is traversed with a usage of extra stack and each triangle frustum is checked until at least one frustum contains the checked point. The TOP tree query cost is given as O (log n). The maximum stack size here must be carefully tuned because it has a major impact on the Fragment Shader register usage and performance.

The Partitioned Shadow Volumes is a new and promising method of rendering shadows on the modern GPU pipeline.

Recently also the Stackless and Deep Partitioned Shadow Volumes algorithm appeared [13]. It introduces various optimizations to the second stage of Partitioned Shadow Volumes algorithm, the TOP tree traversal, including stackless and hybrid approaches. The TOP tree generation is also slightly modified but only in order to add extra information, like node ancestor and the depth. Although those changes improve the various parts of PSV algorithm, they still relies on the random insertion order of the triangles into the tree.

4 Disadvantages of the PSV Method

The PSV rendering algorithm is free of most of the common shadow volumes drawbacks, but it's also introducing several new problems.

The fragment shader stack size is one of the main PSV algorithm "bottlenecks", as it has a large impact on rendering speed due to the nature of fragment shaders. Dynamic arrays are not supported by the shaders and the maximum stack size must be determined at the compile time. The larger stack size consumes more registers and decreases the efficiency. On the other hand, the smaller stack size will crash the application in case of stack overflow. Maximum stack size for the second PSV pass must be carefully tuned in order not to avoid unnecessary overheat.

The PSV method is suited for fully dynamic scenes where light position or model silhouette changes every frame. The TOP tree is entirely recomputed every frame which is not necessary in many real applications.

The PSV TOP tree structure is generated by inserting the scene triangles into the TOP structure in a random order. While this is in fact the fastest method of the tree generation, it has several drawbacks. Usually the generated tree is far from optimal and it takes much time to traverse. Also, the insertion order that changes every frame might lead to small visible artifacts and flickering in certain systems.

The Partitioned Shadow Volumes leaves a very big room for optimization of the static scenes. Many real-time rendering engines tend to reduce the rendering overheat by caching the lighting/shadows information of the static parts of the scene. This is called precomputed lighting. That caching is usually done once at the scene design time or at the load time. The redundant TOP tree calculation is not necessary in most cases and the TOP tree itself could be optimized to speed up the shadow rendering process itself and reduce the required maximum stack size. The computed TOP tree could be stored in memory reused many times as long as the light position and the shadow caster does not change. The precomputing of the TOP tree at the load time also gives a great opportunity to improve the TOP tree structure itself. It could be more balanced so the traversal time is smaller. The TOP tree can be improved by using the BSP split plane selection heuristics in order to determine the best triangle insertion order. The possible TOP tree improvements will be discussed in the next section of this paper.

5   Balanced Partitioned Shadow Volumes (bPSV)

The proposed PSV modification was inspired by the observation that the TOP tree generated by inserting triangles in the random order is not optimal. The subtree depths are not balanced in many cases and take much time to traverse. Furthermore, such unbalanced TOP trees require larger stack size during the traversal and larger stack requires more fragment shader registers and decreases the efficiency. Fragment shaders are working in a slightly different way than CPU programs and the allocation of larger stack size makes a noticeable difference in a performance. The more balanced TOP tree can be generated in a very similar way to BSP tree generation by using a heuristic split plane selection to determine preferred order of triangle insertions. We start with a single set of triangles, then we calculate the heuristic score for each triangle plane. The plane heuristic score is based on the number of triangles in front, behind, and intersecting that plane. Then the triangle with the best score is chosen and used to divide space into three sets: front, back, and intersection. All remaining triangles are moved into those sets and then the algorithm starts again. It runs until there are not triangles left.

5.1   Plane in 3D Graphics

A plane concept is widely used in computer graphics. A plane is usually represented by the following equation:

$$Ax + By + Cz + D = 0 \tag{1}$$

where A, B, C are the nonzero plane normal vectors determining which side is plane facing, and D is the plane distance to the origin, and the (x, y, z) is the any point lying on the plane.

5.2   Plane-Point Relations in 3D Space

The distance of a given point to the plane is given by the following equation:

$$d = Ax + By + Cz + D \tag{2}$$

which is in fact a sum of plane distance D and the dot product between plane normal and the given point: $d = N \cdot p + D$.

Such value of point to plane distance can be used to classify the point relations to the plane to one of three sets:

- front – the point is on the positive side of the plane, when $d >$ epsilon,
- back – the point is on the negative side of the plane, when $d < -$epsilon,
- on – those points are lying on the plane, so they satisfy the equation $|d| \leq$ epsilon.

Epsilon value is used here because of the rounding errors that occur in the floating point arithmetic. It should be chosen carefully for the specific application.

5.3   Plane-Triangle Relations in the 3D Space

In order to determine the relation between the triangle and the plane in the 3D space three distances to triangle vertices must be calculated. The proposed PSV modification requires detection of three cases:

- triangle is in front of the plane, when all distances of the nodes to the plane are positive,
- triangle is behind the plane, when all distances of the nodes to the plane are negative,
- in the remaining cases the triangle is intersecting the plane.

6   Modified TOP Tree Generation Algorithm

The TOP tree in the PSV algorithm is generated by inserting the shadow caster triangles in the random order. This is the fasted method of TOP tree generation but the structure generated in such a way is not optimal and takes long time to traverse. We propose to make the TOP tree generation algorithm more similar to BSP while retaining the TOP tree advantages. Instead of inserting the triangles in random order, a heuristic function is used to choose the best split plane/triangle at given time and make the TOP tree more balanced. It has been observed that this better TOP tree takes less time to traverse and requires smaller stack size.

It should be noted that our approach is using heuristics only to improve the PSV tree. The PSV tree generated by our method is not the best possible PSV tree for the given set of triangles, but in most cases it is better than TOP tree generated by inserting triangles in random order.

The modified TOP tree generation algorithm is defined as follows:

(1) start with set of all the scene triangles,
(2) for each triangle plane in the list:
    (2.1) count the number of triangles in front of the current plane, behind the current plane and intersecting it,
    (2.2) calculate the heuristic score for the current plane,
(3) select the split plane with the best score,
(4) divide the current triangles set into three separate sets – front, back, and intersect,
(5) recursively, for each all of those sets start again 2.

The algorithm is run until all the triangles are inserted into the TOP tree, although the early-stop condition could be introduced with the random insertion of the remaining triangles. Such hybrid approach could give more control over the balanced TOP creation time.

The proposed heuristic function is used to select the best triangle to insert to the TOP tree for the given triangles set. The heuristic score is calculated for each triangle plane by checking the relation of every other triangle with the current plane. The number of triangles in the front, back, and intersecting the plane is calculated for each plane. Then the triangle with the lowest heuristic value is chosen and used to split the current triangles set into next three sets: front set, back set, and intersect set.

A heuristics formula can be defined as follows:

$$\text{score} = \alpha|\text{ front-back}| + \beta \text{ intersect} \tag{3}$$

where the score is a heuristic split value for a given plane, whereas, front and back are the numbers of triangles in front of and behind the split plane, intersect is the number of triangles intersecting the plane (crossing the plane or being on it), and α, β are the weights used to tune the influence of given factor.

The Eq. (3) reflects a balance of the number of nodes in the right and left child of the TOP tree node. Also the number of nodes in the intersection child of the tree should be minimized while maintaining the left-right balance because the intersection child is visited in more cases than right or left child. The α is the weight of the left-right subtrees balance. It can be changed to alter the influence of the left-right subtree balance. The β is the weight of the intersecting triangles count. Generally, intersecting a triangle should not be preferred because the intersection subtree of the node has to be visited in both cases and thus increases the stack size. The α and β factors are used to decide whether the left-right subtree balance is more important than the number of intersecting triangles.

7 Experiments

The tests have been performed on nVidia® GeForce 820 M. The TOP tree building and rendering time have been separately measured for the original PSV algorithm and the proposed improved PSV algorithm (bPSV). The second pass (rendering) algorithm was the same in both cases, only the first pass (top tree build) method was changed. The test scene consisted a single light source, the tested model, and the ground plane for the reference. Three meshes have been used in the tests: Venus, Teapot, and Elephant with different numbers of faces.

These three freely available 3D meshes are:

- Venus Mesh – 11 563 faces
  (http://graphics.csie.ntu.edu.tw/ ~ robin/courses/cg04/model/index.html),
- Teapot Mesh – 6 332 faces
  (https://people.sc.fsu.edu/ ~ jburkardt/data/obj/teapot.obj),
- Elephant Mesh – 10 150 faces
  (http://graphics.csie.ntu.edu.tw/ ~ robin/courses/cg04/model/index.html).

The commonly available meshes were chosen instead of providing our own meshes.

The speedup of rendering time is significant (Tables 1, 2 and 3). The TOP tree build time was much longer, but it is not a problem as we are working under assumption that the modification is dedicated for precomputed lighting, so the rendering stage speedup is worth the investment.

Table 1. Rendering speedup for Venus Mesh – 11 563 faces

	Pass 1: TOP tree build [ms]	Pass 2: Rendering [ms]	Rendering speedup
PSV	36.6	8.4	1
bPSV	10895	5.8	1.45

Table 2. Rendering speedup for Teapot Mesh – 6 332 faces

	Pass 1: TOP tree build [ms]	Pass 2: Rendering [ms]	Rendering speedup
PSV	25	3.8	1
bPSV	5556.4	2.6	1.46

Table 3. Rendering speedup for Elephant Mesh – 10 150 faces

	Pass 1: TOP tree build [ms]	Pass 2: Rendering [ms]	Rendering speedup
PSV	30.3	3.5	1
bPSV	8682.7	2.6	1.34

The received results show that the more optimized TOP tree makes the rendering faster. It is worth noting that all tests were made with the stack size of maximum 32 nodes, which was unnecessarily large in the improved PSV case. The reducing stack size in the second case could lead to more extra performance gain, but we left it without change for the reference.

Of course, the amount of performance gain also depends on various factors like mesh topology and the light position. The further performance gain could be achieved by decreasing the maximum stack size in the shader.

Clearly, there is a trade-off between the amount of time spent on constructing the TOP tree and the rendering time. The long balanced TOP tree computation time makes the proposed improvement not suited for fully dynamic lighting where the shadow caster or light position changes every frame. On the other hand, almost 170% speedup is clearly worth the improved PSV investment for the static scenes. The performance gain is easily noticeable in the cases where the TOP structure is not recomputed every frame.

Static (precomputed) lighting is a common optimization method used in computer graphics [14, 15]. It is based on the observation that the most parts of the scene do not change during the runtime, so the lighting information can be computed once during scene creation time and then stored on the hard disk for later usage. Then the application is just loading the stored data and use it to speed up the rendering. Our improved PSV algorithm is perfect for such static lighting approach, because it tends to move the amount of work from the second PSV pass (rendering) to the first PSV pass (top tree generation). The TOP tree is generated only once and then just reused for faster rendering.

Fully static lighting is not the only possible usage of proposed optimization. The non-changing lighting position and shadow caster combination could be detected at real-time. Then the Balanced TOP Tree can be also computed during runtime in a separate CPU thread without blocking the rendering thread (in the opposition to the original approach, which has used compute shaders for TOP tree creation) and used when it's ready, providing a smooth transition between classic PSV rendering and the BSP-optimized PSV rendering. It can be used the optimize rendering of various scenes, not only the entire static ones.

The proposed solution is fully compatible with the existing PSV rendering algorithm. Our modification is related only to the TOP tree generation algorithm. The second pass, the fragment shader method and the TOP tree data structure itself does not change, so it's easy to integrate our modification into existing PSV pipeline. Both PSV variations could be used for the single scene, original PSV for the moving lights and objects but the improved PSV for the stationary environment. Such improvement makes the entire PSV rendering system more flexible for both static and dynamic scenes.

8   Conclusions

In the paper we proposed an improved version of the Partitioned Shadows Volumes algorithm for static scenes where the light position and shadow casters silhouette do not change.

Our modification is based on the observation that more balanced TOP tree takes less time to traverse. We have improved the TOP tree creation by using heuristic function similar to one used in BSP trees. The triangle insertion order in the proposed method is no longer random.

There is a trade-off between the amount of time spent on constructing the TOP tree and the rendering time. We try to optimize the rendering time by spending more time constructing a better TOP tree.

Our method is dedicated for the static scenes. Our experiments shows that using the heuristic to determine optimal triangle insertion order into the TOP tree increases drastically the TOP tree generation time, but it also significantly decreases the rendering time.

The presented method can be easily integrated with the existing PSV rendering pipeline and used along the original algorithm. The balanced TOP tree can be computed in background and used instead of the original PSV approach. Such a hybrid system can easily allow fast rendering of complex scenes made of both static and dynamic types of objects, which are common in the most of the applications like simulations, computer games, or virtual reality systems.

References

1. Akenine-Möller, T., Haines, E., Hoffman, N.: Real-Time Rendering. CRC Press, London (2008)
2. Choroś, K., Kaczyński, K.: Time and quality of 3D rendering process using programming code optimisation techniques. Int. J. Intell. Inf. Database Syst. **2**(3), 309–319 (2008)
3. Petring, R., Eikel, B., Jähn, C., Fischer, M., auf der Heide, F.M.: Real-time 3D rendering of heterogeneous scenes. In: Bebis, G., et al. (eds.) ISVC 2013. LNCS, vol. 8033, pp. 448–458. Springer, Heidelberg (2013). doi:10.1007/978-3-642-41914-0_44
4. Kronander, J., Banterle, F., Gardner, A., Miandji, E., Unger, J.: Photorealistic rendering of mixed reality scenes. Comput. Graph. Forum **34**(2), 643–665 (2015)
5. Eisemann, E., Assarsson, U., Schwarz, M., Valient, M., Wimmer, M.: Efficient real-time shadows. In: ACM SIGGRAPH 2013 Courses, pp. 1–46. ACM (2013)
6. Crow, F.C.: Shadow algorithms for computer graphics. ACM SIGGRAPH Comput. Graph. **11**(2), 242–248 (1977)
7. Williams, L.: Casting curved shadows on curved surfaces. ACM SIGGRAPH Comput. Graph. **12**(3), 270–274 (1978)
8. Dimitrov, R.: Cascaded Shadow Maps. Developer Documentation, NVIDIA Corp. (2007)
9. Tang, B., Luo, J., Ni, G., Duan, W., Gao, Y.: Light space partitioned shadow maps. IEICE Trans. Inf. Syst. **100**(1), 234–237 (2017)
10. Slater, M., Steed, A., Chrysanthou, Y.: Computer Graphics and Virtual Environments: From Realism to Real-Time. Pearson Education, London (2002)
11. Chin, N., Feiner, S.: Near real-time shadow generation using BSP trees. ACM SIGGRAPH Comput. Graph. **23**(3), 99–106 (1989)
12. Gerhards, J., Mora, F., Aveneau, L., Ghazanfarpour, D.: Partitioned shadow volumes. Comput. Graph. Forum **34**(2), 549–559 (2015)
13. Mora, F., Gerhards, J., Aveneau, L., Ghazanfarpour, D.: Stackless and deep partitioned shadow volumes. In: EGSR16 Eurographics Symposium on Rendering (2016)
14. Sloan, P.P., Kautz, J., Snyder, J.: Precomputed radiance transfer for real-time rendering in dynamic, low-frequency lighting environments. ACM Trans. Graph. **21**(3), 527–536 (2002)
15. Zhou, K., Hu, Y., Lin, S., Guo, B., Shum, H.Y.: Precomputed shadow fields for dynamic scenes. ACM Trans. Graph. **24**(3), 1196–1201 (2005)

Online Comparison System with Certain and Uncertain Criteria Based on Multi-criteria Decision Analysis Method

Paweł Ziemba[1]([⊠]), Jarosław Jankowski[2], and Jarosław Wątróbski[2]

[1] The Jacob of Paradies University,
Teatralna 25, 66-400 Gorzów Wielkopolski, Poland
pziemba@ajp.edu.pl
[2] West Pomeranian University of Technology,
Żołnierska 49, 71-210 Szczecin, Poland
{jjankowski,jwatrobski}@wi.zut.edu.pl

Abstract. The Internet makes it possible to find and analyse information about goods and services which can be purchased online. An abundance of e-commerce services and available products makes customers disorientated. Shoppers usually devote a great deal of time to browse offers and select their optimal products and they often are in need of advisers who could share their expert knowledge on products customers know very little about. That is why, on the Internet the comparison services, such as Google shopping, Shopzilla, PriceGrabber are being developed. The applications support customers in decision-making by letting them compare many similar products and offers of online shops. This article is focused on the issue related to the support of product evaluation in comparison services. Presently applied comparison algorithms include only the product price but do not concentrate on other criteria of evaluation. In relation to the above-mentioned problem, it is suggested that the number of evaluation criteria should be increased and an adequate multi-criteria decision analysis (MCDA) method ought to be applied. The comparison system proposed in the paper is based on MCDA method and is able to use certain and uncertain criteria. As a result, it increases the functionality of this type of Internet service.

Keywords: Comparison services · Multi-criteria decision analysis methods · Fuzzy TOPSIS · Certain and uncertain criteria · Intelligent web systems

1 Introduction

There has been a continuous dynamic development of e-commerce for many years [1]. One can state that recently the Internet has become a major channel for delivering products and services [2]. The value of e-commerce in the world in 2016 amounted to $975 billion [3], whereas in Poland it was over $8 billion (€7.6 billion) [4]. The biggest share in e-commerce have internet shops and online auction sites, however, comparison services, e.g. Google Shopping, PriceGrabber and Shopzilla also play an important role. They are particularly popular in Asia and Great Britain [5], and in Poland as well.

© Springer International Publishing AG 2017
N.T. Nguyen et al. (Eds.): ICCCI 2017, Part II, LNAI 10449, pp. 579–589, 2017.
DOI: 10.1007/978-3-319-67077-5_56

On the Polish market, there are two mostly recognized comparison services, i.e. Ceneo and Skąpiec [6]. This type of services are mostly used for comparing electronics industry products [5], but there are also dedicated services for comparing finance [7], insurance [8], tourist [9, 10] and other services.

One can list essential reasons for the popularity of comparison services: novelty effect, promotional activities of comparison services, increasing awareness of possibilities created by online shopping and possibility of saving money [11]. Indeed, comparison services work as specialized search engines making it possible for users to save time and money by gathering offers of many sellers in one place [12]. Search engines allow users to define the requirements of a specific product and then to present a list of sellers who can deliver this product [13]. The user does not know all online companies selling the product, however, comparison services can present a complete list of offers increasing competition on the market [14]. The services have changed the way consumers look for and buy goods [12].

The main negative aspect related to comparison services is the fact that they only use the price of a product in various shops as a comparison criterion and they do not take into consideration other criteria, such as delivery time, delivery cost, the quality of customer service, etc. Therefore, comparison services allow considering only a single-criterion decision problem, while the problem comprises more criteria, not just its price. Moreover, comparison services are based on the assumption that the user knows what product he or she wants to buy and he or she is looking for the product in various shops. Nevertheless, a large number of different but similar products available in e-commerce services may confuse customers who must devote a lot of their time to browsing offers and choosing the optimum product. What customers often need is an adviser who could share their expert knowledge on the characteristics of products in the field in which the customer is a layman. Consequently, comparison services are not able to capture a decision-making process which the user makes while shopping. At the onset of the decision-making process, there should be considered different sellers as well as different products in which the customer is interested, from which he or she will select a preferred decision variant. The products ought to be considered with regard to many criteria.

The aim of this article is to formulate a model of a comparison system which is capable of capturing a multi-criteria decision problem and a full course of the decision-making process by introducing the multi-criteria decision analysis (MCDA) method. The method would constitute an engine of the comparison system, which would support the user in making a shopping decision in a broader aspect than the present comparison services.

Section 2 discusses the course of the decision-making process and points out to stages of the process which are not supported by the present comparison services. Additionally, Sect. 2 contains the characteristics of the MCDA methods as well as the recommendation of a method applied in a proposed system. Section 3 presents a model of the proposed comparison system taking into account its elements and realized stages of the decision process. Section 4 contains an analysis of a case in which the prepared model taking into consideration certain and uncertain data has been used. Conclusions and suggestions for further research can be found in the final part of the article.

2 Decision Process and Multi-criteria Decision Analysis Methods

2.1 Decision Process

A decision process can be defined as a set of actions and activities which directly or indirectly support the decision maker [15]. In the model of a course of a decision process, according to Roy [16], there are the following stages: (I) determining the object of the decision and defining the set of decision variants as well as determining the multi-criteria problematic, (II) the analysis of consequences and development of a consistent set of criteria, (III) modeling preferences and aggregating performances of the variants, (IV) developing the recommendation on the basis of the results of evaluation of variants. Roy [16] notices that the decision process cannot be simplified by eliminating individual stages.

As it has been noted in the introduction, comparison services do not fully comprise the first two stages, of the decision process, concentrating on structuring the decision process and indicating the considered decision variants and indicating many criteria with regard to which variants will be considered. The operation of comparison services, however, comprises Stages III and IV, that is attributing evaluations to decision variants, aggregation of the evaluations and indicating a preferred variant. Naturally, the stages in comparison services are simplified indeed, since only one criterion is considered, that is the price of a product.

2.2 Multi-criteria Decision Analysis Methods

In the MCDA methods, it is essential to determine the decision-maker's preferences, which are expressed by means of binary relations. While comparing decision variants, there might be two basic relations: (1) variants are indifferent $-a_i I a_j$, (2) one of the variants has strict preference with regard to another one $-a_i P a_j$. Furthermore, the set of basic preference relations can be extended by: (3) weak preference of one variant with regard to another one $-a_i Q a_j$, (4) incomparability of the variants $-a_i R a_j$ [17]. The characterized preference situations are related to thresholds used in many multi-criteria methods. They are: indifference (q), preference (p) and veto (v) thresholds [16].

In the literature, there are many MCDA methods [18]. They can be divided on the basis of the operational approach, which was applied, used for aggregating the decision-maker's preferences. One can distinguish two basic approaches: (I) use of a single synthesizing criterion without incomparabilities, (II) synthesis by outranking with incomparabilities [17]. The application of the method based on a single synthesized criterion makes it possible to differentiate relations I and P. On the other hand, in outranking methods relations Q and R are also used.

The MCDA methods are intended for solving various reference problematics. One can list the following problematics [19]: choosing of a small number of good variants $(P \cdot \alpha)$; sorting variants for specified classes $(P \cdot \beta)$; determining a ranking of variants $(P \cdot \gamma)$; describing variants and their consequences $(P \cdot \delta)$.

Depending on the kind of input information, there are deterministic and non-deterministic methods. The deterministic methods assume that input information is certain. Therefore, it takes the character of quantitative data (crisp). On the other hand, non-deterministic methods have uncertain input data which are usually ex-pressed on a quantitative scale (linguistic) [20]. Many non-deterministic methods are actually methods operating on fuzzy data [21].

It should be assumed that the decision-maker solving a decision problem with the use of the proposed system does not have knowledge referring to modelling the decision problem, the course of the decision process or the MCDA methods. That is why, the MCDA method used in a comparison system should require the decision-maker to a minimum extent to deal with the aspects of modelling the problem. Therefore, the use of the MCDA methods based on the outranking relation seems questionable, since the decision-maker may find it difficult to correctly interpret thresholds p, q and v, which are used in this group of methods. Moreover, the incomparability relation, applied in outranking methods, would hinder decision-makers to make a decision. For this reason, in authors' opining, it should not be found in the prepared comparison system. Therefore, it is necessary to consider the use of methods based on a single synthesized criterion. The method ought to consider the problematics of ranking $(P \cdot \gamma)$. Additionally, the selected MCDA method should be flexible with regard to input data, or more precisely, it should make it possible to consider certain and uncertain data. It results from the fact that the decision-maker does not need to be an expert in the area in which he or she examines products or services and he or she can have different degrees of certainty regarding the parameters of compared variants. All the assumptions are met by the Fuzzy TOPSIS [22]. Its additional advantage is the calculation simplicity as well as the easiness of interpreting a solution. That is why, in the proposed comparison system, the authors decided to apply the Fuzzy TOPSIS.

2.3   Fuzzy TOPSIS Method

In the Fuzzy TOPSIS [22, 23], the evaluation of an i-th variant with regard to a j-th criterion is expressed as a trapezoidal fuzzy number $\dot{x}_{ij} = (a_{ij}, b_{ij}, c_{ij}, d_{ij})$, whereas the weight of the j-the criterion is a fuzzy number $\dot{w}_j = (w_{j1}, w_{j2}, w_{j3}, w_{j4})$. A set of criteria can be divided into criteria such as profit and cost. For the criteria of profit, normalized fuzzy values are obtained by means of the formula (1):

$$\dot{r}_{ij} = \left(\frac{a_{ij}}{d_j^*}, \frac{b_{ij}}{d_j^*}, \frac{c_{ij}}{d_j^*}, \frac{d_{ij}}{d_j^*}\right), \quad \text{where} \quad d_j^* = \max_i d_{ij} \tag{1}$$

Whereas for the criteria of cost, normalization is calculated by the formula (2):

$$\dot{r}_{ij} = \left(\frac{a_j^-}{d_{ij}}, \frac{a_j^-}{c_{ij}}, \frac{a_j^-}{b_{ij}}, \frac{a_j^-}{a_{ij}}\right), \quad \text{where} \quad a_j^- = min_i a_{ij} \tag{2}$$

After considering the weights of criteria, weighted normalized values of variants $\tilde{v}_{ij} = \dot{r}_{ij}(\bullet)\dot{w}_j$ are obtained. On the basis of the weighted normalized values of variants, the following solutions are determined: a fuzzy positive-ideal solution $A^* = \left(\tilde{v}_1^*, \tilde{v}_2^*, \ldots, \tilde{v}_n^*\right)$ and a fuzzy negative-ideal solution $A^- = \left(\tilde{v}_1^-, \tilde{v}_2^-, \ldots, \tilde{v}_n^-\right)$ where $\tilde{v}_j^* = \max_i \left\{\tilde{v}_{ij4}\right\}$ and

$\tilde{v}_j^- = \min_i \{\tilde{v}_{ij1}\}$. The distance of the i-th variant from the fuzzy positive-ideal solution d_i^* and the fuzzy negative-ideal one d_i^- is determined on the basis of the formula (3):

$$d_i^* = \sum_{j=1}^{n} d_v\left(\tilde{v}_{ij}, \tilde{v}_j^*\right) \wedge d_i^- = \sum_{j=1}^{n} d_v\left(\tilde{v}_{ij}, \tilde{v}_j^-\right) \tag{3}$$

where $d_v(\bullet, \bullet)$ is the measure of a distance between two fuzzy numbers. In the Fuzzy TOPSIS, to determine the distance, the vertex method [22, 24] is applied, which uses the Euclidean distance, according to the formula (4):

$$d_v(\tilde{m}, \tilde{n}) = \sqrt{\frac{1}{4}\left[(m_1 - n_1)^2 + (m_2 - n_2)^2 + (m_3 - n_3)^2 + (m_4 - n_4)^2\right]} \tag{4}$$

A global evaluation of every variant is expressed by a closeness coefficient between the positive-ideal solution and the negative-ideal one. The coefficient is described in the formula (5):

$$CC_i = \frac{d_i^-}{d_i^* + d_i^-} \tag{5}$$

In the Fuzzy TOPSIS one can use fuzzy and crisp sets, however, the crisp sets are presented in the form of singletons, where $a_{ij} = b_{ij} = c_{ij} = d_{ij}$.

3 The Framework of the Proposed Comparison System

In the proposed system, a knowledge base, containing information about basic criteria of evaluations of products belonging to individual categories, is indispensable. Such a base ought to contain information about the criteria themselves and dependencies between them. In general, the evaluation criteria should include: technical, functional and economical aspects of products. In the authors' opinion, apart from the criteria referring to products, the applicability of criteria describing shops (e.g. delivery time, shipping costs) should be also taken into account. In fact, every decision variant can be described by a structure containing parameters of a product and parameters of its supplier, since when two same products having the same price are available, data referring its supplier will usually influence a given choice. For evaluation criteria, one can also use some features specific to some group of products, however, the selection of such criteria must be carried out with due caution. It is due to the fact that an objectively higher value of a measure of a variant' given feature makes the variant better than another one with regard to this feature. For instance, a 3 GHz-processor is not always more efficient than a 2.5 GHz-processor. Apart from the knowledge base in the system, there is also an indispensable database storing information about the values of criterial evaluations for individual products.

When the decision-maker has the data about the products, they must initially select the products (decision variants) from one category they are interested in. In this con-text, a comparison system ought to take into consideration two situations: (1) when complete data about all compared decision variants are written in the database, (2) when data about all decision variants are not written in the database.

The data written in the database have the quantitative nature and it can be considered as certain (e.g. for a product such as a "central processing unit", the data is, among other things, the price, the number of cores, thermal design power). The data can be expressed in the form of singletons. When the data about products are not written in the database, the decision-maker should provide information about the data. Such an evaluation depends on the decision-maker's knowledge and skills, therefore, it ought to be assumed that the evaluation is uncertain. In that case, the criterial evaluations of the variants provided by the decision-maker should be expressed on a fuzzy scale.

Before inputting the evaluations of variants or obtaining them from the database, the decision-maker should select the criteria with regard to which individual variants will evaluated. When the database and knowledge base are used, the decision-maker will obtain a predefined set of criteria for the evaluation of products of a given type. This set might be limited by the decision-maker to the criteria which are really essential to him or her. If the database and knowledge base are not used, then the decision-maker should provide their own criteria. Apart from selecting criteria, the decision-maker needs to attribute weights to them. Owing to the fact that the weights are attributed by the decision-maker, they should be regarded as uncertain and expressed in the form of fuzzy sets. The authors decided to use the fuzzy scale and their corresponding linguistic values proposed by Chen et al. [22] for the Fuzzy TOPSIS. The scale and their values are presented in Fig. 1.

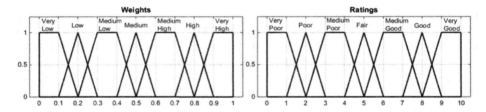

Fig. 1. Linguistic values for weights of criteria and uncertain evaluations of variants

Owing to the fact that there might be different degrees of the decision-maker's uncertainty, the authors of this article suggest that the decision-maker is given the possibility of selecting more than one linguistic values describing the evaluation of a variant or a criterial weight. For example, the decision-maker could simultaneously assume the values "Medium" and "Medium High" as the weight of an j-th criterion. The aggregation of such k fuzzy weight will be similar to the aggregation of many experts' opinions in the Fuzzy TOPSIS [22]. On the basis of the selected linguistic values and their corresponding fuzzy sets, a new trapezoidal fuzzy number $\tilde{w}_j = (w_{j1}, w_{j2}, w_{j3}, w_{j4})$ will be generated, where (6):

$$w_{j1} = min_k\{w_{jk1}\}, \quad w_{j2} = \frac{1}{K}\sum_{k=1}^{K} w_{jk2}, \quad w_{j3} = \frac{1}{K}\sum_{k=1}^{K} w_{jk3}, \tag{6}$$
$$w_{j4} = max_k\{w_{jk4}\}$$

Analogically, the aggregation of several linguistic values describing uncertain evaluations of variants will be carried out.

In the next stage, the aggregation of the decision-maker's preferences into a ranking of variants with the use of the Fuzzy TOPSIS is carried out. A graphic diagram of the operation of the proposed system is comprised in Fig. 2. Figure 2 presents the system's feeding sources (the database and the decision-maker) and the course of the decision process with relation to the stages expressed in the Roy model.

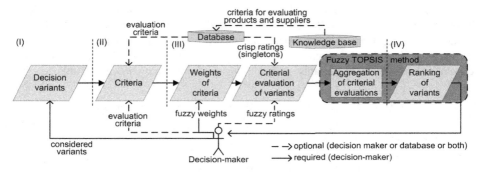

Fig. 2. The framework of the proposed comparison system

4 Results

In order to verify the proposed solution, an example decision problem consisting in selecting a central processing unit was solved. The selection was carried out on the basis of certain criteria which were hypothetically contained in the database and uncertain criteria whose values were provided by the user. Evaluations of variants for the uncertain criteria and the weights of criteria were expressed by means of linguistic values described by fuzzy sets.

Six decision variants were considered. The variants comprised three products and each of the products was offered by two online shops (A1 – Core i3-5005U Seller 1, A2 – Core i3-5005U Seller 2, A3 – Pentium 3825U Seller 1, A4 – Pentium 3825U Seller 2, A5 – Pentium N3710 Seller 1, A6 – Pentium N3710 Seller 2). The variants were considered with regard to three certain criteria (C1 – price, C3 – thermal design power, C4 – delivery time) and two uncertain ones (C2 – performance, C5 – customer service quality). Individual decision variants along with the criteria, preference directions, weights of criteria and evaluation of variants are presented in Table 1. It should be noted that in the case of the criterion C5, two linguistic weights were applied (Medium Low, and Medium). It reflects a higher degree of the decision-maker's uncertainty with regard to the importance of the criterion C5. A similar situation took

Table 1. Weights of criteria and evaluations of decision variants

Criterion	Weight	A1	A2	A3	A4	A5	A6
C1 (min)	H	$275	$261	$180	$172	$170	$161
C2 (max)	VH	F	F	MP	MP	VP, P, MP	VP, P, MP
C3 (min)	L	15 W	15 W	15 W	15 W	6 W	6 W
C4 (min)	L	2 days	3 days	2 days	3 days	2 days	3 days
C5 (max)	ML, M	F	MG	F	MG	F	MG

place while evaluating variants A5 and A6 with regard to the criterion C2. In this case, because of a high degree of the decision-maker's uncertainty with regard to evaluations of variants, the decision-maker applied three linguistic evaluations (Very Poor, Poor, Medium Poor).

A fuzzy-decision matrix and fuzzy weights of variants, obtained on the basis of Table 1 are shown in Table 2. Table 3 depicts weighted normalized values of variants. Table 4 presents fuzzy positive-ideal and fuzzy negative-ideal solutions. Obtained distances of individual variants from the positive-ideal and negative-ideal solutions, closeness coefficients of variants, and ranking of variants are shown in Table 5.

Table 2. Fuzzy-decision matrix and fuzzy weights of variants

	C1	C2	C3	C4	C5
A1	[275, 275, 275, 275]	[4, 5, 5, 6]	[15, 15, 15, 15]	[2, 2, 2, 2]	[4, 5, 5, 6]
A2	[261, 261, 261, 261]	[4, 5, 5, 6]	[15, 15, 15, 15]	[3, 3, 3, 3]	[5, 6, 7, 8]
A3	[180, 180, 180, 180]	[2, 3, 4, 5]	[15, 15, 15, 15]	[2, 2, 2, 2]	[4, 5, 5, 6]
A4	[172, 172, 172, 172]	[2, 3, 4, 5]	[15, 15, 15, 15]	[3, 3, 3, 3]	[5, 6, 7, 8]
A5	[170, 170, 170, 170]	[0, 1.667, 2.333, 5]	[6, 6, 6, 6]	[2, 2, 2, 2]	[4, 5, 5, 6]
A6	[161, 161, 161, 161]	[0, 1.667, 2.333, 5]	[6, 6, 6, 6]	[3, 3, 3, 3]	[5, 6, 7, 8]
Weight	[0.7, 0.8, 0.8, 0.9]	[0.8, 0.9, 1, 1]	[0.1, 0.2, 0.2, 0.3]	[0.1, 0.2, 0.2, 0.3]	[0.2, 0.4, 0.45, 0.6]

Table 3. Weighted normalized fuzzy-decision matrix

	C1	C2	C3	C4	C5
A1	[0.41, 0.468, 0.468, 0.527]	[0.533, 0.75, 0.833, 1]	[0.04, 0.08, 0.08, 0.12]	[0.1, 0.2, 0.2, 0.3]	[0.1, 0.25, 0.281, 0.45]
A2	[0.432, 0.493, 0.493, 0.555]	[0.533, 0.75, 0.833, 1]	[0.04, 0.08, 0.08, 0.12]	[0.067, 0.133, 0.133, 0.2]	[0.125, 0.3, 0.394, 0.6]
A3	[0.626, 0.716, 0.716, 0.805]	[0.267, 0.45, 0.667, 0.833]	[0.04, 0.08, 0.08, 0.12]	[0.1, 0.2, 0.2, 0.3]	[0.1, 0.25, 0.281, 0.45]
A4	[0.655, 0.749, 0.749, 0.842]	[0.267, 0.45, 0.667, 0.833]	[0.04, 0.08, 0.08, 0.12]	[0.067, 0.133, 0.133, 0.2]	[0.125, 0.3, 0.394, 0.6]
A5	[0.663, 0.758, 0.758, 0.852]	[0, 0.25, 0.389, 0.833]	[0.1, 0.2, 0.2, 0.3]	[0.1, 0.2, 0.2, 0.3]	[0.1, 0.25, 0.281, 0.45]
A6	[0.7, 0.8, 0.8, 0.9]	[0, 0.25, 0.389, 0.833]	[0.1, 0.2, 0.2, 0.3]	[0.067, 0.133, 0.133, 0.2]	[0.125, 0.3, 0.394, 0.6]

The results analysis indicates that the best variant, with regard to the considered criteria, is A4, the second position in the ranking takes the variant A6. Moreover, it can

Table 4. Fuzzy positive-ideal and fuzzy negative-ideal solutions

	C1	C2	C3	C4	C5
A^*	[0.9, 0.9, 0.9, 0.9]	[1, 1, 1, 1]	[0.3, 0.3, 0.3, 0.3]	[0.3, 0.3, 0.3, 0.3]	[0.6, 0.6, 0.6, 0.6]
A^-	[0.41, 0.41, 0.41, 0.41]	[0, 0, 0, 0]	[0.04, 0.04, 0.04, 0.04]	[0.067, 0.067, 0.067, 0.067]	[0.1, 0.1, 0.1, 0.1]

Table 5. d_i^*, d_i^-, CC_i values and ranking of variants

	A1	A2	A3	A4	A5	A6
d_i^*	1.408	1.381	1.386	1.354	1.455	1.418
d_i^-	1.28	1.329	1.317	1.377	1.367	1.437
CC_i	0.476	0.49	0.487	0.504	0.484	0.503
Ranking	6	3	4	1	5	2

be noticed that three first positions in the ranking are taken by products from Seller 2. Also, the prepared multi-criteria procedure, which is based on the Fuzzy TOPSIS and taking into consideration different degrees of uncertainty of data, works properly, what is proved by the presented decision problem and its solution.

5 Conclusion

The comparison system will considerably develop the functionality of comparison services. Consequently, their operation will be able to comprise all stages of a decision process, starting from structuring a problem to preparing a multiple-criteria ranking. The Fuzzy TOPSIS method, used in the proposed comparison system, allows introducing multi-criteria decision analysis in comparison services. The authors' idea of including evaluations characterized by a different degree of uncertainty in the Fuzzy TOPSIS method allows decision-makers, who have incomplete knowledge on their examined decision problems, to use the comparison system.

Further research into the comparison system should deal with constructing a knowledge base on criteria of evaluations of sellers and products belonging to various categories [25]. The knowledge base could be implemented in the form of ontology. As a result, an automatic reasoning of new knowledge on the basis of the knowledge written in the base could be possible [26]. Also, in the future, an agent system obtaining information about products from e-commerce services ought to be worked out.

References

1. Jankowski, J., Wątróbski, J., Ziemba, P.: Modeling the impact of visual components on verbal communication in online advertising. In: Núñez, M., Nguyen, N.T., Camacho, D., Trawiński, B. (eds.) ICCCI 2015. LNCS, vol. 9330, pp. 44–53. Springer, Cham (2015). doi:10.1007/978-3-319-24306-1_5

2. Kawaf, F., Tagg, S.: The construction of online shopping experience: a repertory grid approach. Comput. Hum. Behav. **72**, 222–232 (2017)
3. Global B2C E-commerce Report 2016, Ecommerce Foundation. https://www.ecommerce wiki.org/wikis/www.ecommercewiki.org/images/5/56/Global_B2C_Ecommerce_Report_2016.pdf. Accessed 30 Apr 2017
4. European B2C E-commerce Report 2016, Ecommerce Europe. https://www.ecommerce-europe.eu/app/uploads/2016/07/European-B2C-E-commerce-Report-2016-Light-Version-FINAL.pdf. Accessed 30 Apr 2017
5. Yelken, Y.: yStats.com: Popularity of Online Comparison Shopping Services Varies Worldwide, 23 Mar 2016. https://www.linkedin.com/pulse/ystatscom-popularity-online-comparison-shopping-services-yelken. Accessed 30 Apr 2017
6. Gemius & Izba Gospodarki Elektronicznej. https://ecommercepolska.pl/files/9414/6718/9485/E-commerce_w_polsce_2016.pdf. Accessed 30 Apr 2017
7. Laffey, D., Gandy, A.: Comparison websites in UK retail financial services. J. Financ. Serv. Mark. **14**(2), 173–186 (2009)
8. Marano, P.: The EU regulation on comparison websites of insurance products. In: Marano, P., Rokas, I., Kochenburger, P. (eds.) The "Dematerialized" Insurance. Distance Selling and Cyber Risks from an International Perspective, pp. 59–84. Springer, Cham (2016). doi:10.1007/978-3-319-28410-1_3
9. Holland, C.P., Jacobs, J.A., Klein, S.: The role and impact of comparison websites on the consumer search process in the US and German airline markets. Inf. Technol. Tour. **16**(1), 127–148 (2016)
10. Paraskeyas, A., Kontoyiannis, K.: Travel comparison websites: an old friend with new clothes. In: Frew, A.J. (ed.) Information and Communication Technologies in Tourism 2005, pp. 486–496. Springer, Vienna (2005). doi:10.1007/3-211-27283-6_44
11. Chmielarz, W., Zborowski, M.: The application of a conversion method in a confrontational pattern-based design method used for the evaluation of it systems. Ann. Comput. Sci. Inf. Syst. **2**, 1227–1234 (2014)
12. Laffey, D., Gandy, A.: Applying Stabell and Fjeldstad's value configurations to E-commerce: a cross-case analysis of UK comparison websites. J. Strateg. Inf. Syst. **18**(4), 192–204 (2009)
13. Laffey, D.: Comparison websites: evidence from service sector. Serv. Ind. J. **30**(12), 1939–1954 (2010)
14. Ronayne, D.: Price Comparison Websites. Warwick Economics Research Paper Series, no. 1056 (2015)
15. Guitouni, A., Wheaton, K., Wood, D.: An essay to characterise models the military decision-making processes. In: The 11th International Command and Control Research and Technology Symposium, Cambridge (2006)
16. Roy, B.: Multicriteria Methodology for Decision Aiding. Springer, Dordrecht (1996)
17. Guitouni, A., Martel, J.M.: Tentative guidelines to help choosing an appropriate MCDA method. Eur. J. Oper. Res. **109**(2), 501–521 (1998)
18. Wątróbski, J., Jankowski, J., Ziemba, P.: Multistage performance modelling in digital marketing management. Econ. Sociol. **9**(2), 101–125 (2016)
19. Jacquet-Lagreze, E., Siskos, Y.: Preference disaggregation: 20 years of MCDA experience. Eur. J. Oper. Res. **130**(2), 233–245 (2001)
20. Roy, B.: Paradigms and challenges. In: Figueira, J., Greco, S., Ehrogott, M. (eds.) Multiple Criteria Decision Analysis: State of the Art Surveys. International Series in Operations Research & Management Science, vol. 78, pp. 3–24. Springer, New York (2005). doi:10.1007/0-387-23081-5_1

21. Ozturk, M., Tsoukias, A., Vincke, P.: Preference modelling. In: Figueira, J., Greco, S., Ehrgott, M. (eds.) Multiple Criteria Decision Analysis: State of the Art Surveys. International Series in Operations Research & Management Science, vol. 78, pp. 27–59. Springer, New York (2005). doi:10.1007/0-387-23081-5_2

22. Chen, C.T., Lin, C.T., Huang, S.F.: A fuzzy approach for supplier evaluation and selection in supply chain management. Int. J. Prod. Econ. **102**(2), 289–301 (2006)

23. Kutlu, A.C., Ekmekcioglu, M.: Fuzzy failure modes and effects analysis by using fuzzy TOPSIS-based fuzzy AHP. Expert Syst. Appl. **39**(1), 61–67 (2012)

24. Chen, C.T.: Extensions of the TOPSIS for group decision-making under fuzzy environment. Fuzzy Sets Syst. **114**(1), 1–9 (2000)

25. Ziemba, P., Jankowski, J., Wątróbski, J., Piwowarski, M.: Web projects evaluation using the method of significant website assessment criteria detection. In: Nguyen, N.T., Kowalczyk, R. (eds.) Transactions on Computational Collective Intelligence XXII. LNCS, vol. 9655, pp. 167–188. Springer, Heidelberg (2016). doi:10.1007/978-3-662-49619-0_9

26. Ziemba, P., Wątróbski, J., Jankowski, J., Wolski, W.: Construction and restructuring of the knowledge repository of website evaluation methods. In: Ziemba, E. (ed.) Information Technology for Management. LNBIP, vol. 243, pp. 29–52. Springer, Cham (2016). doi:10.1007/978-3-319-30528-8_3

Assessing and Improving Sensors Data Quality in Streaming Context

Rayane El Sibai[1]([⊠]), Yousra Chabchoub[1], Raja Chiky[1], Jacques Demerjian[2], and Kablan Barbar[2]

[1] LISITE Laboratory, ISEP, 92130 Issy Les Moulineaux, France
rayane.el_sibai@etu.upmc.fr, {yousra.chabchoub,raja.chiky}@isep.fr
[2] LARIFA-EDST Laboratory, Lebanese University, Fanar, Lebanon
{jacques.demerjian,kbarbar}@ul.edu.lb

Abstract. An environmental monitoring process consists of a regular collection and analysis of sensors data streams. It aims to infer new knowledge about the environment, enabling the explorer to supervise the network and to take right decisions. Different data mining techniques are then applied to the collected data in order to infer aggregated statistics useful for anomalies detection and forecasting. The obtained results are closely dependent on the collected data quality. In fact, the data are often dirty, they contain noisy, erroneous and missing values. Poor data quality leads to defective and faulty results. One solution to overcome this problem will be presented in this paper. It consists of evaluating and improving the data quality, to be able to obtain reliable results. In this paper, we first introduce the data quality concept. Then, we discuss the existing related research studies. Finally, we propose a complete sensors data quality management system.

Keywords: Data quality · Sensor errors · Streaming data

1 Introduction

An environmental monitoring process is continuously collecting and analyzing data streams generated by environmental sensors. Several observables can be addressed such as temperature, humidity, pressure, etc. The objective of the monitoring process is to filter useful and reliable information and to infer new knowledge which helps the environment explorer to make quickly right decisions.

This whole process, from the data collection to the data analysis, will lead to two key problems: the data volume and the data quality.

In fact, data streams generated by sensors have an always increasing rate, leading to a huge volume of data, sent to the monitoring system in a continuous way. The arrival rate of the data is very high compared to the available treatment and storage capacities of the monitoring system. Thus a permanent and exhaustive storage is very expensive and sometimes impossible. That is why we need to treat the data stream in a single pass, without storing it. Due to this

© Springer International Publishing AG 2017
N.T. Nguyen et al. (Eds.): ICCCI 2017, Part II, LNAI 10449, pp. 590–599, 2017.
DOI: 10.1007/978-3-319-67077-5_57

problem, new challenges related to data streams analysis have appeared. One solution to overcome this problem is to save some of this data in a compact structure called *summary*. This challenge is discussed in more details in [1,2].

On the other hand, in a real-world such as sensors environments, the data are often dirty, they contain noisy, erroneous and missing values. This is due to many factors: local interferences, malicious nodes, network congestion, limited sensor precision, harsh environment, sensor breakdown, sensor malfunction, miscalibration and insufficient battery power of the sensor.

As in the data analysis process, the conclusions and decisions are based on the data, this leads to defective and faulty results. One solution to overcome this problem is to use sensors with high precision to could assume that the arising errors are small, and to deploy redundant sensors to cover the breakdown of a sensor. Nevertheless, this approach is very expensive as it requires very high costs for the sensors. That is why we opted for a software solution where we evaluate and improve data quality using several complementary methods.

The purpose of this paper is to study data quality challenges in dynamic environments, especially, in the sensors environments. We focus on data quality metrics that are first presented for a static dataset, in section two, then in a dynamic context, in section three. In this latter section, we discuss in particular the existing research studies about the data quality in sensors environments. Finally, we present in section four a new and complete sensors data quality management system.

2   Data Quality Concept

Dirty data are reported to cost US industry billions of dollars each year [3]. Several reasons are behind such a problem, especially, the malfunction of the tools that record and collect the data. Data quality has been addressed in many research fields, initiated by the mathematicians in 1960. They proposed a mathematical theory to manage data duplication. In 1980, management researchers opted to control the data to detect and eliminate those with poor quality. It was only from 1990 that the computer science researchers have begun to study the quality of the warehouse's data.

In [4,5], a set of data quality dimensions in static environments was defined. Four data quality categories have been proposed, each one of them is divided into several dimensions.

Intrinsic data quality category measures the quality of the data value itself. *Believability* and *Reputation* dimensions represent the degree to which the user considers the data as correct and trustworthy. *Accuracy* dimension qualifies the difference between the value stored in the database and the real value that the data aims to represent. The last dimension in the intrinsic category is *Data Objectivity*. It represents the degree to which the data are equitable and unbiased.

Contextual data quality category considers the context in which the data will be used. *Relevancy* dimension also known as Helpfulness describes the satisfaction degree of the user's needs and tasks. *Timeliness* dimension also known as

Freshness represents the age of the data. It can be exploited in several ways. On the one hand, it can mean the time between the date of data creation and the current date. On the other hand, it may represent the time between the last verification or updating of the data and the current date. Finally, *Completeness* dimension is given by the ratio of the size of the data registered in the database and the size of real-world data.

The third data quality category is the *Representational* category. It is used to capture the quality of the data representation. *Interpretability* dimension represents the degree to which the data is clear, simple and appropriate for the user. The degree to which the semantic relation between the different information is understandable by the user is depicted by the *Ease of Understanding* dimension. *Representational Consistency* dimension, also known as Homogeneity and Value Consistency represents the degree to which the data are compatible with the previous data. The last dimension is *Concise Representation*, also known as Structural Consistency. It is the degree to which the data structure is suitable to the data itself.

The last data quality category is *Accessibility*. It is related to the accessibility of the data and their security level. The first dimension is *Availability*. It measures the probability that a user's query is answered within a specific time range. The second dimension is *Access Security*, known as Privacy. It incorporates technical security aspects such as data encryption/decryption, user login, anonymization of the user and authentication of the data source.

3   Sensors Data Qualification Systems

In this section, we extend the quality dimensions for sensors data. Such data have an important characteristic: they are temporally and spatially correlated. In fact, for a particular sensor, the current values is often dependent on the past ones. Moreover, values issued from sensors geographically close to each other are expected to be closely correlated. The definition, assessment, and improvement of the quality dimensions for sensors data must take into account these characteristics.

Nowadays, no such study is provided. The domain of sensors data quality is not enough explored. Few works have addressed the issues of the evaluation and improvement of sensors data quality, and especially, in a streaming context. In this section, we discuss these works, and we evaluate their efficiency based on two metrics:

- The ability of the system to evaluate and improve the sensors data quality while taking into account the most important quality dimensions (in view of their impact on the data analysis results): precision, accuracy and completeness. These dimensions are discussed in details in the next section.
- The ability of the system to adapt to the streaming context.

In 2006, Shawn et al. proposed a sensor data cleaning framework *Extensible Sensor stream Processing* (ESP) [6] to clean sensor data. The framework uses

Cassandra Query Language (CQL) as a declarative language to clean the data based on the user's rules. The cleaning process aims to detect outliers, replace missing data and remove duplicate data. The detection of erroneous data is achieved through two steps. First, the framework ESP considers that the data is erroneous if it is largely higher than the expected value defined by the user. Second, ESP uses the spatial correlation between the data. It compares each sensor value with the ones recorded by its neighboring sensors. If the difference between the value and the mean is two times higher than the standard deviation, the data is considered as erroneous. For the missing data problem, ESP replaces each missing value using interpolation. This framework is interesting but it remains incomplete. In fact, a deviated data can be an erroneous data or an event (a real phenomenon). The outliers detection process of ESP does not make this differentiation. Also, no detection and removal for noisy data were performed.

The origin of the data is used to assess the accuracy dimension of both sensors and data in [7]. The proposed framework computes the sensor and data confidence degrees and updates them over time. A filtering process is performed based on a confidence interval defined by the user. The objective is to select only the data having a confidence degree belonging to the confidence interval. For the computation of the data item confidence degree, two similarity measures are used: the value and the provenance similarities. To evaluate the value similarity, the authors assume that the theoretical data distribution follows the normal distribution. The confidence degree is then calculated using the difference between the experimental and the theoretical distribution. The provenance similarity is based on the fact that if several sensors record the same value, this increases the confidence degree of this value. However, in many real use cases, no a priori knowledge about the data distribution can be defined. Moreover, the framework does not present any way to improve the data accuracy.

In [8,9], an environmental phenomena monitoring system for volcanic data analysis is proposed. The system includes three data management layers. In the acquisition layer, the sensors' observations are recorded. Then the data are analyzed in the processing layer. They are filtered, reduced or aggregated. Finally, the discovery layer allows the user to exploit the data in real time. The data quality characteristics are displayed via a graphical user interface. The evaluation criteria of the data quality are the accuracy, the completeness and the time-related aspect. To evaluate the accuracy dimension, no differentiation between errors and real events has been taken into account.

A data cleaning system based on machine learning algorithms was proposed in [10]. The system aims to evaluate the accuracy dimension of the data generated from the environmental sensor network Jornada Experimental Range (JER). For every sensor, each value is predicted using three machine learning algorithms. If the recorded value is far from that predicted, the value is declared as an error. It will be then replaced by the expected value, or by the average of the erroneous value and the expected value, or using the linear interpolation. According to the author, this approach is very expensive and so it will be implemented as a post-processing phase. It needs to save all the recorded values, which is not suitable for the streaming environment constraints. Just like previous works, no differentiation between errors and real events has been made. Also, the completeness dimension was not addressed, while it is quite important in sensors environments as the transmission conditions are very variable and data is very likely to be missed.

Three works have elaborated the quality of sensors data in streaming context. In [11], the authors proposed a model to evaluate and store information about data accuracy and completeness dimensions. This information is recorded over jumping windows instead of each item in order to reduce the storage size of the quality information. According to the authors, a data is considered as inaccurate only when it exceeds the highest sensor range. This study does not address the detection of the anomalies related to a sudden change in the process measured by the sensor. Moreover, no improvement of data quality is proposed.

The previous work [11] was extended in [12,13]. In [12] the authors studied the impact of several data stream operators on the quality of the data. Data stream operators aim to manage the data by applying modifying, generating, reducing and merging actions. [13] provided the information collected about the quality of the data to the user. Data quality on each window is evaluated, then a filter is applied to keep only the data having a quality degree that meets the requirements. However, just like the initial work, no data quality improvement was proposed. Moreover, the rules used to evaluate the accuracy of the data do not differentiate errors and events.

4   Proposition: A Complete Sensors Data Quality Management System

Errors identification in the raw data is a very important issue, it separates the erroneous data and the real important events, and thus ensures a good data accuracy. In the most of the studies presented above, erroneous data are always linked to the precision of the sensor (do not confuse with the *precision* dimension detailed in the following, which is related to noise), and without making any difference between the real events and the erroneous data.

Also, we recall that, currently, few data quality management systems address at the same time, the dimensions: precision, accuracy and completeness. In the literature, very few studies plan to enhance the quality of the data while taking into account these three quality dimensions, as well as the dynamic nature of the data. The proposed definitions and evaluation methods of these dimensions are detailed in Subsect. 4.2. We first provide in Subsect. 4.1 a global view of the architecture of the Complete Sensors Data Quality Management System that we propose.

4.1   Sensors Data Quality Management System, Architecture

We propose the Complete Sensors Data Quality Management System, presented in Fig. 1.

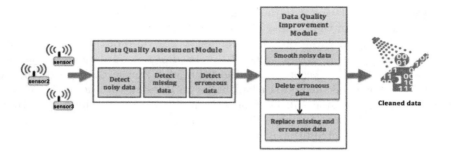

Fig. 1. Sensors data quality management system

After the data acquisition from each sensor, the Sensors Data Quality Management System will:

- Detect and improve noisy data to enhance the precision dimension. Smoothing algorithms will be used to remove the noise from the data.
- Detect and remove errors to improve the accuracy dimension. In order to ensure a good accuracy of the data, it is necessary to separate the true data from the erroneous ones. So that, only the true data will be kept. In order to detect the erroneous data, we will study in Sect. 4.2 the various types of errors that can occur in sensors environments.
- Replace missing data and erroneous data which have been deleted, in order to improve the completeness dimension.

The proposed Sensors Data Quality Management System addresses the main data quality dimensions: precision, accuracy and completeness. The precise definition and evaluation methods of these dimensions are detailed in the following subsection.

4.2 Sensor Data Quality Metrics

Precision Dimension: The precision of data is a measure of random noise [14]. It depicts how close are the data to each other. The potential causes of noisy data are the fluctuations and interferences in the environment, the low battery of the sensor, the hardware failure and the poor calibration of the sensor. As a result, the data can be influenced by random errors also called noise, which makes them slightly deviated from the true values.

Random errors are always present in the data and cannot be controlled. They impact the variability of the data around the average without affecting it. Thus, the measurements will be scattered and dispersed around the true values.

The commonly used statistical measure for noise identification is the *variance* [15,16]. Actually, a high variance is a sign of noisy data. According to [14], the precision of the data can also be measured using the signal-to-noise ratio, or equivalently, the coefficient of variation CV. Evaluating the variance of the data is useless without relating it to the mean. That is why the coefficient of variation is more efficient than the standard deviation for noise detection.

The coefficient of variation CV is defined as the ratio of the standard deviation σ to the mean μ. A low value of CV implies a good data precision (see [14] for more details). According to the value of CV, the data precision can be considered as good, medium or bad based on decision thresholds. Only data with a medium precision will be denoised. In this case, smoothing algorithms will be applied. In fact, the smoothing process reduces the data variance, and so, attenuates the contribution of the noise in the data.

Accuracy Dimension: The accuracy dimension describes how close are the observations values to the true values. It represents the difference between the observation value and the true value which the sensor reading aims to represent. Due to the instrumental, physical and human limitations, malfunction and mis-calibration of the sensor, the observations values can deviate significantly from the true values. These deviated values called errors are abnormal compared to other data and they affect the average of the data.

We know that each value exceeding the maximum range of the sensor is inaccurate and erroneous, however, a value less than the maximum sensor range can also be wrong. Actually, a deviant value compared to other data or to the expected value is considered as abnormal. This comparison is often based on sensor's temporal and spatial neighbor's observations.

The abnormal data may represent an anomaly in the sensor or an anomaly in the environment. In the first case, it is a false measure called systematic error and must be removed and replaced to avoid its impact on the data analysis process. In the second case, it is a real measure called event that describes a real-world phenomenon which must be exploited.

The consideration of abnormal values as errors and their systematic rejection has a significant impact on the data analysis results, and can result in missing important information for the network supervisor. Our objective, in this section, is to study the source of erroneous values, in order to classify the abnormal data into errors and true events. By using a set of rules, and by exploiting the characteristics of these erroneous data, the separation can be made, and therefore, the accuracy of the abnormal data can be judged.

Table 1 describes the faults that can occur in sensors data. For each fault, we will provide its characteristics, potential causes, and the relevant features for modeling it.

- Outliers: A sudden and temporary increase or decrease in the sensor values is manifested by the appearance of outliers. Both erroneous data and events generate outliers. The use of the spatial correlation between the sensors makes it possible to identify the source of the outliers. This is based on the fact that the sensor errors are spatially uncorrelated, whereas the events are correlated.
- Spikes: Spikes error is a combination of several outliers. It increases the gradient and decreases the temporal correlation with the previous data. After the evaluation of the gradient, data classification will be based on the human

Table 1. Taxonomy of sensor data faults from a data centric view

Fault	Form of occurrence	Potential cause(s)	Detection features
Outlier	Single point	Unknown	Rate of change
Spikes	Successive points	Low battery Battery failure Sensor failure Connection failure	Gradient
Stuck-at x	Successive points	Low battery Dead sensor Sensor ADC malfunction	Variation Variance
Clipping	Set of points	Bad calibration	Variation Variance

knowledge and the environment's context. For instance, the temperature during a day is not expected to have a large gradient.

- Stuck-at: A stuck-at x error occurs when the sensor is stuck on an incorrect value x. During such a situation, a set of successive values will have the same value $x \pm \epsilon$. The error may last for a long time, and the sensor may or not return to its normal behavior. This type of error is also called *constant* in [15]. As during the stuck-at error, the variance of the data drops significantly, the variance can be modeled in order to detect this type of error. However, the variance modeling requires the use of a temporal window with a specific size, which is not a trivial task. If the size of the window is very large, the error can be missed. On the contrary, if the size of the window is very small, this can engender false positive detections.
- Clipping: Stuck-at data are not always errors. This is the case of clipping situation. The data that exhibit a clipping fault, still hold some real and important information, and should not be discarded. Clipping errors occur when the environment data are outside the maximum sensitivity range of the sensor. In this case, we observe consecutive data having the maximum value of the sensor sensitivity range. The sensor is thereby saturated, and the data exhibiting this fault have a very small variance.

Completeness Dimension: The loss of data has serious consequences for the environmental monitoring system, and leads to reduced information, and so, to erroneous and distorted results. One solution to deal with this problem is to use a reliable transmission protocol [17], which requires data retransmission in case of failure. However, this will be costly in terms of battery consumption. Another way to avoid data loss is to deploy multiple sensors in the same region to be monitored. This approach is not only expensive in terms of hardware, but it will also create a problem of data redundancy. In fact, it requires an efficient policy to merge the data and will increase the preparation time of data before their analysis. We can thus conclude that the most efficient way to deal with the problem of missing data is to estimate and regenerate them using statistical methods. Time series models and other algorithms based on the moving average can be used to impute missing data.

5   Conclusion

In this paper, we focus on the data quality assessment and improvement in sensors networks. Dirty data engender a faulty data analysis process leading to ineffective results and decisions. So, it is very important to guarantee the quality of the data. Three main data quality dimensions are addressed in this paper: precision, accuracy and completeness. We proposed the architecture of a complete sensor data quality management system, composed of two modules. The objective of the first module is to evaluate data quality based on the detection of noisy, erroneous and missing data. Then, in the second module, the data quality will be enhanced using smoothing methods to deal with noise, and estimation algorithms to regenerate missing and erroneous data. An implementation of this framework will be provided in our future works, in the context of the project WAVES[1]. The goal of this project is to design and develop a platform for a real-time supervision of the water distribution network.

References

1. El Sibai, R., Chabchoub, Y., Demerjian, J., Kazi-Aoul, Z., Barbar, K.: A performance study of the chain sampling algorithm. In: Proceedings of the IEEE Seventh International Conference on Intelligent Computing and Information Systems (ICICIS), pp. 487–494. IEEE (2015)
2. El Sibai, R., Chabchoub, Y., Demerjian, J., Kazi-Aoul, Z., Barbar, K.: Sampling algorithms in data stream environments. In: Proceedings of the IEEE First International Conference on Digital Economy Emerging Technologies and Business Innovation (ICDEc), pp. 29–36. IEEE (2016)
3. Fan, W.: Data quality: from theory to practice. ACM SIGMOD Rec. **44**(3), 7–18 (2015)
4. Strong, D.M., Lee, Y.W., Wang, R.Y.: Data quality in context. Commun. ACM **40**, 103–110 (1997)
5. Wang, R.Y., Strong, D.M.: Beyond accuracy: what data quality means to data consumers. J. Manag. Inf. Syst. **12**, 5–33 (1996)
6. Jeffery, S.R., Alonso, G., Franklin, M.J., Hong, W., Widom, J.: Declarative support for sensor data cleaning. In: Fishkin, K.P., Schiele, B., Nixon, P., Quigley, A. (eds.) Pervasive 2006. LNCS, vol. 3968, pp. 83–100. Springer, Heidelberg (2006). doi:10.1007/11748625_6
7. Lim, H.S., Moon, Y.S., Bertino, E.: Research issues in data provenance for streaming environments. In: Proceedings of the 2nd SIGSPATIAL ACM GIS 2009 International Workshop on Security and Privacy in GIS and LBS, pp. 58–62 (2009)
8. Rodriguez, C.G.: Qualité des données capteurs pour les systèmes de surveillance de phénomènes environnementaux. Ph.D. thesis, Villeurbanne, INSA (2010)

---

[1] http://www.waves-rsp.org.

9. Rodriguez, C.G., Servigne, S.: Sensor data quality for geospatial monitoring applications. In: AGILE, 15th Internationale Conference on Geographic Information Science, pp. 1–6 (2012)
10. Ramirez, G., Fuentes, O., Tweedie, C.E.: Assessing data quality in a sensor network for environmental monitoring. In: Fuzzy Information Processing Society (NAFIPS), pp. 1–6. IEEE (2011)
11. Klein, A., Do, H.H., Hackenbroich, G., Karnstedt, M., Lehner, W.: Representing data quality for streaming and static data. In: 23rd International Conference on Data Engineering Workshop, pp. 3–10. IEEE (2007)
12. Klein, A., Lehner, W.: Representing data quality in sensor data streaming environments. J. Data Inf. Qual. (JDIQ) 1, 10–28 (2009)
13. Olbrich, S.: Warehousing and analyzing streaming data quality information. In: AMCIS, p. 159 (2010)
14. Smith, S.: Digital Signal Processing: A Practical Guide for Engineers and Scientists. Newnes, Oxford (2013)
15. Sharma, A.B., Golubchik, L., Govindan, R.: Sensor faults: detection methods and prevalence in real-world datasets. ACM Trans. Sens. Netw. (TOSN) 6(3), 23 (2010)
16. Abuaitah, G.R., Wang, B.: A taxonomy of sensor network anomalies and their detection approaches. In: Technological Breakthroughs in Modern Wireless Sensor Applications, pp. 172–206. IGI Global (2015)
17. Pang, Q., Wong, V.W.S.: Reliable data transport and congestion control in wireless sensor networks. Int. J. Sens. Netw. 3(1), 16–24 (2008)

Neural Network Based Eye Tracking

Pavel Morozkin[1,2(✉)], Marc Swynghedauw[1], and Maria Trocan[2]

[1] SuriCog, 130 Rue de Lourmel, 75015 Paris, France
{pmor,ms}@suricog.com
[2] Institut Supérieur d'Electronique de Paris,
28 Rue Notre Dame des Champs, 75006 Paris, France
maria.trocan@isep.fr

Abstract. The EyeDee embedded eye tracking solution developed by SuriCog is the world's first solution using the eye as a real-time mobile digital cursor, while maintaining full mobility. In order to reduce the time of eye image transmission, image compression techniques can be employed. Being hardware implemented, several standard image coding systems (JPEG and JPEG2000) were evaluated for their potential use in the next generation device of the EyeDee product line. In order to satisfy low-power, low-heat, low-MIPS requirements several non-typical approaches have been considered. One example consists in the complete replacement of currently used eye tracking algorithm based on image processing coupled with geometric eye modeling by a precisely tuned and perfectly trained neural network, which directly transforms wirelessly transmitted floating-point values of decimated eye image (result of the 3D perspective projection of a model of rotating pupil disk) into five floating-point parameters of pupil's ellipse (result of the eye tracking). Hence implementation of the eye tracking algorithm is reduced to a known challenge of neural network construction and training, preliminary results of which are presented in the paper.

Keywords: Eye tracking · Human–machine interaction · Neural networks

1 Introduction

In order to understand the impact of human's vision during daily activities it is common practice to use eye tracking solutions, which are implemented as human–machine interaction (HMI) devices, which collect and exchange data with some processing units. The EyeDee (Fig. 1) embedded eye tracking solution developed by SuriCog is the world's first solution using the eye as a real-time mobile digital cursor for industry-grade and multimedia applications (interaction with objects placed in a known environment), decision critical applications (control centers), ergonomic assessment and training applications (e.g., cockpit of an aircraft or a helicopter), while maintaining full mobility, which is resulted in total freedom of user's movements.

Since the eye movements are the fastest movements that the human body is able to produce, eye tracking solutions must be especially responsive. This implies restrictions to the algorithms used to measure eye rotation, and also hard constraints on the implementation of these algorithms on physical devices. For example, sending an eye

© Springer International Publishing AG 2017
N.T. Nguyen et al. (Eds.): ICCCI 2017, Part II, LNAI 10449, pp. 600–609, 2017.
DOI: 10.1007/978-3-319-67077-5_58

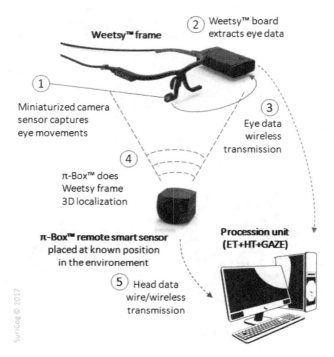

Fig. 1. EyeDee eye tracking solution: Weetsy™ frame, Weetsy™ board, π-Box™ remote smart sensor and processing unit (ET + HT + GAZE).

image taken from a miniaturized camera sensor [1] over a wireless media has to be done in a minimum time. In order to reduce the time of image transmission, image compression techniques can be employed. To most efficiently deploy an implementation of an image coding system, a particular profile has to be selected, based on the result of the complex tradeoff between compressed image size vs. image compression/decompression time vs. computational/implementation complexity and also memory requirements.

Being hardware implemented, several standard image coding systems (JPEG and JPEG2000) were evaluated [2, 3] for their potential use in the next generation device of the EyeDee product line. However, in order to reach low-power, low-heat, low-MIPS requirements, such a device has to be designed as an extra-small, wearable remotely controlled IoT-device, which is able to perform data exchange with a processing unit targeted for further data analysis, which in turn complicates the usage of the image coding systems currently available on the market. To satisfy these factors several non-typical approaches have been considered. One potentially applicable example consists in the complete replacement of currently used eye tracking algorithm based on image processing coupled with geometric eye modeling by a precisely tuned and perfectly trained neural network, which directly transforms wirelessly transmitted floating-point values of decimated eye image (result of the 3D perspective projection of a model of rotating pupil disk) into five floating-point parameters of pupil's ellipse (result of the eye tracking). Hence the implementation of the eye tracking algorithm is reduced to a neural network construction and training approach, which is proposed in this paper.

2   Eye Tracking Approaches

In order to create eye images for experimentation we have developed a simulator based on a 3D eye model projected (Fig. 2) onto a plane (CMOS sensor).

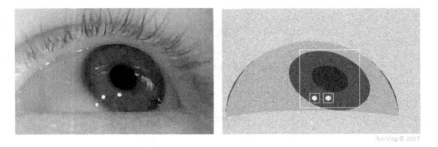

Fig. 2. Eye images: real one and generated by simulator.

It can be seen that the 3D projection of the rotating pupil disk has a geometrical shape of an ellipse. The reconstruction of the gaze (i.e. direction in which the user is looking) is based on the center of rotation of the eye coupled with the coordinates of pupil's center, which are calculated based on five ellipse parameters (Fig. 3).

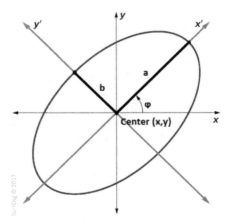

Fig. 3. Five ellipse parameters: center (x, y), major/minor (a/b) axis, rotation angle (φ).

The eye tracking can be interpreted as a multivariate function, which unambiguously associates an ROI image (region that contains the centered image of the pupil) with five ellipse parameters. This function can be based on:

- Image processing (Fig. 4), i.e. filtering the eye image aimed on ellipse shape preservation, followed by its measuring;
- Training of the neural network (Fig. 5) (wherein the inputs are the decimated ROI coupled with the decimated ROI edges – both can be the high-frequency subbands of the 2D wavelet transform [4] – and outputs are the ellipse parameters), followed by loss function optimization (minimization of the average ellipse reconstruction error).

In this paper the proposed eye tracking solution is based on the second approach.

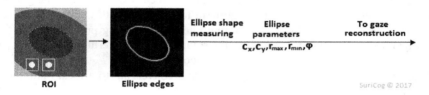

Fig. 4. Eye tracking: image processing based.

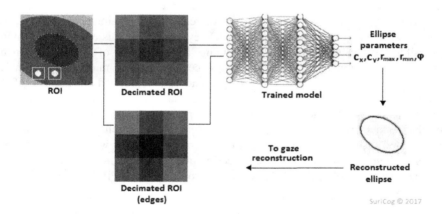

Fig. 5. Eye tracking: neural network based.

3 Neural Network Construction and Training

The neural network [5] is aimed on finding a correlation between the floating-point values of a decimated ROI image (decimated in order to reduce the amount of wirelessly transferred data) and the five floating-point parameters of the pupil's ellipse (necessary for gaze reconstruction). In order to estimate the relationships among these variables, a regression analysis function is applied using Torch7 [6] software (neural network 'nn' and optimization 'optim' packages). This function is further integrated into the EyeDee eye tracking software running on Windows platform. During an initial testing we decided to keep only one layer for the network, and since the decimation reduces 120×120 pixels 8 bpp ROI image into $3 \times 3/5 \times 5$ pixels 32 bpp image, the neural network has 9/25 inputs (18–50 inputs in case of using ellipse edges) and 5 outputs.

The training of the neural network is based on the well-known back-propagation approach [7] coupled with a gradient descent optimization method [8]. The output of the network is compared to the desired output using a loss function. Therefore, the training can be interpreted as the loss function optimization (error minimization). The training is done with the following hyper-parameters:

- Number of epochs – the number of iterations over the training dataset;
- Learning rate – the size of the step taken at each stochastic estimate of the gradient;
- Learning rate decay – allows the algorithm to converge with high precision (it is often recommended to lower the learning rate as the training progresses);
- Weight decay – used to L2-regularize the solution (model overfitting reduction [9]), which prevents the weights from growing too large;
- Momentum – used to prevent the system from converging to a local minimum or saddle point.

It should be noted that since there are no strict rules (only practices) on the selection of the hyper-parameters, we use reinforcement learning [10], i.e., tuning of the hyper-parameters until expected output result is reached.

It should be noted also that to ensure correct ellipse parameters reconstruction based on decimated ROI images of all possible eye positions, it is essential to provide all possible input–output pairs for the training. Since number of such pairs can be sufficiently large (number of all possible ellipse centers c_x and c_y multiplied by number of number of all possible ellipse sizes r_{max} and r_{min} multiplied by number of all possible ellipse rotation angles φ) training can take an amount of time. To reduce this complexity, it is possible to reduce number of input–output pairs by removing such a pairs that cannot practically appear, because ellipse sizes and ellipse center positions are fluctuated (user moves the eye) in a certain range. Another way is to use several modern techniques aimed on training acceleration, such as dropout [11] or batch normalization [12], use of ReLU [13] (or its improved versions PReLU [14] or RReLU [15]).

4   Experimental Results

After the integration of the Torch7 software into the EyeDee eye tracking solution several experiments were completed (Fig. 6).

In a first time we compute two ellipses – one calculated using image processing (Fig. 4) and one obtained using the trained neural network (Fig. 5), and we measure their similarity, based on three coefficients: distance (ε_d), shape (ε_s) and orientation (ε_0) defined below.

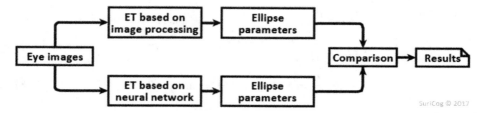

Fig. 6. Testing scheme for two eye tracking approaches: image processing based vs. neural network based.

1. Distance coefficient (ε_d):

$$\varepsilon_d = e^{-\frac{|c_1 - c_2|}{|c_1||c_2|}},\tag{1}$$

where c denotes the ellipse center.

2. Shape coefficient (ε_s):

   For each ellipse equation:

$$ax^2 + 2bxy + cy^2 + 2dx + 2ey + 1 = 0,\tag{2}$$

   we can form the matrix M:

$$M = \begin{bmatrix} a & b & d \\ b & c & e \\ d & e & 1 \end{bmatrix},\tag{3}$$

   and calculate the 3 eigen values:

   d_1, d_2, d_3 for the first ellipse and $\delta_1, \delta_2, \delta_3$ for the second one, which are further used for vectors v and w computation:

$$v = \left(\frac{sign(d_3)}{\sqrt{|d_3|}}, \frac{sign(d_2)}{\sqrt{|d_2|}}, \frac{sign(d_1)}{\sqrt{|d_1|}} \right), \quad |d_3| \geq |d_2| \geq |d_1| > 0,\tag{4}$$

$$w = \left(\frac{sign(\delta_3)}{\sqrt{|\delta_3|}}, \frac{sign(\delta_2)}{\sqrt{|\delta_2|}}, \frac{sign(\delta_1)}{\sqrt{|\delta_1|}} \right), \quad |\delta_3| \geq |\delta_2| \geq |\delta_1| > 0,\tag{5}$$

   Finally, the shape coefficient is given by:

$$\varepsilon_s = e^{-\frac{|w - v|}{|w||v|}},\tag{6}$$

3. Orientation coefficient (ε_0)

$$R_{1,2} = \begin{bmatrix} \cos \varphi_{1,2} & -\sin \varphi_{1,2} \\ \sin \varphi_{1,2} & \cos \varphi_{1,2} \end{bmatrix}, \quad R = R_1^{-1} \times R_2, \tag{7}$$

$$\Theta_0 = \arccos(R_{00}), \quad \Theta_1 = \arccos(R_{11}), \tag{8}$$

$$\varepsilon_0 = e^{-\sqrt{\sin^2 \Theta_0 + \sin^2 \Theta_1}}, \tag{9}$$

where $\varphi_{1,2}$ denotes the ellipse rotation angles, $R_{1,2}$ the rotation matrices, R_{00} and R_{11} the first two components of the diagonal matrix R used to obtain the orientation angles Θ_0 and Θ_1.

According to our preliminary experimentation results, the use of a trained neural network has a good potential. Increasing the number of training epochs results in increasing the average accuracy of the reconstructed ellipse (see Table 1). In our experimental framework we use the following hyper-parameters: the learning rate is set to 1e−5, learning rate decay to 1e−4 and both weight decay and momentum are set to 0.

Table 1. Average ellipse similarity for different ROI sizes.

ROI size	Epochs	ε_d	ε_o	ε_s
3 × 3	1000	0.999985	0.697736	1.000000
	4000	0.999983	0.772567	1.000000
	10,000	0.999984	0.833981	1.000000
4 × 4	1000	0.999991	0.659846	1.000000
	4000	0.999997	0.736138	1.000000
	10,000	0.999997	0.699567	1.000000
5 × 5	1000	0.999991	0.694999	1.000000
	4000	0.999995	0.775739	1.000000
	10,000	0.999993	0.755159	1.000000

Increasing the number of decimation points results into lower average error (see Fig. 7) i.e., computed as cumulated difference between the original (expected) result and the approximated one (generated by the neural network during the training), measured on each epoch. For example, changing the decimated ROI size from 3 × 3 to 4 × 4 immediately decreased the reconstruction error, providing better quality of the output results c_x, c_y, r_{max}, r_{min} and φ, and thus increasing the similarity between the original and reconstructed ellipses (Table 1).

Increasing the number of training iterations (epochs) results into lower average orientation coefficient (see Fig. 8), computed as a result of comparison of original and reconstructed ellipses for all samples used in the training dataset. Distance coefficient (ε_d) and shape coefficient (ε_s) are fluctuating in more narrow range (see Table 1).

It can be shown that the best average error of orientation coefficient (ε_o) over all sizes of decimated ROI is reached, when the number of training iterations is equal to 40,000. To find an optimal number of hidden layers (Fig. 9) we kept this number of

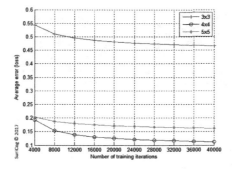

Fig. 7. Neural network training (average error decrease), 40,000 training iterations.

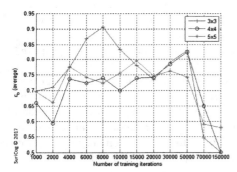

Fig. 8. Trained neural network: average orientation coefficient (ε_o) vs. number of training iterations.

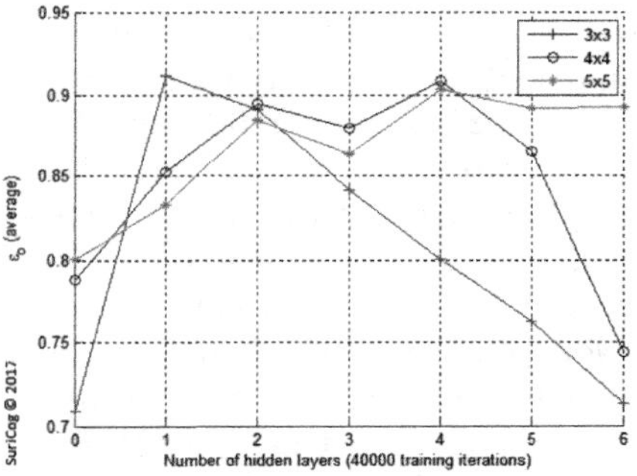

Fig. 9. Trained neural network: average orientation coefficient (ε_o) vs. number of hidden layers.

iterations. It can be shown that best average error of the orientation coefficient (ε_o) is reached, when the number of hidden layers is equal to 2.

The visual comparison of ellipse reconstruction (see Fig. 10) shows that there are some accuracy issues, i.e., some ellipses generated by the trained neural network do not perfectly fit into the original ones generated by the simulator (although some are very close to, see Fig. 10a, b). Even more, for new eye positions, not present in the training data set, unexpected results are generated by the trained neural network (see Fig. 10c, d). In order to improve the accuracy, it is necessary to find the best neural network configuration and define the training strategy to avoid common issues such as overtraining, finding a local minimum (instead of global one) etc.

From the computational point of view (computational complexity) the eye tracking approach, which is based on image processing, usually requires significant computing resources to perform the processing (eye image filtering, pupil features extraction, etc.) even if some of it is done in parallel (image blurring, histogram calculation, etc.). Also the processing time of eye image is not constant (due to conditional logic, and, in consequence, not constant number of operations executed in run-time), which can be

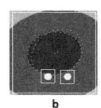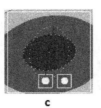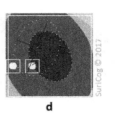

a   b   c   d

Fig. 10. Visual comparison of ellipse reconstruction (increasing degradation order, from best (a) to worst (d)).

essential for time-critical applications. The neural network based approach is expected to be much faster (average speed-up of use of neural network over image processing is ×125), because of the constant-time response. However, the use of neural network has a well known challenge of its training (hyper-parameters tuning, training time reduction, output results quality maximization).

From the implementation point of view (implementation complexity), the proposed neural-network eye tracking approach reduces drastically the amount of wireless transmitted data, (max. 25 floating-point values of decimated ROI image vs. 14,400 bytes for the original ROI image). Even more, if the next coming version of the Weetsy™ board will have higher memory space, enough to store the trained neural network, it will be possible to keep the network inside the device and send wirelessly five ellipse parameters, which results in the elimination of any remote eye tracking software (hence making Weetsy™ frame and Weetsy™ board a standalone wireless eye tracking solution).

From the conceptual point of view, employing a neural network can be interpreted as a move of the eye image processing power from *online* (run-time) to *offline* (pre-trained network).

5   Conclusion and Future Work

In this paper we propose a new eye tracking approach based on neural networks instead of using classical image processing (i.e., filtering the ROI image aimed on preservation of edges of pupil's ellipse, followed by measuring of its shape). A neural network was implemented in Torch7 software, which was completely integrated into EyeDee software running on the Windows. During an initial testing, the number of neural network layers was limited to only one layer, which takes 9/25 inputs (values of decimated ROI images) and produces 5 outputs (pupil's ellipse parameters). According to our preliminary experimentations, the use of a trained neural network has a good potential for eye tracking. However, the visual quality of ellipse reconstruction shows that some ellipses generated by the trained neural network do not perfectly fit into the original ones, generated by the simulator. Therefore, there is a need of further investigation in order to achieve correct construction of the network, followed by its training in a more efficient manner in order to avoid common failures such as overtraining, fining of local minimum (instead of global one), etc.

Future work consists in statistical experimentation with all parameters: adding more noise to simulated eye images, adding the ROI offset (with use of real eye images, the pupil is not necessarily centered in the found ROI) and training of the neural network to produce correct results with non-centered ROI, and finally testing the trained neural network on real images.

References

1. Morozkin, P., Swynghedauw, M., Trocan, M.: Design of an embedded image acquisition system. In: 2015 IEEE International Conference on Electronics, Circuits, and Systems (ICECS), pp. 504–505 (2015)
2. Morozkin, P., Swynghedauw, M., Trocan, M.: An image compression for embedded eye-tracking applications. In: 2016 International Symposium on INnovations in Intelligent SysTems and Applications (INISTA), article 33 (2016)
3. Morozkin, P., Swynghedauw, M., Trocan, M.: Image quality impact for eye tracking systems accuracy. In: 2016 IEEE International Conference on Electronics, Circuits and Systems (ICECS), pp. 429–431 (2016)
4. Antonini, M., Barlaud, M., Mathieu, P., Daubechies, I.: Image coding using wavelet transform. IEEE Trans. Image Process. **1**(2), 205–220 (1992)
5. Demuth, H.B., Beale, M.H., De Jesus, O., Hagan, M.T.: Neural Network Design. Martin Hagan, Oklahoma (2014)
6. Torch framework. www.torch.ch. Accessed 17 May 2017
7. Vogl, T.P., Mangis, J.K., Rigler, A.K., Zink, W.T., Alkon, D.L.: Accelerating the convergence of the back-propagation method. Biol. Cybern. **59**(4), 257–263 (1988)
8. Mason, L., Baxter, J., Bartlett, P.L., Frean, M.R.: Boosting algorithms as gradient descent. In: NIPS, pp. 512–518 (1999)
9. Ng, A.Y.: Feature selection, L1 vs. L2 regularization, and rotational invariance. In: Proceedings of the Twenty-First International Conference on Machine Learning, p. 78. ACM (2004)
10. Sutton, R.S., Barto, A.G.: Reinforcement Learning: An Introduction, vol. 1. MIT Press, Cambridge (1998)
11. Srivastava, N., Hinton, G.E., Krizhevsky, A., Sutskever, I., Salakhutdinov, R.: Dropout: a simple way to prevent neural networks from overfitting. J. Mach. Learn. Res. **15**(1), 1929–1958 (2014)
12. Ioffe, S., Szegedy, C.: Batch normalization: accelerating deep network training by reducing internal covariate shift. arXiv preprint arXiv:1502.03167 (2015)
13. Krizhevsky, A., Sutskever, I., Hinton, G.E.: Imagenet classification with deep convolutional neural networks. In: Advances in Neural Information Processing Systems, pp. 1097–1105 (2012)
14. He, K., Zhang, X., Ren, S., Sun, J.: Delving deep into rectifiers: surpassing human-level performance on imagenet classification. In: Proceedings of the IEEE International Conference on Computer Vision, pp. 1026–1034 (2015)
15. Xu, B., Wang, N., Chen, T., Li, M.: Empirical evaluation of rectified activations in convolutional network. arXiv preprint arXiv:1505.00853 (2015)

Author Index

Abduali, Balzhan   II-491
Abedin, Mahmudul   I-298
Achilleos, Achilleas P.   I-484
Adnan, Foysal Amin   II-479
Ahmed, Muyeed   II-479
Albishry, Nabeel   II-469
Alimzhanov, Yermek   II-509
Aloui, Nadia   I-233
Amirova, Dina   II-491
Aubakirov, Sanzhar   II-509
Augustynek, Martin   II-541

Bac, Maciej   I-113
Bădică, Costin   I-331, I-381
Bąk, Jarosław   I-93
Balabanov, Kristiyan   I-223, I-361
Barbar, Kablan   II-590
Bartuskova, Aneta   I-63, I-452, I-462
Basher, Sheikh Faisal   I-298
Bernas, Marcin   II-119
Blecha, Petr   I-548
Blinkiewicz, Michał   I-93
Bobrowski, Leon   I-73
Boryczka, Urszula   II-76, II-107
Bosse, Tibor   I-125
Bourgne, Gauvain   I-202
Bregulla, Markus   II-195, II-205, II-249
Bruha, Radek   II-305
Bryjova, Iveta   II-182, II-541
Bytniewski, Andrzej   I-34

Cerri, Stefano A.   I-212
Chabchoub, Yousra   II-590
Charfi, Nesrine   I-538
Chau, Nguyen Hai   I-266
Chebil, Raoudha   I-212
Chiky, Raja   I-137, II-590
Chłopaś, Łukasz   II-292
Chohra, Amine   II-32
Chojnacka-Komorowska, Anna   I-342
Choroś, Kazimierz   II-569
Choudhury, Deboshree   I-307
Cimler, Richard   I-528, II-315, II-335, II-345
Cimr, Dalibor   I-528

Ciorbaru, Vicentiu-Marian   I-192
Crick, Tom   II-469
Cupek, Rafał   II-238, II-272, II-282, II-292

Danicek, Matej   II-345
Deepthi, P.S.   II-129
Demerjian, Jacques   II-590
Diakou, Chrysostomi Maria   I-569
Djaghloul, Younes   I-172
Dolezal, Rafael   II-171
Doroz, Rafal   II-161
Draszawka, Karol   II-438
Drewniak, Marek   II-195, II-227, II-272, II-282, II-292
Du Nguyen, Van   I-83
Du, Phuong-Hanh   I-148
Duda, Jakub   II-292
Dworak, Kamil   II-107
Dziędziel, Grzegorz   II-292

El Sibai, Rayane   II-590
Ellouze, Mehdi   I-172
Esmaili, Parvaneh   I-497

Faltynova, Kamila   II-541
Fathalla, Said   I-14
Fernandes de Mello Araújo, Eric   II-386
Fietz, Robinson Guerra   I-223
Filonenko, Alexander   II-549, II-558
Fojcik, Marcin   II-249, II-272, II-282
Formolo, Daniel   I-160
Fougères, Alain-Jérôme   I-389
Foulonneau, Muriel   I-172
Franke, Annelore   II-386

Gargouri, Faiez   I-233
Gataullin, Ramil   II-519, II-529
George, K.M.   II-417
Georgiou, Kyriaki   I-484
Gilmullin, Rinat   II-519, II-529
Gładysz, Barbara   I-579
Gogoglou, Antonia   I-244
Gouider, Mohamed Salah   II-448

Grzechca, Damian   II-215, II-260
Guinand, Frédéric   I-422, I-442
Gumede, Andile M.   I-257
Gwetu, Mandlenkosi V.   I-257
Gwizdałła, Tomasz M.   II-66

Hajdú-Szücs, Katalin   I-401
Hamem, Sihem   II-448
Haron, Habibollah   I-497
Hasan, Syeda Shabnam   I-288
Hernes, Marcin   I-34, I-113, I-342
Himel, Ahsan Habib   I-298
Hlioui, Fedia   I-233
Hoang, Dinh Tuyen   I-182
Hosain, Rukshar Wagid   II-386
Hwang, Dosam   I-182

Imtiaz, Mir Tahsin   II-479
Ivanović, Mirjana   I-381
Ivković, Jovana   I-381

Jankowski, Jarosław   II-579
Jędrzejowicz, Joanna   II-3, II-357
Jędrzejowicz, Piotr   II-3
Jo, Kang-Hyun   II-549, II-558
Jodłowiec, Marcin   I-24
Juanals, Brigitte   II-376

Kadery, Ivan   II-479
Kannot, Yaman   I-14
Karaskova, Natalie   II-171
Karibayeva, Aidana   II-491
Karwowski, Jan   I-518
Kempa, Olgierd   I-317
Khakimov, Bulat   II-529
Khan, Haymontee   I-288
Khan, Raiyan   II-479
Khisha, Joytu   I-307
Khusainov, Aidar   II-407
Kiss, Attila   I-401
Kłak, Sebastian   II-249
Kleanthous, Styliani   I-569
Klein, Michel C.A.   I-473
Kokkinaki, Angelika I.   I-569
Kolar, Karel   II-171
Komarek, Ales   II-325
Konstantinidis, Andreas   I-484
Korczak, Jerzy   I-113
Kotelnikova, Anastasiia   I-433

Kovarnik, Jaroslav   I-548
Kowalczyk, Mateusz   I-411
Kozierkiewicz-Hetmańska, Adrianna   I-44, I-103
Krejcar, Ondrej   II-335, II-345
Krenek, Jiri   II-171
Kriz, Pavel   II-305
Krótkiewicz, Marek   I-24
Kubicek, Jan   II-182, II-541
Kuca, Kamil   II-171, II-182
Kuhnova, Jitka   I-528
Kurnianggoro, Laksono   II-549, II-558
Kutrzyński, Marcin   I-317

Labidi, Mohamed   II-459
Laki, Sándor   I-401
Lasota, Tadeusz   I-317
Latif, Asiful Haque   II-479
Le, Hong-Quang   II-22
Lejouad Chaari, Wided   I-212
Leon, Florin   I-331
Lis, Robert   I-433
Logofătu, Doina   I-223, I-361
Lu, Dang-Nhac   II-22
Luckner, Marcin   I-518
Lupu, Andrei-Ştefan   I-331

Madani, Kurosh   II-32
Madiyeva, Aigerim   II-501
Madiyeva, Gulmira   II-509
Maia, Nuno   I-558
Małecki, Krzysztof   II-56
Maleszka, Bernadetta   II-428
Maleszka, Marcin   I-54
Maltsevskaya, Nadezhda V.   II-171
Mannan, Noel   I-288
Manolopoulos, Yannis   I-244
Mansurova, Madina   II-509
Manzoor, Adnan   I-473
Maraoui, Mohsen   II-459
Maresova, Petra   II-541
Mariano, Manuel   I-558
Markides, Christos   I-484
Marreiros, Goreti   I-558
Mars, Ammar   II-448
Mashbu, Ruhul   I-298
Matouk, Kamal   I-342
Matyska, Jan   II-335, II-345
Medeiros, Lenin   I-125

Melikova, Michaela   II-171
Meltzer, Flavian   II-205
Merayo, Mercedes G.   I-83
Mercik, Jacek   II-13
Mercl, Lubos   II-325
Métais, Elisabeth   I-137
Meziane, Farid   I-137
Mikitiuk, Artur   I-411, I-422
Minel, Jean-Luc   II-376
Mollee, Julia S.   I-473
Molnarova, Kristyna   II-182
Moni, Jebun Nahar   I-288
Morozkin, Pavel   II-600
Mukhamedshin, Damir   II-407

Nachazel, Tomas   I-371
Nalepa, Marek   II-238
Nemcova, Zuzana   I-371
Neves, José   I-558
Nevzorova, Olga   II-407
Ngo, Thi-Thu-Trang   II-22
Nguyen, Manh-Hai   II-22
Nguyen, Ngoc Thanh   I-83
Nguyen, Ngoc-Hoa   I-148
Nowak-Brzezińska, Agnieszka   II-139,
    II-150
Nowakowski, Arkadiusz   II-45

Ogorodnikov, Nikita   II-398

Pałka, Dariusz   II-97
Papadopoulos, George A.   I-484
Paryani, Jyotsna   II-417
Paszek, Krzysztof   II-260
Paul, Anirudha   II-479
Pavlik, Jakub   II-325
Penhaker, Marek   II-182, II-541
Peter, Lukas   II-182
Pham, Hai-Dang   I-148
Piekarz, Jakub   II-195
Pietranik, Marcin   I-44
Płaczek, Bartłomiej   II-119
Poloczek, Dawid   II-260
Porwik, Piotr   II-161
Pozo, Manuel   I-137
Procházka, Jan   I-351
Przybyła-Kasperek, Małgorzata   II-139,
    II-150
Pscheidl, Pavel   II-315

Racakova, Veronika   II-171
Rahman, Rashedur M.   I-288, I-298, I-307,
    II-479
Rahman, Rashida   I-288
Rakhimova, Diana   II-501
Rebedea, Traian   I-192
Rędziński, Michał   II-227
Rybka, Paweł   II-215
Rybotycki, Tomasz   II-150

Safaverdi, Hossein   II-161
Sapek, Alicja   II-119
Schmitt, Ulrich   I-3
Schrittenloher, Sebastian   II-195
Sec, David   II-335
Seredynski, Franciszek   I-442
Siemiński, Andrzej   I-277
Sikder, Tonmoy   I-298
Simiński, Roman   II-139, II-150
Sitarczyk, Mateusz   I-103
Skinderowicz, Rafał   II-87
Sobeslav, Vladimir   II-325
Sobol, Gil   I-361
Solaiman, Basel   I-538
Soldano, Henry   I-202
Soukal, Ivan   I-63, I-452, I-462
Stamate, Daniel   I-361
Štekerová, Kamila   I-351
Stepan, Jan   II-335, II-345
Strąk, Łukasz   II-45
Suder, Tomasz   II-569
Suleymanov, Dzhavdet   II-529
Sundetova, Aida   II-491
Swynghedauw, Marc   II-600
Szuba, Tadeusz   I-507
Szwarc, Krzysztof   II-76
Szymański, Julian   II-438

T.K., Ashwin Kumar   II-417
Tamanna, Nusrat Jahan   I-298
Telec, Zbigniew   I-317
Thampi, Sabu M.   II-129
Tokarz, Krzysztof   II-227, II-260
Tomášková, Hana   I-528, II-315
Toshimasa, Yamanaka   II-367
Tran, Thi-Thu-Hien   II-22
Tran, Van Cuong   I-182
Trawiński, Bogdan   I-317
Trejbal, Jan   II-171
Tretyakova, Antonina   I-442

Trichili, Hanene   I-538
Trocan, Maria   II-600
Trojanowski, Krzysztof   I-411, I-422
Tryfonas, Theo   II-469
Tucnik, Petr   I-371, I-548
Tukeyev, Ualsher   II-491
Turki, Slim   I-172

van der Wal, C. Natalie   I-160
van Halteren, Aart T.   I-473
Vanin, Artem   I-433
Veillon, Lise-Marie   I-202
Vicente, Henrique   I-558
Viriri, Serestina   I-257

Wahyono,   II-549
Wąs, Jarosław   II-97
Wątróbski, Jarosław   II-56, II-579

Wesolowski, Tomasz Emanuel   II-161
Wieczorek, Wojciech   II-45
Wojtkiewicz, Krystian   I-24
Wolski, Waldemar   II-56
Wrobel, Krzysztof   II-161
Wypych, Michał   I-422

Xanat, Vargas Meza   II-367

Yermekov, Zhantemir   II-509

Zakrzewska, Magdalena   II-357
Zerin, Naushaba   I-307
Zhumanov, Zhandos   II-501
Ziębiński, Adam   II-215, II-238, II-249,
    II-272, II-282
Ziemba, Paweł   II-579
Zonenberg, Dariusz   II-292
Zrigui, Mounir   II-459

Printed in the United States
By Bookmasters